# Sustainable Agriculture Reviews

Volume 50

**Series Editor**
Eric Lichtfouse
CNRS, IRD, INRAE, Coll France, CEREGE
Aix-Marseille University
Aix-en-Provence, France

*Other Publications by Dr. Eric Lichtfouse*

**Books**
Scientific Writing for Impact Factor Journals
https://www.novapublishers.com/catalog/product_info.php?products_id=42242

Environmental Chemistry
http://www.springer.com/978-3-540-22860-8

Sustainable Agriculture
Volume 1: http://www.springer.com/978-90-481-2665-1
Volume 2: http://www.springer.com/978-94-007-0393-3

**Book series**
Environmental Chemistry for a Sustainable World
http://www.springer.com/series/11480

Sustainable Agriculture Reviews
http://www.springer.com/series/8380

**Journal**
Environmental Chemistry Letters
http://www.springer.com/10311

Sustainable agriculture is a rapidly growing field aiming at producing food and energy in a sustainable way for humans and their children. Sustainable agriculture is a discipline that addresses current issues such as climate change, increasing food and fuel prices, poor-nation starvation, rich-nation obesity, water pollution, soil erosion, fertility loss, pest control, and biodiversity depletion.

Novel, environmentally-friendly solutions are proposed based on integrated knowledge from sciences as diverse as agronomy, soil science, molecular biology, chemistry, toxicology, ecology, economy, and social sciences. Indeed, sustainable agriculture decipher mechanisms of processes that occur from the molecular level to the farming system to the global level at time scales ranging from seconds to centuries. For that, scientists use the system approach that involves studying components and interactions of a whole system to address scientific, economic and social issues. In that respect, sustainable agriculture is not a classical, narrow science. Instead of solving problems using the classical painkiller approach that treats only negative impacts, sustainable agriculture treats problem sources.

Because most actual society issues are now intertwined, global, and fast-developing, sustainable agriculture will bring solutions to build a safer world. This book series gathers review articles that analyze current agricultural issues and knowledge, then propose alternative solutions. It will therefore help all scientists, decision-makers, professors, farmers and politicians who wish to build a safe agriculture, energy and food system for future generations.

More information about this series at http://www.springer.com/series/8380

Vipin Kumar Singh • Rishikesh Singh
Eric Lichtfouse
Editors

# Sustainable Agriculture Reviews 50

## Emerging Contaminants in Agriculture

Springer

*Editors*
Vipin Kumar Singh
Department of Botany, Centre of Advanced Study in Botany, Institute of Science
Banaras Hindu University
Varanasi, India

Rishikesh Singh
Institute of Environment and Sustainable Development
Banaras Hindu University
Varanasi, India

Eric Lichtfouse
Aix Marseille University, CNRS, IRD, INRAE, Coll France, CEREGE
Aix-en-Provence, France

ISSN 2210-4410 ISSN 2210-4429 (electronic)
Sustainable Agriculture Reviews
ISBN 978-3-030-63248-9 ISBN 978-3-030-63249-6 (eBook)
https://doi.org/10.1007/978-3-030-63249-6



This Springer imprint is published by the registered company Springer Nature Switzerland AG
The registered company address is: Gewerbestrasse 11, 6330 Cham, Switzerland

# Preface

> The soil is the great connector of lives, the source and destination of all. It is the healer and restorer and resurrector, by which disease passes into health, age into youth, death into life. Without proper care for it we can have no community, because without proper care for it we can have no life.
>
> Wendell Berry

Agriculture is essential for the survival and livelihood of a continuously rising human population. Increasing industrialization and urbanization have unexpectedly produced huge quantities of contaminants of organic and inorganic nature, exerting undesirable effects on human health and the environment. Many contaminants considered as emerging contaminants, such as antibiotics, antibiotic resistance genes, nanomaterials, hormones, personal-care products, and toxic byproducts, have been identified from agriculture fields, and proper management practices are required to safeguard the agroecosystem. Surprisingly, to date, no regulatory limits have been set for most emerging contaminants. Food chain contamination, transfer of emerging contaminants from agroecosystems to waters, and accumulation in diverse flora and fauna are potential threats perturbing the environmental homoeostasis. As a consequence, there is a need for identification and precise determination of emerging contaminants, development of methods for contaminant extraction from different soil matrices, appraisal of toxicological impacts in agriculture, and remediation methods.

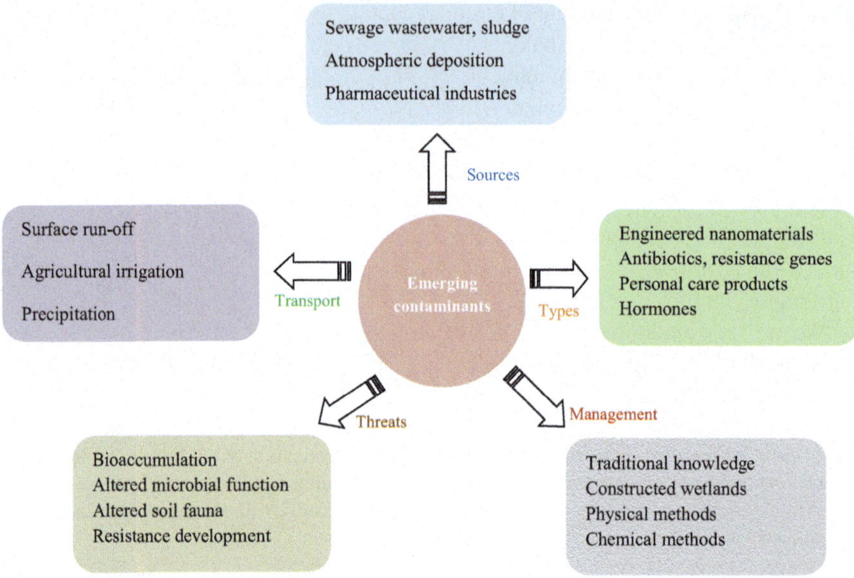


Sources, types, transport mechanism, threats, and management of emerging contaminants in agroecosystems

This book, entitled *Emerging Contaminants in Agriculture*, contains 15 chapters reviewing the latest advances regarding sources, measurement, consequences in agroecosystems, fate and transport, and strategies applied for management of emerging contaminants in agriculture. The first chapter by Hasnat presents emerging contaminants and their effects on the environment and human health. Implications of sewage wastewater and sludge as sources of contaminants with focus on source, fate, transport, and ecological toxicity are discussed by Singh et al. in Chap. 2. In Chap. 3, Nath discusses the contribution of dew as a medium for deposition of emerging contaminants in agriculture. The process of dew formation, chemistry, and contribution in soil biogeochemical processes is also elaborated. Chapter 4, about particulate material contamination with negative consequences on agricultural productivity, is authored by Das et al. Dyes as emerging contaminants in agriculture, their fate, transport, toxic impacts, and remediation are presented in Chap. 5 by Singh et al. Kaur in Chap. 6 sheds light on environmental modifications, transfer, negative consequences on soil processes, and important strategies to combat pharmaceutical contamination.

Sources, determination and fate of antibiotics, and antibiotic resistance genes are presented in Chap. 7 by Singh et al. Sources, soil contamination, and impacts of polycyclic aromatic hydrocarbons are explored in Chap. 8 by Yadav et al. In Chap. 9, Mishra and Khare present nano-formulation of agrochemicals, their benefits for agriculture, environmental hazards, and mitigation strategies. Chapter 10 by Gani et al. discusses transport of emerging contaminants from agroecosystems to native groundwater, along with factors influencing fate and transportation of agricultural soils. Crop application and fate of engineered nanomaterials are elaborated by Kaur in Chap. 11. Chapter 12 by Shakya and Ahmad reviews sources, fate, and exposure pathways of nanomaterials in agroecosystems. Singh and Patel in Chap. 13 review the effects of engineered nanomaterials with emphasis on toxicity at morphological, physiological, and molecular level; they also discuss defense responses and detoxification strategies in crop plants. Chapter 14 by Mishra et al. presents pesticide contamination and associated hazards in agriculture with case studies both from global and Indian perspectives. In the final chapter, Borthakur and Singh discuss the application of traditional knowledge in sustainable management of agro-environment.

We are very much thankful to all the learned contributors for their valuable time and contribution. The completion of this volume would have not been possible without the strenuous efforts of eminent reviewers and colleagues. Finally, we extend our sincere thanks to the technical staff at Springer Nature for typesetting and efficient production of the book.

Varanasi, Uttar Pradesh, India Vipin Kumar Singh

Varanasi, Uttar Pradesh, India Rishikesh Singh

Aix-en-Provence, France Eric Lichtfouse

# Contents

# About the Editors

**Vipin Kumar Singh** is currently working as CSIR research Associate in the Department of Botany, Centre of Advanced Study, Banaras Hindu University, Varanasi, Uttar Pradesh, India. He obtained his Ph.D. degree in botany from the Banaras Hindu University in 2015. He has worked on the UGC project entitled "Detoxification of arsenic contaminated groundwater by transgenic ecofriendly microbes." Dr. Singh is recipient of prestigious UGC research fellowship in science for meritorious students. He has more than 6 years of research experience in the area of groundwater hydro-geochemistry and bioremediation approaches to clean the contaminated groundwater sites. His area of interest includes use of fast-growing plants and microbes to clean the contaminated soil and water system. Dr. Singh has published 6 chapters, 14 experimental research articles, and 2 review articles in journals of national and international repute. He is lifetime member of the Indian Association of Hydrologists (IAH) and Association of Microbiologists of India (AMI). Dr. Singh has actively participated in many national and international seminars, symposia, conferences, and workshop related to application of microbes in management of contaminated sites.

**Rishikesh Singh** has obtained his master's degree in Environmental Science from the Department of Botany, Institute of Science, Banaras Hindu University, India. He is actively engaged in research after completion of Ph.D. at the Institute of Environment and Sustainable Development, Banaras Hindu University. The area of his research is management and conservation of polluted sites and agroecosystem. He has published more than 35 papers in international journals in the fields of soil nutrient dynamics, agroecosystem restoration, environmental contamination, and waste management.

**Eric Lichtfouse** is a biogeochemist at the French National Institute for Agricultural Research, CEREGE, Aix en Provence, France. He has invented carbon-13 dating, a method allowing to measure the dynamics of organic molecules encapsulated in complex media. Dr. Lichtfouse is professor of scientific communication and author of the book *Scientific Writing for Impact Factor Journals*. He is founder and chief editor of the journal *Environmental Chemistry Letters* and the book series *Sustainable Agriculture Reviews and Environmental Chemistry for a Sustainable World*. Dr. Lichtfouse has won awards in analytical chemistry and science editing.

# Contributors

**Faraz Ahmad** ICMR-National Institute of Pathology, New Delhi, India

**Muntjeer Ali** Environmental Engineering Group, Department of Civil Engineering, Indian Institute of Technology Roorkee, Roorkee, Uttrakhand, India

**Rahul Arya** Institute of Environmental Studies, Kurukshetra University, Kurukshetra, Haryana, India

**Rahul Bhadouria** Department of Botany, University of Delhi, New Delhi, India

**Anwesha Borthakur** Leuven International and European Studies (LINES), KU Leuven, Leuven, Belgium

**Faizal Bux** Institute for Water and Wastewater Technology, Durban University of Technology, Durban, South Africa

**Anamika Das** Department of Paramedical Sciences, Guru Kashi University, Bathinda, India

**Sujit Das** Laboratory of Applied Stress Biology, Department of Botany, University of Gour Banga, Malda, West Bengal, India

**Monika Dubey** Environmental Engineering Group, Department of Civil Engineering, Indian Institute of Technology Roorkee, Roorkee, Uttrakhand, India

**Khalid Muzamil Gani** Institute for Water and Wastewater Technology, Durban University of Technology, Durban, South Africa

**Dipti Grover** Institute of Environmental Studies, Kurukshetra University, Kurukshetra, Haryana, India

**Parnika Gupta** Department of Environmental Science and Technology, Central University of Punjab, Bathinda, India

**Jaskiran Kaur** Department of Biotechnology, Doaba College, Jalandhar, Punjab, India

**Tamanreet Kaur** Department of Zoology, Kanya MahaVidyalaya, Jalandhar, Punjab, India

**Absar Ahmad Kazmi** Environmental Engineering Group, Department of Civil Engineering, Indian Institute of Technology Roorkee, Roorkee, Uttrakhand, India

**Puja Khare** Central Institute of Medicinal and Aromatic Plants (CIMAP), Lucknow, India

**Abhishek Kumar** Soil Ecosystem and Restoration Ecology Lab, Department of Botany, Panjab University, Chandigarh, India

**Ajay Kumar** Agriculture Research Organization (ARO), Volcani Center, Rishon LeZion, Israel

**Sheena Kumari** Institute for Water and Wastewater Technology, Durban University of Technology, Durban, South Africa

**Ajay Kumar Mishra** Terrestrial Ecosystem Management Laboratory, Kyoto University, Kyoto, Japan

International Rice Research Institute, New Delhi, India

**Disha Mishra** Central Institute of Medicinal and Aromatic Plants (CIMAP), Lucknow, India

**Jyotiprakash Mishra** Department of Agronomy, OUAT, Bhubaneswar, Odisha, India

**Sarita Mishra** Department of Chemistry, UdaiPratap College, Varanasi, Uttar Pradesh, India

**Supriya Nath** Central Water and Power Research Station, Khadakwasla, Pune, Maharashtra, India

**Kin Israel Notarte** Faculty of Medicine and Surgery, University of Santo Tomas, Manila, Philippines

**Debanjana Pal** Laboratory of Applied Stress Biology, Department of Botany, University of Gour Banga, Malda, West Bengal, India

**Shilpi Pandey** Department of Botany, Centre of Advanced Study in Botany, Institute of Science, Banaras Hindu University, Varanasi, India

**Kajal Patel** Natural Resources Management Laboratory, Department of Botany, University of Delhi, New Delhi, India

**Abhijit Sarkar** Laboratory of Applied Stress Biology, Department of Botany, University of Gour Banga, Malda, West Bengal, India

**Amita Shakya** Department of Agriculture and Environmental Sciences, National Institute of Food Technology Entrepreneurship and Management, Kundli, Sonipat, Haryana, India

**Sheetal Sharma** International Rice Research Institute, New Delhi, India

**Anand Narain Singh** Soil Ecosystem and Restoration Ecology Lab, Department of Botany, Panjab University, Chandigarh, India

**Jatinder Singh** Department of Environmental Science and Technology, Central University of Punjab, Bathinda, India

**Pardeep Singh** Department of Environmental Science, PGDAV College, University of Delhi, New Delhi, India

**Ranjana Singh** Department of Botany, Government Model Degree College, Arniya-Bulandsahr, UP, India

**Rishikesh Singh** Institute of Environment and Sustainable Development, Banaras Hindu University, Varanasi, India

**Vipin Kumar Singh** Department of Botany, Centre of Advanced Study in Botany, Institute of Science, Banaras Hindu University, Varanasi, India

**G. N. Tanjina Hasnat** Institute of Forestry and Environmental Sciences, University of Chittagong, Chittagong, Bangladesh

**Diksha Tokas** Soil Ecosystem and Restoration Ecology Lab, Department of Botany, Panjab University, Chandigarh, India

**Rakhi Tyagi** Institute of Environmental Studies, Kurukshetra University, Kurukshetra, Haryana, India

**Shobhit Raj Vimal** Department of Environmental Microbiology, Babasaheb Bhimao Ambedkar University, Lucknow, Uttar Pradesh, India

**Rajni Yadav** Soil Ecosystem and Restoration Ecology Lab, Department of Botany, Panjab University, Chandigarh, India

# Part I
# Sources of Emerging Contaminants in Agroecosystems

# Chapter 1
# Sources and Impacts of Emerging Contaminants in Agroecosystems

**G. N. Tanjina Hasnat**

**Abstract** Emerging contaminants present in the environment for the long time at a tolerable level have increased nowadays in significance level due to human activities. Throwing of man-made hazardous chemicals and equipments into the soil and water increases the percentage of chemicals in the adjacent agricultural fields, water, and air. Crops and fishes produced in the contaminated agricultural fields and water bear the higher percentage of hazardous chemicals that people and livestock uptake through food chain. Emerging contaminants have potential threats to human and environmental health. Among the natural and man-made sources, the contaminants that produced through man-made process received high concerns recently because of its diverse, persistence and hazardous nature.

In this chapter, I had tried to find out the possible sources of emerging contaminants and their consequences on the agro-ecosystem as well as on the whole environmental system. Different kinds of pharmaceuticals and personal care products, bio-terrorism agents, persistent organic pollutants, nanomaterials, metabolites and transformation products of man-made chemicals, flame retardants are the major sources of man-made emerging contaminants. A number of diseases and abnormalities in humans and animals, environmental degradation, and ecosystem imbalance are the important outcomes of these emerging contaminants. This is a chain of reaction, and exposure to these agents makes the way for further exposure. High contamination sometimes creates radioactivity that causes cancer. Long term plan with strict policies and legislations could be helpful in future reduction of emerging contaminants from environment. For betterment of environment and for a livable future world a high concentration should be given on emerging contaminants and more research works are recommended.

**Keywords** Agricultural environment · Diseases · Ecosystems · Food chain · Health effects · Naturally produced contaminants · Man-made chemicals

---

G. N. Tanjina Hasnat (✉)
Institute of Forestry and Environmental Sciences, University of Chittagong, Chittagong, Bangladesh
e-mail: gnthasnat@gmail.com



V. Kumar Singh et al. (eds.), *Sustainable Agriculture Reviews 50*, Sustainable Agriculture Reviews 50, https://doi.org/10.1007/978-3-030-63249-6_1

## Abbreviations

| | |
|---|---|
| 4MBC | 4-methyl-benzilidine-camphor |
| BBB | blood brain barrier |
| CECs | contaminants of emerging concerns |
| DEET | N,N-diethyl-m-toluamide |
| EHMC | 2-ethyl-hexyl-4-trimethoxycinnamate |
| GIS | Geographic Information System |
| HBCD | hexabromocyclododecane |
| HHCB | 1,3,4,6,7,8-hexahydro-4,6,6,7,8,8,-hexamethyl-cyclopenta[*g*]benzopyran |
| NSAIDS | nonsteroidal anti-inflammatory drug |
| OPFRs | organophosphate flame retardants |
| PBDE | polybrominated diphenyl ethers |
| PCDDs | polychlorinated dibenzo-para-dioxins |
| PCDFs | polychlorinated dibenzofurans |
| RS | Remote Sensing |
| TBBPA | tetrabromobisphenol A |
| TCDD | 2,3,7,8-tetrachlorodibenzo-p-dioxin |
| UNEP | United Nations Environment Programme |

## 1.1 Introduction

Emerging contaminants are the naturally occurring and man-made chemical substances which are not newly created rather than existed in nature at a tolerable level since long. The tolerable level of these chemicals in nature is essential. But, the presence of these substances are recognized very recently and created concerns as these chemicals or substances have increased in significance level in nature due to human activities or natural disasters (Rosenfeld and Feng 2011; Boxall 2012; Sauvé and Desrosiers 2014; Singh et al. 2018; Dey et al. 2019). These contaminants are found in the food sources such as crops, plants, livestock and fishes; and in water sources such as groundwater, surface water, municipal water, drinking water, etc. Thus the contaminant agents readily entered into humans or animals body, and cause different kinds of diseases and abnormalities (Pal et al. 2014; Birch et al. 2015; Careghini et al. 2015; Clarke and Cummins 2015; Richardson and Kimura 2017). Emerging contaminants have potential threats to humans, livestock as well as environmental health.

Emerging contaminants are also known as emerging organic contaminants (Jurado et al. 2012; Lapworth et al. 2012) and contaminants of emerging concerns (CECs) (Dulio et al. 2018). New contaminants are emerging recently with the discovery of new sources and methods or technologies to identify it (DoD 2010; Rosenfeld and Feng 2011). The contaminants are increased in nature by throwing of pharmaceutical materials, medication equipment, nanomaterials, industrial materials, veterinary medicines, solid wastes, fertilizers, biosolids, etc. into the soil and water that makes it more vulnerable and increase the percentage of chemicals in the adjacent agricultural fields, water, and air (Rosenfeld and Feng 2011). Thus humans and animals are directly and indirectly exposed to these emerging contaminants. From chemical analysis point of view, emerging contaminants are classified into two major parts – naturally produced compounds and man-made chemicals. From my personal point of view, for maintaining environmental balances there are many chemical substances naturally existing in the environment. The optimum level of these chemical substances is very important for sustaining environment and human health. But increase in excessive rates create imbalance in nature. Due to natural phenomenon or human induced causes, these contaminants increased from its threshold level and sometimes modified that act adversely.

Emerging contaminants are the most time-oriented research issue and recently has attracted the immense concern of researchers, environmentalists and policy makers. These contaminants emerge the awareness nowadays. The main purpose of this chapter is to discuss about the major sources of these agents and their impacts on environment as well as on human health.

## 1.2 Naturally Produced Compounds

Naturally occurring chemicals or microorganisms which have the potential to cause great harm to human and environmental health are termed as naturally produced emerging contaminants (Rosenfeld and Feng 2011). Toxins are produced by various natural contaminants such as bacteria, viruses, insects, fungi and plants (Sultana and Hanif 2009; Boxall 2012; Suleiman and Kurt 2015). These toxins are separated according to their origin such as mycotoxins (fungal origin) (Sultana and Hanif 2009; Suleiman and Kurt 2015), bacterial toxins (bacterial origin) (Alouf 2006), phycotoxins (algal origin) (Rossini and Hess 2010; Anfossi et al. 2013; Botana and Alfonso 2015), phytotoxins (plant origin) (Strobel 1974; Duke and Dayan 2011; Cimmino et al. 2014), and zootoxins (animal origin) (Gwaltney-Brant 2017; Gwaltney-Brant et al. 2018). Persistent organic pollutants are generally found in the nature due to human activities such as use of pharmaceuticals, insecticides, pesticides, industrial chemicals and are resistant to the environmental degradation (Lesage 2015), but many persistent organic pollutants arise naturally at the time of natural phenomena like volcanoes and other biosynthetic pathways (El-Shahawi et al. 2010).

Tungsten, a naturally occurring element is water-soluble and found above safety limits in water sources. This contaminant agent has toxic health effects (Koutsospyros et al. 2011; Witten et al. 2012). Excessive exposure to it affects the eyes, skin, respiratory system and blood (NIOSH 2010) as well as has the potential of acting like carcinogen (Shukla et al. 2014; Laulicht et al. 2015). Polycyclic aromatic hydrocarbons are naturally produced at the time of incomplete combustion of organic matter or wildfire (Ramesh et al. 2011). The agents are also found during chemical transformation of organic sediments into fossil fuels, burning of vegetation due to lightning strikes, and decaying of organic matter and volcanic activities (Lee 2010; Abdel-Shafy and Mansour 2016). In garden compost polychlorinated dibenzo-para-dioxins (PCDDs) and polychlorinated dibenzofurans (PCDFs) grow naturally (Öberg et al. 1993).

## 1.3 Man-Made Chemicals

Man-made emerging contaminants are emerging concern now as it is detected in living and non-living environmental substances as well as food chain, sometimes more than the threshold level (Lei et al. 2015). The adverse effects of man-made emerging contaminants are diverse on human and ecosystem health, though many negative impacts are still unknown and under research. The broadly classified of most common and adverse man-made emerging contaminants are – pharmaceuticals and personal care products, bio-terrorism agents, persistent organic pollutants, nanomaterials, and metabolites and transformation products.

### *1.3.1 Pharmaceuticals and Personal Care Products*

Pharmaceutical products are the therapeutic drugs used to prevent and treat human and animal diseases. Personal care products are those chemicals used to cleanse, alter, or enhance the appearance of the body as well as improve the quality of daily life (Boxall et al. 2012; Paulsen 2015; Stefanakis and Becker 2016). Pharmaceutical products basically consider antibiotics, hormones and drugs (Table 1.1), whereas personal care products are considered as both drugs such as mouthwashes, acne medications, steroid creams, sunscreens, lip balms, hand sanitizers, soaps, etc. and cosmetics such as lotions, make-up, perfume, etc. and contain a large and diverse group of organic compounds (Table 1.2) (Paulsen 2015). Pharmaceuticals and personal care products have chemical properties making them environmentally persistent (Clarke and Smith 2011). The pharmaceuticals and personal care products have various known and unknown side-effects on ecosystems and environment. The metabolites and transformation products of pharmaceuticals and personal care products are found in wastewater and aquatic environments throughout the world (Liu and Wong 2013; Dai et al. 2014; De García et al. 2014; Kosma et al. 2014; Ebele et al. 2017; Yang et al. 2017a) thus known as effluent derived contaminants (Oulton et al. 2010; Thomaidis et al. 2012). Even after treatment of wastewater containing pharmaceuticals and personal care products, almost all of the categories of antibiotics and most of the pharmaceuticals and personal care products are not eliminated, therefore, fully or partially persist in the treated water (Lin et al. 2010a; Tabe et al. 2010; Kosma et al. 2014). This treated water sometimes serve as

**Table 1.1** Most common pharmaceuticals products

| Contaminant groups | Representative compounds |
|---|---|
| Human antibiotics and veterinary | Trimethoprim, erytromycine, amoxicillin, lincomycin, sulfamethaxozole, chloramphenicol, clarithromycin, erythromycin, sulfamethoxazole, sulfadimethoxine, ciprofloxacin, norfloxacin, chloramphenicol, roxithromycin |
| Analgesics, anti-inflammatory drugs | Ibuprofene, diclofenac, paracetamol, codein, acetaminophen, acetylsalicilyc acid, fenoprofen, diclofenac, ibuprofen, acetaminophen, acetylsalicylic acid |
| Psychiatric drugs | Diazepam, carbamazepine, primidone, salbutamol |
| β-blockers | Metoprolol, propanolol, timolol, atenolol, sotalol, metoprolol, propanolol |
| Lipid regulators | Bezafibrate, clofibric acid, fenofibric acid, etofibrate, gemfibrozil |
| X-ray contrasts | Iopromide, iopamidol, diatrizoate |
| Hormones | Estrone (E1), estradiol (E2), ethinylestradiol (EE2), 17β-estradiol, estrone, estriol, 17α-ethinyloestradiol, mestranol |
| Antiepileptic drugs | Carbamazepine, primidone |
| Blood lipid regulators | Clofibrate, gemfibrozil |
| Contrast media | Diatrizoate, iopromide |
| Cytostatic drugs | Ifosfamide, cyclophosphamide |

**Sources:** Daughton and Ternes (1999), Liu and Wong (2013), Stefanakis and Becker (2016)

**Table 1.2** List of some of the most commonly used personal care products

| Contaminant groups | Representative compounds |
|---|---|
| Fragrances | Nitro, polycyclic and macrocyclic musks, phthalates |
| Sun-screen agents/ ultraviolet filters | Benzophenone, methylbenzylidene camphor; 2-ethyl-hexyl-4-trimethoxycinnamate (EHMC), 4-methyl-benzilidine-camphor (4MBC) |
| Insect repellents | N,N-diethyltoluamide; N,N-diethyl-m-toluamide (DEET) |
| Antimicrobial agents/ disinfectants | Triclosan, Triclocarban |
| Synthetic musks/ fragrances | Galaxolide (HHCB), Toxalide (AHTN) |
| Preservatives | Parabens (alkyl-p-hydroxybenzoates) |

**Sources:** Daughton and Ternes (1999), Liu and Wong (2013), Stefanakis and Becker (2016)

recreational and drinking water (Oulton et al. 2010; Wells et al. 2010; Clarke and Smith 2011; Hernando et al. 2011; Reif et al. 2011). The pharmaceuticals and personal care products are receiving attention in recent days as important emerging contaminants in the environment (Thomaidis et al. 2012; Liu and Wong 2013; Solé and Sanchez-Hernandez 2018). The globally used high quantities of pharmaceuticals and personal care products create serious pollution (Liu and Wong 2013). These contaminants are found to accumulate in liver, brain, muscle, and blood tissues of fish species (Escarrone et al. 2016). Some of the most common pharmaceuticals and personal care products are enlisted in Tables 1.1 and 1.2.

Sewage treatment plants have been found as the main source of pharmaceuticals and personal care products throughout the world (Li 2014; Subedi et al. 2017; Yang et al. 2017a). In China, pharmaceuticals and personal care products are found in livestock sewage, hospital sewage, swine nursery sewage, slaughter house sewage, municipal sewage, swine sewage, sewage of a contraceptives producing factory (Liu and Wong 2013), and sewage influents and effluents (Peng et al. 2012). The report from other countries such as Korea, UK, Sweden, Finland, US, Japan, Portugal, Canada, Finland and Norway have also revealed the existence of pharmaceuticals and personal care products in sewage influent and effluent (Behera et al. 2011; Liu and Wong 2013). The most important group of organic compounds found in sewage treatment plants are antibiotics (Peng et al. 2012), antifungal drugs (Peng et al. 2012), hormones (Liu et al. 2012), pharmaceuticals (Sui et al. 2011), carbamazepine (Zhou et al. 2011), antimicrobial agents (Zhao et al. 2010), synthetic musks (Nakata and Shinohara 2010), insect repellents (Sui et al. 2011), ultraviolet filters (Nakata and Shinohara 2010), preservatives (Kasprzyk-Hordern et al. 2009), and triclosan (Liu and Wong 2013).

Pharmaceuticals and personal care products also contaminate the water bodies if treated water discharges from sewage treatment plants to the aquatic environment (Liu and Wong 2013). Antibiotics (Kasprzyk-Hordern et al. 2009), antifungal drugs (Huang et al. 2010), hormones (Ferguson et al. 2013), pharmaceuticals (Ramaswamy et al. 2011), carbamazepine (Zhou et al. 2011), triclosan (Ramaswamy et al. 2011), antimicrobial agents, preservatives, polycyclic musks, ultra-violet (UV) filters (Kasprzyk-Hordern et al. 2009), synthetic musks, galaxolide (HHCB) and tonalide (AHTN) (Parolini et al. 2015), N,N-diethyl-meta-toluamide (DEET) (Liu and Wong 2013) are the major chemical compounds found in China, Vietnam, France, UK, Finland, US, Korea, Spain, Netherlands, Australia, Brazil, India as contaminants of surface water bodies (Liu and Wong 2013). Surface water bodies including drinking water sources, rivers, lakes, pore water, sea water, off-shore seawater, as well as raw water sources are reported to be contaminated with pharmaceuticals and personal care products (Yang et al. 2018). In addition, the contaminant agents are also detected in the sediments of surface water (Ashfaq et al. 2019). Pharmaceuticals and personal care products in sediments are less studied yet, but the chemical compounds like antibiotics, antifungal drugs, hormones, antimicrobial agents, polycyclic musks and UV filters have been found in the sediments of river and coastal water (Conkle et al. 2012).

Soil is affected by pharmaceuticals and personal care products through the application of sludge residuals into it. Throwing wastes into the soil or landfill and livestock wastes used in the agricultural fields as fertilizers, and used surface or underground water for irrigation purposes (Liu and Wong 2013) affect the agricultural production as well as food chain. Pollutants in the soil spread in two ways – (i) uptake by plants, and (ii) leaching to the ground water or runoff to the surface water (Liu and Wong 2013).

Pharmaceuticals and personal care products are the most important and commonly used products by human, thus people are easily exposed to the contaminants. Human beings are exposed to pharmaceuticals and personal care products in three different ways – (i) daily use of personal care products, (ii) inhalation, and (iii) dietary intake (Liu and Wong 2013). Chemical groups of pharmaceuticals and personal care products having relatively high volatility are found in the dusts of indoor air (Rostkowski et al. 2019; Wang et al. 2016) that people inhale. Pharmaceuticals and personal care products are also accessible to humans through drinking water and food (Qiao et al. 2011). These contaminants have also been found in the tap water, bottled water and even in baby bottles (Li et al. 2010a). Due to daily inhalation or uptake by human beings the contaminants may accumulate in human milk, blood, and urine (Liu and Wong 2013).

Pharmaceuticals and personal care products have different effects on ecosystems and human health. Antibiotics in environment increase the resistance of bacteria and promote the emergence of antibiotic resistance genes (Kumar et al. 2012). Growth of algae is inhibited by antibiotics such as roxithromycin, clarithromycin and tylosin (Han et al. 2020); blood lipid regulator such as gemfibrozil (Araujo et al. 2011); and antimicrobial agents such as triclosan and triclocarban (Yang et al. 2008). Caffeine, the stimulant drug causes endocrine disruption of goldfish (Li et al. 2012). Diclofenac, a nonsteroidal anti-inflammatory drug (NSAIDS) causes renal lesions, gill alterations, and immune system changes of rainbow trout (Derakhsh et al. 2020). Carbamazepine, an antiepileptic drug leads to oxidative stress in rainbow trout (Li et al. 2010b). The propranolol ($\beta$-blocker) reduces the viable eggs of Japanese medaka (Massarsky et al. 2011). The synthetic musk 1,3,4,6,7,8-hexahydro-4,6,6,7,8,8,-hexamethyl-cyclopenta[*g*]benzopyran (HHCB) creates oxidative stress in goldfish (Chen et al. 2012). Hormones induce adverse effects on reproductive and development systems. These effects may also lead to reduced fertility, feminization of males, induction of vitellogenesis in male, and generation of intersex organisms (Golshan 2015; Talib and Randhir 2016).

### 1.3.2 Bio-terrorism Agents

Bio-terrorism is the intentional release or dissemination of biological agents in order to injure or kill people, animals, and plants (Balali-Mood et al. 2013; Christian 2013; Jansen et al. 2014; Sengupta and Agrawal 2014; Pal et al. 2017). Modification of biological agents in the laboratory makes these agents more dangerous. All these agents are although phylogenetically diverse but have potentiality of dissemination in form of aerosol (Greenfield et al. 2002). There are quite large numbers of recognized biological agents. Human modified form of biological agents includes bacteria, viruses, insects, fungi, and toxins (Jansen et al. 2014; Sengupta and Agrawal 2014; Pal et al. 2017) and have been potentially used in bioterrorism. These biological agents are released from a point source. Some of these agents are transmitted from person to person while others are not transmitted. In the transmitted agents the infection generally takes hours to days to appear the symptoms. Smallpox, viral fever, and plague are highly transmissible from person to person through the respiratory droplets (CDC 2002; Darling et al. 2002; Horton et al. 2002). Isolating the infected person is the best solution to prevent infection. The U.S. Centre for Disease Control and Prevention (CDC) has developed a classification system for biological terror agents (Mahy 2003). The CDC categorizes biological agents as Category A, B and C (CDC 2002; Darling et al. 2002; Moran 2002; Rotz et al. 2002; Pappas et al. 2006) which are elaborated as:

#### 1.3.2.1 Category A Agents and Diseases

These are the highest priority agents. These diseases are transmitted from person to person. These agents are potential to cause serious public health effects, create social disaster and need special attention to prevent the effects. The agents have the ability to create high illness and mortality rate. Terrorism experts prefer this category as these agents have capability to cause great harm to civilians and environment (CDC 2002; Darling et al. 2002; Horton et al. 2002; He et al. 2009). Six diseases are caused by category A agents. The biological agents along with associated diseases are listed in Table 1.3.

#### 1.3.2.2 Category B Agents and Diseases

Category B agents are comparatively less easily disseminated from person to person. The causal diseases create less illness and low mortality rate than category A agents (CDC 2002; Horton et al. 2002; Moran 2002; Pappas et al. 2006). Diseases and causal biological agents of category B are listed in Table 1.4.

**Table 1.3** Category A bio-terrorism agents and associated diseases

| Biological agent | Diseases |
|---|---|
| *Bacillus anthracis* | Anthrax |
| *Clostridium botulinum* toxin | Botulism |
| Ebola virus (Filoviruses) | Viral hemorrhagic fevers |
| *Francisella tularensis* | Tularemia |
| Lassa virus (Arenaviruses) | Viral hemorrhagic fevers |
| Machupo (Arenaviruses) | Viral hemorrhagic fevers |
| Marburg virus (Filoviruses) | Viral hemorrhagic fevers |
| *Variola major* | Smallpox |
| *Yersinia pestis* | Plague |

**Source:** Anderson and Bokor (2012), CDC (2002), Darling et al. (2002), Horton et al. (2002)

**Table 1.4** Category B bio-terrorism agents and resultant diseases

| Biological agents and products | Diseases |
|---|---|
| *Brucella* species | Brucellosis |
| *Burkholderia mallei* | Glanders |
| *Burkholderia pseudomallei* | Melioidosis |
| *Chlamydia psittaci* | Psittacosis |
| *Clostridium botulinum* toxin | Botulism |
| *Clostridium perfringens* | Epsilon toxin |
| *Cryptosporidium parvum* | Cryptosporidiosis |
| *Coxiella burnetii* | Q fever |
| Dengue fever (Flaviviruses) | Viral encephalitis |
| Eastern equine encephalitis (Alphaviruses) | Viral encephalitis |
| *Escherichia coli* O157:H7 | Bacterial and clinical infection |
| Mycotoxins | Hepatotoxicity, cancer |
| *Ricinus communis* (castor beans) | Ricin toxin poisoning |
| Saint Louis encephalitis (Flaviviruses) | Viral encephalitis |
| *Salmonella* bacteria | Salmonellosis |
| *Shigella* dysenteriae | Shigellosis |
| *Staphylococcus* enterotoxin B. | Epsilon toxin |
| Venezuelan equine encephalitis (Alphaviruses) | Viral encephalitis |
| *Vibrio parahaemolyticus* | Cholera |
| *Vibrio vulnificus* | Infections |
| West Nile encephalitis (Flaviviruses) | Viral encephalitis |
| Western equine encephalitis (Alphaviruses) | Viral encephalitis |

**Source:** Anderson and Bokor (2012), CDC (2002), Horton et al. (2002), Moran (2002), Pappas et al. (2006)

#### 1.3.2.3 Category C Agents and Diseases

Category C type of viruses are easily available in nature, so they can be easily multiplicated under laboratory conditions and consequently used for mass dissemination. These viruses cause major effects on human health and have potentiality in high rate of illness and mortality (CDC 2002; Horton et al. 2002; Moran 2002). Diseases and causal biological agents of category C are listed in Table 1.5.

### *1.3.3 Persistent Organic Pollutants*

Globally, man-made persistent organic pollutants are now emerging issue and increased in environment due to current and past uses in the form of pesticides, solvents, pharmaceuticals and industrial chemicals. Most common persistent organic pollutants compounds adopted by the United Nations Environment Programme (UNEP) in the Stockholm Convention are – polycyclic aromatic hydrocarbon, aldrin, dieldrin, endrin, dichlorodiphenyltrichloroethane, chlordane,

**Table 1.5** Category C bio-terrorism agents and resultant diseases

| Biological agents | Diseases |
|---|---|
| Hantavirus | Infectious disease |
| *Mycobacterium tuberculosis* | Infectious disease |
| Nipah virus | Infectious disease |
| Tickborne encephalitis viruses | Infectious disease |
| Tickborne hemorrhagic fever viruses | Hemorrhagic fever |
| *Rickettsia prowazekii* | Typhus fever |
| Yellow fever virus | Hemorrhagic disease |

**Source:** Anderson and Bokor (2012), CDC (2002), Horton et al. (2002), Moran (2002)

heptachlor, mirex, toxaphene, hexachlorobenzene, polychlorinated biphenyls, PCDDs and PCDFs, chlordecone, hexabromobiphenyl, and hexachlorocyclohexanes (WHO 1999; Wong and Poon 2003). Persistent organic pollutants are detected in river water as well as river and marine sediment. A high concentration of dieldrin and dichlorodiphenyltrichloroethane was noticed in agricultural soil where vegetables were cultivated intensively (Muendo et al. 2012; Olatunji 2019).

Polycyclic aromatic hydrocarbons are discharged from industrial and wastewater treatment plants, attached with solid particles and settled at the bottom of lakes, rivers and sea (Bouloubassi et al. 2012). These agents are mainly generated by anthropogenic activities (Lee 2010) in residential and industrial areas (Lee 2010; Abdel-Shafy and Mansour 2016). The contaminants produced during incomplete combustion of organic materials such as coal, oil, petrol, gas, garbage, and wood in domestic area (Kim et al. 2013; Rengarajan et al. 2015). Human induced sources of polycyclic aromatic hydrocarbons are residential heating, motor vehicle exhaust, coal gasification and liquefying plants, carbon black, coal-tar pitch and asphalt production, petrochemical industries, rubber tire and cement manufacturing companies, bitumen and asphalt industries, wood preservation, commercial heat and power generation, waste incineration, coke and aluminum production, catalytic cracking towers and related activities in petroleum refineries (Lee 2010; Abdel-Shafy and Mansour 2016). Polycyclic aromatic hydrocarbons have toxic, mutagenic and carcinogenic properties (Ghosal et al. 2016). The global distribution of persistent organic pollutants generally occurs through atmospheric media due to its widespread sources and persistent characteristics (Wang et al. 2012). People are exposed to polycyclic aromatic hydrocarbons in gaseous or particulate phases in ambient air. The agents are also highly soluble in lipid and are readily absorbed by human beings and other animals (Yebra-Pimentel et al. 2015; Abdel-Shafy and Mansour 2016). Exposure to high concentration of polycyclic aromatic hydrocarbons for long time creates serious health risks, particularly risks of lung cancer (Lee 2010; Abdel-Shafy and Mansour 2016).

Polychlorinated biphenyls, another persistent organic pollutant basically are a group of man-made chemicals and no any natural sources are found yet. Polychlorinated biphenyls are used as adhesive, paints, flame-retardants, surface coating, coolants and lubricants in transformers, capacitors, and other electrical equipment because of their insulating and heat absorbing capacity (Hong et al. 2018). The agents are very much harmful to human beings and persistence in nature (Winneke 2011). Polychlorinated biphenyls have been banned since 1974, but are still found in the environment as a contaminant. It is estimated that the polychlorinated biphenyls entered into the air, water, and soil before 1974 at the time of manufacturing and used, and persisted in the natural environment (Worm et al. 2017). These agents are transported at the time of accidental spills and leaks during their transport, from products containing polychlorinated biphenyls, hazardous waste sites, illegal disposal, and burning of industrial wastes. In remote areas, urban or industrial sites and in rural areas, outdoor polychlorinated biphenyl level in the air is comparatively lesser than the indoor level in air (Melymuk et al. 2012). The agents can travel long distances in air and deposited in areas far away from source, followed by binding to soil strongly and sticking to organic particles and sediments (Gioia et al. 2013) facilitating uptake by organisms, fishes, marine mammals and other animals (Beyer and Biziuk 2009). Lipophilic polychlorinated biphenyls then build up in the fatty tissues of invertebrates, birds, fishes, mammals, and human's body through food chain. Polychlorinated biphenyls are also taken up through normal breathing (Xing et al. 2011). The agents affect fertility, child development, and immune system. The other major health effects found are acne, skin rashes, liver damage in adults, pubertal disorders in female, neurobehavioral and immunological changes in children upon exposure (Mantovani and Fucic 2014). Polychlorinated biphenyls are identified as carcinogens and cause certain kinds of human cancers like liver and biliary tract cancer (Liu et al. 2010; Man et al. 2011; Zani et al. 2017). In animals, polychlorinated biphenyls exposure leads to decreased response to pain, reduced production of antibody, negative effects on fertility, reproductive organs, and female hormonal activity, damage lung, kidney, pancreas, liver, cause diarrhea, breathing difficulties, dehydration, coma, anemia, stomach or thyroid gland injuries as well as cancer (Sauer et al. 1994; Pavuk et al. 2004).

Man-made PCDDs and PCDFs create greater environmental problem (Al-Alawi 2008; Chalmers et al. 2019). PCDDs and/or PCDFs are generally produced through combustion process (Fiedler 2003). PCDDs and/or PCDFs are emitted to the atmosphere from two different sources – industrial and non-industrial (Cheruiyot et al. 2016). Municipal waste combustion, coal combustion, landfill gas combustion, cement manufacturing, chemical waste incineration plant, and clinical waste incinerator are major industrial sources. Crematoria, animal carcass incinerator, cars, vans, buses, motorcycles, etc. are the main non-industrial sources of PCDDs and PCDFs in environment. Among all sources municipal solid waste incineration contribute maximum for release of PCDDs and PCDFs (Vassura et al. 2011). These agents are also produced at the time of iron and steel production (Li et al. 2011), chlorine bleaching of pulp and paper (Wang et al. 2014), graphite electrodes (Yamamoto et al. 2018), car exhausts and emissions from cars (Rey et al. 2014). The intake of PCDDs and PCDFs occur through inhalation (Yang et al. 2017b). In indoor, PCDDs and PCDFs concentration are found high when particles get coated with polychlorinated biphenyls (Jacobs et al. 2013) and pentachlorophenol are used as a wood preservative (Lee et al. 2006a, b). Smoking cigarettes also increase the intake of PCDDs and PCDFs (Wohlfahrt-Veje et al. 2014). According to WHO, the intake of PCDDs and PCDFs through drinking water is negligible, but intake through food causes major exposure to general people (Lu et al. 2016). PCDDs and PCDFs are lipophilic and animal origin foods are the most important sources (Fernandez-Gonzalez et al. 2015); in addition, plants too contain very low level of these contaminants. Particularly, daily intake of fatty fishes and crabs from contaminated waters increase PCDDs and PCDFs levels in human body significantly (Lee et al. 2006a, b). Infants are also vulnerable as in human breast milk PCDDs and PCDFs have been detected at high concentrations (Chovancová et al. 2011; Tai et al. 2013).

2,3,7,8-tetrachlorodibenzo-p-dioxin (TCDD) and related compounds causes lethality, body weight loss, immunosuppression, oedema, dermal, developmental, reproductive, and neurobehavioural alterations, liver and endocrine toxicity, alterations in lipid metabolism and gluconeogenesis, modulation of responsiveness to hormones and growth factors, alterations in circulating levels of hormones, thymic atrophy, induction of various enzyme activities, tumour promotion and increased risk of cancer (Ciftci et al. 2011; Goodale 2013).

### *1.3.4 Nanomaterials*

Nanomaterials are small nanoscale substances, having 1–100 nm structural components, nanomaterials contain thermal stability, high strength and conductivity, low permeability, and microwave absorption capacity (Chen et al. 2014; Măruţescu et al. 2019; Prabha et al. 2020). Three different kinds of nanomaterials are generally present in the environment – natural, incidental, and engineered (Boxall 2012; Stuart and Compton 2015). Engineered nanomaterials are more harmful to environment and human health. These nanomaterials are produced intentionally and used in a various commercial purposes such as medical services, drug delivery, water purification, photovoltaics, contaminated area remediation, soil restoration problems; industrial products manufacturing purposes such as cosmetics, paints, cement, cleansing materials, sporting goods, cloths, tires, foods; and energy applications such as insulation materials, batteries, supercapacitors, energy storage and conversion, catalyst, and sensors (Edebali et al. 2018; Park et al. 2018; Ahmad et al. 2019; Bandala and Berli 2019). Due to lack of proper handling techniques and faulty instrumentation in commercial and industrial uses these contaminated agents sometimes released into different environmental matrices (Nowack et al. 2012).

Carbon-based (natural or engineered) nanomaterials are used for biomedical sensors, supercapacitors, and semiconductors (Kuzma et al. 2020). Metal oxide (natural or engineered) nanomaterials are used in tumor cell treatment, drug delivery applications, wastewater treatment, environmental clean-up, improving agricultural crops, etc. (Rasmussen et al. 2010; Xu et al. 2012; Rosa et al. 2017). Zero-valent metals (engineered) have been used in water filtration, iron wall remediation, removal of metal pollution such as chlorinated organic and nitroaromatic compounds, arsenic, nitrate, dyes, phenols etc. (Fu et al. 2014). Quantum dots (engineered) are applied in medicine, biological imaging, cancer therapy, photovoltaics, telecommunication, quantum computing, lasers, LEDs, solar cells, sensors, transistors, etc. (Jin et al. 2011; Frigerio et al. 2012). Dendrimers (engineered) are used for drug delivery, gene delivery, biomedical applications, tissue engineering, anticancer drugs, dendritic sensor, enhancing solubility, Photodynamic therapy, detecting dye molecule, etc. (Abbasi et al. 2014; Noriega-Luna et al. 2014). Composite nanomaterials (engineered) have a multivariate use such as catalysts, modifiers, optical sensors, semiconductor junctions, cancer detector, etc. (Calandra et al. 2013; Wu et al. 2015). Nanosilver particles (engineered) are most attractive nanomaterials and used mostly in biomedical applications and personal health care. These nanomaterials also reveal antimicrobial, antiviral, antifungal, and anti-inflammatory properties (Ge et al. 2014).

Nanomaterial contaminants may be released from point and non-point sources. These contaminants are mixed with water through different ways such as solid wastes or wastewater effluents may discharged into water or soil, direct discharges into water or soil, or accidental spillages from industries and eventually contaminate water (Patnaik 2017). Nanomaterials are persistent in nature and are transported to a greater distance through different media such as water, air as well as organisms (Lin et al. 2010b). Researches revealed that nanomaterials have negative effects on wildlife health (Poma et al. 2014). Aquatic organisms are much more affected by nanomaterials as they rapidly accumulate carbon black, titanium dioxide and polystyrene (Molins-Delgado et al. 2016). Nanomaterials cause the cell death of bacteria, impact soil microbial function and processes (Pawlett et al. 2013), and induce oxidative stress in the fish brains (Millán-Chiu et al. 2020). These contaminants also inhibit plant growth (Millán-Chiu et al. 2020).

Humans are generally exposed to nanomaterials through ingestion, inhalation, injection, and dermal exposure (WHO 2019). People working in the contaminated environment, are exposed to nanomaterials (Li et al. 2018). Due to small size, nanomaterials have the potential to reach the circulatory system (Tasso et al. 2020) and pass through both the blood brain barrier (BBB) and the placenta (Pietroiusti et al. 2013; Ali and Rytting 2014). Nanomaterials also affect the lung, liver, spleen, lymph nodes, nervous systems, cardiovascular systems and pulmonary arteries (Pietroiusti et al. 2013; Arami et al. 2015; De Matteis 2017; WHO 2019; Tasso et al. 2020). The dermal exposure occurs basically when sunscreen products are used (Tasso et al. 2020). Exposure via ingestion occurs through contaminated drinking water and food (Johnson et al. 2015; Omlor et al. 2015). Moreover, Engineered nanomaterials causes immunosuppression, immunostimulation, hypersensitivity or autoimmunity (Bonner 2011; Inoue and Takano 2011; WHO 2019).

### 1.3.5 Metabolites and Transformation Products of Man-Made Chemicals

Liver is the primary site for drug metabolism and most of the drugs pass through it and are subjected to multiple transformations in the human body. Drug is a complex chemical mixture and the frequent conversion reduces the general effects of drugs on living organisms (Michael et al. 2014; Aronson 2015). In human body, the metabolism of drugs occurs in two steps. In first step, oxidation, reduction or hydrolysis occurs and second step involves conjunction reactions (Guengerich and Yoshimoto 2018). Some pharmaceutical metabolites of human are detected in the environment including salicylic acid, gentisic acid, nortriptyline, o-hydroxy atorvastatin, p-hydroxy atorvastatin, 2-hydroxycarbamazepine, 3-hydroxycarbamazepine, 10,11-dihydro-10,11-epoxycarbamazepine, 10,11-dihydro-10,11-dihydroxycarbamazepine, 10,11-dihydro-10-hydroxycarbamazepine, clofibric acid, benzoylecgonine, erythromycin-$H_2O$, norfluoxetine, hydroxy-ibuprofen, carboxy-ibuprofen, 2-ethylene-1,5-dimethyl-3,3-diphenylpyrrolidine, ritalinic acid, 1-acetyl-1-methyl-2-dimethyloxamoyl-2-phenylhydrazine, dimethyloxalamide acid-(N0-methyl-N-phenyl)-hydrazide, 1-acetyl-1-methyl-2- phenylhydrazide, desmethylsertraline, desmethylvenlafaxine, and N4-acetylsulfamethoxazole (Michael et al. 2014). These metabolites are detected in influents and effluents, groundwater, river water, lake water, drinking water, biosolids, seawater, pharmaceutical effluents, and hospital effluents. People are exposed not only to one or two chemicals rather they are frequently affected by a complex mixture of chemicals (Paiva et al. 2011). These contaminants affect male reproductive system, cause infertility, and induce toxicity in humans. In animals, the contaminants cause decreased sperm count and reduced male reproductive organs (Paulsen 2015).

The metabolites and transformation products of fragrance chemicals cause anxiety, depression, headaches, migraines, phobias, panic attacks, obsessive-compulsive disorder, hormone disruption, and other neurological disorders. These products also create allergies, skin irritation, asthma, sinus, eczema, respiratory diseases, and multiple chemical sensitivity (Nigen et al. 2003; Monteiro and Boxall 2010; Jain 2012; Belsito et al. 2013; Park and Igarashi 2013). In addition, man-made metabolites and transformation products are also the important sources of indoor air pollution (Paulsen 2015).

### 1.3.6 Flame Retardants

Flame retardants are a group of man-made chemical contaminants used in preventing ignition or slowing fire. Flame retardants are toxic to the environment and persistent in nature (Segev et al. 2009). Flame retardants are commonly used in furnishings materials, electronics and electrical devices, building and construction materials, and transportation products. The most common furnishing materials in which flame retardants used are foams, cushions, chair coverings, sofas, curtains, wall hanging, beddings, mattresses, doormats, carpets, bean bags, etc.; electronics and electrical devices are desktops, laptops, fans, cameras, transformers, phones, ovens, televisions, generators, printers motors, etc.; building and construction materials are electrical wires, cables fibre glass, cellulose, polystyrene, polyurethane, insulation facings, etc.; and transportation products are parts of automobiles, airplanes and trains (Stubbings and Harrad 2014; Wei et al. 2015). Flame retardants are persistent in nature and remain in the environment (Xiao-Ju et al. 2012). Most common flame retardants are brominated flame retardants, polybrominated diphenyl ethers (PBDE's), tetrabromobisphenol A (TBBPA), hexabromocyclododecane (HBCD), and organophosphate flame retardants (OPFRs) (Kowalski and Mazur 2014; Noyes et al. 2015).

These contaminants build up in people and animal body over time affect the health. People exposed to flame retardants through three ways – inhalation, food and skin (Cao et al. 2014; Xu et al. 2016). The adverse health effects in animals and humans are – endocrine and thyroid disruption, problems in brain development and learning, abnormal mental and physical development, negative impacts on immune system and reproductive organs, harmful effects on neurological functions as well as causes cancer. These contaminants have adverse effects on fetus and child development. Children are more vulnerable to the toxic effects of flame retardants than adults as their body organs are still under development (Babrauskas et al. 2011; Wikoff and Birnbaum 2011; Lyche et al. 2015).

Environment is the source and sink of emerging contaminants, not only of a single contaminant rather than a matrix of numbers of natural and man-made contaminants (Zhang et al. 2013a). Basically, the man-made contaminants directly and indirectly effects on agro-ecosystems and human health. Different types of contaminants cause different types of diseases to livestock and human. Excessive exposure sometimes causes cancer. Due to adverse health impacts many man-made chemical compounds are banned in past but the persistent nature of those chemicals in environment and continual adding of new invented chemical compounds creates more disastrous situation. Due to this, contamination increased in recent days above the danger level, and thus, receiving more concerns of researchers, environmentalists and policy makers within very short time.

## 1.4 Agricultural Environment and Emerging Contaminants

Agricultural environments are more susceptible to emerging contaminants. Agricultural systems are badly affected by these contaminants in two ways i.e., directly and indirectly. These agents are directly released to the agricultural environment when fertilizers and pesticides are used in the agricultural fields and livestock are treated with veterinary medicines or their metabolites. Indirect release involves different ways. The most feasible way is soil fertilization with the residues of livestock, composting materials, and application of pesticides (Boxall 2012). Moreover, pharmaceuticals and personal care products, nanomaterials, flame retardants, and metabolites are released to the environment when irrigated with contaminated water and fertilized with biosolids (Boxall 2012; Yu et al. 2013; Qin et al. 2015).

Natural toxins that are released from plants, algae, and fungi have high potentiality to contaminate water as well as agricultural environment (Sharma et al. 2011). Veterinary medicines used for livestock are released to soil from compost, manure and sludge applied as fertilizer and have high contamination probability to the water and land (Kim et al. 2011). Release of natural and synthetic hormones by animal in the lands directly or through indirect processes such as manuring or sludge application to land impact the water resources and agricultural lands (Adeel et al. 2017). Man-made chemicals that are applied directly to the agricultural lands or irrigation water contaminated with some transformation products possess the characteristic tendency to contaminate the water resources and agricultural lands (Qin et al. 2015). Nanomaterials are released to the agricultural environment when nanopesticides, veterinary residues, and sludge are applied to crops and cultivable fields are irrigated with contaminated water. Bio-terrorism agents also possess high contamination potentiality consequently affecting agricultural environment, water resources, crops and livestock (Boxall 2012). Veterinary antibacterial from agriculture and aquaculture sectors create negative effects on soil. These agents reduce soil sulfate, thus decomposition of dung organic matter inhibited (Bakiu and Durmishaj 2018) and antibacterial resistance of soil increased (FEEDAP 2012). Addition of pesticides and metabolites of antibiotics significantly reduce the rate of degradation of human drugs in soil (Boxall 2012).

Emerging contaminants are transported to the agricultural system in different ways and finally enter into the food chain. Pharmaceuticals and personal care products and other emerging contaminants are degraded by physical, chemical and biological processes and stick to the soil particles of the agricultural lands as well as are acquired by plants (Clarke and Cummins 2015; Unuofin 2020). Some particles are leached into the groundwater while others via run-off and drainage find their ways to the receiving water bodies (Tran et al. 2019). Emerging contaminants are also accumulated in the water sediment (Luigi et al. 2015). Geographic information system (GIS) and remote sensing (RS) are using now-a-days to monitor the contaminated soil, air, water and sediment status under water, and are also using to apply appropriate remedy to restore the quality (Ramadas and Samantaray 2018).

## 1.5 Effects of Emerging Contaminants on Ecosystems and Human Health

Invertebrates are covering 95% of all animal species and affected severely by emerging contaminants (Canesi and Fabbri 2015). In invertebrates, insecticides show a number of risks such as biological effects, endocrine disrupting effects, inhibition of metamorphosis, reduction in growth rate, detrimental impacts on reproduction, prevention of adult emergence, disruption of water balance, even the larval and adult mortality. The mortality of invertebrates (prey) induces the mortality of vertebrates (predators) (Canesi and Fabbri 2015; Barbosa et al. 2018). In vertebrates, emerging contaminants effect on animal behaviors such as alter neural signaling or cognitive abilities, reduce the courtship and nest buildings behaviors in fishes, change animal movements, alter migration behavior, create aggression and ineffective communication, modify the adaptation nature and fear responses (Saaristo et al. 2018). Hormones used for human and livestock pose negative effects on the reproductive biology in lower vertebrates such as fishes, amphibians and other aquatic wildlife (Orlando and Ellestad 2014). Diclofenac, a non-steroidal anti-inflammatory drug used as veterinary and human pharmaceuticals reduced the population of three vulture species in Asia. Fluoroquinolone antibiotics are responsible for fatal embryo chondral damage (Boxall 2012).

Humans, are generally exposed to the emerging contaminants through food chain, particularly, the consumption of contaminated plants or plant parts, livestock, fishes and drinking water. People are affected by using of contaminated ground water, surface water, and inhaling indoor and outdoor air dusts (Lei et al. 2015; Gwenzi et al. 2018). Emerging contaminants cause great impacts on human health broadly on nervous system, neurological behaviors, reproduction system, and immune system. Exposure to these contaminants for long time may cause cancers (Lei et al. 2015; Gwenzi et al. 2018).

Emerging contaminants are investigated in the fresh and marine aquatic ecosystems. Fishes, amphibians, mollusks, reptiles, aves and invertebrates have been found seriously affected by these contaminants (Asif et al. 2018; Pittinger and Pecquet 2018). Maximum accumulation of pharmaceuticals and personal care products is reported in liver, gills and muscles of affected fishes (Malarvizhi et al. 2012). Emerging contaminants also have negative impacts on plant growth, algae and soil microbial activities (Wei et al. 2016; Snow et al. 2018). Pharmaceuticals affect follicular development, hinder oocytes and ovarian development, hamper egg maturation and testicular maturation, and cause sterility in animals (Fent et al. 2006; Kvarnryd et al. 2011; Zhang et al. 2013b). Pharmaceuticals, pesticides, parasiticides and herbicides effect on insect, crustaceans, and earthworms physiology and behaviour, as well as animal dung decomposition (González-Tokman et al. 2017; Gill et al. 2018; Villa et al., 2018). Hormonal effect in crabs and crayfishes occurs due to anorexic treatments (Boxall 2010). Estrogens induce endocrine disrupting effects on fishes (Söffker and Tyler 2012). Antibacterial and anti-inflammatory substances are reported to modulate structure of soil microbial communities, inhibit the growth of blue-green algae and aquatic plants, and biochemical effects in exposed fishes. Lipid regulators, analgesics, and beta blockers have also been described to accelerate biochemical effects in fishes. In addition, antianxiety drugs, cardiac glycosides and calcium channel blockers affect the development of invertebrates (Ding and He 2010). Hence, it can be said in word that all the components of ecosystems are vulnerable to be affected by emerging contaminants.

## 1.6 Conclusion and Recommendations

Emerging contaminants have received major concerns in recent days. Through manufacturing process, use and intake by livestock, emerging contaminants easily enter into the agricultural environment, thus into food chain. Organisms are eventually affected by a diverse type of diseases and the contaminant agents cause significant harm to environment and human health. Emerging contaminants can be compared with double sided blade. For maintaining regular food chain and regular life, use of these contaminants is crucial. Daily use and exposure to these contaminants cause severe health hazards, and as remedies of these health effects more medicines are needed, again increasing the contents of emerging contaminants in receiving environment. The contaminant agents are found to cause catastrophic hazards in ecosystems. Currently, it is difficult to identify the non-point sources of emerging contaminants. Though it is late to resolve the whole problem in a short time, but a long-term plan could minimize the present condition. Some recommendations for management of emerging contaminants are listed below:

- Finding out the each and every point source, and development of strategies for reducing the effects of emerging contaminants by proper recycling or processing
- Maintenance of strict rules for using safety masks and appearances when working in the emerging contaminants exposed areas
- Complete ban of disastrous emerging contaminants and invention of more environment friendly ones
- Long-term plan (100 years) to reduce emerging contaminants in nature
- Strict policies and legislations
- Compulsory education (partly) in schools and colleges
- Regular researches in universities and research centers
- Regular interval trainings to farmers, and
- Development of geographic information system (GIS) and remote sensing (RS) applications for regular monitoring of emerging contaminants production, use, disposal areas, legislative applications, etc.

## References

Abbasi E, Aval SF, Akbarzadeh A, Milani M, Nasrabadi HT, Joo SW, Hanifehpour Y, Nejati-Koshki K, Pashaei-Asl R (2014) Dendrimers: synthesis, applications, and properties. Nanoscale Res Lett 9(1):247. https://doi.org/10.1186/1556-276X-9-247

Abdel-Shafy HI, Mansour MS (2016) A review on polycyclic aromatic hydrocarbons: source, environmental impact, effect on human health and remediation. Egypt J Pet 25(1):107–123. https://doi.org/10.1016/j.ejpe.2015.03.011

Adeel M, Song X, Wang Y, Francis D, Yang Y (2017) Environmental impact of estrogens on human, animal and plant life: a critical review. Environ Int 99:107–119. https://doi.org/10.1016/j.envint.2016.12.010

Ahmad A, Lokhat D, Rafatullah M (2019) Survey of graphene-based nanotechnologies. In: Jawaid M, Ahmad A, Lokhat D (eds) Graphene-based nanotechnologies for energy and environmental applications. Elsevier, pp 23–39. https://doi.org/10.1016/B978-0-12-815811-1.00002-8

Al-Alawi MT (2008) Dioxin characterisation, formation, and minimisation. In: Mehmetli E, Koumanova B (eds) The fate of persistent organic pollutants in the environment, NATO Science for Peace and Security Series. Springer, Dordrecht, pp 269–282. https://doi.org/10.1007/978-1-4020-6642-9_21

Ali S, Rytting E (2014) Influences of nanomaterials on the barrier function of epithelial cells. In: Capco D, Chen Y (eds) Nanomaterial. Advances in experimental medicine and biology, vol 811. Springer, Dordrecht, pp 45–54. https://doi.org/10.1007/978-94-017-8739-0_3

Alouf JE (2006) Bacterial protein toxins. In: Holst O (ed) Bacterial toxins. Methods in molecular biology, vol 145. Springer Nature, Switzerland AG, Cham, pp 1–26. https://doi.org/10.1385/1592590527

Anderson PD, Bokor G (2012) Bioterrorism: pathogens as weapons. J Pharm Pract 25(5):521–529. https://doi.org/10.1177/0897190012456366

Anfossi L, Baggiani C, Giovannoli C, D'Arco G, Giraudi G (2013) Lateral-flow immunoassays for mycotoxins and phycotoxins: a review. Anal Bioanal Chem 405:467–480. https://doi.org/10.1007/s00216-012-6033-4

Arami H, Khandhar A, Liggitt D, Krishnan KM (2015) In vivo delivery, pharmacokinetics, biodistribution and toxicity of iron oxide nanoparticles. Chem Soc Rev 44(23):8576–8607. https://doi.org/10.1039/C5CS00541H

Araujo L, Villa N, Camargo N, Bustos M, García T, de Jesus Prieto A (2011) Persistence of gemfibrozil, naproxen and mefenamic acid in natural waters. Environ Chem Lett 9:13–18. https://doi.org/10.1007/s10311-009-0239-5

Aronson JK (ed) (2015) Meyler's side effects of drugs: the international encyclopedia of adverse drug reactions and interactions, 16th edn. Elsevier, New York, pp 1–7674

Ashfaq M, Li Y, Rehman MS, Zubair M, Mustafa G, Nazar MF, Yu CP, Sun Q (2019) Occurrence, spatial variation and risk assessment of pharmaceuticals and personal care products in urban wastewater, canal surface water, and their sediments: a case study of Lahore, Pakistan. Sci Total Environ 688:653–663. https://doi.org/10.1016/j.scitotenv.2019.06.285

Asif MB, Hai FI, Price WE, Nghiem LD (2018) Impact of pharmaceutically active compounds in marine environment on aquaculture. In: Hai F, Visvanathan C, Boopathy R (eds) Sustainable aquaculture. Applied environmental science and engineering for a sustainable future. Springer, Cham, pp 265–299. https://doi.org/10.1007/978-3-319-73257-2_9

Babrauskas V, Blum A, Daley R, Birnbaum L (2011) Flame retardants in furniture foam: benefits and risks. Fire Saf Sci 10:265–278. https://doi.org/10.3801/IAFSS.FSS.10-265

Bakiu R, Durmishaj S (2018) Medical waste effects and management: overview and future directions. SF J Environ Earth Sci 1(2):1019

Balali-Mood M, Moshiri M, Etemad L (2013) Medical aspects of bio-terrorism. Toxicon 69:131–142. https://doi.org/10.1016/j.toxicon.2013.01.005

Bandala ER, Berli M (2019) Engineered nanomaterials (ENMs) and their role at the nexus of food, energy, and water. Mater Sci Energy Technol 2(1):29–40. https://doi.org/10.1016/j.mset.2018.09.004

Barbosa PR, Oliveira MD, Barros EM, Michaud JP, Torres JB (2018) Differential impacts of six insecticides on a mealybug and its coccinellid predator. Ecotox Environ Safe 147:963–971. https://doi.org/10.1016/j.ecoenv.2017.09.021

Behera SK, Kim HW, Oh JE, Park HS (2011) Occurrence and removal of antibiotics, hormones and several other pharmaceuticals in wastewater treatment plants of the largest industrial city of Korea. Sci Total Environ 409(20):4351–4360. https://doi.org/10.1016/j.scitotenv.2011.07.015

Belsito D, Bickers D, Bruze M, Calow P, Dagli ML, Fryer AD, Greim H, Miyachi Y, Saurat JH, Sipes IG, Panel TR (2013) A toxicological and dermatological assessment of alkyl cyclic ketones when used as fragrance ingredients. Food Chem Toxicol 62(1):S1–S44. https://doi.org/10.1016/j.fct.2013.09.033

Beyer A, Biziuk M (2009) Environmental fate and global distribution of polychlorinated biphenyls. In: Whitacre DM (ed) Reviews of environmental contamination and toxicology (continuation of residue reviews), vol 201. Springer, Boston, pp 137–158. https://doi.org/10.1007/978-1-4419-0032-6_5

Birch GF, Drage DS, Thompson K, Eaglesham G, Mueller JF (2015) Emerging contaminants (pharmaceuticals, personal care products, a food additive and pesticides) in waters of Sydney estuary, Australia. Mar Pollut Bull 97(1–2):56–66. https://doi.org/10.1016/j.marpolbul.2015.06.038

Bonner JC (2011) Carbon nanotubes as delivery systems for respiratory disease: do the dangers outweigh the potential benefits? Expert Rev Respir Med 5(6):779–787. https://doi.org/10.1586/ers.11.72

Botana LM, Alfonso A (eds) (2015) Phycotoxins: chemistry and biochemistry, 2nd edn. John Wiley & Sons, New York, pp 1–544

Bouloubassi I, Roussiez V, Azzoug M, Lorre A (2012) Sources, dispersal pathways and mass budget of sedimentary polycyclic aromatic hydrocarbons (PAH) in the NW Mediterranean margin, Gulf of Lions. Mar Chem 142:18–28. https://doi.org/10.1016/j.marchem.2012.07.003

Boxall A (2010) Impacts of emerging contaminants on the environment. In: Batty LC, Hallberg KB (eds) Ecology of industrial pollution, Ecological reviews. Cambridge University Press, Cambridge, pp 101–125. https://doi.org/10.1017/CBO9780511805561.006

Boxall ABA (2012) New and emerging water pollutants arising from agriculture. Organisation for Economic Co-operation and Development. OECD, London, pp 1–49

Boxall AB, Rudd MA, Brooks BW, Caldwell DJ, Choi K, Hickmann S, Innes E, Ostapyk K, Staveley JP, Verslycke T, Ankley GT (2012) Pharmaceuticals and personal care products in the environment: what are the big questions? Environ Health Perspect 120(9):1221–1229. https://doi.org/10.1289/ehp.1104477

Calandra P, La Parola V, Turco Liveri V, Lidorikis E, Finocchi F (2013) Composite nanoparticles. Hindawi 2013:1–2. https://doi.org/10.1155/2013/536341

Canesi L, Fabbri E (2015) Environmental effects of BPA: focus on aquatic species. Dose-Response 13(3):1–14. https://doi.org/10.1177/1559325815598304

Cao Z, Xu F, Covaci A, Wu M, Wang H, Yu G, Wang B, Deng S, Huang J, Wang X (2014) Distribution patterns of brominated, chlorinated, and phosphorus flame retardants with particle size in indoor and outdoor dust and implications for human exposure. Environ Sci Technol 48(15):8839–8846. https://doi.org/10.1021/es501224b

Careghini A, Mastorgio AF, Saponaro S, Sezenna E (2015) Bisphenol A, nonylphenols, benzophenones, and benzotriazoles in soils, groundwater, surface water, sediments, and food: a review. Environ Sci Pollut Res 22(8):5711–5741. https://doi.org/10.1007/s11356-014-3974-5

CDC (2002) Guidance for protecting building environments from airborne chemical, biological, or radiological attacks, vol 139. Department of Health and Human Services (DHHS), Center for Disease Control and Prevention (CDC). National Institute for Occupational Safety and Health Publication (NIOSH), Cincinnati, pp 1–40. http://www.ncjrs.gov/App/publications/abstract.aspx?ID=201419

Chalmers G, Adams R, Bustin A, Bustin M (2019) The environmental significance of sediment surface area as a controlling factor in the preservation of Polychlorinated Dibenzo-P-Dioxins and Dibenzofurans (PCDD/PCDF) in sediments adjacent to woodfibre pulp mill, Howe Sound, British Columbia. Minerals 9(11):711. https://doi.org/10.3390/min9110711

Chen F, Gao J, Zhou Q (2012) Toxicity assessment of simulated urban runoff containing polycyclic musks and cadmium in Carassius auratus using oxidative stress biomarkers. Environ Pollut 162:91–97. https://doi.org/10.1016/j.envpol.2011.10.016

Chen Y, Liu X, Mao X, Zhuang Q, Xie Z, Han Z (2014) γ-Fe 2 O 3–MWNT/poly (p-phenylenebenzobisoxazole) composites with excellent microwave absorption performance and thermal stability. Nanoscale 6(12):6440–6447. https://doi.org/10.1039/C4NR00353E

Cheruiyot NK, Lee WJ, Yan P, Mwangi JK, Wang LC, Gao X, Lin NH, Chang-Chien GP (2016) An overview of PCDD/F inventories and emission factors from stationary and mobile sources:

what we know and what is missing. Aerosol Air Qual Res 16(12):2965–2988. https://doi.org/10.4209/aaqr.2016.10.0447

Chovancová J, Čonka K, Kočan A, Sejáková ZS (2011) PCDD, PCDF, PCB and PBDE concentrations in breast milk of mothers residing in selected areas of Slovakia. Chemosphere 83(10):1383–1390. https://doi.org/10.1016/j.chemosphere.2011.02.070

Christian MD (2013) Biowarfare and bioterrorism. Crit Care Clin 29(3):717–756. https://doi.org/10.1016/j.ccc.2013.03.015

Ciftci O, Ozdemir I, Tanyildizi S, Yildiz S, Oguzturk H (2011) Antioxidative effects of curcumin, β-myrcene and 1, 8-cineole against 2, 3, 7, 8-tetrachlorodibenzo-p-dioxin-induced oxidative stress in rats liver. Toxicol Ind Health 27(5):447–453. https://doi.org/10.1177/0748233710388452

Cimmino A, Andolfi A, Evidente A (2014) Phytotoxic terpenes produced by phytopathogenic fungi and allelopathic plants. Nat Prod Commun 9(3):401–408. https://doi.org/10.1177/1934578X1400900330

Clarke RM, Cummins E (2015) Evaluation of "classic" and emerging contaminants resulting from the application of biosolids to agricultural lands: a review. Hum Ecol Risk Assess: Int J 21(2):492–513. https://doi.org/10.1080/10807039.2014.930295

Clarke BO, Smith SR (2011) Review of 'emerging' organic contaminants in biosolids and assessment of international research priorities for the agricultural use of biosolids. Environ Int 37(1):226–247. https://doi.org/10.1016/j.envint.2010.06.004

Conkle JL, Gan J, Anderson MA (2012) Degradation and sorption of commonly detected PPCPs in wetland sediments under aerobic and anaerobic conditions. J Soils Sediments 12(7):1164–1173. https://doi.org/10.1007/s11368-012-0535-8

Dai G, Huang J, Chen W, Wang B, Yu G, Deng S (2014) Major pharmaceuticals and personal care products (PPCPs) in wastewater treatment plant and receiving water in Beijing, China, and associated ecological risks. Bull Environ Contam Toxicol 92(6):655–661. https://doi.org/10.1007/s00128-014-1247-0

Darling RG, Catlett CL, Huebner KD, Jarrett DG (2002) Threats in bioterrorism I: CDC category A agents. Emerg Med Clin N Am 20(2):273–309. https://doi.org/10.1016/S0733-8627(02)00005-6

Daughton CG, Ternes TA (1999) Pharmaceuticals and personal care products in the environment: agents of subtle change? Environ Health Perspect 107:907–938. https://doi.org/10.1289/ehp.99107s6907

De García SA, Pinto GP, García-Encina PA, Irusta-Mata R (2014) Ecotoxicity and environmental risk assessment of pharmaceuticals and personal care products in aquatic environments and wastewater treatment plants. Ecotoxicology 23(8):1517–1533. https://doi.org/10.1007/s10646-014-1293-8

De Matteis V (2017) Exposure to inorganic nanoparticles: routes of entry, immune response, biodistribution and in vitro/in vivo toxicity evaluation. Toxics 5(4):29. https://doi.org/10.3390/toxics5040029

Derakhsh MP, Mashinchian Moradi A, Sharifpour I, Jamili S (2020) Toxic effects of diclofenac on gills, liver and kidney of Cyprinus carpio (Linnaeus, 1758). Iran J Fish Sci 19(2):735–747. URL: http://jifro.ir/article-1-2682-en.html

Dey S, Bano F, Malik A (2019) Pharmaceuticals and personal care product (PPCP) contamination—a global discharge inventory. In: Prasad MN, Vithanage M, Kapley A (eds) Pharmaceuticals and personal care products: waste management and treatment technology: emerging contaminants and micro pollutants. Butterworth-Heinemann, pp 1–26. https://doi.org/10.1016/B978-0-12-816189-0.00001-9

Ding C, He J (2010) Effect of antibiotics in the environment on microbial populations. Appl Microbiol Biotechnol 87(3):925–941. https://doi.org/10.1007/s00253-010-2649-5

Duke SO, Dayan FE (2011) Modes of action of microbially-produced phytotoxins. Toxins 3(8):1038–1064. https://doi.org/10.3390/toxins3081038

Dulio V, van Bavel B, Brorström-Lundén E, Harmsen J, Hollender J, Schlabach M, Slobodnik J, Thomas K, Koschorreck J (2018) Emerging pollutants in the EU: 10 years of NORMAN in support of environmental policies and regulations. Environ Sci Eur 30(1):1–13. https://doi.org/10.1186/s12302-018-0135-3

Ebele AJ, Abdallah MA, Harrad S (2017) Pharmaceuticals and personal care products (PPCPs) in the freshwater aquatic environment. Emerg Contam 3(1):1–16. https://doi.org/10.1016/j.emcon.2016.12.004

Edebali S, Oztekin Y, Arslan G (2018) Metallic engineered nanomaterial for industrial use. In: Hussain CM (ed) Handbook of nanomaterials for industrial applications. Elsevier, pp 67–73. https://doi.org/10.1016/B978-0-12-813351-4.00004-3

El-Shahawi MS, Hamza A, Bashammakh AS, Al-Saggaf WT (2010) An overview on the accumulation, distribution, transformations, toxicity and analytical methods for the monitoring of persistent organic pollutants. Talanta 80(5):1587–1597. https://doi.org/10.1016/j.talanta.2009.09.055

Escarrone ALV, Caldas SS, Primel EG, Martins SE, Nery LEM (2016) Uptake, tissue distribution and depuration of triclosan in the guppy Poecilia vivipara acclimated to freshwater. Sci Total Environ 560:218–224. https://doi.org/10.1016/j.scitotenv.2016.04.039

EFSA panel on additives and products or substances used in animal feed (FEEDAP) (2012) Guidance on the assessment of bacterial susceptibility to antimicrobials of human and veterinary importance. EFSA J 10(6):2740. https://doi.org/10.2903/j.efsa.2012.2740

Fent K, Weston AA, Caminada D (2006) Ecotoxicology of human pharmaceuticals. Aquat Toxicol 76(2):122–159. https://doi.org/10.1016/j.aquatox.2005.09.009

Ferguson EM, Allinson M, Allinson G, Swearer SE, Hassell KL (2013) Fluctuations in natural and synthetic estrogen concentrations in a tidal estuary in south-eastern Australia. Water Res 47(4):1604–1615. https://doi.org/10.1016/j.watres.2012.12.020

Fernandez-Gonzalez R, Yebra-Pimentel I, Martinez-Carballo E, Simal-Gandara J (2015) A critical review about human exposure to polychlorinated dibenzo-p-dioxins (PCDDs), polychlorinated dibenzofurans (PCDFs) and polychlorinated biphenyls (PCBs) through foods. Crit Rev Food Sci Nutr 55(11):1590–1617. https://doi.org/10.1080/10408398.2012.710279

Fiedler H (2003) Dioxins and furans (PCDD/PCDF). In: Fiedler H (ed) Persistent organic pollutants. Springer, Berlin, Heidelberg, pp 123–201. https://doi.org/10.1007/10751132_6

Frigerio C, Ribeiro DS, Rodrigues SS, Abreu VL, Barbosa JA, Prior JA, Marques KL, Santos JL (2012) Application of quantum dots as analytical tools in automated chemical analysis: a review. Anal Chim Acta 735:9–22. https://doi.org/10.1016/j.aca.2012.04.042

Fu F, Dionysiou DD, Liu H (2014) The use of zero-valent iron for groundwater remediation and wastewater treatment: a review. J Hazard Mater 267:194–205. https://doi.org/10.1016/j.jhazmat.2013.12.062

Ge L, Li Q, Wang M, Ouyang J, Li X, Xing MM (2014) Nanosilver particles in medical applications: synthesis, performance, and toxicity. Int J Nanomedicine 9:2399–2407. https://doi.org/10.2147/IJN.S55015

Ghosal D, Ghosh S, Dutta TK, Ahn Y (2016) Current state of knowledge in microbial degradation of polycyclic aromatic hydrocarbons (PAHs): a review. Front Microbiol 7:1369. https://doi.org/10.3389/fmicb.2016.01369

Gill JP, Sethi N, Mohan A, Datta S, Girdhar M (2018) Glyphosate toxicity for animals. Environ Chem Lett 16(2):401–426. https://doi.org/10.1007/s10311-017-0689-0

Gioia R, Dachs J, Nizzetto L, Lohmann R, Jones KC (2013) Atmospheric transport, cycling and dynamics of polychlorinated biphenyls (PCBs) from source regions to remote oceanic areas. In: LL MC, Dachs J, Hapeman CJ (eds) Occurrence, fate and impact of atmospheric pollutants on environmental and human health, ACS Symp Ser Am Chem Soc, pp 3–18. https://doi.org/10.1021/bk-2013-1149.ch001

Golshan M (2015) Environmental contaminants and endocrine associated male infertility in fish, Vodňany, pp 1–170. https://doi.org/10.1016/j.aquatox.2014.06.004

González-Tokman D, Martínez I, Villalobos-Ávalos Y, Munguía-Steyer R, del Rosario O-ZM, Cruz-Rosales M, Lumaret JP (2017) Ivermectin alters reproductive success, body condition and sexual trait expression in dung beetles. Chemosphere 178:129–135. https://doi.org/10.1016/j.chemosphere.2017.03.013

Goodale BC (2013) Developmental toxicity of polycyclic aromatic hydrocarbons: defining mechanisms with systems-based transcriptional profiling (Graduate Thesis), Oregon State University. URL: http://hdl.handle.net/1957/42865

Greenfield RA, Drevets DA, Machado LJ, Voskuhl GW, Cornea P, Bronze MS (2002) Bacterial pathogens as biological weapons and agents of bioterrorism. Am J Med Sci 323(6):299–315. https://doi.org/10.1097/00000441-200206000-00003

Guengerich FP, Yoshimoto FK (2018) Formation and cleavage of C–C bonds by enzymatic oxidation–reduction reactions. Chem Rev 118(14):6573–6655. https://doi.org/10.1021/acs.chemrev.8b00031

Gwaltney-Brant SM (2017) Zootoxins. In: Gupta RC (ed) Reproductive and developmental toxicology, 2nd edn. Academic Press, New York, pp 963–972. https://doi.org/10.1016/B978-0-12-804239-7.00049-4

Gwaltney-Brant SM, Dunayer E, Youssef H (2018) Terrestrial zootoxins. In: Gupta RC (ed) Veterinary toxicology, 3rd edn. Academic Press, New York, pp 781–801. https://doi.org/10.1016/B978-0-12-811410-0.00058-1

Gwenzi W, Mangori L, Danha C, Chaukura N, Dunjana N, Sanganyado E (2018) Sources, behaviour, and environmental and human health risks of high-technology rare earth elements as emerging contaminants. Sci Total Environ 636:299–313. https://doi.org/10.1016/j.scitotenv.2018.04.235

Han Q, Zheng Y, Qi Q, Peng J, Song J, Guo J, Guo J (2020) Involvement of oxidative stress in the sensitivity of two algal species exposed to roxithromycin. Ecotoxicology 29:625–633. https://doi.org/10.1007/s10646-020-02192-6

He J, Kraft AJ, Fan J, Van Dyke M, Wang L, Bose ME, Khanna M, Metallo JA, Henrickson KJ (2009) Simultaneous detection of CDC category "A" DNA and RNA Bioterrorism agents by use of multiplex PCR & RT-PCR enzyme hybridization assays. Viruses 1(3):441–459. https://doi.org/10.3390/v1030441

Hernando MD, Rodríguez A, Vaquero JJ, Fernández-Alba AR, García E (2011) Environmental risk assessment of emerging pollutants in water: Approaches under horizontal and vertical EU legislation. Crit Rev Env Sci Tec 41(7):699–731. https://doi.org/10.1080/10643380903140224

Hong WJ, Jia H, Ding Y, Li WL, Li YF (2018) Polychlorinated biphenyls (PCBs) and halogenated flame retardants (HFRs) in multi-matrices from an electronic waste (e-waste) recycling site in Northern China. J Mater Cycles Waste Manag 20(1):80–90. https://doi.org/10.1007/s10163-016-0550-8

Horton HH, Misrahi JJ, Matthews GW, Kocher PL (2002) Critical biological agents: disease reporting as a tool for determining bioterrorism preparedness. J Law Med Ethics 30(2):262–266. https://doi.org/10.1111/j.1748-720X.2002.tb00392.x

Huang Q, Yu Y, Tang C, Peng X (2010) Determination of commonly used azole antifungals in various waters and sewage sludge using ultra-high performance liquid chromatography–tandem mass spectrometry. J Chromatogr A 1217(21):3481–3488. https://doi.org/10.1016/j.chroma.2010.03.022

Inoue KI, Takano H (2011) Aggravating impact of nanoparticles on immune-mediated pulmonary inflammation. Sci World J 11:382–390. https://doi.org/10.1100/tsw.2011.44

Jacobs MM, Massey RI, Clapp RW (2013) The burden of cancer from organic chemicals. In: Carpenter DO (ed) Effects of persistent and bioactive organic pollutants on human health. John Wiley & Sons, Inc, pp 26–56. https://doi.org/10.1002/9781118679654.ch3

Jain KK (2012) Drug-induced neurological disorders, 3rd edn. Hogrefe Publishing, Cambridge, MA. Pp: x+12

Jansen HJ, Breeveld FJ, Stijnis C, Grobusch MP (2014) Biological warfare, bioterrorism, and biocrime. Clin Microbiol Infect 20(6):488–496. https://doi.org/10.1111/1469-0691.12699

Jin S, Hu Y, Gu Z, Liu L, Wu HC (2011) Application of quantum dots in biological imaging. J Nanomater 2011. https://doi.org/10.1155/2011/834139

Johnson DS, Duncan SE, Bianchi LM, Chang HH, Eigel WN, O'Keefe SF (2015) Packaging modifications for protecting flavor of extended-shelf-life milk from light. J Dairy Sci 98(4):2205–2214. https://doi.org/10.3168/jds.2014-8857

Jurado A, Vàzquez-Suñé E, Carrera J, de Alda ML, Pujades E, Barceló D (2012) Emerging organic contaminants in groundwater in Spain: a review of sources, recent occurrence and fate in a European context. Sci Total Environ 440:82–94. https://doi.org/10.1016/j.scitotenv.2012.08.029

Kasprzyk-Hordern B, Dinsdale RM, Guwy AJ (2009) The removal of pharmaceuticals, personal care products, endocrine disruptors and illicit drugs during wastewater treatment and its impact on the quality of receiving waters. Water Res 43(2):363–380. https://doi.org/10.1016/j.watres.2008.10.047

Kim KR, Owens G, Kwon SI, So KH, Lee DB, Ok YS (2011) Occurrence and environmental fate of veterinary antibiotics in the terrestrial environment. Water, Air, & Soil Pollution 214(1–4):163–174. https://doi.org/10.1007/s11270-010-0412-2

Kim KH, Jahan SA, Kabir E, Brown RJ (2013) A review of airborne polycyclic aromatic hydrocarbons (PAHs) and their human health effects. Environ Int 60:71–80. https://doi.org/10.1016/j.envint.2013.07.019

Kosma CI, Lambropoulou DA, Albanis TA (2014) Investigation of PPCPs in wastewater treatment plants in Greece: occurrence, removal and environmental risk assessment. Sci Total Environ 466–467:421–438. https://doi.org/10.1016/j.scitotenv.2013.07.044

Koutsospyros AD, Koutsospyros DA, Strigul N, Braida W, Christodoulatos C (2011) Tungsten: environmental pollution and health effects. In: Nriagu JO (ed) Encyclopedia of environmental health. Elsevier Inc, pp 418–426. https://doi.org/10.1016/B978-0-444-52272-6.00650-4

Kowalski B, Mazur M (2014) The simultaneous determination of six flame retardants in water samples using SPE pre-concentration and UHPLC-UV method. Water Air Soil Pollut 225(3):1866. https://doi.org/10.1007/s11270-014-1866-4

Kumar RR, Lee JT, Cho JY (2012) Fate, occurrence, and toxicity of veterinary antibiotics in environment. Korean Soc Appl Biol Chem 55(6):701–709. https://doi.org/10.1007/s13765-012-2220-4

Kuzma M, Gerhard E, Shan D, Yang J (2020) Advances in bioresorbable electronics and uses in biomedical sensing. In: Cao H, Coleman T, Hsiai T, Khademhosseini A (eds) Interfacing bioelectronics and biomedical sensing. Springer, Cham, pp 29–72. https://doi.org/10.1007/978-3-030-34467-2_2

Kvarnryd M, Grabic R, Brandt I, Berg C (2011) Early life progestin exposure causes arrested oocyte development, oviductal agenesis and sterility in adult Xenopus tropicalis frogs. Aquat Toxicol 103(1–2):18–24. https://doi.org/10.1016/j.aquatox.2011.02.003

Lapworth DJ, Baran N, Stuart ME, Ward RS (2012) Emerging organic contaminants in groundwater: a review of sources, fate and occurrence. Environ Pollut 163:287–303. https://doi.org/10.1016/j.envpol.2011.12.034

Laulicht F, Brocato J, Cartularo L, Vaughan J, Wu F, Kluz T, Sun H, Oksuz BA, Shen S, Peana M, Medici S (2015) Tungsten-induced carcinogenesis in human bronchial epithelial cells. Toxicol Appl Pharm 288(1):33–39. https://doi.org/10.1016/j.taap.2015.07.003

Lee BK (2010) Sources, distribution and toxicity of polyaromatic hydrocarbons (PAHs) in particulate matter. In: Villanyi V (ed) Air pollution. IntechOpen. https://doi.org/10.5772/10045

Lee CC, Guo YL, Kuei CH, Chang HY, Hsu JF, Wang ST, Liao PC (2006a) Human PCDD/PCDF levels near a pentachlorophenol contamination site in Tainan, Taiwan. Chemosphere 65(3):436–448. https://doi.org/10.1016/j.chemosphere.2006.01.063

Lee CC, Lin WT, Liao PC, Su HJ, Chen HL (2006b) High average daily intake of PCDD/Fs and serum levels in residents living near a deserted factory producing pentachlorophenol (PCP) in Taiwan: influence of contaminated fish consumption. Environ Pollut 141(2):381–386. https://doi.org/10.1016/j.envpol.2005.08.032

Lei M, Zhang L, Lei J, Zong L, Li J, Wu Z, Wang Z (2015) Overview of emerging contaminants and associated human health effects. BioMed Res Int 2015. https://doi.org/10.1155/2015/404796

Lesage G (2015) Persistent organic pollutant. In: Drioli E, Giorno L (eds) Encyclopedia of membranes. Springer, Berlin, Heidelberg. https://doi.org/10.1007/978-3-642-40872-4
Li WC (2014) Occurrence, sources, and fate of pharmaceuticals in aquatic environment and soil. Environ Pollut 187:193–201. https://doi.org/10.1016/j.envpol.2014.01.015
Li X, Ying GG, Su HC, Yang XB, Wang L (2010a) Simultaneous determination and assessment of 4-nonylphenol, bisphenol A and triclosan in tap water, bottled water and baby bottles. Environ Int 36(6):557–562. https://doi.org/10.1016/j.envint.2010.04.009
Li ZH, Velisek J, Zlabek V, Grabic R, Machova J, Kolarova J, Randak T (2010b) Hepatic antioxidant status and hematological parameters in rainbow trout, Oncorhynchus mykiss, after chronic exposure to carbamazepine. Chem Biol Interact 183(1):98–104. https://doi.org/10.1016/j.cbi.2009.09.009
Li X, Li Y, Zhang Q, Wang P, Yang H, Jiang G, Wei F (2011) Evaluation of atmospheric sources of PCDD/Fs, PCBs and PBDEs around a steel industrial complex in northeast China using passive air samplers. Chemosphere 84(7):957–963. https://doi.org/10.1016/j.chemosphere.2011.06.004
Li Z, Lu G, Yang X, Wang C (2012) Single and combined effects of selected pharmaceuticals at sublethal concentrations on multiple biomarkers in *Carassius auratus*. Ecotoxicology 21(2):353–361. https://doi.org/10.1007/s10646-011-0796-9
Li Z, Cong H, Yan Z, Liu A, Yu B (2018) The potential human health and environmental issues of nanomaterials. In: Hussain CM (ed) Handbook of nanomaterials for industrial applications. Elsevier, pp 1049–1054. https://doi.org/10.1016/B978-0-12-813351-4.00060-2
Lin AY, Lin CF, Tsai YT, Lin HH, Chen J, Wang XH, Yu TH (2010a) Fate of selected pharmaceuticals and personal care products after secondary wastewater treatment processes in Taiwan. Water Sci Technol 62(10):2450–2458. https://doi.org/10.2166/wst.2010.476
Lin D, Tian X, Wu F, Xing B (2010b) Fate and transport of engineered nanomaterials in the environment. J Environ Qual 39(6):1896–1908. https://doi.org/10.2134/jeq2009.0423
Liu JL, Wong MH (2013) Pharmaceuticals and personal care products (PPCPs): a review on environmental contamination in China. Environ Int 59:208–224. https://doi.org/10.1016/j.envint.2013.06.012
Liu S, Li S, Du Y (2010) Polychlorinated biphenyls (PCBs) enhance metastatic properties of breast cancer cells by activating Rho-associated kinase (ROCK). PLoS One 5(6):e11272. https://doi.org/10.1371/journal.pone.0011272
Liu S, Ying GG, Zhao JL, Zhou LJ, Yang B, Chen ZF, Lai HJ (2012) Occurrence and fate of androgens, estrogens, glucocorticoids and progestagens in two different types of municipal wastewater treatment plants. J Environ Monitor 14(2):482–491. https://doi.org/10.1039/C1EM10783F
Lu F, Jiang Y, Wu D, Zhou J, Li S, Zhang J (2016) Levels and profiles of polychlorinated dibenzo-p-dioxin and dibenzofurans in raw and treated water from water treatment plants in Shenzhen, China. Environ Pollut 211:233–240. https://doi.org/10.1016/j.envpol.2015.12.062
Luigi V, Giuseppe M, Claudio R (2015) Emerging and priority contaminants with endocrine active potentials in sediments and fish from the River Po (Italy). Environ Sci Pollut Res 22(18):14050–14066. https://doi.org/10.1007/s11356-015-4388-8
Lyche JL, Rosseland C, Berge G, Polder A (2015) Human health risk associated with brominated flame-retardants (BFRs). Environ Int 74:170–180. https://doi.org/10.1016/j.envint.2014.09.006
Mahy BWJ (2003) An overview on the use of a viral pathogen as a bioterrorism agent: why smallpox? Antiviral Res 57(1):1–5. https://doi.org/10.1016/s0166-3542(02)00194-8
Malarvizhi A, Kavitha C, Saravanan M, Ramesh M (2012) Carbamazepine (CBZ) induced enzymatic stress in gill, liver and muscle of a common carp, Cyprinus carpio. J King Saud Univ Sci 24(2):179–186. https://doi.org/10.1016/j.jksus.2011.01.001
Man YB, Lopez BN, Wang HS, Leung AO, Chow KL, Wong MH (2011) Cancer risk assessment of polybrominated diphenyl ethers (PBDEs) and polychlorinated biphenyls (PCBs) in former agricultural soils of Hong Kong. J Hazard Mater 195:92–99. https://doi.org/10.1016/j.jhazmat.2011.08.010

Mantovani A, Fucic A (2014) Puberty dysregulation and increased risk of disease in adult life: possible modes of action. Reprod Toxicol 44:15–22. https://doi.org/10.1016/j.reprotox.2013.06.002

Măruţescu L, Chifiriuc MC, Postolache C, Pircalabioru GG, Bolocan A (2019) Nanoparticles' toxicity for humans and environment. In: Grumezescu AM (ed) Nanomaterials for drug delivery and therapy. William Andrew Publishing, pp 515–535. https://doi.org/10.1016/B978-0-12-816505-8.00012-6

Massarsky A, Trudeau VL, Moon TW (2011) β-blockers as endocrine disruptors: the potential effects of human β-blockers on aquatic organisms. J Exp Zool Part A: Ecol Genet Physiol 315A(5):251–265. https://doi.org/10.1002/jez.672

Melymuk L, Robson M, Helm PA, Diamond ML (2012) PCBs, PBDEs, and PAHs in Toronto air: spatial and seasonal trends and implications for contaminant transport. Sci Total Environ 429:272–280. https://doi.org/10.1016/j.scitotenv.2012.04.022

Michael I, Vasquez MI, Hapeshi E, Haddad T, Baginska E, Kümmerer K, Fatta-Kassinos D (2014) Metabolites and transformation products of pharmaceuticals in the aquatic environment as contaminants of emerging concern. In: Lambropoulou DA, Nollet LM (eds) Transformation products of emerging contaminants in the environment: analysis, processes, occurrence, effects and risks, 1st edn. New York, John Wiley & Sons Ltd, pp 425–470

Millán-Chiu BE, del Pilar R-TM, Loske AM (2020) Nanotoxicology in plants. In: Patra J, Fraceto L, Das G, Campos E (eds) Green nanoparticles. Nanotechnology in the life sciences. Springer, Cham, pp 43–76. https://doi.org/10.1007/978-3-030-39246-8_3

Molins-Delgado D, Gago-Ferrero P, Díaz-Cruz MS, Barceló D (2016) Single and joint ecotoxicity data estimation of organic UV filters and nanomaterials toward selected aquatic organisms. Urban groundwater risk assessment. Environ Res 145:126–134. https://doi.org/10.1016/j.envres.2015.11.026

Monteiro SC, Boxall AB (2010) Occurrence and fate of human pharmaceuticals in the environment. In: Whitacre DM (ed) Reviews of environmental contamination and toxicology, vol 202. Springer, New York, pp 53–154. https://doi.org/10.1007/978-1-4419-1157-5_2

Moran GJ (2002) Threats in bioterrorism. II: CDC category B and C agents. Emerg Med Clin N Am 20(2):311–330. https://doi.org/10.1016/s0733-8627(01)00003-7

Muendo BM, Lalah JO, Getenga ZM (2012) Behavior of pesticide residues in agricultural soil and adjacent River Kuywa sediment and water samples from Nzoia sugarcane belt in Kenya. Environmentalist 32(4):433–444. https://doi.org/10.1007/s10669-012-9407-4

Nakata H, Shinohara RI (2010) Concentrations of benzotriazole UV stabilizers and polycyclic musks in wastewater treatment plant samples in Japan. In: Isobe T, Nomiyama K, Subramanian A, Tanabe S (eds) Interdisciplinary studies on environmental chemistry — environmental specimen bank. TERRAPUB, Tokyo, pp 51–58

Nigen S, Knowles SR, Shear NH (2003) Drug eruptions: approaching the diagnosis of drug-induced skin diseases. J Drugs Dermatol: JDD 2(3):278–299

NIOSH (2010) NIOSH pocket guide to chemical hazards: tungsten. Link: www.cdc.gov/niosh/npg/npgd0645.html. Viewed on: 08.06.20.

Noriega-Luna B, Godínez LA, Rodríguez FJ, Rodríguez A, Zaldívar-Lelo de Larrea G, Sosa-Ferreyra CF, Mercado-Curiel RF, Manríquez J, Bustos E (2014) Applications of dendrimers in drug delivery agents, diagnosis, therapy, and detection. J Nanomater. https://doi.org/10.1155/2014/507273

Nowack B, Ranville JF, Diamond S, Gallego-Urrea JA, Metcalfe C, Rose J, Horne N, Koelmans AA, Klaine SJ (2012) Potential scenarios for nanomaterial release and subsequent alteration in the environment. Environ Toxicol Chem 31(1):50–59. https://doi.org/10.1002/etc.726

Noyes PD, Haggard DE, Gonnerman GD, Tanguay RL (2015) Advanced morphological—behavioral test platform reveals neurodevelopmental defects in embryonic zebrafish exposed to comprehensive suite of halogenated and organophosphate flame retardants. Toxicol Sci 145(1):177–195. https://doi.org/10.1093/toxsci/kfv044

Öberg LG, Wagman N, Andersson R, Rappe C (1993) De novo formation of PCDD/Fs in compost and sewage sludge–a status report. Organohalogen Compd 11:297–302

Olatunji OS (2019) Evaluation of selected polychlorinated biphenyls (PCBs) congeners and dichlorodiphenyltrichloroethane (DDT) in fresh root and leafy vegetables using GC-MS. Sci Rep 9(1):1–10. https://doi.org/10.1038/s41598-018-36996-8

Omlor AJ, Nguyen J, Bals R, Dinh QT (2015) Nanotechnology in respiratory medicine. Respir Res 16(1):64. https://doi.org/10.1186/s12931-015-0223-5

Orlando EF, Ellestad LE (2014) Sources, concentrations, and exposure effects of environmental gestagens on fish and other aquatic wildlife, with an emphasis on reproduction. Gen Comp Endocrinol 203:241–249. https://doi.org/10.1016/j.ygcen.2014.03.038

Oulton RL, Kohn T, Cwiertny DM (2010) Pharmaceuticals and personal care products in effluent matrices: a survey of transformation and removal during wastewater treatment and implications for wastewater management. J Environ Monit 12(11):1956–1978. https://doi.org/10.1039/C0EM00068J

Paiva JC, Cabral IO, Soares BM, Sombra CM, Ferreira JR, Moraes MO, Cavalcanti BC, Pessoa C (2011) Biomonitoring of rural workers exposed to a complex mixture of pesticides in the municipalities of Tianguá and Ubajara (Ceará state, Brazil): genotoxic and cytogenetic studies. Environ Mol Mutagen 52(6):492–501. https://doi.org/10.1002/em.20647

Pal A, He Y, Jekel M, Reinhard M, Gin KY (2014) Emerging contaminants of public health significance as water quality indicator compounds in the urban water cycle. Environ Int 71:46–62. https://doi.org/10.1016/j.envint.2014.05.025

Pal M, Tsegaye M, Girzaw F, Bedada H, Godishala V, Kandi V (2017) An overview on biological weapons and bioterrorism. Am J Biomed Res 5(2):24–34. https://doi.org/10.12691/ajbr-5-2-2

Pappas G, Panagopoulou P, Christou L, Akritidis N (2006) Category B potential bioterrorism agents: bacteria, viruses, toxins, and foodborne and waterborne pathogens. Infect Dis Clin N Am 20(2):395–421. https://doi.org/10.1016/j.idc.2006.02.002

Park MH, Igarashi K (2013) Polyamines and their metabolites as diagnostic markers of human diseases. Biomol Ther 21(1):1–9. https://doi.org/10.4062/biomolther.2012.097

Park CM, Wang D, Su C (2018) Recent developments in engineered nanomaterials for water treatment and environmental remediation. In: Hussain CM (ed) Handbook of nanomaterials for industrial applications. Elsevier, pp 849–882. https://doi.org/10.1016/B978-0-12-813351-4.00048-1

Parolini M, Magni S, Traversi I, Villa S, Finizio A, Binelli A (2015) Environmentally relevant concentrations of galaxolide (HHCB) and tonalide (AHTN) induced oxidative and genetic damage in *Dreissena polymorpha*. J Hazard Mater 285:1–10. https://doi.org/10.1016/j.jhazmat.2014.11.037

Patnaik P (2017) Handbook of environmental analysis: chemical pollutants in air, water, soil, and solid wastes (3rd edition). CRC Press, Boca Raton, pp 1–1420. https://doi.org/10.1080/03067319.2017.1421129

Paulsen L (2015) The health risks of chemicals in personal care products and their fate in the environment. Chemistry Honors Papers 15. Link: http://digitalcommons.conncoll.edu/chemhp/15

Pavuk M, Cerhan JR, Lynch CF, Schecter A, Petrik J, Chovancova J, Kocan A (2004) Environmental exposure to PCBs and cancer incidence in eastern Slovakia. Chemosphere 54(10):1509–1520. https://doi.org/10.1016/j.chemosphere.2003.10.038

Pawlett M, Ritz K, Dorey RA, Rocks S, Ramsden J, Harris JA (2013) The impact of zero-valent iron nanoparticles upon soil microbial communities is context dependent. Environ Sci Pollut Res 20(2):1041–1049. https://doi.org/10.1007/s11356-012-1196-2

Peng X, Huang Q, Zhang K, Yu Y, Wang Z, Wang C (2012) Distribution, behavior and fate of azole antifungals during mechanical, biological, and chemical treatments in sewage treatment plants in China. Sci Total Environ 426:311–317. https://doi.org/10.1016/j.scitotenv.2012.03.067

Pietroiusti A, Campagnolo L, Fadeel B (2013) Interactions of engineered nanoparticles with organs protected by internal biological barriers. Small 9(9–10):1557–1572. https://doi.org/10.1002/smll.201201463

Pittinger CA, Pecquet AM (2018) Review of historical aquatic toxicity and bioconcentration data for the brominated flame retardant tetrabromobisphenol A (TBBPA): effects to fish, invertebrates, algae, and microbial communities. Environ Sci Pollut 25(15):14361–14372. https://doi.org/10.1007/s11356-018-1998-y

Poma A, Colafarina S, Fontecchio G, Chichiriccò G (2014) Transgenerational effects of NMs. In: Capco D, Chen Y (eds) Nanomaterial. Advances in experimental medicine and biology, vol 811. Springer, Dordrecht, pp 235–254. https://doi.org/10.1007/978-94-017-8739-0_12

Prabha AS, Dorothy R, Jancirani S, Rajendran S, Singh G, Kumaran SS (2020) Recent advances in the study of toxicity of polymer-based nanomaterials. In: Rajendran S, Mukherjee A, Shukla RK (eds) Nanotoxicity: prevention and antibacterial applications of nanomaterials. Elsevier, pp 143–165. https://doi.org/10.1016/B978-0-12-819943-5.00007-5

Qiao T, Yu Z, Zhang X, Au DW (2011) Occurrence and fate of pharmaceuticals and personal care products in drinking water in southern China. J Environ Monit 13(11):3097–3103. https://doi.org/10.1039/C1EM10318K

Qin Q, Chen X, Zhuang J (2015) The fate and impact of pharmaceuticals and personal care products in agricultural soils irrigated with reclaimed water. Crit Rev Env Sci Tec 45(13):1379–1408. https://doi.org/10.1080/10643389.2014.955628

Ramadas M, Samantaray AK (2018) Applications of remote sensing and GIS in water quality monitoring and remediation: a state-of-the-art review. In: Bhattacharya S, Gupta A, Gupta A, Pandey A (eds) Water remediation. Energy, environment, and sustainability. Springer, Singapore, pp 225–246. https://doi.org/10.1007/978-981-10-7551-3_13

Ramaswamy BR, Shanmugam G, Velu G, Rengarajan B, Larsson DJ (2011) GC–MS analysis and ecotoxicological risk assessment of triclosan, carbamazepine and parabens in Indian rivers. J Hazard Mater 186(2–3):1586–1593. https://doi.org/10.1016/j.jhazmat.2010.12.037

Ramesh A, Archibong AE, Hood DB, Guo Z, Loganathan BG (2011) Global environmental distribution and human health effects of polycyclic aromatic hydrocarbons. In: Loganathan BG, Lam PK (eds) Global contamination trends of persistent organic chemicals. CRC Press, New York, pp 95–124

Rasmussen JW, Martinez E, Louka P, Wingett DG (2010) Zinc oxide nanoparticles for selective destruction of tumor cells and potential for drug delivery applications. Expert Opin Drug Del 7(9):1063–1077. https://doi.org/10.1517/17425247.2010.502560

Reif R, Santos A, Judd SJ, Lema JM, Omil F (2011) Occurrence and fate of pharmaceutical and personal care products in a sewage treatment works. J Environ Monit 13(1):137–144. https://doi.org/10.1039/C0EM00175A

Rengarajan T, Rajendran P, Nandakumar N, Lokeshkumar B, Rajendran P, Nishigaki I (2015) Exposure to polycyclic aromatic hydrocarbons with special focus on cancer. Asian Pac J Trop Biomed 5(3):182–189. https://doi.org/10.1016/S2221-1691(15)30003-4

Rey MD, Font R, Aracil I (2014) PCDD/F emissions from light-duty diesel vehicles operated under highway conditions and a diesel-engine based power generator. J Hazard Mater 278:116–123. https://doi.org/10.1016/j.jhazmat.2014.05.101

Richardson SD, Kimura SY (2017) Emerging environmental contaminants: Challenges facing our next generation and potential engineering solutions. Environ Technol Innov 8:40–56. https://doi.org/10.1016/j.eti.2017.04.002

Rosa DLG, García-Castañeda C, Vázquez-Núñez E, Alonso-Castro ÁJ, Basurto-Islas G, Mendoza Á, Cruz-Jiménez G, Molina C (2017) Physiological and biochemical response of plants to engineered NMs: implications on future design. Plant Physiol Biochem 110:226–235. https://doi.org/10.1016/j.plaphy.2016.06.014

Rosenfeld PE, Feng LGH (2011) 16 – Emerging contaminants. In: Rosenfeld PE, Feng LGH (eds) Risks of hazardous wastes. Elsevier, Amsterdam, pp 215–222. https://doi.org/10.1016/B978-1-4377-7842-7.00016-7

Rossini GP, Hess P (2010) Phycotoxins: chemistry, mechanisms of action and shellfish poisoning. In: Luch A (ed) Molecular, clinical and environmental toxicology, Experientia Supplementum, vol 100. Birkhäuser, Basel, pp 65–122. https://doi.org/10.1007/978-3-7643-8338-1_3

Rostkowski P, Haglund P, Aalizadeh R, Alygizakis N, Thomaidis N, Arandes JB, Nizzetto PB, Booij P, Budzinski H, Brunswick P, Covaci A (2019) The strength in numbers: comprehensive characterization of house dust using complementary mass spectrometric techniques. Anal Bioanal Chem 411(10):1957–1977. https://doi.org/10.1007/s00216-019-01615-6

Rotz LD, Khan AS, Lillibridge SR, Ostroff SM, Hughes JM (2002) Public health assessment of potential biological terrorism agents. Emerg Infect Dis 8(2):225. https://doi.org/10.3201/eid0802.010164

Saaristo M, Brodin T, Balshine S, Bertram MG, Brooks BW, Ehlman SM, McCallum ES, Sih A, Sundin J, Wong BB, Arnold KE (2018) Direct and indirect effects of chemical contaminants on the behaviour, ecology and evolution of wildlife. Proc Roy Soc B-Biol Sci 285(1885):20181297. https://doi.org/10.1098/rspb.2018.1297

Sauer PJ, Huisman M, Koopman-Esseboom C, Morse DC, Smits-van Prooije AE, Van de Berg KJ, Tuinstra LT, Van der Paauw CG, Boersma ER, Weisglas-Kuperus N, Lammers J (1994) Effects of polychlorinated biphenyls (PCBs) and dioxins on growth and development. Hum Exp Toxicol 13(12):900–906. https://doi.org/10.1177/096032719401301213

Sauvé S, Desrosiers M (2014) A review of what is an emerging contaminant. Chem Cent J 8(1):1–7. https://doi.org/10.1186/1752-153X-8-15

Segev O, Kushmaro A, Brenner A (2009) Environmental impact of flame retardants (persistence and biodegradability). Int J Env Res Pub He 6(2):478–491. https://doi.org/10.3390/ijerph6020478

Sengupta P, Agrawal A (2014) Emerging threat of bio-terrorism: an international perspective. J Politics Govern 3(1):81–88

Sharma NK, Tiwari SP, Tripathi K, Rai AK (2011) Sustainability and cyanobacteria (blue-green algae): facts and challenges. J Appl Phycol 23(6):1059–1081. https://doi.org/10.1007/s10811-010-9626-3

Shukla S, Chaudhary S, Umar A, Chaudhary GR, Mehta SK (2014) Tungsten oxide (WO3) nanoparticles as scaffold for the fabrication of hydrazine chemical sensor. Sensor Actuat B-Chem 196:231–237. https://doi.org/10.1016/j.snb.2014.02.016

Singh VK, Kumar A, Singh R (2018) Biochar amendment to soil for sustainable agriculture. In: Lichtfouse E (ed) Sustainable agriculture reviews, vol 32. Springer, Cham, pp 207–227. https://doi.org/10.1007/978-3-319-98914-3_8

Snow DD, Cassada DA, Biswas S, Shafieifini M, Li X, D'Alessio M, Carter L, Sallach JB (2018) Detection, occurrence and fate of emerging contaminants in agricultural environments. Water Environ Res 90(10):1348–1370. https://doi.org/10.2175/106143018X15289915807254

Söffker M, Tyler CR (2012) Endocrine disrupting chemicals and sexual behaviors in fish–a critical review on effects and possible consequences. Crit Rev Toxicol 42(8):653–668. https://doi.org/10.3109/10408444.2012.692114

Solé M, Sanchez-Hernandez JC (2018) Elucidating the importance of mussel carboxylesterase activity as exposure biomarker of environmental contaminants of current concern: an in vitro study. Ecol Indic 85:432–439. https://doi.org/10.1016/j.ecolind.2017.10.046

Stefanakis AI, Becker JA (2016) A review of emerging contaminants in water: classification, sources, and potential risks. In: McKeown AE (ed) Impact of water pollution on human health and environmental sustainability. IGI Global, pp 55–80. https://doi.org/10.4018/978-1-4666-9559-7.ch003

Strobel GA (1974) Phytotoxins produced by plant parasites. Annu Rev Plant Physiol 25(1):541–566. https://doi.org/10.1146/annurev.pp.25.060174.002545

Stuart EJ, Compton RG (2015) Nanoparticles-emerging contaminants. In: Moretto L, Kalcher K (eds) Environmental analysis by electrochemical sensors and biosensors. nanostructure science and technology. Springer, New York, pp 855–878. https://doi.org/10.1007/978-1-4939-1301-5_8

Stubbings WA, Harrad S (2014) Extent and mechanisms of brominated flame-retardant emissions from waste soft furnishings and fabrics: a critical review. Environ Int 71:164–175. https://doi.org/10.1016/j.envint.2014.06.007

Subedi B, Balakrishna K, Joshua DI, Kannan K (2017) Mass loading and removal of pharmaceuticals and personal care products including psychoactives, antihypertensives, and antibiotics in two sewage treatment plants in southern India. Chemosphere 167:429–437. https://doi.org/10.1016/j.chemosphere.2016.10.026

Sui Q, Huang J, Deng S, Chen W, Yu G (2011) Seasonal variation in the occurrence and removal of pharmaceuticals and personal care products in different biological wastewater treatment processes. Environ Sci Technol 45(8):3341–3348. https://doi.org/10.1021/es200248d

Suleiman RA, Kurt RA (2015) Current maize production, postharvest losses and the risk of mycotoxins contamination in Tanzania. In: 2015 ASABE Annual International Meeting. Am Soc Agri Biol Eng. https://doi.org/10.13031/aim.20152189434

Sultana N, Hanif NQ (2009) Mycotoxin contamination in cattle feed and feed ingredients. Pakistan Vet J 29(4):211–213

Tabe S, Yang P, Zhao X, Hao C, Seth R, Schweitzer L, Jamal T (2010) Occurrence and removal of PPCPs and EDCs in the Detroit River watershed. Water Pract Tech 5(1):wpt2010015. https://doi.org/10.2166/wpt.2010.015

Tai PT, Nishijo M, Anh NTN, Maruzeni S, Nakagawa H, Van Luong H, Anh TH, Honda R, Kido T, Nishijo H (2013) Dioxin exposure in breast milk and infant neurodevelopment in Vietnam. Occup Environ Med 70(9):656–662. https://doi.org/10.1136/oemed-2012-101021

Talib A, Randhir TO (2016) Managing emerging contaminants: status, impacts, and watershed-wide strategies. Expo Health 8(1):143–158. https://doi.org/10.1007/s12403-015-0192-4

Tasso M, Huvelle MA, Bessone ID, Picco AS (2020) Toxicity assessment of nanomaterials. In: Sharma S, Javed Y (eds) Magnetic nanoheterostructures. Nanomedicine and nanotoxicology. Springer, Cham, pp 383–446. https://doi.org/10.1007/978-3-030-39923-8_13

Thomaidis NS, Asimakopoulos AG, Bletsou AA (2012) Emerging contaminants: a tutorial mini-review. Global NEST J 14(1):72–79

Tran NH, Reinhard M, Khan E, Chen H, Nguyen VT, Li Y, Goh SG, Nguyen QB, Saeidi N, Gin KY (2019) Emerging contaminants in wastewater, stormwater runoff, and surface water: application as chemical markers for diffuse sources. Sci Total Environ 676:252–267. https://doi.org/10.1016/j.scitotenv.2019.04.160

Unuofin JO (2020) Garbage in garbage out: the contribution of our industrial advancement to wastewater degeneration. Environ Sci Pollut Res 27:22319–22335. https://doi.org/10.1007/s11356-020-08944-5

US Department of Defense (DoD) (2010) Emerging chemical & material risks. Link: https://www.denix.osd.mil/portal/page/portal/CMRMD/ECMR. Accessed: 08.06.2020

Vassura I, Passarini F, Ferroni L, Bernardi E, Morselli L (2011) PCDD/Fs atmospheric deposition fluxes and soil contamination close to a municipal solid waste incinerator. Chemosphere 83(10):1366–1373. https://doi.org/10.1016/j.chemosphere.2011.02.072

Villa S, Di Nica V, Pescatore T, Bellamoli F, Miari F, Finizio A, Lencioni V (2018) Comparison of the behavioural effects of pharmaceuticals and pesticides on *Diamesa zernyi* larvae (Chironomidae). Environ Pollut 238:130–139. https://doi.org/10.1016/j.envpol.2018.03.029

Wang XP, Sheng JJ, Gong P, Xue YG, Yao TD, Jones KC (2012) Persistent organic pollutants in the Tibetan surface soil: Spatial distribution, air–soil exchange and implications for global cycling. Environ Pollut 170:145–151. https://doi.org/10.1016/j.envpol.2012.06.012

Wang X, Zhang H, Ni Y, Du Q, Zhang X, Chen J (2014) Kinetics of PCDD/Fs formation from non-wood pulp bleaching with chlorine. Environ Sci Tech 48(8):4361–4367. https://doi.org/10.1021/es404347h

Wang B, Zhao Y, Lan Z, Yao Y, Wang L, Sun H (2016) Sampling methods of emerging organic contaminants in indoor air. Trends Environ Anal Chem 12:13–22. https://doi.org/10.1016/j.teac.2016.11.001

Wei GL, Li DQ, Zhuo MN, Liao YS, Xie ZY, Guo TL, Li JJ, Zhang SY, Liang ZQ (2015) Organophosphorus flame retardants and plasticizers: sources, occurrence, toxicity and human exposure. Environ Pollut 196:29–46. https://doi.org/10.1016/j.envpol.2014.09.012

Wei L, Wang S, Zuo Q, Liang S, Shen S, Zhao C (2016) Nano-hydroxyapatite alleviates the detrimental effects of heavy metals on plant growth and soil microbes in e-waste-contaminated soil. Environ Sci: Processes Impacts 18(6):760–767. https://doi.org/10.1039/C6EM00121A
Wells MJ, Bell KY, Traexler KA, Pellegrin ML, Morse A (2010) Emerging pollutants. Water Environ Res 82(10):2095–2170. https://doi.org/10.2175/106143010X12756668802292
World Health Organization (WHO) (1999) Persistent organic pollutants. Programme for promotion of chemical safety division of control of tropical diseases, and food safety unit. Link: https://www.who.int/foodsafety/areas_work/chemical-risks/pops/en/. Viewed on: 08.06.20
World Health Organization (WHO) (2019) Principles and methods to assess the risk of immunotoxicity associated with exposure to nanomaterials (Environmental Health Criteria: n. 244). International programme on chemical safety (IPCS). Inter-organization programme for the sound management of chemicals a cooperative agreement among FAO, ILO, UNEP, UNIDO, UNITAR, WHO, World Bank and OECD
Wikoff DS, Birnbaum L (2011) Human health effects of brominated flame retardants. In: Eljarrat E, Barceló D (eds) Brominated flame retardants. The handbook of environmental chemistry, vol 16. Springer, Berlin, Heidelberg, pp 19–53. https://doi.org/10.1007/698_2010_97
Winneke G (2011) Developmental aspects of environmental neurotoxicology: lessons from lead and polychlorinated biphenyls. J Neurol Sci 308(1–2):9–15. https://doi.org/10.1016/j.jns.2011.05.020
Witten ML, Sheppard PR, Witten BL (2012) Tungsten toxicity. Chem-Biol Interact 196(3):87–88. https://doi.org/10.1016/j.cbi.2011.12.002
Wohlfahrt-Veje C, Audouze K, Brunak S, Antignac JP, Le Bizec B, Juul A, Skakkebæk NE, Main KM (2014) Polychlorinated dibenzo-p-dioxins, furans, and biphenyls (PCDDs/PCDFs and PCBs) in breast milk and early childhood growth and IGF1. Reproduction 147:391–399. https://doi.org/10.1530/REP-13-0422
Wong MH, Poon BH (2003) Sources, fates and effects of persistent organic pollutants in China, with emphasis on the Pearl River Delta. In: Fiedler H (ed) Persistent organic pollutants. The handbook of environmental chemistry (Vol. 3 Series: Anthropogenic Compounds), vol 3O. Springer, Berlin, Heidelberg, pp 355–369. https://doi.org/10.1007/10751132_13
Worm B, Lotze HK, Jubinville I, Wilcox C, Jambeck J (2017) Plastic as a persistent marine pollutant. Annu Rev Environ Resour 42:1–26. https://doi.org/10.1146/annurev-environ-102016-060700
Wu W, Jiang C, Roy VA (2015) Recent progress in magnetic iron oxide–semiconductor composite nanomaterials as promising photocatalysts. Nanoscale 7(1):38–58. https://doi.org/10.1039/C4NR04244A
Xiao-Ju Y, Huan HE, Ying P, Xiao-Meng W, Zhan-Qi G, Shao-Gui Y, Cheng S (2012) Determination of organophosphorus flame retardants in surface water by solid phase extraction coupled with gas chromatography-mass spectrometry. Chinese J Anal Chem 40(11):1693–1697. https://doi.org/10.1016/S1872-2040(11)60586-0
Xing GH, Liang Y, Chen LX, Wu SC, Wong MH (2011) Exposure to PCBs, through inhalation, dermal contact and dust ingestion at Taizhou, China–A major site for recycling transformers. Chemosphere 83(4):605–611. https://doi.org/10.1016/j.chemosphere.2010.12.018
Xu P, Zeng GM, Huang DL, Feng CL, Hu S, Zhao MH, Lai C, Wei Z, Huang C, Xie GX, Liu ZF (2012) Use of iron oxide nanomaterials in wastewater treatment: a review. Sci Total Environ 424:1–10. https://doi.org/10.1016/j.scitotenv.2012.02.023
Xu F, Giovanoulis G, Van Waes S, Padilla-Sanchez JA, Papadopoulou E, Magnér J, Haug LS, Neels H, Covaci A (2016) Comprehensive study of human external exposure to organophosphate flame retardants via air, dust, and hand wipes: the importance of sampling and assessment strategy. Environ Sci Technol 50(14):7752–7760. https://doi.org/10.1021/acs.est.6b00246
Yamamoto T, Higashino K, Abe T, Takasuga T, Takemori H, Weber R, Sasaki Y (2018) PCDD/PCDF formation in the chlor-alkali process—laboratory study and comparison with patterns from contaminated sites. Environ Sci Pollut Res 25(32):31874–31884. https://doi.org/10.1007/s11356-017-0777-5

Yang LH, Ying GG, Su HC, Stauber JL, Adams MS, Binet MT (2008) Growth-inhibiting effects of 12 antibacterial agents and their mixtures on the freshwater microalga *Pseudokirchneriella subcapitata*. Environ Toxicol Chem: Int J 27(5):1201–1208. https://doi.org/10.1897/07-471.1

Yang Y, Ok YS, Kim KH, Kwon EE, Tsang YF (2017a) Occurrences and removal of pharmaceuticals and personal care products (PPCPs) in drinking water and water/sewage treatment plants: a review. Sci Total Environ 596–597:303–320. https://doi.org/10.1016/j.scitotenv.2017.04.102

Yang L, Liu G, Zheng M, Jin R, Zhu Q, Zhao Y, Zhang X, Xu Y (2017b) Atmospheric occurrence and health risks of PCDD/Fs, polychlorinated biphenyls, and polychlorinated naphthalenes by air inhalation in metallurgical plants. Sci Total Environ 580:1146–1154. https://doi.org/10.1016/j.scitotenv.2016.12.071

Yang YY, Zhao JL, Liu YS, Liu WR, Zhang QQ, Yao L, Hu LX, Zhang JN, Jiang YX, Ying GG (2018) Pharmaceuticals and personal care products (PPCPs) and artificial sweeteners (ASs) in surface and ground waters and their application as indication of wastewater contamination. Sci Total Environ 616–617:816–823. https://doi.org/10.1016/j.scitotenv.2017.10.241

Yebra-Pimentel I, Fernández-González R, Martínez-Carballo E, Simal-Gándara J (2015) A critical review about the health risk assessment of PAHs and their metabolites in foods. Crit Rev Food Sci Nutr 55(10):1383–1405. https://doi.org/10.1080/10408398.2012.697497

Yu Y, Liu Y, Wu L (2013) Sorption and degradation of pharmaceuticals and personal care products (PPCPs) in soils. Environ Sci Pollut Res 20(6):4261–4267. https://doi.org/10.1007/s11356-012-1442-7

Zani C, Ceretti E, Covolo L, Donato F (2017) Do polychlorinated biphenyls cause cancer? A systematic review and meta-analysis of epidemiological studies on risk of cutaneous melanoma and non-Hodgkin lymphoma. Chemosphere 183:97–106. https://doi.org/10.1016/j.chemosphere.2017.05.053

Zhang X, Brar SK, Yan S, Tyagi RD, Surampalli RY (2013a) Fate and transport of fragrance materials in principal environmental sinks. Chemosphere 93(6):857–869. https://doi.org/10.1016/j.chemosphere.2013.05.055

Zhang XF, Zhang LJ, Li L, Feng YN, Chen B, Ma JM, Huynh E, Shi QH, De Felici M, Shen W (2013b) Diethylhexyl phthalate exposure impairs follicular development and affects oocyte maturation in the mouse. Environ Mol Mutagen 54(5):354–361. https://doi.org/10.1002/em.21776

Zhao JL, Ying GG, Liu YS, Chen F, Yang JF, Wang L (2010) Occurrence and risks of triclosan and triclocarban in the Pearl River system, South China: from source to the receiving environment. J Hazard Mater 179(1–3):215–222. https://doi.org/10.1016/j.jhazmat.2010.02.082

Zhou XF, Dai CM, Zhang YL, Surampalli RY, Zhang TC (2011) A preliminary study on the occurrence and behavior of carbamazepine (CBZ) in aquatic environment of Yangtze River Delta, China. Environ Monit Assess 173(1–4):45–53. https://doi.org/10.1007/s10661-010-1369-8

# Chapter 2
# Sewage Wastewater and Sludge as Source of Traditional and Emerging Contaminants in Agroecosystems

**Vipin Kumar Singh, Rishikesh Singh, Ajay Kumar, Rahul Bhadouria, and Shilpi Pandey**

**Abstract** The enormous release of untreated sewage wastewater (330 $km^3$) and sludge containing contaminants like antibiotics, antibiotic resistance genes, heavy metals, metalloids, personal care products, infectious microorganisms including coliforms and hormones into agroecosystem has been documented to pose severe risks to food chain and crop productivity. The sludge amount produced in European Union would rise to approximately 12 million tonnes by 2020. Surprisingly, nearly 30–60% of antibiotics get concentrated in treatment facilities, serving as source of environmental contaminants and 30–90% of antibiotics used in livestock may release in nature without any modification. The heavy metal content in sewage sludge with respect to dry weight may reach upto 0.5–2%. The recently rising interest of farmers in organic farming has accelerated the employment of sewage wastewater and sludge as an innovative source of irrigation water and fertilizers, respectively so as to augment crop production, minimize the agrochemical exploitation and overcome the associated threats to agricultural productivity. Although, multifaceted beneficial effects of sewage water and sludge deployment in agriculture are reported worldwide, the introduction of traditional as well as emerging

V. K. Singh (✉) · S. Pandey
Department of Botany, Centre of Advanced Study in Botany, Institute of Science, Banaras Hindu University, Varanasi, India
e-mail: vipinks85@gmail.com; pandeyshilpi35@gmail.com

R. Singh
Institute of Environment and Sustainable Development, Banaras Hindu University, Varanasi, India
e-mail: rishikesh.iesd@gmail.com

A. Kumar
Agriculture Research Organization (ARO), Volcani Center, Rishon LeZion, Israel
e-mail: ajaykumar_bhu@yahoo.com

R. Bhadouria
Department of Botany, University of Delhi, New Delhi, India
e-mail: rahulbhadouriya2@gmail.com



V. Kumar Singh et al. (eds.), *Sustainable Agriculture Reviews 50*, Sustainable Agriculture Reviews 50, https://doi.org/10.1007/978-3-030-63249-6_2

contaminants like antibiotics, personal care products, and hormones in food chain via utilization of sewage wastewater and sludge cannot be denied. Therefore, the proper treatment is imperative to nullify the unforeseen consequences of sewage wastewater and sludge utilization in agroecosystem.

We review recent information regarding the sewage application in agriculture, commonly observed traditional and emerging contaminants, major bottlenecks during application, potential benefits and risks of using sewage water and sludge for management of crop productivity and techniques currently being practiced for sequestration of contaminants. The information on various organic and inorganic contaminants present in sewage wastewater and sludge could be helpful in predicting the toxicity to agroecosystem, threats of accumulation in crop produce and hence human health. In depth information pertinent to source fingerprinting, assessment, ecotoxicology along with different aspects of fate and transport of important contaminants would help foster in the design and application of sustainable strategies and improvement in currently employed tools for the management of globally produced huge amounts of sewage wastewater and sludge.

**Keywords** Agrochemicals · Agroecosystem · Antibiotics · Crop productivity · Organic agriculture · Personal care products

## Abbreviations

| | |
|---|---|
| ELISA | Enzyme linked immunosorbent assay |
| GC-MS | Gas chromatography-mass spectrometry |
| US-EPA | United states environmental protection agency |

## 2.1 Introduction

Enhancement in crop productivity based on limited available natural resources like water and land is chief constraint under changing climatic conditions. Deterioration of soil characteristics and contamination of water with hazardous substances galvanized by massive anthropogenic interferences (Emadodin and Bork 2012; Singh and Singh 2018; Duttagupta et al. 2020) are considered as crucial factors responsible for significant modulation in food productivity across the globe. Moreover, substantial changes in environmental conditions due to anthropogenic activities have restricted the agricultural productivity multiple folds posing threats to food security (Liu et al. 2019; Jiang et al. 2020a). Environmental stresses of both biotic and abiotic nature are reported worldwide to potentially inhibit the yield of numerous economically important crops (Mamphogoro et al. 2020; Razzaq et al. 2020). Accessibility to sufficient nutritional food to ever rising global population could only be made possible through sustainable utilization of existing natural resources integrated with increased crop production. United Nations has projected two billion expansions in human population during upcoming 30 years mostly in developing regions of the world, indicating need for improving global food productivity both qualitatively and quantitatively. In this connection, the crop productivity needs to be doubled in order to achieve the goals of food security (Prosekov and Ivanova 2018).

Current world agricultural practices are largely dependent on deployment of different types of agrochemicals including herbicides, pesticides, insecticides, fungicides and fertilizers in order to raise the crop productivity (Zhang et al. 2020). As per an estimate, the application of nitrogen and phosphorus based fertilizers in agricultural activities may reach upto 120 and 50 million tonnes, respectively, by the forthcoming decade 2030 to 2040 (Vance et al. 2003). Very little amount of chemical fertilizers are utilized by crop plants (López-Arredondo et al. 2014; Raun and Johnson 1999) and considerably large amounts of the agrochemicals applied to the agricultural fields are released into the surrounding environment leading to contamination of air, water and soil (Nag et al. 2020; Kafaei et al. 2020). Apart from being very much expensive, the deployment of chemical fertilizers disturbs the natural attributes of soil physico-chemical and biological characteristics. Modifications in soil productivity, microbiological characteristics, soil fauna and associated biological processes, resistance development, nitrate accumulation as well as toxicity to non-target organisms (Geisseler and Scow 2014; Zhu et al. 2018; Ye et al. 2020; Guo et al. 2020; Li et al. 2020a; Cai et al. 2020; Dzul-Caamal et al. 2020) are chief outcomes of indiscriminate application of agrochemicals in agro-ecosystem. The above-mentioned negative consequences of chemical fertilizers necessitates for the investigation of innovative sources of fertilizers posing minimal impact on agricultural ecosystems.

Currently, organic agriculture has received considerable attention among farmers and scientists globally because of augmenting effect on soil characteristics and inherent potential in improving crop productivity (Badgley et al. 2007; Emadodin et al. 2020) without interrupting ecosystem homoeostasis. Sewage wastewater and sludge may serve as an important source of agriculturally important nutrients to meet the rising demand of chemical fertilizers (Fang et al. 2020; Vogel et al. 2020). Globally, huge quantities of sewage sludge are generated and consequently utilized for the purpose of land filling as well as in different agricultural practices as fertilizer (Lundin et al. 2004; Christodoulou and Stamatelatou 2016; Praspaliauskas et al. 2020). They are important source of essential nutrients and trace elements like nitrogen, phosphorus, potassium, sulfur, calcium and magnesium required for vital plant processes (Fang et al. 2018). The contribution of sewage sludge towards the cultivation of *Sorghum* for enhanced biomass generation is illustrated by Kołodziej et al. (2015). The increase in biomass was found to be directly associated with the increasing amount of sewage sludge. Furthermore, incorporation of sewage sludge at increased amounts modified the soil attributes as expressed in terms of nutrient composition and enzyme activities.

Nevertheless, the exploitation of untreated sewage wastewater and sludge containing unacceptable level of heavy metals, metalloids, microplastics, organic contaminants, endocrine disruptors, personal care products, antibiotics, antibiotic resistance genes, organic materials and pathogenic microbes has been prevented because of harmful impact on different environmental components and human health (Qian et al. 2016; Abril et al. 2020; Arun et al. 2020; Lopes et al. 2020; Praspaliauskas et al. 2020; Rolsky et al. 2020; Wang et al. 2020c, d). The harmful effects linked with application of sewage wastewater and sludge could be reduced to a great extent by incorporating different treatment technologies including incineration with the resultant of lowered sludge volume and energy generation (Lundin et al. 2004; Raheem et al. 2018).

So far as the treatment of sewage wastewater and sludge is concerned, different physical, chemical and biological strategies like hydrothermal treatment, alkalization, microwave coupled with fenton reaction, biological digestion, composting, and supercritical liquids have been applied to eliminate pathogenic microbes and environmentally hazardous substances of organic and inorganic nature together with the concomitant generation of energy and useful products like char and tar (Qian et al. 2016; Lopes et al. 2020; Lopez et al. 2020; Wang et al. 2020a, b). Nonetheless, most of the technologies employed till date for the treatment of sewage wastewater and sludge are not proved absolutely efficient in sequestering environmentally hazardous entities suggesting intensive process optimization as well as small scale field studies to assess the possible risks in agroecosystems.

The present chapter has dealt with sewage wastewater and sludge as source of traditional as well as emerging contaminants such as metals, metalloids, antibiotics, antibiotic resistance genes, hormones, endocrine disruptors, and personal care products. Relevant information pertaining to advantages like increment in soil micronutrient and macronutrient content, organic material, improved soil productivity and disadvantages like increased heavy metals/metalloids content, modulation in microbial community structure and function, food chain contamination, and negative consequences on human health and environment are briefly presented. A brief insight on fate of sewage wastewater and sludge contaminants in agricultural field under natural condition has also been given in order to understand the environmental processes and factors responsible for adsorption, transformation and biodegradation of contaminants. In addition, currently used methodologies for treatment of both organic and inorganic contaminants present in sewage wastewater and sludge affecting agricultural ecosystem so as to attain the natural characteristics of cultivated ecosystems and eliminate the limitations of presently deployed technologies have also been described. The contribution of sewage wastewater and sludge as a potential source of various traditional and emerging contaminants is presented in the following sections.

## 2.2 Sewage Wastewater and Sludge as Source of Contamination in Agroecosystems

### *2.2.1 Traditional Contaminants*

#### 2.2.1.1 Sewage Wastewater and Sludge as Source of Heavy Metals and Metalloids

Anthropogenic activities are known to add large amounts of hazardous metals and metalloids in sewage wastewater and sludge eventually posing considerable risks to human health and sustainability of agricultural ecosystem. Being soluble in nature, metals and metalloids including zinc, chromium, nickel, lead, arsenic and mercury are readily absorbed by different life forms inducing toxic effects (Singh et al. 2013; Younus et al. 2020). After entry into various environmental matrices, metals and metalloids may contaminate the food chain and accumulate into organisms of higher trophic levels (Barakat 2011). The metals and metalloids concentration in soil ecosystem beyond the recommended limits has often been documented to exert negative effects on plants, microbes, and associated biological processes (Zemanová et al. 2020; Diaconu et al. 2020; Humberto et al. 2020), signifying the regular monitoring and application of suitable strategies for decontamination of affected agroecosystems.

At lower concentrations, some of the metals like copper and zinc may contribute considerably in vital physiological processes of plants and humans (Magu et al. 2016; Alamer et al. 2020), however, the increased availability hinders the important biological processes. Although, application of sewage wastewater and sludge to land areas are noticed to enhance the soil fertility and important characteristics, they are frequently laden with noxious metals and metalloids (Yang et al. 2020a, b), serving as end points for different contaminants. The presence of heavy metals such as manganese, nickel, zinc, copper and lead in sewage wastewater is demonstrated by Singh et al. (2012). The highest content was recorded for iron and least for zinc. The introduction of wastewater into land led not only to rise in organic carbon content, available nutrients but also in the level of hazardous metals implying future risks to soil microbes and grown vegetables. The content of heavy metals like zinc, copper, lead and cadmium in sewage effluents above the World Health Organization recommended limit is described recently by Sayo et al. (2020). The utilization of wastewater caused accumulation of noxious metals in soils with implications in transfer to agricultural produce and threat to food safety. Unacceptable level of heavy metals including copper, zinc and manganese in sewage sludge (Wang et al. 2020d) could potentially endanger the soil characteristics and crop productivity. The heavy metals content in collected sludge samples were found to be ranged from 42.7 to 423.0 mg $kg^{-1}$. The presence of arsenic in hydrothermally treated sewage sludge determined as 28.76 mg $kg^{-1}$ is demonstrated recently (Li et al. 2020b). The contents of chromium, copper, zinc, cadmium and lead were recorded as 3696.54,

3856.72, 4137.98, 1.94 and 105.21 mg $kg^{-1}$, respectively, thereby indicating non-suitability for agricultural applications.

The continuous deployment of sewage effluents for agricultural irrigation could impose toxicity to soil beneficial microbes, microbial processes together with the heavy metals and metalloids accumulation in crop produce, requiring the assessment of spatial and temporal variations and factors influencing mobility in soil environment for effective management of undesirable environmental contaminants. In addition, the mechanisms underlying the mobilization of metals and metalloids could be helpful in sequestering the noxious contaminants.

#### 2.2.1.2 Sewage Wastewater and Sludge as Source of Organic Contaminants

The huge application of pesticides and herbicides in stored product preservation and agricultural activities, respectively are important sources of agroecosystem contamination consequently affecting the productivity of economically important crops. The household burning of coal, wood and natural gas are considered as important sources of diverse polycyclic aromatic hydrocarbons (Khillare et al. 2020). The presence of polybrominated diphenyl ethers in polyurethane foam used for making household furniture gets entry into sewage after degeneration (Wang et al. 2007). The presence of different organochlorine pesticides in sewage wastewater of treatment facilities and impact of irrigation on soil and groundwater contamination is elaborated by Zhang et al. (2013). The important organochlorine pesticides contaminants identified in impacted area was hexachlorocyclohexane, dichlorodiphenyltrichloroethylene and endosulfan, suggesting development of sustainable technology for cost effective elimination of environmental contaminants. Sewage sludge containing numbers of organochlorine pesticides and polychlorinated biphenyl has been demonstrated by Clarke et al. (2010). The respective concentration of selected organochlorine pesticides like dieldrin, chlordane and dichlorodiphenyldichloroethylene was found to be 770, 290 and 270 μg $kg^{-1}$. The increased concentrations of polycyclic aromatic hydrocarbons in sewage sludge obtained from treatment plants in Delhi, India, infringing the regulatory limits are illustrated recently by Khillare et al. (2020). The average content of polycyclic aromatic hydrocarbons was determined as 20.67 mg $kg^{-1}$. Among the detected organic contaminants, the proportion of elevated molecular weight polycyclic aromatic hydrocarbons was 81% in contrast to the 19% of reduced molecular weight polycyclic aromatic hydrocarbons.

The pesticides and herbicides are generally of persistent nature and exert negative consequences on soil inhabiting microbes, microbial processes and diversity of fauna, resulting in perturbations in soil productivity. The presence of pesticides in cultivated crops such as grains, vegetables and feed samples may impose health hazards in humans and animals. The precise determination of different organic contaminants, factors affecting mobilization and effect of degradation products on various environmental matrices is necessary for managing the contamination of farm lands caused by irrigation with treated sewage wastewater and input of sludge as fertilizer.

### 2.2.2 *Emerging Contaminants*

#### 2.2.2.1 Sewage Wastewater and Sludge as Source of Antibiotics and Antibiotic Resistance Genes

Antibiotics are common in practice to cure various human and livestock diseases in order to improve life quality. The exploitation of antibiotics in developing nations is known to increase 36% during the year 2000–2010 (Thung et al. 2016) pointing towards the rise in global consumption (Borghi and Palma 2014). A large proportion of antibiotics (25–75%) taken by humans and animals are excreted in unmodified forms (Watkinson et al. 2009), consequently finding their ways in sewage wastewater and sludge, thereby posing risks to receiving aquatic and terrestrial ecosystem. The continued exposure to even low concentrations of antibiotics inducing resistance development in bacteria (Liu et al. 2016) and spread of antibiotic resistance genes via horizontal gene transfer and mutation (Summers 2006; Ding and He 2010) is of global concern. The introduction of antibiotics in potable water may induce allergenic responses (Kümmerer 2009) in humans. Furthermore, the presence of antibiotics in aquatic environment may hinder the life forms affecting ecosystem functioning. For instance, the modulating effect of chlortetracycline on alga *Pseudokirchneriella subcapitata* is investigated by Yang et al. (2008). The presence of antibiotics and antibiotic resistance genes in agroecosystem has been recognized as emerging contaminants in agriculture, necessitating the integration of advanced technologies for successful elimination from affected sites. Sewage wastewaters laden with increased concentrations of non metaboilized antibiotic residues are potential threat not only for humans but also for agricultural ecosystem (Shi et al. 2020a).

The scarcity of water for agricultural irrigation in different parts of globe can be fulfilled by utilization of sewage wastewater, however, the high concentration of antibiotics may influence the beneficial microorganisms, microbial community structure and function along with microbially driven processes including cycling of important nutrients and fixation of nitrogen (Gomes et al. 2018; Ma et al. 2019; Urra et al. 2019; Santás-Miguel et al. 2020; Tong et al. 2020). The prevalence of antibiotics in sewage wastewater containing marbofloxacin, climbazole, and norfloxacin with the dominance of fluoroquinolones is explored recently by Shi et al. (2020a). The occurrence of antibiotics associated with class quinolone, tetracycline, and macrolide in urban wastewater is presented by Dan et al. (2020). The investigation depicted seasonal variations in the concentrations of antibiotics. The concentration of selected antibiotics in winter ranged from 20 to 265 ng $l^{-1}$, while the concentration ranged from 34 to 2034 ng $l^{-1}$ in summer. The identification of antibiotics falling in the category of sulfonamides, fluoroquinolones, macrolides, tetracyclins, beta lactams, etc. in sewage sludge has been investigated currently (Ajibola et al. 2020). The maximum concentration of antibiotics in sewage sludge from different treatment facilities was determined as 4689 ng $g^{-1}$ and 1201 ng $g^{-1}$ for oxytetracycline and ciprofloxacin, respectively on dry weight basis. The occurrence of large numbers of antibiotics indicated the possible threat to agroecosystem rendered by introduction of sewage sludge to cultivated land areas. The assessment of fluoroquniolone antibiotics like norfloxacin, ciprofloxacin, and ofloxacin in sewage sludge procured from treatment plant is recently conducted by Arun et al. (2020). The mean concentration of selected antibiotics was determined as 26 μg $g^{-1}$ and order of contribution in sludge was ciprofloxacin> norfloxacin> ofloxacin. Most notably, the concentrations of antibiotics were high in wet sludge in contrast to dry sludge suggesting solubility as one of the main parameters responsible for increased content.

Dissemination of antibiotic resistance genes from sewage wastewater and sludge into agroecosystem through horizontal gene transfer is regarded as the potential future risks to human health and environmental compartments. The incessant application of sewage sludge as manure into agricultural fields may enhance the abundance of antibiotic resistance genes (Chen et al. 2016). The presence of antibiotic resistance genes in sewage wastewater, and sludge containing 18 antibiotic resistance genes based on metagenomics approaches is demonstrated by Yang et al. (2014). The abundance of antibiotic resistance genes was documented in the order of influent>effluent>digested sludge>activated sludge. Reports are available supporting the presence of antibiotic resistance genes in sewage wastewater (Volkmann et al. 2004) as determined through real time polymerase chain reaction. The identified antibiotic resistance genes included *vanA* and *ampC* in 21 and 78%, respectively of the analyzed samples. The identification of antibiotic resistance genes belonging to tetracycline (*tetM*, *tetO*, *tetQ* and *tetW*) and sulfonamide (*sulI* and *sulII*) class of antibiotics in municipal wastewater and rural sewage treatment facilities quantified through q-PCR with differences in abundance is described by Chen and Zhang (2013). The precise determination and early identification of resistance genes in contaminated agroecosystem could help in restricting the spread of antibiotic resistance genes from soil dwelling microorganisms to humans essentially needed for health management. Latest identification of antibiotic resistance genes conferring resistance to antibiotics belonging to the class fluoroquinolones ((*aac-(6′)-1b-cr, qnrB, qnrS*), beta-lactams (*ampC*) and sulfonamide (*sul1*) is illustrated currently by Kumar et al. (2020). The effluents heavily laden with different antibiotic resistance genes at the time of treatment could exert profound influence under natural environmental condition upon interaction with inorganic contaminants like heavy metals. The occurrence of antibiotic resistance genes such as $bla_{TEM}$, *ermB*, *sulf1*, *sulf2*, *tetA*, and *tetB* in sludge even after treatment under alkaline condition is reported recently (Lopes et al. 2020), indicating inefficiency of currently applied treatment procedures. Thus, farmland application of sewage wastewater and sludge as well as sewage treatment plants are considered as prominent sources of antibiotic resistance genes conferring resistance to multitude of commonly used antibiotics (Yang et al. 2014).

Although most of the traditional processes for elimination of antibiotics (Bao et al. 2020) and antibiotic resistance genes present in sewage wastewater and sludge are not effective, the application of technologically advanced strategies like oxidation, biological treatment and surface adsorption (Bao et al. 2020) could help in sequestration to a safer limit. Moreover, any single treatment methodology cannot be applied equally because of considerable differences in composition of sewage wastewater and sludge probably due to variation in source and environmental phenomena. The information regarding the characteristics of antibiotics, antibiotic resistance genes, fate and transport as well as ecotoxicity in given environmental compartments could provide an insight for sustainable management of contaminated agroecosystem.

#### 2.2.2.2 Sewage Wastewater and Sludge as Source of Human Pathogens

Rapid increase in anthropogenic activities is responsible for generation of enormous quantity of sewage wastewater and sludge, eventually affecting the receiving aquatic and terrestrial ecosystems. Although, microorganisms are ubiquitous miniature entities with characteristic potential to perform important ecological processes and potential to modulate environmental integrity, the prevalence in wastewater and sludge produced even after treatment via different approaches is of rising concern to human and animal health globally (Dumontet et al. 2001). The large numbers of microorganisms such as Gram positive and Gram negative bacterial species, viruses, and yeasts present in sewage wastewater and sludge (Jiang et al. 2020b) may compromise the human health due to pathogenic nature. The commonly observed bacteria are those belonging to genus *Escherichia coli* (Jiang et al. 2018a), *Salmonella* (Jiang et al. 2018b), *Camplylobacter* (Guimaraes et al. 2016), and *Enterococcus* (Jiang et al. 2018a). The viral pathogens encountered in sewage wastewater and sludge belonging to rotavirus, hepatitis virus, and adenovirus is presented by different researchers worldwide (Magri et al. 2015; El-Senousy and Abou-Elela 2017). Therefore, the sewage wastewater and sludge heavy loaded with pathogenic microbes and parasites necessitate the development of sustainable technology for reducing the pathogen to acceptable limits set by World Health Organization.

Occurrence of pathogenic microbes designated as *Enterococcus*, *Pesudomonas aeruginosa*, and *Escherichia coli* in sewage wastewater is elaborated currently by Racar et al. (2020), indicating risks to agroecosystem if applied directly, therefore, requiring the deployment of apposite treatment technology before being used for irrigation in agricultural practices. The presence of pathogenic bacteria such as *Enterococcus*, *Escherichia coli*, *Pseuodomonas* and *Staphylococcus* in sewage sludge obtained from sewage treatment plants in Spain is illustrated recently by Lopez et al. (2020) suggesting thorough characterization and treatment of sewage sludge based on currently existing physical, chemical and biological techniques. The prevalence of *Cryptosporidium* and *Giardia* cysts have been reported in nearly 90% of untreated sewage sludge samples and in all treated samples procured from wastewater treatment facilities (Amorós et al. 2016) implying the further improvement in pathogen removal efficiency. The assessment of sewage sludge indicating enrichment with bacterial pathogen *Salmonella* in both treated and untreated samples is well documented (Sahlström et al. 2004).

The employment of sewage wastewater and sludge into agroecosystem may introduce the pathogenic microbes and favor the abundance as well as diversity of antibiotic resistance genes (Chen et al. 2016; Lopez et al. 2020), threatening native soil microbial characteristics. The increased concentrations of pathogenic bacteria in wastewater and sludge facilitated by abundance of nutrients required for microbial growth and multiplication needs effective, economical and ecologically sound techniques for elimination of target life threatening human pathogens. If left untreated, the pathogenic microbes occurring in wastewater and sludge may be transferred into water, soil, crop produce and live stocks imposing severe threat to human health (Bradford et al. 2013).

#### 2.2.2.3 Sewage Wastewater and Sludge as Source of Hormones, Endocrine Disruptors and Personal Care Products

Various emerging contaminants like hormones along with conjugates, endocrine disrupting chemicals and personal care products derived from different sources are ultimately discharged into sewage wastewater and sludge responsible for undesirable consequences in exposed life forms and receiving bodies including aquatic and terrestrial ones. Presence of increased as well as even minute concentrations of emerging contaminants poses deleterious impact on environment (Leal et al. 2020) and organisms because of considerable changes in physiological and biochemical processes of organisms coming in direct or indirect contact. Most of the hormones are easily dissolved in aqueous environment and differ significantly from other classes of drugs with respect to structure and physico-chemical characteristics (Nieto et al. 2010). Endocrine disruptors also referred as "endocrine disrupting chemicals" are chemically synthesized molecules and mixture thereof interfering with the normal biological functions of natural hormone (Zoeller et al. 2012; Yang et al. 2020a, b) existing in human and other animals, consequently affecting growth, development, reproductive and other hormone dependent vital physiological and biochemical processes (Gonsioroski et al. 2020) via interaction with receptors generally employed by endogenous hormones. Some of the important endocrine disruptors frequently observed in environmental matrices include ethyl *p*-hydrobenzoate, bisphenol A and dibutyl phthalate (Wang et al. 2020c) and are also detected in personal care products. Generally personal care products are described in conjunction with pharmaceuticals. The continued occurrence of personal care products in different environmental media is of growing concern due to incessant utilization by human population globally (Sadutto et al. 2020). The introduction of non-persistent personal care products like scents, preservatives, disinfectants, sunscreen UV filters and health care based substances from different sources with the concentration ranging from ng $l^{-1}$ to μg $l^{-1}$ (Yang et al. 2017; Jiang et al. 2019) needs thorough monitoring of affected ecosystem, especially agroecosystem for alleviating the hazardous impacts (Zhang et al. 2017; Ali et al. 2018).

The occurrence and quantification of natural hormone 17β-estradiol and testosterone in sewage wastewater based on enzyme linked immunosorbent assay (ELISA) and estrogen receptor binding assays is registered recently by Aziz and Ojumu (2020). The maximum concentration of 17β-estradiol in collected raw wastewater was equivalent to 80.22 ng $l^{-1}$ and that of testosterone was found as 281.3 ng $l^{-1}$, comparatively very high from the set trigger value of 0.7 and 1.0 ng $l^{-1}$ as laid down by USEPA and World Health Organization, and European Union, respectively. The presence of estrogen hormones in sewage sludge as determined through Gas Chromatography/Mass spectroscopy (GC/MS) is demonstrated by Ternes et al. (2002). The concentrations of estrone, 17β-estradiol, and 17α-ethinylestradiol were determined to reach the level of 37, 49, and 17 ng $g^{-1}$, respectively. The prevalence of estrogen hormones even after the sludge digestion suggested persistent nature, requiring further treatment to eliminate traces of contaminants.

The perturbing nature of endocrine disruptors modulating reproductive and neuronal system in humans and animals indicates threat of such emerging contaminants to agroecosystem upon releases from sources like sewage. Presence of environmentally noxious endocrine disruptors including androstenedione and metabolized products in influents as well as discharges from wastewater treatment facilities is illustrated currently by Zwart et al. (2020), displaying possibility of contaminant transfer into agroecosystem. Experimental investigation indicating the prevalence of endocrine disruptors such as nonylphenol, nonylphenol ethoxylates, triclosan and bisphenol A in sewage sludge employing the technique of gas chromatography integrated with mass spectroscopy is described by Gatidou et al. (2007). The widespread occurrence of endocrine disruptors in sewage sludge exerting ecotoxicological impacts on natural environment (Kutluyer et al. 2020; De la Parra-Guerra and Olivero-Verbel 2020) and human health (Sprague et al. 2013) calls for the investigation of different factors influencing contaminant transport, degradation, and uptake by different life forms together with deployment of economical tools for rapid assessment and identification in complex media.

Studies have been conducted to assess and quantify the level of different personal care products in sewage wastewater based on chromatographic tools integrated with mass spectrometry (Fan et al. 2020). The comprehensive review on prevalence of personal care products in potable water and wastewater from sewage treatment facilities is presented by Yang et al. (2017). Monitoring of sludge samples at different treatment stages have been conducted in order to reveal the occurrence of personal care products like preservatives, sunscreen filters, and disinfectants including triclosan and triclocarban (Abril et al. 2020). The contents of disinfectants and UV sunscreen filters were determined upto 512 ng $g^{-1}$ and 662 ng $g^{-1}$, respectively based on dry mass basis. The wide occurrence in wastewater may impose undesirable consequences on human health and environment once entered into the agroecosystem. Careful monitoring and precise determination, therefore is necessary for management of contaminated sites.

The treatment processes currently being used are not much efficient in either completely eliminating or reducing the contaminants present in sewage wastewater and sludge, suggesting optimization of operational conditions in both conventional and new technologies meant for removal of emerging contaminants to the level of USEPA and World Health Organization trigger value. Furthermore, the elimination of contaminants by a particular treatment technique largely depends on characteristics of wastewater and sludge characteristics such as pH, organic matter, inorganic and organic contaminants, type of microorganisms present, and sources. The successful elimination of hormones, endocrine disruptors and personal care products is necessary as during farmland application of sewage wastewater and sludge, there are ample chances of food chain contamination (Smith 2009) and consequently harmful impact on human health. Some of the important traditional and emerging contaminants present in sewage wastewater and sludge are illustrated in Fig. 2.1.

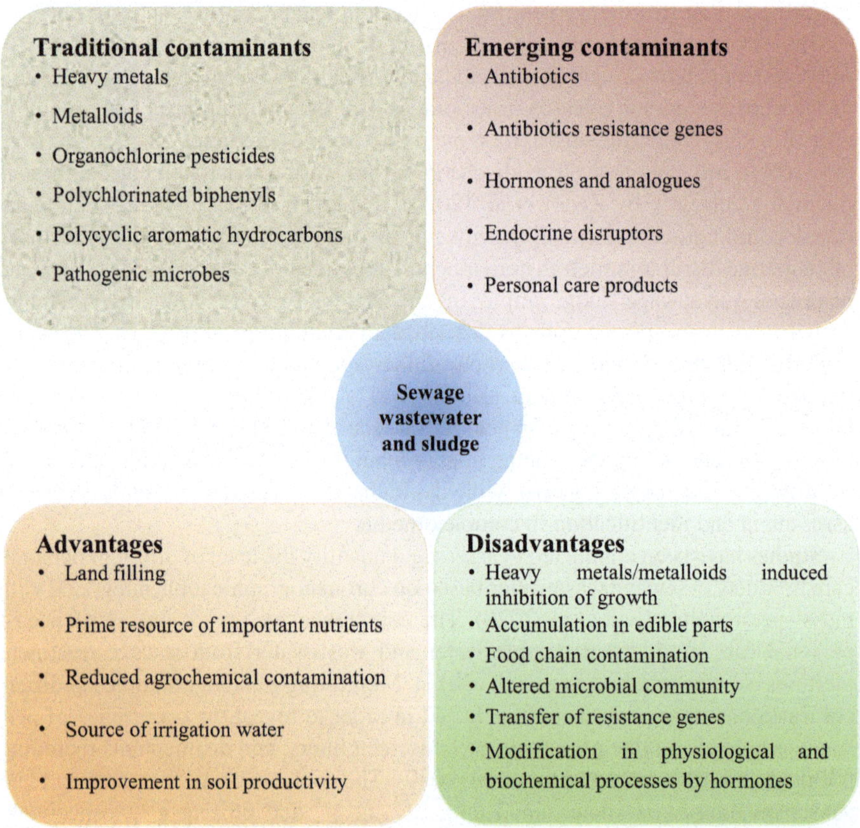


**Fig. 2.1** Traditional and emerging contaminants present in sewage wastewater and sludge along with advantages and disadvantages. (Source: Sayo et al. 2020; Zhang et al. 2013; Khillare et al. 2020; Dan et al. 2020; Ajibola et al. 2020; Yang et al. 2014; Lopes et al. 2020; Guimaraes et al. 2016; Jiang et al. 2019; Aziz and Ojumu 2020; Fang et al. 2020; Rolsky et al. 2020; Santás-Miguel et al. 2020)

## 2.3 Fate of Sewage Wastewater and Sewage Sludge in Agroecosystem

Treated sewage wastewater and sludge components may undergo natural processes of surface binding and transformation after getting entry into the agroecosystem. Adsorption on soil matrices (Pan and Chu 2016) may reduce the availability of traditional and emerging contaminants to microbes, fauna, and cultivated plants (Fig. 2.2). The adherence of inorganic and organic chemical entities onto soil components is characteristically modulated not only by physico-chemical characteristics of contaminants itself but also by soil physical, chemical and biological attributes (Shi et al. 2020b; Umeh et al. 2020). The exposure of wastewater and sludge contaminants to sunlight under open agricultural fields may lead to substantial modification in structure, degradation (Sturini et al. 2012; Torres-Palma et al. 2020) and biological effects exerted therein. Furthermore, increased temperature effect rendered by sunlight may facilitate the volatilization of selected organic and inorganic residues. The contaminants could be transported and accumulated in different parts of economically important crop plants (Eid et al. 2020b) including edible portions. The commencement of seasonal precipitation may lead to leaching and run-off of different contaminants finally ending up in terrestrial and aquatic ecosystems.

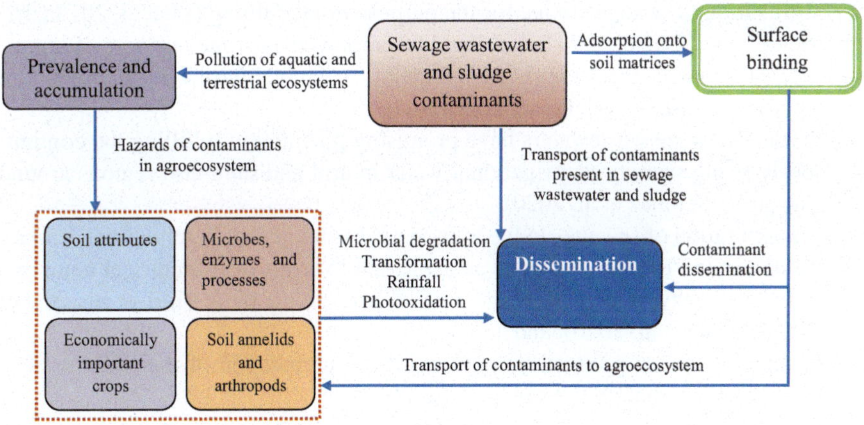


**Fig. 2.2** Fate, transport and effect of contaminants present in sewage wastewater and sludge in agroecosystem. The contaminants are transferred, and accumulated to nearby aquatic and terrestrial system affecting natural environment. The entry into agroecosystem may modulate the soil characteristics, microbiological properties, crop plants and soil fauna. The soil surface bound contaminants may be transformed biologically and non-biologically followed by dissemination to vicinity

## 2.4 Advantages and Disadvantages of Sewage Wastewater and Sludge Application in Agroecosystem

Sewage wastewater and sludge after proper treatment may be used for irrigation as well as for soil fertilization purposes. Nevertheless, most of the currently used practices for treatment are not efficient in eliminating the various organic and inorganic contaminants to a safer level, thereby indicating risks to agroecosystem. The studies conducted so far have presented both advantageous and disadvantageous impacts of sewage wastewater and sludge application in cultivated ecosystems. Important benefits and risks are briefly presented in following section.

### *2.4.1 Advantages of Sewage Wastewater and Sludge Application in Agriculture*

Worldwide, enormous amounts of sewage wastewater and sludge are generated. They are important source of essential nutrients and trace elements like nitrogen, phosphorus, potassium, sulfur, calcium and magnesium required for vital plant processes (Fang et al. 2018). The judicious application could have important benefits including

(a) Utilization of sewage sludge for the purpose of land filling (Li et al. 2014a, b).
(b) Sewage wastewater and sludge could be employed as prime resource of important nutrients to meet the rising demand of chemical fertilizers in agriculture (Fang et al. 2020; Vogel et al. 2020).
(c) Treated sewage wastewaters have promising potential in fulfilling the continuously rising scarcity of irrigational water in arid and semi-arid regions around the globe (Yılmaz et al. 2020).
(d) Incorporation of sewage sludge at increased amounts has potential opportunity in modifying the soil attributes as expressed in terms of soil nutrient composition, soil enzyme activities and overall soil productivity (Eid et al. 2020a; Skowrońska et al. 2020), and
(e) Sludge possesses promising opportunities in preparation of materials used in building construction (Chang et al. 2020)

### *2.4.2 Disadvantages of Sewage Wastewater and Sludge Application in Agriculture*

Apart from beneficial nutrients, sewage wastewater and sludge also contain unacceptable level of various environmental contaminants like heavy metals, metalloids, microplastics, organic contaminants, personal care products, antibiotics, antibiotic resistance genes, organic materials and pathogenic microbes bringing harmful impact on different environmental components and human health (Qian et al. 2016; Abril et al. 2020; Arun et al. 2020; Lopes et al. 2020; Praspaliauskas et al. 2020; Rolsky et al. 2020; Wang et al. 2020d). Following negative consequences of utilizing sewage wastewater and sludge could be expected.

(a) Presence of heavy metals and metalloids in sewage wastewater and sludge (Eid et al. 2020b) beyond the recommended limits may limit the growth and development not only of plants but also of exposed humans. The raised contents of heavy metals and metalloids are described to impose oxidative stress in soil microorganisms resulting into alteration in biological processes conducted. The accumulation in different plant parts including edible ones may directly influence the human and livestock health.
(b) The presence of antibiotics even in traces could modify the microbiological community structure and function to a greater extent (Santás-Miguel et al. 2020), leading to considerable changes in soil characteristics consequently the overall crop productivity.
(c) Dissemination of antibiotic resistance genes through application of sewage wastewater and sludge has long been documented by various researchers across the globe (Christou et al. 2017; Piña et al. 2020). The transfer of antibiotic resistance genes to beneficial soil microbes could modulate the native soil microbiological characteristics and associated ecological functions. Additionally, the transfer of antibiotic resistance genes to clinically important microbes may threaten the human health severely, and
(d) The introduction of hormones, endocrine disrupting chemicals and personal care products in agroecosystem (Pedersen et al. 2005) beyond the safety limit may challenge the normal physiological and biochemical processes of exposed humans and animals rendered by interaction with hormone receptors. The prevalence of increased concentrations of mutagenic and carcinogenic compounds present in sewage wastewater and sludge are another largely unnoticed risk to human health and environment.

Although, the negative consequences associated with the deployment of sewage wastewater and sludge could be reduced to a great extent by incorporating different treatment technologies including incineration with the resultant of lowered sludge volume and energy generation (Lundin et al. 2004; Raheem et al. 2018), none of the technologies deployed to date could have proven success in eliminating all the chemical hazards to an acceptable level set by different regulatory bodies. The key advantages and disadvantages of employing sewage wastewater and sludge in agroecosystem are represented in Fig. 2.1.

## 2.5 Treatment of Sewage Wastewater and Sludge for Application in Agroecosystem

The most commonly employed techniques for management of sewage sludge involve land filling, high temperature burning, hydrothermal carbonization and application to cultivated ecosystem (Pradel et al. 2016; Wang et al. 2017; Chu et al. 2020). The process of land filling, however, leads to release of greenhouse gas, engagement of large land areas and phenomenon of secondary pollution. The high temperature burning (incineration) is regarded as the most viable option for treatment due to potential in minimizing the sludge content. The disposal of sludge to cultivated land sites after treatment is another important and sustainable strategy for management of huge sewage sludge produced globally (Tarpani and Azapagic 2018; Gherghel et al. 2019). Composting as another less expensive and alternative strategy for treating sewage sludge as well as for reutilization in intended purposes is reported worldwide (Duan et al. 2019). The composting may sufficiently facilitate breakdown of organic contaminants, nevertheless, chances of rise in content of heavy metals also prevails (Zheng et al. 2018; Yu et al. 2019). Additionally, the process of composting may also accelerate the metal mobilization at the time of composting (Liu et al. 2007).

Constructed wetlands (CWs) are among the environment friendly and cost effective techniques integrating physical, chemical and biological processes for eliminating the antibiotics present in sewage wastewater (Dan et al. 2020). The association of conventional treatment processes with wetland system harbors characteristic potential in removing antibiotics and antibiotic resistance genes (García et al. 2020). The important mechanisms underlying sequestration of target antibiotics in constructed wetlands may involve biological degradation, sorption, acquisition by plants, photooxidation, and loss through volatilization depending on the physicochemical attributes of contaminants (Imfeld et al. 2009; Li et al. 2014a, b; Verlicchi and Zambello 2014). Since, the treatment of sewage wastewater and sludge is governed by different parameters like pH, nutrient composition, organic material's content, contaminant characteristics, and microbiological features, so far, none of the treatment technologies could have shown efficiency in eliminating contaminants absolutely, implying the need of further research and development for ensuring successful contaminant sequestration.

## 2.6 Conclusion and Future Perspectives

Rapid rise in anthropogenic activities has introduced large amount of sewage wastewater and sludge in natural ecosystem, ultimately finding their ways in different aquatic and terrestrial environment. The prevalence of exceedingly high content of various inorganic and organic contaminants of both traditional and emerging nature such as heavy metals, metalloids, pesticides, antibiotics, antibiotic resistance genes, hormones and personal care products from varied sources, after introduction into agroecosystem may substantially modulate the soil physical, chemical and biological attributes, thereby productivity of multiple economically important crops, putting the risks to global food security. Different components of farmland applied sewage wastewater and sludge may undergo adsorption, transformation and uptake by plants and soil microorganisms. Although, there are multifarious advantages of applying sewage wastewater and sludge into agroecosystem because of increased proportions of essential micro- and macro-nutrients and organic materials, the negative consequences to receiving environments require the management in sustainable manner. The currently used techniques of sewage wastewater and sludge treatment comprise of nano-filtration, aerobic and anerobic digestion, pyrolysis, constructed wetlands, and wastewater treatment plants, but are not effective in completely eliminating the contaminants present, suggesting the improvement in current practice of management. Some of the future prospects in this area are provided below.

- The investigations are needed to precisely determine the concentration of different contaminants present in sewage wastewater and sludge in order to decipher the fate and transport as well as accumulation in different plant parts especially, the edible portion to avoid the toxicity.
- The assessment of toxicity of different organic and inorganic contaminants together with the impact of using sewage wastewater and sludge as a whole on agriculturally important microorganisms associated biological processes such as nutrient cycling and fixation of nitrogen, soil fauna and crop plants is imperative to safeguard the gradually declining productivity of cultivable land areas.
- Through understanding of the processes influencing the fate and transport of contaminants existing in sewage wastewater and sludge is inevitable for controlling the food chain contamination and hazardous impact on human health.
- Development of low cost, ecologically sound innovative technologies for treatment of sewage wastewater and sludge could help not only in removing contaminants but would also provide important opportunities in simultaneous energy generation and production of valuable materials like char and fertilizers for agricultural field applications.

## References

Abril C, Santos JL, Martín J, Aparicio I, Alonso E (2020) Occurrence, fate and environmental risk of anionic surfactants, bisphenol A, perfluorinated compounds and personal care products in sludge stabilization treatments. Sci Total Environ 711:135048. https://doi.org/10.1016/j.scitotenv.2019.135048

Ajibola AS, Tisler S, Zwiener C (2020) Simultaneous determination of multiclass antibiotics in sewage sludge based on QuEChERS extraction and liquid chromatography-tandem mass spectrometry. Anal Methods 12:576–586. https://doi.org/10.1039/C9AY02188D

Alamer K, Ali E, Al-Thubaiti M, Al-Ghamdi M (2020) Zinc nutrition and its activated roles on growth, inflorescences attributes and some physiological parameters of *Tagetes erecta* L. Plants. Pak J Biol Sci 23:35. https://doi.org/10.3923/pjbs.2020.35.44

Ali AM, Rønning HT, Sydnes LK, Alarif WM, Kallenborn R, Al-Lihaibi SS (2018) Detection of PPCPs in marine organisms from contaminated coastal waters of the Saudi Red Sea. Sci Total Environ 621:654–662. https://doi.org/10.1016/j.scitotenv.2017.11.298

Amorós I, Moreno Y, Reyes M, Moreno-Mesonero L, Alonso JL (2016) Prevalence of *Cryptosporidium* oocysts and *Giardia* cysts in raw and treated sewage sludges. Environ Technol 37:2898–2904. https://doi.org/10.1080/09593330.2016.1168486

Arun S, Kumar RM, Raj J, Mukhopadhyay M, Ilango K, Chakrabort P (2020) Occurrence, source and risk assessment of fluoroquinolones in dumpsite soil and sewage sludge from Chennai, India. Environ Toxicol Pharmacol:103410. https://doi.org/10.1016/j.etap.2020.103410

Aziz M, Ojumu T (2020) Exclusion of estrogenic and androgenic steroid hormones from municipal membrane bioreactor wastewater using UF/NF/RO membranes for water reuse application. Membranes 10:37. https://doi.org/10.3390/membranes10030037

Badgley C, Moghtader J, Quintero E, Zakem E, Chappell MJ, Aviles-Vazquez K, Samulon A, Perfecto I (2007) Organic agriculture and the global food supply. Renewable Agric Food Syst 22:86–108. https://doi.org/10.1017/S1742170507001640

Bao LP, Chong CO, Mohamed SMS, Jo-Shu C, Tau CL, Su SL, Joon CJ (2020) Conventional and emerging technologies for removal of antibiotics from wastewater. J Hazard Mater:122961. https://doi.org/10.1016/j.jhazmat.2020.122961

Barakat MA (2011) New trends in removing heavy metals from industrial wastewater. Arab J Chem 4:361–377. https://doi.org/10.1016/j.arabjc.2010.07.019

Borghi AA, Palma MSA (2014) Tetracycline: production, waste treatment and environmental impact assessment. Braz J Pharm Sci 50:25–40. https://doi.org/10.1590/S1984-82502011000100003

Bradford SA, Morales VL, Zhang W, Harvey RW, Packman AI, Mohanram A, Welty C (2013) Transport and fate of microbial pathogens in agricultural settings. Crit Rev Environ Sci Technol 43:775–893. https://doi.org/10.1080/10643389.2012.710449

Cai S, Wang J, Lv W, Xu S, Zhu H (2020) Nitrogen fertilization alters the effects of earthworms on soil physicochemical properties and bacterial community structure. Appl Soil Ecol 150:103478. https://doi.org/10.1016/j.apsoil.2019.103478

Chang Z, Long G, Zhou JL, Ma C (2020) Valorization of sewage sludge in the fabrication of construction and building materials: a review. Resour Conserv Recycl 154:104606. https://doi.org/10.1016/j.resconrec.2019.104606

Chen H, Zhang M (2013) Occurrence and removal of antibiotic resistance genes in municipal wastewater and rural domestic sewage treatment systems in eastern China. Environ Int 55:9–14. https://doi.org/10.1016/j.envint.2013.01.019

Chen Q, An X, Li H, Su J, Ma Y, Zhu YG (2016) Long-term field application of sewage sludge increases the abundance of antibiotic resistance genes in soil. Environ Int 92:1–10. https://doi.org/10.1016/j.envint.2016.03.026.

Christodoulou A, Stamatelatou K (2016) Overview of legislation on sewage sludge management in developed countries worldwide. Water Sci Technol 73:453–462. https://doi.org/10.2166/wst.2015.521

Christou A, Agüera A, Bayona JM, Cytryn E, Fotopoulos V, Lambropoulou D, Manaia CM, Michael C, Revitt M, Schröder P, Fatta-Kassinos D (2017) The potential implications of reclaimed wastewater reuse for irrigation on the agricultural environment: the knowns and unknowns of the fate of antibiotics and antibiotic resistant bacteria and resistance genes–a review. Water Res 123:448–467. https://doi.org/10.1016/j.watres.2017.07.004

Chu Q, Xue L, Singh BP, Yu S, Müller K, Wang H, Feng Y, Pan G, Zheng X, Yang L (2020) Sewage sludge-derived hydrochar that inhibits ammonia volatilization, improves soil nitrogen retention and rice nitrogen utilization. Chemosphere 245:125558. https://doi.org/10.1016/j.chemosphere.2019.125558

Clarke BO, Porter NA, Marriott PJ, Blackbeard JR (2010) Investigating the levels and trends of organochlorine pesticides and polychlorinated biphenyl in sewage sludge. Environ Int 36:323–329. https://doi.org/10.1016/j.envint.2010.01.004

Dan A, Zhang X, Dai Y, Chen C, Yang Y (2020) Occurrence and removal of quinolone, tetracycline, and macrolide antibiotics from urban wastewater in constructed wetlands. J Clean Prod 252:119677. https://doi.org/10.1016/j.jclepro.2019.119677

De la Parra-Guerra A, Olivero-Verbel J (2020) Toxicity of nonylphenol and nonylphenol ethoxylate on *Caenorhabditis elegans*. Ecotoxicol Environ Saf 187:109709. https://doi.org/10.1016/j.ecoenv.2019.109709

Diaconu M, Pavel LV, Hlihor RM, Rosca M, Fertu DI, Lenz M, Corvini PX, Gavrilescu M (2020) Characterization of heavy metal toxicity in some plants and microorganisms–a preliminary approach for environmental bioremediation. New Biotechnol 56:130–139. https://doi.org/10.1016/j.nbt.2020.01.003

Ding C, He J (2010) Effect of antibiotics in the environment on microbial populations. Appl Microbiol Biotechnol 87:925–941. https://doi.org/10.1007/s00253-010-2649-5

Duan Y, Awasthi SK, Liu T, Zhang Z, Awasthi MK (2019) Evaluation of integrated biochar with bacterial consortium on gaseous emissions mitigation and nutrients sequestration during pig manure composting. Bioresour Technol 291:121880. https://doi.org/10.1016/j.biortech.2019.121880

Dumontet S, Scopa A, Kerje S, Krovacek K (2001) The importance of pathogenic organisms in sewage and sewage sludge. J Air Waste Manag Assoc 51:848–860. https://doi.org/10.1080/10473289.2001.10464313

Duttagupta S, Mukherjee A, Das K, Dutta A, Bhattacharya A, Bhattacharya J (2020) Groundwater vulnerability to pesticide pollution assessment in the alluvial aquifer of Western Bengal basin, India using overlay and index method. Geochemistry 125601. DOI: https://doi.org/10.1016/j.chemer.2020.125601.

Dzul-Caamal R, Vega-López A, Osten JRV (2020) Distribution of heavy metals in crop soils from an agricultural region of the Yucatan Peninsula and biochemical changes in earthworm *Eisenia foetida* exposed experimentally. Environ Monit Assess 192:1–15. https://doi.org/10.1007/s10661-020-08273-7

Eid EM, Hussain AA, Taher MA, Galal TM, Shaltout KH, Sewelam N (2020a) Sewage sludge application enhances the growth of *Corchorus olitorius* plants and provides a sustainable practice for nutrient recirculation in agricultural soils. J Soil Sci Plant Nutr 20:149–159. https://doi.org/10.1007/s42729-019-00113-z

Eid EM, Shaltout KH, Abdallah SM, Galal TM, El-Bebany AF, Sewelam NA (2020b) Uptake prediction of ten heavy metals by *Eruca sativa* Mill. cultivated in soils amended with sewage sludge. Bull Environ Contam Toxicol 104:134–143. https://doi.org/10.1007/s00128-019-02746-3

El-Senousy WM, Abou-Elela SI (2017) Assessment and evaluation of an integrated hybrid anaerobic–aerobic sewage treatment system for the removal of enteric viruses. Food Environ Virol 9:287–303. https://doi.org/10.1007/s12560-017-9286-4

Emadodin I, Bork HR (2012) Degradation of soils as a result of long-term human-induced transformation of the environment in Iran: an overview. J Land Use Sci 7:203–219. https://doi.org/10.1080/1747423X.2011.560292

Emadodin I, Reinsch T, Rotter A, Orlando-Bonaca M, Taube F, Javidpour J (2020) A perspective on the potential of using marine organic fertilizers for the sustainable management of coastal ecosystem services. Environ Sustain:1–11. https://doi.org/10.1007/s42398-020-00097-y

Fan X, Gao J, Li W, Huang J, Yu G (2020) Determination of 27 pharmaceuticals and personal care products (PPCPs) in water: the benefit of isotope dilution. Front Environ Sci Eng 14:8. https://doi.org/10.1007/s11783-019-1187-3

Fang L, Li JS, Guo MZ, Cheeseman CR, Tsang DC, Donatello S, Poon CS (2018) Phosphorus recovery and leaching of trace elements from incinerated sewage sludge ash (ISSA). Chemosphere 193:278–287. https://doi.org/10.1016/j.chemosphere.2017.11.023

Fang L, Li JS, Donatello S, Cheeseman CR, Poon CS, Tsang DC (2020) Use of Mg/Ca modified biochars to take up phosphorus from acid-extract of incinerated sewage sludge ash (ISSA) for fertilizer application. J Clean Prod 244:118853. https://doi.org/10.1016/j.jclepro.2019.118853

García J, García-Galán MJ, Day JW, Boopathy R, White JR, Wallace S, Hunter RG (2020) A review of emerging organic contaminants (EOCs), antibiotic resistant bacteria (ARB), and antibiotic resistance genes (ARGs) in the environment: increasing removal with wetlands and reducing environmental impacts. Bioresour Technol:123228. https://doi.org/10.1016/j.biortech.2020.123228

Gatidou G, Thomaidis NS, Stasinakis AS, Lekkas TD (2007) Simultaneous determination of the endocrine disrupting compounds nonylphenol, nonylphenol ethoxylates, triclosan and bisphenol A in wastewater and sewage sludge by gas chromatography–mass spectrometry. J Chromatogr A 1138:32–41. https://doi.org/10.1016/j.chroma.2006.10.037

Geisseler D, Scow KM (2014) Long-term effects of mineral fertilizers on soil microorganisms–a review. Soil Biol Biochem 75:54–63. https://doi.org/10.1016/j.soilbio.2014.03.023

Gherghel A, Teodosiu C, De Gisi S (2019) A review on wastewater sludge valorisation and its challenges in the context of circular economy. J Clean Prod 228:244–263. https://doi.org/10.1016/j.jclepro.2019.04.240

Gomes MP, de Brito JCM, Carneiro MMLC, da Cunha MRR, Garcia QS, Figueredo CC (2018) Responses of the nitrogen-fixing aquatic fern *Azolla* to water contaminated with ciprofloxacin: impacts on biofertilization. Environ Pollut 232:293–299. https://doi.org/10.1016/j.envpol.2017.09.054

Gonsioroski A, Mourikes VE, Flaws JA (2020) Endocrine disruptors in water and their effects on the reproductive system. Int J Mol Sci 21:1929. https://doi.org/10.3390/ijms21061929

Guimaraes JR, Guadagnini RA, Franco RMB, dos Santos LU (2016) Inactivation of *Clostridium perfringens*, total coliforms, and *Escherichia coli* by UV/$H_2O_2$ in wastewater treatment plant effluent. J Adv Oxid Technol 19:93–97. https://doi.org/10.1515/jaots-2016-0112

Guo Z, Wan S, Hua K, Yin Y, Chu H, Wang D, Guo X (2020) Fertilization regime has a greater effect on soil microbial community structure than crop rotation and growth stage in an agroecosystem. Appl Soil Ecol 149:103510. https://doi.org/10.1016/j.apsoil.2020.103510

Humberto A, Paula M, Butler B, Jorge P, Matus F, Merino C, Cornejo P, Yakov K (2020) Meta-analysis of heavy metal effects on soil enzyme activities. Sci Total Environ:139744. https://doi.org/10.1016/j.scitotenv.2020.139744

Imfeld G, Braeckevelt M, Kuschk P, Richnow H (2009) Monitoring and assessing processes of organic chemicals removal in constructed wetlands. Chemosphere 74:349–362. https://doi.org/10.1016/j.chemosphere.2008.09.062

Jiang Y, Dennehy C, Lawlor PG, Hu Z, Zhan X, Gardiner GE (2018a) Inactivation of enteric indicator bacteria and system stability during dry co-digestion of food waste and pig manure. Sci Total Environ 612:293–302. https://doi.org/10.1016/j.scitotenv.2017.08.214

Jiang Y, Dennehy C, Lawlor PG, Hu Z, Yang Q, McCarthy G, Tan SP, Zhan X, Gardiner GE (2018b) Inactivation of *Salmonella* during dry co-digestion of food waste and pig manure. Waste Manag 82:231–240. https://doi.org/10.1016/j.wasman.2018.10.037

Jiang X, Qu Y, Liu L, He Y, Li W, Huang J, Yang H, Yu G (2019) PPCPs in a drinking water treatment plant in the Yangtze river delta of China: occurrence, removal and risk assessment. Front Environ Sci Eng 13:27. https://doi.org/10.1007/s11783-019-1109-4

Jiang Q, Qi Z, Xue L, Bukovsky M, Madramootoo CA, Smith W (2020a) Assessing climate change impacts on greenhouse gas emissions, N losses in drainage and crop production in a subsurface drained field. Sci Total Environ 705:135969. https://doi.org/10.1016/j.scitotenv.2019.135969

Jiang Y, Xie SH, Dennehy C, Lawlor PG, Hu ZH, Wu GX, Zhan XM, Gardiner GE (2020b) Inactivation of pathogens in anaerobic digestion systems for converting biowastes to bioenergy: a review. Renew Sustain Energy Rev 120:109654. https://doi.org/10.1016/j.rser.2019.109654

Kafaei R, Arfaeinia H, Savari A, Mahmoodi M, Rezaei M, Rayani M, Sorial GA, Fattahi N, Ramavandi B (2020) Organochlorine pesticides contamination in agricultural soils of southern Iran. Chemosphere 240:124983. https://doi.org/10.1016/j.chemosphere.2019.124983

Khillare PS, Sattawan VK, Jyethi DS (2020) Profile of polycyclic aromatic hydrocarbons in digested sewage sludge. Environ Technol 41:842–851. https://doi.org/10.1080/09593330.2018.1512654

Kołodziej B, Antonkiewicz J, Stachyra M, Bielińska EJ, Wiśniewski J, Luchowska K, Kwiatkowski C (2015) Use of sewage sludge in bioenergy production—a case study on the effects on *Sorghum* biomass production. Eur J Agron 69:63–74. https://doi.org/10.1016/j.eja.2015.06.004

Kumar M, Ram B, Sewwandi H, Honda R, Chaminda T (2020) Treatment enhances the prevalence of antibiotic-resistant bacteria and antibiotic resistance genes in the wastewater of Sri Lanka, and India. Environ Res 183:109179. https://doi.org/10.1016/j.envres.2020.109179

Kümmerer K (2009) Antibiotics in the aquatic environment–a review–part II. Chemosphere 75:435–441. https://doi.org/10.1016/j.envint.2016.06.025.

Kutluyer F, Çakir Sahilli Y, Kocabaş M, Aksu Ö (2020) Sperm quality and oxidative stress in chub *Squalius orientalis* and *Padanian barbell, Barbus plebejus* (Teleostei: Cyprinidae) after *in vitro* exposure to low doses of bisphenol A. Drug Chem Toxicol:1–6. https://doi.org/10.1080/01480545.2020.1726379

Leal CS, Mesquita DP, Amaral AL, Amaral AM, Ferreira EC (2020) Environmental impact and biological removal processes of pharmaceutically active compounds: the particular case of sulfonamides, anticonvulsants and steroid estrogens. Crit Rev Environ Sci Technol 50:698–742. https://doi.org/10.1080/10643389.2019.1642831

Li Y, Zhu G, Ng WJ, Tan SK (2014a) A review on removing pharmaceutical contaminants from wastewater by constructed wetlands: design, performance and mechanism. Sci Total Environ 468:908–932. https://doi.org/10.1016/j.scitotenv.2013.09.018

Li YL, Liu JW, Chen JY, Shi YF, Mao W, Liu H, Li Y, He S, Yang JK (2014b) Reuse of dewatered sewage sludge conditioned with skeleton builders as landfill cover material. Int J Environ Sci Technol 11:233–240. https://doi.org/10.1007/s13762-013-0199-y

Li Y, Tremblay J, Bainard LD, Cade-Menun B, Hamel C (2020a) Long-term effects of nitrogen and phosphorus fertilization on soil microbial community structure and function under continuous wheat production. Environ Microbiol 22:1066–1088. https://doi.org/10.1111/1462-2920.14824

Li C, Xie S, Wang Y, Pan X, Yu G, Zhang Y (2020b) Simultaneous heavy metal immobilization and antibiotics removal during synergetic treatment of sewage sludge and pig manure. Environ Sci Pollut Res:1–10. https://doi.org/10.1007/s11356-020-09230-0.

Liu Y, Ma L, Li Y, Zheng L (2007) Evolution of heavy metal speciation during the aerobic composting process of sewage sludge. Chemosphere 67:1025–1032. https://doi.org/10.1016/j.chemosphere.2006.10.056

Liu YY, Wang Y, Walsh TR, Yi LX, Zhang R, Spencer J, Doi Y, Tian G, Dong B, Huang X, Yu LF (2016) Emergence of plasmid-mediated colistin resistance mechanism MCR-1 in animals and human beings in China: a microbiological and molecular biological study. Lancet Infect Dis 16:161–168. https://doi.org/10.1016/S1473-3099(15)00424-7

Liu B, Martre P, Ewert F, Porter JR, Challinor AJ, Müller C, Ruane AC, Waha K, Thorburn PJ, Aggarwal PK, Ahmed M (2019) Global wheat production with 1.5 and 2.0° C above pre-industrial warming. Glob Chang Biol 25:1428–1444. https://doi.org/10.1111/gcb.14542

Lopes BC, Machado EC, Rodrigues HF, Leal CD, Araújo JCD, Teixeira de Matos A (2020) Effect of alkaline treatment on pathogens, bacterial community and antibiotic resistance genes in different sewage sludges for potential agriculture use. Environ Technol 41:529–538. https://doi.org/10.1080/09593330.2018.1505960

Lopez A, Rodríguez-Chueca J, Mosteo R, Gomez J, Ormad MP (2020) Microbiological quality of sewage sludge after digestion treatment: a pilot scale case of study. J Clean Prod:120101. https://doi.org/10.1016/j.jclepro.2020.120101

Lopez-Arredondo DL, Leyva-González MA, González-Morales SI, López-Bucio J, Herrera-Estrella L (2014) Phosphate nutrition: improving low-phosphate tolerance in crops. Annu Rev Plant Biol 65:95–123. https://doi.org/10.1146/annurev-arplant-050213-035949

Lundin M, Olofsson M, Pettersson GJ, Zetterlund H (2004) Environmental and economic assessment of sewage sludge handling options. Resour Conserv Recycl 41:255–278. https://doi.org/10.1016/j.resconrec.2003.10.006

Ma J, Zhu D, Chen QL, Ding J, Zhu YG, Sheng GD, Qiu YP (2019) Exposure to tetracycline perturbs the microbiome of soil oligochaete *Enchytraeus crypticus*. Sci Total Environ 654:643–650. https://doi.org/10.1016/j.scitotenv.2018.11.154

Magri ME, Fidjeland J, Jönsson H, Albihn A, Vinnerås B (2015) Inactivation of adenovirus, reovirus and bacteriophages in fecal sludge by pH and ammonia. Sci Total Environ 520:213–221. https://doi.org/10.1016/j.scitotenv.2015.03.035

Magu MM, Keriko JM, Kareru PG, Chege CW (2016) Burdens of selected heavy metals in common fish species from specific Kenyan freshwaters. Int J Fish Aquat Stud 4:173–179

Mamphogoro TP, Babalola OO, Aiyegoro OA (2020) Sustainable management strategies for bacterial wilt of sweet peppers (*Capsicum annuum*) and other Solanaceous crops. J Appl Microbiol. https://doi.org/10.1111/jam.14653

Nag SK, Saha K, Bandopadhyay S, Ghosh A, Mukherjee M, Raut A, Raman RK, Suresh VR, Mohanty SK (2020) Status of pesticide residues in water, sediment, and fishes of Chilika Lake, India. Environ Monit Assess 192:1–10. https://doi.org/10.1007/s10661-020-8082-z

Nieto A, Borrull F, Pocurull E, Marcé RM (2010) Occurrence of pharmaceuticals and hormones in sewage sludge. Environ Toxicol Chem 29:1484–1489. https://doi.org/10.1002/etc.188

Pan M, Chu LM (2016) Adsorption and degradation of five selected antibiotics in agricultural soil. Sci Total Environ 545:48–56. https://doi.org/10.1016/j.scitotenv.2015.12.040

Pedersen JA, Soliman M, Suffet IH (2005) Human pharmaceuticals, hormones, and personal care product ingredients in runoff from agricultural fields irrigated with treated wastewater. J Agric Food Chem 53:1625–1632. https://doi.org/10.1021/jf049228m

Piña B, Bayona JM, Christou A, Fatta-Kassinos D, Guillon E, Lambropoulou D, Michael C, Polesel F, Sayen S (2020) On the contribution of reclaimed wastewater irrigation to the potential exposure of humans to antibiotics, antibiotic resistant bacteria and antibiotic resistance genes–NEREUS COST Action ES1403 position paper. J Environ Chem Eng 8:102131. https://doi.org/10.1016/j.jece.2018.01.011

Pradel M, Aissani L, Villot J, Baudez JC, Laforest V (2016) From waste to added value product: towards a paradigm shift in life cycle assessment applied to wastewater sludge–a review. J Clean Prod 131:60–75. https://doi.org/10.1016/j.jclepro.2016.05.076

Praspaliauskas M, Žaltauskaitė J, Pedišius N, Striūgas N (2020) Comprehensive evaluation of sewage sludge and sewage sludge char soil amendment impact on the industrial hemp growth performance and heavy metal accumulation. Ind Crop Prod 150:112396. https://doi.org/10.1016/j.indcrop.2020.112396

Prosekov AY, Ivanova SA (2018) Food security: the challenge of the present. Geoforum 91:73–77. https://doi.org/10.1016/j.geoforum.2018.02.030

Qian L, Wang S, Xu D, Guo Y, Tang X, Wang L (2016) Treatment of municipal sewage sludge in supercritical water: a review. Water Res 89:118–131. https://doi.org/10.1016/j.watres.2015.11.047

Racar M, Dolar D, Karadakić K, Čavarović N, Glumac N, Ašperger D, Košutić K (2020) Challenges of municipal wastewater reclamation for irrigation by MBR and NF/RO: Physico-chemical and microbiological parameters, and emerging contaminants. Sci Total Environ:137959. https://doi.org/10.1016/j.scitotenv.2020.137959

Raheem A, Sikarwar VS, He J, Dastyar W, Dionysiou DD, Wang W, Zhao M (2018) Opportunities and challenges in sustainable treatment and resource reuse of sewage sludge: a review. Chem Eng J 337:616–641. https://doi.org/10.1016/j.cej.2017.12.149

Raun WR, Johnson GV (1999) Improving nitrogen use efficiency for cereal production. Agron J 91:357–363

Razzaq A, Ali A, Safdar LB, Zafar MM, Rui Y, Shakeel A, Shaukat A, Ashraf M, Gong W, Yuan Y (2020) Salt stress induces physiochemical alterations in rice grain composition and quality. J Food Sci 85:14–20. https://doi.org/10.1111/1750-3841.14983

Rolsky C, Kelkar V, Driver E, Halden RU (2020) Municipal sewage sludge as a source of microplastics in the environment. Curr Opin Environ Sci Health 14:16–22. https://doi.org/10.1016/j.coesh.2019.12.001

Sadutto D, Álvarez-Ruiz R, Picó Y (2020) Systematic assessment of extraction of pharmaceuticals and personal care products in water and sediment followed by liquid chromatography–tandem mass spectrometry. Anal Bioanal Chem:1–15. https://doi.org/10.1007/s00216-019-02207-0.

Sahlström L, Aspan A, Bagge E, Danielsson-Tham ML, Albihn A (2004) Bacterial pathogen incidences in sludge from Swedish sewage treatment plants. Water Res 38:1989–1994. https://doi.org/10.1016/j.watres.2004.01.031

Santás-Miguel V, Arias-Estévez M, Díaz-Raviña M, Fernández-Sanjurjo MJ, Álvarez-Rodríguez E, Núñez-Delgado A, Fernández-Calviño D (2020) Interactions between soil properties and tetracycline toxicity affecting to bacterial community growth in agricultural soil. Appl Soil Ecol 147:103437. https://doi.org/10.1016/j.apsoil.2019.103437

Sayo S, Kiratu JM, Nyamato GS (2020) Heavy metal concentrations in soil and vegetables irrigated with sewage effluent: a case study of Embu sewage treatment plant, Kenya. Sci Afr 8:e00337. https://doi.org/10.1016/j.sciaf.2020.e00337

Shi Y, Liu J, Zhuo L, Yan X, Cai F, Luo W, Ren M, Liu Q, Yu Y (2020a) Antibiotics in wastewater from multiple sources and surface water of the Yangtze River in Chongqing in China. Environ Monit Assess 192:159. https://doi.org/10.1007/s10661-020-8108-6

Shi Z, Peng S, Lin X, Liang Y, Lee SZ, Allen HE (2020b) Predicting Cr (VI) adsorption on soils: the role of the competition of soil organic matter. Environ Sci Process Impacts 22:95–104. https://doi.org/10.1039/C9EM00477G

Singh AL, Singh VK (2018) Assessment of groundwater quality of Ballia district, Uttar Pradesh, India, with reference to arsenic contamination using multivariate statistical analysis. Appl Water Sci 8(3):95. https://doi.org/10.1007/s13201-018-0737-3

Singh PK, Deshbhratar PB, Ramteke DS (2012) Effects of sewage wastewater irrigation on soil properties, crop yield and environment. Agric Water Manag 103:100–104. https://doi.org/10.1016/j.agwat.2011.10.022

Singh AL, Singh VK, Srivastava A (2013) Effect of arsenic contaminated drinking water on human chromosome: a case study. Indian J Clin Biochem 28:422–425. https://doi.org/10.1007/s12291-013-0330-3

Skowrońska M, Bielińska EJ, Szymański K, Futa B, Antonkiewicz J, Kołodziej B (2020) An integrated assessment of the long-term impact of municipal sewage sludge on the chemical and biological properties of soil. Catena 189:104484. https://doi.org/10.1016/j.catena.2020.104484

Smith SR (2009) Organic contaminants in sewage sludge (biosolids) and their significance for agricultural recycling. Philos Trans R Soc A 367:4005–4041. https://doi.org/10.1098/rsta.2009.0154

Sprague BL, Trentham-Dietz A, Hedman CJ, Wang J, Hemming JD, Hampton JM, Buist DS, Bowles EJA, Sisney GS, Burnside ES (2013) Circulating serum xenoestrogens and mammographic breast density. Breast Cancer Res 15:R45. https://doi.org/10.1186/bcr3432

Sturini M, Speltini A, Maraschi F, Profumo A, Pretali L, Fasani E, Albini A (2012) Sunlight-induced degradation of soil-adsorbed veterinary antimicrobials marbofloxacin and enrofloxacin. Chemosphere 86:130–137. https://doi.org/10.1016/j.chemosphere.2011.09.053

Summers AO (2006) Genetic linkage and horizontal gene transfer, the roots of the antibiotic multi-resistance problem. Anim Biotechnol 17:125–135. https://doi.org/10.1080/10495390600957217

Tarpani RRZ, Azapagic A (2018) Life cycle costs of advanced treatment techniques for wastewater reuse and resource recovery from sewage sludge. J Clean Prod 204:832–847. https://doi.org/10.1016/j.jclepro.2018.08.300

Ternes TA, Andersen H, Gilberg D, Bonerz M (2002) Determination of estrogens in sludge and sediments by liquid extraction and GC/MS/MS. Anal Chem 74:3498–3504. https://doi.org/10.1021/ac015717z

Thung I, Aramin H, Vavinskaya V, Gupta S, Park JY, Crowe SE, Valasek MA (2016) The global emergence of Helicobacter pylori antibiotic resistance. Aliment Pharmacol Ther 43:514–533. https://doi.org/10.1111/apt.13497

Tong XN, Wang XZ, He XJ, Wang Z, Li WX (2020) Effects of antibiotics on microbial community structure and microbial functions in constructed wetlands treated with artificial root exudates. Environ Sci Process Impacts 22:217–226. https://doi.org/10.1039/C9EM00458K

Torres-Palma RA, Serna-Galvis EA, Ávila-Torres YP (2020) Photochemical and photocatalytical degradation of antibiotics in water promoted by solar irradiation. In: Singh P, Borthakur A, Mishra PK, Tiway D (eds) Nano-materials as photocatalysts for degradation of environmental pollutants. Elsevier, pp 211–243. https://doi.org/10.1016/B978-0-12-818598-8.00012-2.

Umeh C, Asegbeloyin JN, Akpomie KG, Oyeka EE, Ochonogor AE (2020) Adsorption properties of tropical soils from Awka North Anambra Nigeria for lead and cadmium ions from aqueous media. Chem Afr 3:199–210. https://doi.org/10.1007/s42250-019-00109-3

Urra J, Alkorta I, Lanzén A, Mijangos I, Garbisu C (2019) The application of fresh and composted horse and chicken manure affects soil quality, microbial composition and antibiotic resistance. Appl Soil Ecol 135:73–84. https://doi.org/10.1016/j.apsoil.2018.11.005

Vance CP, Uhde-Stone C, Allan DL (2003) Phosphorus acquisition and use: critical adaptations by plants for securing a nonrenewable resource. New Phytol 157:423–447. https://doi.org/10.1046/j.1469-8137.2003.00695.x

Verlicchi P, Zambello E (2014) How efficient are constructed wetlands in removing pharmaceuticals from untreated and treated urban wastewaters? A review. Sci Total Environ 470:1281–1306. https://doi.org/10.1016/j.scitotenv.2013.10.085

Vogel C, Hoffmann MC, Taube MC, Krüger O, Baran R, Adam C (2020) Uranium and thorium species in phosphate rock and sewage sludge ash based phosphorus fertilizers. J Hazard Mater 382:121100. https://doi.org/10.1016/j.jhazmat.2019.121100

Volkmann H, Schwartz T, Bischoff P, Kirchen S, Obst U (2004) Detection of clinically relevant antibiotic-resistance genes in municipal wastewater using real-time PCR (TaqMan). J Microbiol Methods 56:277–286. https://doi.org/10.1016/j.mimet.2003.10.014

Wang Y, Zhang Q, Lv J, Li A, Liu H, Li G, Jiang G (2007) Polybrominated diphenyl ethers and organochlorine pesticides in sewage sludge of wastewater treatment plants in China. Chemosphere 68:1683–1691. https://doi.org/10.1016/j.chemosphere.2007.03.060

Wang Q, Wei W, Gong Y, Yu Q, Li Q, Sun J, Yuan Z (2017) Technologies for reducing sludge production in wastewater treatment plants: state of the art. Sci Total Environ 587:510–521. https://doi.org/10.1016/j.scitotenv.2017.02.203

Wang G, Deng D, Hu C, Lou L, Luo L, He J, Tian D, Xiao Y, He Y, Zhang S, Deng S (2020a) More effective removal of antibiotic resistance genes from excess sludge by microwave integrated Fenton treatment. Int Biodeterior Biodegradation 149:104920. https://doi.org/10.1016/j.ibiod.2020.104920

Wang C, Fan Y, Hornung U, Zhu W, Dahmen N (2020b) Char and tar formation during hydrothermal treatment of sewage sludge in subcritical and supercritical water: effect of organic matter composition and experiments with model compounds. J Clean Prod 242:118586. https://doi.org/10.1016/j.jclepro.2019.118586

Wang D, Liu Y, Xu Z, Zhao D, Liu Y, Liu Z (2020c) Multitemplate molecularly imprinted polymeric solid-phase microextraction fiber coupled with HPLC for endocrine disruptor analysis in water samples. Microchem J:104802. https://doi.org/10.1016/j.microc.2020.104802

Wang X, Chen T, Zheng G (2020d) Preservation of nitrogen and sulfur and passivation of heavy metals during sewage sludge composting with $KH_2PO_4$ and $FeSO_4$. Bioresour Technol 297:122383. https://doi.org/10.1016/j.biortech.2019.122383

Watkinson AJ, Murby EJ, Kolpin DW, Costanzo SD (2009) The occurrence of antibiotics in an urban watershed: from wastewater to drinking water. Sci Total Environ 407:2711–2723. https://doi.org/10.1016/j.scitotenv.2008.11.059

Yang LH, Ying GG, Su HC, Stauber JL, Adams MS, Binet MT (2008) Growth-inhibiting effects of 12 antibacterial agents and their mixtures on the freshwater microalga *Pseudokirchneriella subcapitata*. Environ Toxicol Chem 27:1201–1208. https://doi.org/10.1897/07-471.1.

Yang Y, Li B, Zou S, Fang HH, Zhang T (2014) Fate of antibiotic resistance genes in sewage treatment plant revealed by metagenomic approach. Water Res 62:97–106. https://doi.org/10.1016/j.watres.2014.05.019

Yang Y, Ok YS, Kim KH, Kwon EE, Tsang YF (2017) Occurrences and removal of pharmaceuticals and personal care products (PPCPs) in drinking water and water/sewage treatment plants: a review. Sci Total Environ 596:303–320. https://doi.org/10.1016/j.scitotenv.2017.04.102

Yang C, Song G, Lim W (2020a) Effects of endocrine disrupting chemicals in pigs. Environ Pollut:114505. https://doi.org/10.1016/j.envpol.2020.114505

Yang W, Song W, Li J, Zhang X (2020b) Bioleaching of heavy metals from wastewater sludge with the aim of land application. Chemosphere 249:126134. https://doi.org/10.1016/j.chemosphere.2020.126134

Ye L, Zhao X, Bao E, Li J, Zou Z, Cao K (2020) Bio-organic fertilizer with reduced rates of chemical fertilization improves soil fertility and enhances tomato yield and quality. Sci Rep 10:1–11. https://doi.org/10.1038/s41598-019-56954-2.

Yılmaz F, Otuzaltı MM, Perendeci NA, Karatay M, Ünşar EK, Ateş M, Akin R, Yıldız O, Orhon D (2020) Potential of aerobic membrane bioreactor for recycling and reuse of domestic wastewater for irrigation. Environ Earth Sci 79:1–11. https://doi.org/10.1007/s12665-020-09006-2.

Younus H, Arsalan A, Alam MF (2020) Arsenic inhibits human salivary aldehyde dehydrogenase: mechanism and a population-based study. Chemosphere 243:125358. https://doi.org/10.1016/j.chemosphere.2019.125358

Yu B, Zheng G, Wang X, Wang M, Chen T (2019) Biodegradation of triclosan and triclocarban in sewage sludge during composting under three ventilation strategies. Front Environ Sci Eng 13:41. https://doi.org/10.1007/s11783-019-1125-4

Zemanová V, Popov M, Pavlíková D, Kotrba P, Hnilička F, Česká J, Pavlík M (2020) Effect of arsenic stress on 5-methylcytosine, photosynthetic parameters and nutrient content in arsenic hyperaccumulator *Pteris cretica* (L.) var. Albo-lineata. BMC Plant Biol 20:1–10. https://doi.org/10.1186/s12870-020-2325-6.

Zhang C, Liao X, Li J, Xu L, Liu M, Du B, Wang Y (2013) Influence of long-term sewage irrigation on the distribution of organochlorine pesticides in soil–groundwater systems. Chemosphere 92:337–343. https://doi.org/10.1016/j.chemosphere.2013.01.020

Zhang Y, Guo W, Yue Z, Lin L, Zhao F, Chen P, Wu W, Zhu H, Yang B, Kuang Y, Wang J (2017) Rapid determination of 54 pharmaceutical and personal care products in fish samples using microwave-assisted extraction—hollow fiber—liquid/solid phase microextraction. J Chromatogr B Anal Technol Biomed Life Sci 1051:41–53. https://doi.org/10.1016/j.jchromb.2017.01.026

Zhang Z, Gao S, Chu C (2020) Improvement of nutrient use efficiency in rice: current toolbox and future perspectives. Theor Appl Genet 133:1365–1384. https://doi.org/10.1007/s00122-019-03527-6

Zheng G, Wang T, Niu M, Chen X, Liu C, Wang Y, Chen T (2018) Biodegradation of nonylphenol during aerobic composting of sewage sludge under two intermittent aeration treatments in a full-scale plant. Environ Pollut 238:783–791. https://doi.org/10.1016/j.envpol.2018.03.112

Zhu Q, Liu X, Hao T, Zeng M, Shen J, Zhang F, De Vries W (2018) Modeling soil acidification in typical Chinese cropping systems. Sci Total Environ 613:1339–1348. https://doi.org/10.1016/j.scitotenv.2017.06.257

Zoeller RT, Brown TR, Doan LL, Gore AC, Skakkebaek NE, Soto AM, Woodruff TJ, Vom Saal FS (2012) Endocrine-disrupting chemicals and public health protection: a statement of principles from The Endocrine Society. Endocrinology 153:4097–4110. https://doi.org/10.1210/en.2012-1422

Zwart N, Jonker W, ten Broek R, de Boer J, Somsen G, Kool J, Hamers T, Houtman CJ, Lamoree MH (2020) Identification of mutagenic and endocrine disrupting compounds in surface water and wastewater treatment plant effluents using high-resolution effect-directed analysis. Water Res 168:115204. https://doi.org/10.1016/j.watres.2019.115204

# Chapter 3
# Dew as Source of Emerging Contaminants in Agricultural System

**Supriya Nath**

**Abstract** Dew water plays crucial role in agriculture in arid regions of various parts of the world and during winter in Indian northern and north-western regions, particularly. Dew is considered to be wet deposition which augments the supply of water during dry periods. It is also a source of moisture as well as nutrients helping in growth of fruits, vegetables and crops like wheat and gram to grow during winter especially at reproductive stage. Dew reduces transpirational loss of water and aids in foliar absorption which helps in survival of natural vegetation in extreme climates. However, dew increases humidity in crop canopy which is favourable for pest growth and spread of diseases in plants that negatively impact yield of crops. Emerging contaminants such as pesticides, biocides and pharmaceutical products dissolved in dew act as input in agricultural field which ultimately affect geochemistry of soil.

In this chapter, the role of dew water in wet deposition of emerging contaminants in agricultural systems has been discussed. In addition, various facets involved in the process of dew formation, its chemical properties, positive and negative impacts on soil biogeochemistry and mechanism of how does it affect the crop productivity by acting as scavengers of surrounding aerosol as well as particles containing emerging contaminants have also been discussed. Fate and transport mechanism of emerging contaminants, their transmission and health effects have also been discussed. These contaminants are suspected to be hazardous for plants as well as consumers of plants. Dew forming on leaves provide medium for dissolution of these emerging contaminants present on leaves causing its direct uptake through absorption or wet deposition to soil when dew falls from leaves onto ground. We need to control these emerging contaminants from entering the environment by banning hazardous pesticides and biocides, minimizing the wastage of pharmaceutical products and replacing them with eco-friendly manures and natural products in place of synthetic ones.

**Keywords** Aerosol · Crop productivity · Hazardous · Pesticides · Wet deposition

---

S. Nath (✉)
Central Water and Power Research Station, Khadakwasla, Pune, Maharashtra, India
e-mail: supriyasesjnu@gmail.com



V. Kumar Singh et al. (eds.), *Sustainable Agriculture Reviews 50*, Sustainable Agriculture Reviews 50, https://doi.org/10.1007/978-3-030-63249-6_3

## 3.1 Introduction

Dew once considered to be a source of pure water having inspired poets of different languages portraying it to be a synonym of purity, has started being subject of worldwide serious concern for ecologists and environmentalists (Lekouch et al. 2011; Beysens et al. 2017; Shohel et al. 2017; Tomaszkiewicz et al. 2017; Nath and Yadav 2018).

Dew formation is a surface phenomenon and dew droplets are formed by condensation of water vapors on any surface such as plastics, metals and plants when the temperature falls below dew point but lies above freezing point. The dew formation and evaporation of dew water affects other atmospheric reactions in ambient conditions (Nath and Yadav 2018). Various environmental conditions in which the dew forms, like meteorological parameters, number and constitution of aerosol particles and gas phase chemical constitution, determine the relative importance and chemical composition of dew water (Lekouch et al. 2011). Dew water chemistry is determined by long range transported atmospheric particles and gases and aerosols of local origin (Beysens et al. 2017). It sequesters aerosols that impact air quality and different ecological and environmental processes. Dew water interacts with inorganic, organic particles and soluble gases present on surface on which dew forms followed by their removal from the atmosphere by drop deposition onto the ground that may act either as nutrients or as pollutants. It may support the development of pathogen causing plant disease as it acts as medium for growth of fungus and bacteria in plants (Singh et al. 2006; Beysens et al. 2017) and may corrode roof of cars in urban areas (Rubio et al. 2002; Beysens et al. 2017).

The dew is being extensively studied across the world for different aspects such as ionic composition, effect of drop size on chemistry, scavenging of pollutants such as mercury (Hg). Dew water is considered as probable water resource in water scarce regions (Tomaszkiewicz et al. 2015) though it provides less quantity of moisture but its frequency helps plant to overcome rain deficiency in such places (Leopold 1952) for most of the time in year. Recently, with the increase in scarcity of potable water, there is surge in groups working on the feasibility of using dew water as source of potable water in Lebanon (Tomaszkiewicz et al. 2017), Bangladesh (Shohel et al. 2017), Jordan (Odeh et al. 2017), Tibet (He and Richards 2015), Morocco (Lekouch et al. 2011), and India (Sharan et al. 2007).

Recently there have been surge in study on emerging contaminants (Dhaka et al. 2019; Galindo-Miranda et al. 2019; Matich et al. 2019; Richardson and Kimura 2019; Riva et al. 2019; Tran et al. 2019). According to the US-EPA (United States – Environmental Protection Agency), emerging contaminants/pollutants are defined as new chemicals that are not regulated and their health and environmental impacts are yet to be assessed completely (Deblonde et al. 2011). Emerging contaminants have existed in the environment and not necessarily are new chemicals but their health hazard was not considered before (Boxall 2012). Sources of emerging contaminants are synthetic compounds like human pharmaceuticals, veterinary medicines, pesticides, cosmetics, steroid hormones, nanomaterials, personal care products, paints and coatings (Boxall 2012; Rashid et al. 2019) or some natural products that get transformed through biochemical process in the environment (Boxall 2012). Some uncommon agricultural chemicals (like many herbicides, pesticides and veterinary medicines) and drugs with their metabolite residues (Rosi-Marshall et al. 2015) are emerging contaminants found in agricultural fields (Wilkinson et al. 2017).

The indiscriminate use of emerging contaminants in modern society in different fields like industry, transport, agriculture and urban lifestyle has allowed these chemicals to enter into the environment as hazardous wastes and non-biodegradable substances (Lei et al. 2015). They enter wastewater from various sources. These emerging contaminants from wastewater treatment plants enter into air as aerosol through air-liquid exchange process and get transported to new places (Hsiao et al. 2019). Emerging contaminants are released to the agricultural environment through different routes. Veterinary medicines may enter soils directly or indirectly when animal waste is used as fertilizer (Boxall et al. 2004; Boxall 2012). Human pharmaceuticals and personal care products enter the agricultural fields when sewage sludge is applied as manure or when wastewater is used for irrigation of land (Boxall 2012).

After Green Revolution, developing countries like India have used pesticides, biocides and fertilizers indiscriminately without knowing the health hazards and negative impact on environment. These pesticides and biocides used for plant protection act as source of emerging contaminants in agriculture directly. They are mishandled and excessively used which causes their deposition on soil and leaves surfaces. Various types of emerging contaminants from agricultural field which are water soluble get dissolved in dew and affects dew water chemistry. Metals like zinc (Zn) get corroded from the metal surface in presence of dew which acts as medium for dissolution (Veleva et al. 2007; Pitt et al. 2013). Quantitatively dew is less in amount to affect plant water budget but it can affect plant health negatively by promoting spread of disease in spite of less quantity (Ben-Asher et al. 2010). In the next section, we have discussed how the dew forms and what are the various factors on which its quantity and quality depends. Sources of water from which dew forms and steps involved in its formation have also been discussed.

## 3.2 Dew Formation

Dew is moisture that condenses on plants, soil or any surface present near ground when the temperature goes below the dew point (Agam and Berliner 2006). Radiative cooling results into condensation of dew on any surface (Lakhani et al. 2012). It forms when atmospheric vapour condensate passively into liquid due to radiation deficit between atmosphere and condensing surface that provides the necessary cooling. Outgoing radiations from the Earth during clear night provides necessary cool temperature near the plant (Bhardwaj et al. 2009). The water vapour present around the plants as a result of evapotranspiration or from soil moisture condenses on these cool surfaces of plants (Bhardwaj et al. 2009). However, the temperature difference between surface and atmosphere reduces during cloudy weather as clouds reduce net long wave radiative flux, and therefore, dew does not form during cloudy weather (Bhardwaj et al. 2009). Dew formation is local phenomenon governed by atmospheric condition and substrate type, structure and chemistry (Beysens 1995; Shohel et al. 2017) as well as the local condition near the substrate surface.

Dew amount depends upon several abiotic environmental factors like wind speed, absolute temperature, atmospheric condition, relative humidity, soil properties and cloudiness (Garratt and Segal 1988).

Dew formation gets water from the following sources (Xu et al. 2015):

1. Dew fall- Water vapor present in the troposphere condenses. This is the major source.
2. Dew rise- It comes from the soil when soil has excess moisture.
3. Guttation- Water and dissolved materials are released by uninjured organs of plants (Jacobs et al. 1999).

Dewfall is more effective in bringing a net gain of water to the soil-plant-water system compared to dew rise and guttation (He and Richards 2015; Xu et al. 2015).

The steps that determine dew chemical composition (Beysens et al. 2017) are:

1. Dew is formed on dry deposition solids.
2. Soluble fraction of the dry deposition is solubilised in dew.
3. Sorption of gases in close proximity into dew solution.

Dew formation does not require aerosol as condensation nuclei rather requires cold surface to condense on the attainment of dew point temperature and high relative humidity during clear night. However, aerosols depositing on the surface on which it forms determines the chemistry of dew water. Dew water composition depends on atmospheric particles coming from long distance transport as well as gas and aerosol of local origin (Beysens et al. 2017). Source of aerosol in agricultural field may be from soil dust, soil crushing, biomass burning and refuse burning. Dew plays significant role in agricultural system by acting positively by providing additional source of water as well as increasing productivity of crops. The role of dew in agricultural systems has been described in the next section.

## 3.3 Role of Dew in Agricultural Systems

Dew plays important role in agriculture in many different ways. It directly provides supplementary source of water in water scarce areas and affects plant productivity indirectly.

### *3.3.1 Role of Dew as Water Source*

Dew water is considered as alternative source of water especially in arid and desert areas for human and agricultural consumption. It helps in recharging the ground water and providing complementary source of portable drinking water in water scarce places like Kutch regions in India (Sharan et al. 2011) as well as various places across different countries like France, Morocco and Chile. In Spain, a single-wall polypropylene tree shelter was installed to harvest dew which resulted in the increment of soil moisture (del Campo et al. 2006). In West Africa, it was reported that dew harvesting was collected up to 0.43 $L/m^2/day$ and was used to counter balance the need of water in maize crop field (Gabin 2015). It is feasible technically to collect dew water on large scale near the seashore for production of fresh water (Rajvanshi 1981; Sharan et al. 2011). Dew harvesting can be economic as it has low initial investment and maintenance cost (Tomaszkiewicz et al. 2017). It helps in limiting the heat gain in buildings located in arid and semi-arid countries (Lekouch et al. 2011). The harvested dew water can also be used for agricultural purposes. During dry seasons or in arid climates, dew acts as a supplement by providing water to plants naturally (Malek et al. 1999).

### *3.3.2 Role of Dew in Crop Productivity*

Dew formation during night covers the plant leaves surfaces with dew water. Stomata present on leaves open after sunrise and water gets assimilated easily without any resistance helping in growth of plants (Slatyer 1967). Dew can be useful to crops as it decreases the vapor-pressure insufficiency, thereby allowing stomatal opening and photosynthesis (Agam et al. 2006). Munné-Bosch and Alegre (1999) found that dew played significant role as it gets absorbed through leaves and rekindles the metabolism of *Melissa officinalis* plants which are water-stressed during the summer. They also reported that dew application on water stressed plants helped in better $CO_2$ assimilation rates and thereby helping the plant to recover its leaf pigment and water budget. Additionally, dew water also reduces rate of evapotranspiration by reducing leaf temperature (Bhardwaj et al. 2009).

Dew provides medium for various kind of emerging contaminants to interact with the plants directly in agricultural environment and form toxins. Various sources of emerging contaminants and their interaction with dew have been discussed in next section.

## 3.4 Sources of Emerging Contaminants in Agricultural Fields

There are various routes through which emerging contaminants enter the agricultural environment (Fig. 3.1). Animal husbandry is associated with agricultural practices, and therefore, veterinary medicines and their metabolites when applied on livestock may enter soils directly (grazing animals) or indirectly when their wastes are used as manure in fields (Boxall et al. 2004). Human medicines and personal care products (Galindo-Miranda et al. 2019) may also enter either by applying animal wastes to land or when wastewater from treatment plant is used in irrigation (Boxall 2012). Emerging contaminants which are applied on plants directly for their protection like pesticides and herbicides or fertilizers with nanomaterials form aerosol during their spray. These aerosols formed while spraying or those fallen on leave surface directly interact with dew during dew formation. This interaction increases the contact time of emerging contaminants with plant and causes direct absorption of released toxins through leaves, exposed surface or through soil during plant uptake.

Nano-scale capsules are being used in addition to existing pesticides may be present in the nanoparticulate form rendering the active ingredient with useful properties for precise, controlled and effective use of pesticides (Lyons and Scrimis 2009; Boxall 2012). Some emerging contaminants are generated in the agricultural

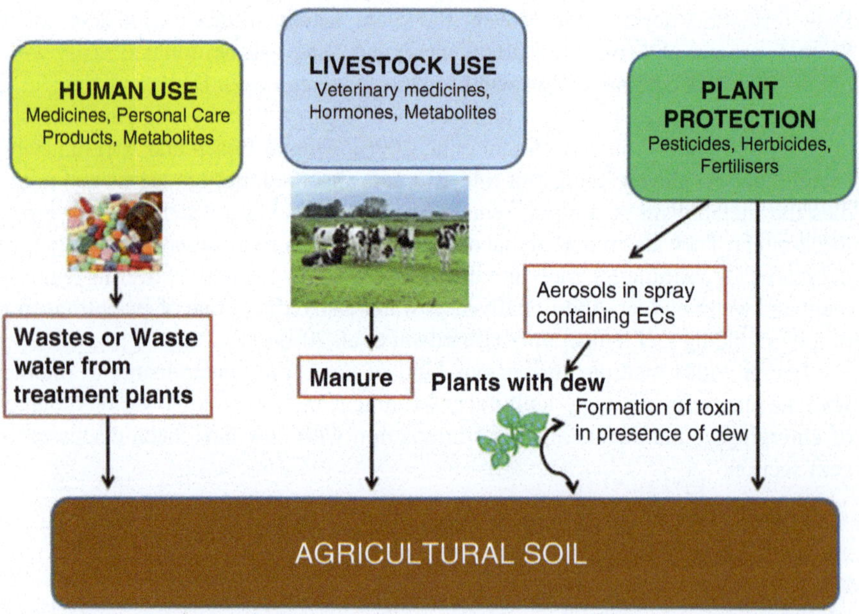


**Fig. 3.1** Schematic representation of sources of emerging contaminants in agricultural field and their interaction with dew. (Modified from Boxall 2012)

environment naturally by fungi, bacteria, algae, plants, and animals (van Egmond 2004) for their defense (Boxall 2012).

Various emerging contaminants from different sources get transported through different ways and end up in the agricultural farmland. These transport mechanism, transmission of emerging contaminants to the atmosphere and interaction of aerosol with dew has been discussed in next section.

## 3.5 Transport Mechanism of Emerging Contaminants in Agricultural Systems

One of the major sources of emerging contaminants in agricultural field is pesticide, fertilizers and herbicides. Their increased use has caused their leaching and they have become a source of emerging contaminants in wastewater (Houtman 2010). Their high concentration and permanence in existence in wastewater has increased research concern on pesticides (Rasheed et al. 2019). They get released as run-off to lakes and rivers (Leu et al. 2004) where eutrophication has increased and percolates in groundwater and surface water (Rasheed et al. 2019) making them unfit for human and animal consumption. Sewage sludge from sewage treatment plants are applied as fertilizers in agricultural fields. It retains pharmaceuticals and personal care products and other emerging contaminants even after treatment (Wilkinson et al. 2017) which results into soil and groundwater contamination from where they are introduced in plants through uptake (Daughton 2006; Barron et al. 2010; Wilkinson et al. 2017).

Various environmental factors like temperature and pH of the soil, saturation degree, percolation rate and physico-chemical properties of both the soil and emerging contaminants affect the rate of adsorption of contaminants (Barron et al. 2010; Wilkinson et al. 2017). Persistence and fate of these emerging contaminants depend on the climate and soil of that place (Wilkinson et al. 2017). Hydrophobicity and soil organic carbon (SOC) content are factors which determine contaminant adsorption on soils (Barron et al. 2010; Wilkinson et al. 2017).

The emerging contaminant undergoes various biological, physical or chemical processes through different transport mechanism as shown in Fig. 3.2. It can either go through soil particles directly, absorbed by plants, percolate to groundwater or may get drained to water system through runoff and drainage water (Boxall 2012). These processes depend on water solubility, volatility of emerging contaminants and how much emerging contaminants are attracted to organic matter (Boxall 2012).

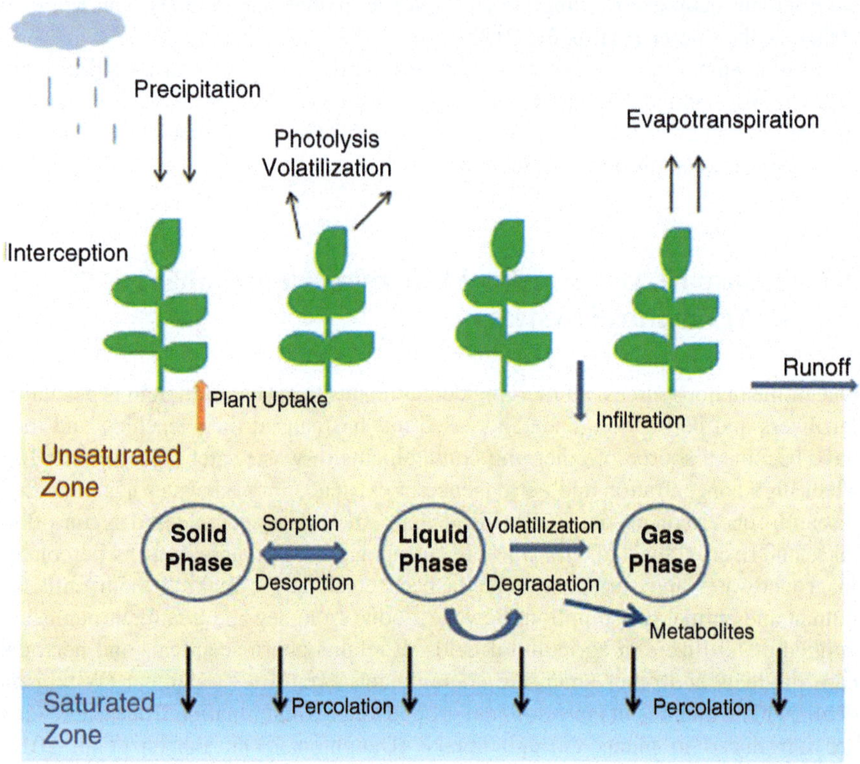


**Fig. 3.2** Transport mechanism of emerging contaminants in agricultural system. (Modified from Boxall 2012)

### *3.5.1 Emerging Contaminants Transmission in Atmosphere*

Various factors which transfer pesticides from the agricultural field to the atmosphere include properties of the pesticides, the environmental situation and agricultural practices used (Bedos et al. 2002). Their solubility and the vapor pressure determine their volatility. As the vapor pressure, the Henry's law constant and the solubility varied over more than five orders of magnitude, therefore physiochemical properties of pesticides in atmosphere varied greatly (Bedos et al. 2002). Pesticides found in the atmosphere may be classified into three categories (Jury et al. 1983; Bedos et al. 2002) which are:

1. Highly volatile pesticides ($K_h \gg 2.65 \times 10^{-5}$)
2. Moderately volatile pesticides
3. Low volatile pesticides ($K_h \ll 2.65 \times 10^{-5}$)

Pesticides can reach atmosphere by three processes:

1. Volatilization (part which are volatile),
2. Evaporation during application in field
3. By long distance transport through wind (Bedos et al. 2002).

These three ways may contaminate the atmosphere for several days after application of pesticides (Bedos et al. 2002).

Once pesticide reaches the atmosphere, it follows mainly three processes (Fig. 3.3):

1. Dispersion- It gets dispersed by turbulence or advection
2. Chemical transformation by undergoing chemical or photochemical processes
3. Washout by dissolution in liquid droplets of fog, clouds or dew

The aerosol formed in these processes may act as condensation nuclei or the pesticides in atmosphere get settled through wet or dry deposition. Wet deposition in form of dew, fog forming near the surface or through rainfall depends on the meteorological conditions.

Herbicide atrazine which is used as spray in agricultural fields are found in precipitation (like dew and rain) far away from their applied source which indicates their atmospheric route of spread (Hayes et al. 2002, 2003; Wilkinson et al. 2017). Rudel et al. (2003) reported some groups of emerging contaminants (like plasticisers, emulsifiers, nonylphenols, and adhesive) in indoor atmosphere homes in the range of 50–1500 ng/m$^3$. Some of emerging contaminants are volatile like agricultural spray drift applied to increase fruit tree productivity are found to spread

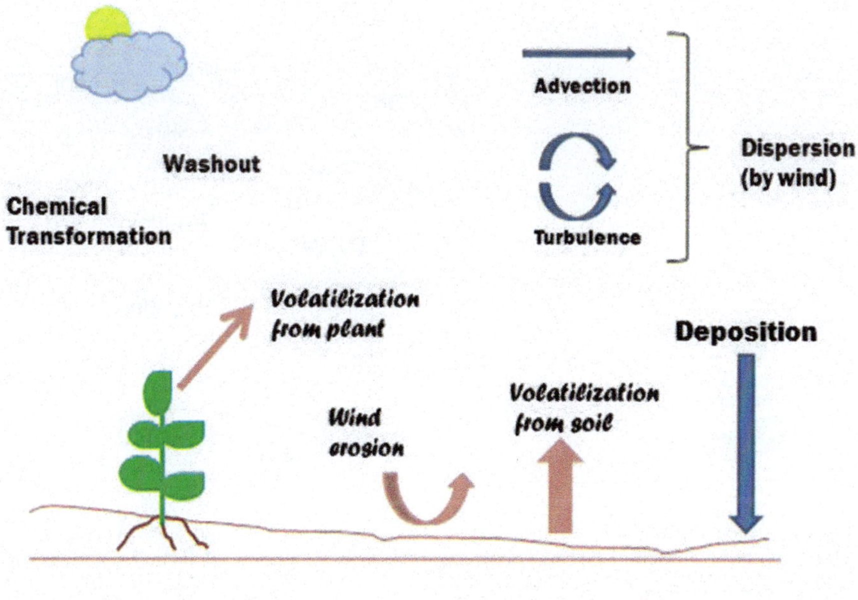


**Fig. 3.3** Processes which are responsible for the transfer of pesticides to the atmosphere. (Modified from: Bedos et al. 2002)

through atmospheric route (Daughton 2006). The rate of transmission and spread of these volatile compounds in the atmosphere has not been determined (Wilkinson et al. 2017), but they get deposited in agricultural fields through dew formation or other forms of precipitation.

### 3.5.2 Chemical Interaction of Dew with Aerosol

Dew water composition not only depends on local atmospheric surrounding, but also gets affected by long range transported aerosols and gases (Beysens et al. 2017). Various complex atmospheric water-gas-particulate matter interactions determine the chemistry of dew water (Rubio et al. 2002; Acker et al. 2008; Yadav and Kumar 2014). It plays important role as processors of carbonaceous species. Liquid water present in dew drops provides sink for soluble gases and reacting medium for aqueous phase chemical reaction (Fig. 3.4). Dew water chemistry gets affected by nature of surface on which it is formed (Beysens et al. 2006; Lekouch et al. 2010; Nath and Yadav 2018). It acts as scavenger of contaminants through interactions and dissolution of various gaseous components as well as other contaminants deposited on leaf surface during the formation of dew and cause their absorption in leaves or wet deposition on soil.

In the schematic diagram (Fig. 3.4), dust carrying emerging contaminants like fertilizer, pesticides and biocides from direct spray or deposition on leaves come

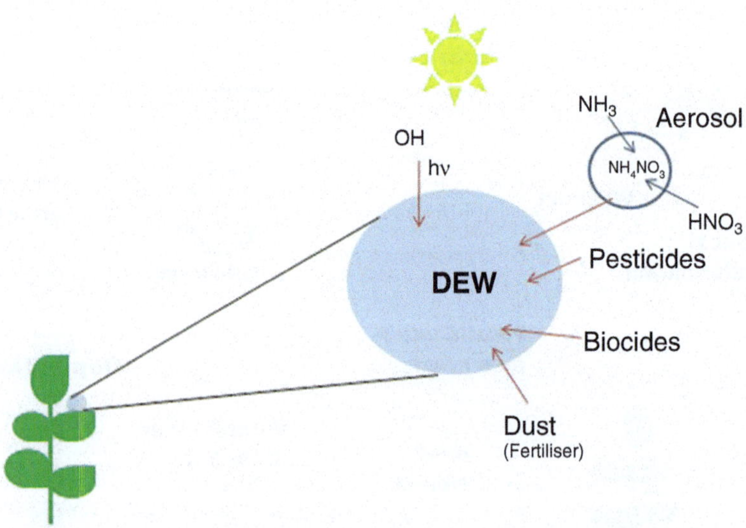


**Fig. 3.4** Schematic diagram of interaction of emerging contaminants and other pollutants from surrounding with dew droplets. Dew droplet has been enlarged to show interaction of various contaminants, aerosol and compounds with dew

inside dew droplet and interact with each other forming toxic substances or carrying toxins present in these chemicals inside the plant. Some reactions take place inside the aerosols forming secondary products which enter dew and further interacts with other substances present in dew droplets. These interactions produce negative impacts which have been discussed in next section.

## 3.6 Negative Impacts of Dew

Dew water has negative effects as well in the form of corrosion and leaching. In watershed regions where dew formation is a predominant phenomenon, dew could be a major source of nutrients as well as pollutants. It may cause plant disease as it acts as medium for growth of fungus and bacteria in plants (Singh et al. 2006; Beysens et al. 2017) as pathogen releases spores under moisture condition. The infection depends on quantity of dew and its sustaining time on plant's surface. Dew acts as medium for growth of foliar pathogens in plants as it provides free water for sporulation and spread of infection (Bhardwaj et al. 2009). Crops like soyabean pod gets affected by moisture from dew and disease spread to its seed which directly reduces their yield. Diseases like corn leaf spot, peanut leaf spot, blight on tomato, tobacco, potato finds dew as source of moisture for development of diseases. The Great Potato Famine in Ireland was widespread due to combined effect of moisture from dew, rainfall and temperature.[1]

Dew may reduce transpiration amount in plants. Acidic dew reduces quality of crops by decreasing its $CO_2$ assimilation efficiency. Formation of acidic dew promotes corrosion processes, also damages surface structures and monuments (Singh et al. 2006). It may corrode roof of cars in urban areas (Rubio et al. 2002; Beysens et al. 2017) and causes economic loss as well. Efficacy of applied pesticide gets diluted in presence of dew on plants (Saab et al. 2017).

Pollutants or emerging contaminants in surrounding gets dissolved in dew causing harm to plants as well as human or animal health which are consumers of those crops. Various health impacts of these emerging contaminantson human and environment has been discussed in next section.

## 3.7 Emerging Contaminants and Their Health Effects

The role of emerging contaminants on human health has not been determined conclusively but experiments have been conducted on animals. So, emerging contaminants are suspected to be mutagen, teratogen, and carcinogen for humans and other animals (Lei et al. 2015). Some endocrine disruptors like Bisphenol A, DDT,

[1] http://agron-www.agron.iastate.edu/courses/Agron541/classes/541/lesson07a/541L7.pdf

**Table 3.1** Various human health effects of different types of emerging contaminants

| Categories of emerging contaminants | Reported effects | References |
|---|---|---|
| Endocrine disrupters | Adverse effect on reproductive, neurological, developmental and immunity in humans | Fisher and Eugster (2014), Gomes et al. (2018) |
| Plasticizers | Toxicity is detected in human blood, adipose tissue, fluids, and breast milk | Wombacher and Hornbuckle (2009) |
| Pesticides metabolites | Interference with endogenous hormones systems | Lapworth and Gooddy (2006), Clausen et al. (2007) |
| Pharmaceutically active compounds | Estrogenic and chronic toxicity | Liu et al. (2015) |
| Polycyclic aromatic contaminants | Carcinogenic, infertility, diabetes, poor fetal development, oxidative stress, cardiovascular disease and inflammation | Jomova et al. (2012), Wang et al. (2015), Ferguson et al. (2017), Yang et al. (2017) |
| Dyes | Toxic, carcinogenic or teratogenic | Bilal et al. (2017), Chatha et al. (2017), Bilal et al. (2018) |

**Table 3.2** Various effects of different types of medicines

| Pharmaceutical categories | Reported effect | References |
|---|---|---|
| Analgesic | Biochemical effects in fish | Lavelle et al.(2004) |
| Anorexic treatments | Hormonal effect in crab and crayfish | Daughton and Ternes (1999) |
| Antianxiety drugs | Effects on development of invertebrates | Pascoe et al.(2003) |
| Antibacterials | Selection of antibacterial resistance, biochemical effects in fish | Westergaard et al. (2001), Sengelov et al. (2003), Lavelle et al. (2004), Pomati et al. (2004) |
| Beta blocker | Biochemical effects in fish | Lavelle et al. (2004) |
| Calcium channel blocker | Effects on development of invertebrates | Pascoe et al. (2003) |
| Cardiac glycoside | Effects on development of invertebrates | Pascoe et al. (2003) |
| Estrogens | Endocrine disrupting effects on fish | Sommer and Bibby (2002), Floate et al. (2005) |
| Lipid regulator | Biochemical effects in fish | Lavelle et al.(2004) |
| Parasiticides | Effects on insect development and physiology, effect on rate of dung decomposition | Sommer and Bibby (2002), Floate et al. (2005) |

Modified from Boxall (2012)

Endosulfan may adversely affect endocrine system (Bilal et al. 2017; Caliman and Gavrilescu 2009; Cook et al. 2012; Preda et al. 2012; Rasheed et al. 2019) and are suspected to produce negative impacts on reproductive system, immunity and nervous system in humans and wildlife (Fisher and Eugster 2014; Gomes et al. 2018;

Rasheed et al. 2019). Various emerging contaminants like plasticizers or musk fragrances have been reported to be found in human blood, adipose tissues, fluids and breast milk (Wombacher and Hornbuckle 2009; Rasheed et al. 2019) but their role has not been determined. Various health impacts of different categories of emerging contaminants of high concern are discussed in Table 3.1.

Different classes of pharmaceutical medicines have been reported to have different effects on aquatic environment and aquatic fauna by various studies as discussed in Table 3.2.

## 3.8 Conclusion and Future Recommendations

Various kinds of emerging contaminants like pharmaceuticals, veterinary medicines, cosmetics, steroid hormones, nanomaterials, personal care products, paints and coatings which have become inevitable part of modern lifestyle enter into groundwater and surface water and indirectly reach agricultural fields. These emerging contaminants react in presence of dew formed on leaves of plants leading to harmful effects on plants as well as those animals and human who are consuming it. Various diseases like cardiovascular disease, nervous disorders, cancer, reproductive disorder, anemia, and hypertension are speculated to be caused by these emerging contaminants. Therefore, we cannot ignore the dangers associated with these emerging contaminants.

Dew is natural phenomena which can enhance crop productivity in pure form. Addition of synthetic products to our water systems or application of pesticides and biocides directly in agricultural fields has contaminated the fields. This problem requires immediate attention and actions. Recommendations for controlling the damage caused by emerging contaminants are:

1. We may control their presence by banning hazardous pesticides and biocides and replacing them with eco-friendly manures and environmentally friendly products.
2. We need to switch to natural products in place of synthetic ones like cosmetics and personal care products.
3. We must minimize the wastage of pharmaceutical products.
4. Farmers need to be informed about the environmental and health risks associated with emerging contaminants. They should be trained and informed through awareness programs so that they may opt for alternatives which are more environment-friendly or use the pesticides judiciously and optimally through better management while applying pesticides.

There have been various studies conducted across the globe to determine the chemistry of dew being affected by pollutants nearby. However, very few studies have focused on its role in connection with emerging pollutants. Further study is required to determine how dew plays role in affecting plants through emerging contaminants. Dew water may be used as source of water instead of acting as agent of deposition of emerging contaminants. For this we require developing ways to collect dew and try to minimize use of synthetic compounds in agricultural fields to reduce input of emerging contaminants.

## References

Acker K, Beysens D, Möller D (2008) Nitrite in dew, fog, cloud and rain water: an indicator for heterogeneous processes on surfaces. Atmos Res 87(3):200–212. https://doi.org/10.1016/j.atmosres.2007.11.002

Agam N, Berliner PR (2006) Dew formation and water vapor adsorption in semi-arid environments—a review. J Arid Environ 65(4):572–590. https://doi.org/10.1016/j.jaridenv.2005.09.004

Barron L, Nesterenko E, Hart K (2010) Holistic visualisation of the multimodal transport and fate of twelve pharmaceuticals in biosolid enriched topsoils. Anal Bioanal Chem 397:287–296. https://doi.org/10.1007/s00216-010-3494-1

Bedos C, Cellier P, Calvet R, Barriuso E (2002) Occurrence of pesticides in the atmosphere in France. Agronomie 22(1):35–49. https://doi.org/10.1051/agro:2001004

Ben-Asher J, Alpert P, Ben-Zvi A (2010) Dew is a major factor affecting vegetation water use efficiency rather than a source of water in the eastern Mediterranean area. Water Resour Res 46(10). https://doi.org/10.1029/2008WR007484

Beysens D (1995) The formation of dew. Atmos Res 39(1):215–237. https://doi.org/10.1016/0169-8095(95)00015-J

Beysens D, Ohayon C, Muselli M, Clus O (2006) Chemical and biological characteristics of dew and rainwater in an urban coastal area (Bordeaux France). Atmos Environ 40:3710–3723. https://doi.org/10.1016/j.atmosenv.2006.03.007

Beysens D, Mongruel A, Acker K (2017) Urban dew and rain in Paris, France: occurrence and physico-chemical characteristics. Atmos Res 189:152–161. https://doi.org/10.1016/j.atmosres.2017.01.013

Bhardwaj J, Singh S, Singh D, Rao V (2009) Dewfall in relation to weather variables in varying environments of mustard crop. J Agrometeorol 11(Special Issue):70

Bilal M, Asgher M, Parra-Saldivar R, Hu H, Wang W, Zhang X, Iqbal HM (2017) Immobilized ligninolytic enzymes: an innovative and environmental responsive technology to tackle dye-based industrial pollutants–a review. Sci Total Environ 576:646–659. https://doi.org/10.1016/j.scitotenv.2016.10.137

Bilal M, Rasheed T, Sosa-Hernández JE, Raza A, Nabeel F, Iqbal H (2018) Biosorption: an interplay between marine algae and potentially toxic elements—a review. Mar Drugs 16(2):65. https://doi.org/10.3390/md16020065

Boxall ABA (2012) New and emerging water pollutants arising from agriculture. Environment Department, University of York

Boxall ABA, Fogg LA, Kay P, Blackwell PA, Pemberton EJ, Croxford A (2004) Veterinary medicines in the environment. Rev Environ Contam Toxicol 180:1–91. https://doi.org/10.1007/0-387-21729-0_1

Caliman FA, Gavrilescu M (2009) Pharmaceuticals, personal care products and endocrine disrupting agents in the environment–a review. CLEAN Soil Air Water 37(4–5):277–303. https://doi.org/10.1002/clen.200900038

Chatha SAS, Asgher M, Iqbal HMN (2017) Enzyme-based solutions for textile processing and dye contaminant biodegradation—a review. Environ Sci Pollut Res 24:14005–14018. https://doi.org/10.1007/s11356-017-8998-1

Clausen L, Arildskov NP, Larsen F, Aamand J, Albrechtsen HJ (2007) Degradation of the herbicide dichlobenil and its metabolite BAM in soils and subsurface sediments. J Contam Hydrol 89(3–4):157–173. https://doi.org/10.1016/j.jconhyd.2006.04.004

Cook SM, VanDuinen BJ, Love NG, Skerlos SJ (2012) Life cycle comparison of environmental emissions from three disposal options for unused pharmaceuticals. Environ Sci Technol 46(10):5535–5541. https://doi.org/10.1021/es203987b

Daughton CG (2006) Origins and fate of PPCPs in the environment. US Environmental Protection Agency

Daughton CG, Ternes TA (1999) Pharmaceuticals and personal care products in the environment: agents of subtle change? Environ Health Perspect 107:907–938. https://doi.org/10.1289/ehp.99107s6907

Deblonde T, Leguille CC, Hartemann P (2011) Emerging pollutants in wastewater: a review of the literature. Int J Hyg Environ Health 214(6):442–448. https://doi.org/10.1016/j.ijheh.2011.08.002

del Campo AD, Navarro RM, Aguilella A, González E (2006) Effect of tree shelter design on water condensation and run-off and its potential benefit for reforestation establishment in semiarid climates. For Ecol Manag 235(1–3):107–115. https://doi.org/10.1016/j.foreco.2006.08.003

Dhaka S, Kumar R, Deep A, Kurade MB, Ji SW, Jeon BH (2019) Metal–organic frameworks (MOFs) for the removal of emerging contaminants from aquatic environments. Coord Chem Rev 380:330–352. https://doi.org/10.1016/j.ccr.2018.10.003

Ferguson KK, McElrath TF, Pace GG, Weller D, Zeng L, Pennathur S, Cantonwine DE, Meeker JD (2017) Urinary polycyclic aromatic hydrocarbon metabolite associations with biomarkers of inflammation, angiogenesis, and oxidative stress in pregnant women. Environ Sci Technol 51(8):4652–4660. https://doi.org/10.1021/acs.est.7b01252

Fisher MM, Eugster EA (2014) What is in our environment that effects puberty? Reprod Toxicol 44:7–14. https://doi.org/10.1016/j.reprotox.2013.03.012

Floate K, Wardaugh K, Boxall ABA, Sherratt T (2005) Faecal residues of veterinary parasiticides: non-target effects in the pasture environment. Annu Rev Entomol 50:153–179. https://doi.org/10.1146/annurev.ento.50.071803.130341

Gabin K (2015) Atmospheric moisture condensing in Benin: a contribution to the correction of water stress in maize in semi-arid environment (in French). Unpublished Ph.D., Universitéd'Abomey-Calavi (Bénin)

Galindo-Miranda JM, Guízar-González C, Becerril-Bravo EJ, Moeller-Chávez G, León-Becerril E, Vallejo-Rodríguez R (2019) Occurrence of emerging contaminants in environmental surface waters and their analytical methodology–a review. Water Supply 19(7):1871–1884. https://doi.org/10.2166/ws.2019.087

Garratt JR, Segal M (1988) On the contribution of atmospheric moisture to dew formation. Bound-Layer Meteorol 45(3):209–236. https://doi.org/10.1007/BF01066671

Gomes IB, Simões LC, Simões M (2018) The effects of emerging environmental contaminants on Stenotrophomonasmaltophilia isolated from drinking water in planktonic and sessile states. Sci Total Environ 643:1348–1356. https://doi.org/10.1016/j.scitotenv.2018.06.263

Hayes TB, Collins A, Lee M, Mendoza M, Noriega N, Stuart AA, Vonk A (2002) Hermaphroditic, demasculinized frogs after exposure to the herbicide atrazine at low ecologically relevant doses. Proc Natl Acad Sci U S A 99(8):5476–5480. https://doi.org/10.1073/pnas.082121499

Hayes T, Haston K, Tsui M, Hoang A, Haeffele C, Vonk A (2003) Atrazine induced hermaphroditism at 0.1 ppb in American leopard frogs (Ranapipiens): laboratory and field evidence. Environ Health Perspect 111(4):568–575. https://doi.org/10.1289/ehp.5932

He S, Richards K (2015) The role of dew in the monsoon season assessed via stable isotopes in an alpine meadow in Northern Tibet. Atmos Res 151:101–109. https://doi.org/10.1016/j.atmosres.2014.02.014

Houtman CJ (2010) Emerging contaminants in surface waters and their relevance for the production of drinking water in Europe. J Integr Environ Sci 7(4):271–295. https://doi.org/10.1080/1943815X.2010.511648

Hsiao TC, Lin AYC, Lien WC, Lin YC (2019) Size distribution, biological characteristics and emerging contaminants of aerosols emitted from an urban wastewater treatment plant. J Hazard Mater:121809. https://doi.org/10.1016/j.jhazmat.2019.121809

Jacobs AF, Heusinkveld BG, Berkowicz SM (1999) Dew deposition and drying in a desert system: a simple simulation model. J Arid Environ 42(3):211–222. https://doi.org/10.1006/jare.1999.0523

Jomova K, Baros S, Valko M (2012) Redox active metal-induced oxidative stress in biological systems. Transit Met Chem 37:127–134. https://doi.org/10.1007/s11243-012-9583-6

Jury WA, Spencer WF, Farmer WJ (1983) Behavior assesment model for trace organics in soil: I. model description. J Environ Qual 12:558–564. https://doi.org/10.2134/jeq1983.00472425001200040025x

Lakhani A, Parmar RS, Prakash S (2012) Chemical composition of dew resulting from radiative cooling at a semi-arid site in Agra, India. Pure Appl Geophys 169(5–6):859–871. https://doi.org/10.1007/s00024-011-0329-8

Lapworth DJ, Gooddy DC (2006) Source and persistence of pesticides in a semi-confined chalk aquifer of Southeast England. Environ Pollut 144(3):1031–1044. https://doi.org/10.1016/j.envpol.2005.12.055

Lavelle N, Ait-Aissa S, Gomez E, Casellas C, Porcher JM (2004) Effects of human pharmaceuticals on cytotoxicity, EROD activity and ROS production in fish hepatocytes. Toxicology 196:41–55. https://doi.org/10.1016/j.tox.2003.11.002

Lei M, Zhang L, Lei J, Zong L, Li J, Wu Z, Wang Z (2015) Overview of emerging contaminants and associated human health effects. Biomed Res Int. https://doi.org/10.1155/2015/404796

Lekouch I, Mileta M, Muselli M, Milimouk-Melnytchouk I, Šojat V, Kabbachi B, Beysens D (2010) Comparative chemical analysis of dew and rain water. Atmos Res 95(2):224–234. https://doi.org/10.1016/j.atmosres.2009.10.002

Lekouch I, Muselli M, Kabbachi B, Ouazzani J, Milimouk IM, Beysens D (2011) Dew, fog, and rain as supplementary sources of water in South-Western Morocco. Energy 36:2257–2265. https://doi.org/10.1016/j.energy.2010.03.017

Leopold LB (1952) Dew as a source of plant moisture. Pac Sci 6(3):259–261

Leu C, Singer H, Stamm C, Müller SR, Schwarzenbach RP (2004) Variability of herbicide losses from 13 fields to surface water within a small catchment after a controlled herbicide application. Environ Sci Technol 38(14):3835–3841. https://doi.org/10.1021/es0499593

Liu A, Goonetilleke A, Egodawatta P (2015) Role of rainfall and catchment characteristics on urban stormwater quality. Springer, Singapore

Lyons K, Scrimis G (2009) Under the regulatory radar? Nanotechnologies and their impacts for rural Australia. In: Merlan F, Raftery D (eds) Tracking rural change: community, policy and Technology in Australia, New Zealand and Europe. Australian National University E Press, Canberra, pp 151–171

Malek E, McCurdy G, Giles B (1999) Dew contribution to the annual water balances in semi-arid desert valleys. J Arid Environ 42(2):71–80. https://doi.org/10.1006/jare.1999.0506

Matich EK, Soria NC, Aga DS, Atilla-Gokcumen GE (2019) Applications of metabolomics in assessing ecological effects of emerging contaminants and pollutants on plants. J Hazard Mater. https://doi.org/10.1016/j.jhazmat.2019.02.084

Munné-Bosch S, Alegre L (1999) Role of dew on the recovery of water-stressed Melissa officinalis L. plants. J Plant Physiol 154(5–6):759–766. https://doi.org/10.1016/S0176-1617(99)80255-7

Nath S, Yadav S (2018) A comparative study on fog and dew water chemistry at New Delhi, India. Aerosol Air Qual Res 18(1):26–36. https://doi.org/10.4209/aaqr.2017.01.0033

Odeh I, Arar S, Al-Hunaiti A, Sa'aydeh H, Hammad G, Duplissy J, Vuollekoski H, Korpela A, Petäjä T, Kulmala M, Hussein T (2017) Chemical investigation and quality of urban dew collections with dust precipitates. Environ Sci Pollut Res 24(13):12312–12318. https://doi.org/10.1007/s11356-017-8870-3

Pascoe D, Karntanut W, Mueller CT (2003) Do pharmaceuticals affect freshwater invertebrates? A study with the cnidarian Hydra vulgaris. Chemosphere 51:521–528. https://doi.org/10.1016/S0045-6535(02)00860-3

Pitt R, Goodson K, Ogburn O, Eppakayala V, Veeravalli SU, Bathi J, Wilson B (2013) Identification and treatment of emerging contaminants in wet weather flows. EPA Report Number July 30, 2013

Pomati F, Netting AG, Calamari D, Neilan BA (2004) Effects of erythromycin and ibuprofen on the growth of Synechocystis sp. and Lemna minor. Aquat Toxicol 67:387–396. https://doi.org/10.1016/j.aquatox.2004.02.001

Preda C, Ungureanu MC, Vulpoi C (2012) Endocrine disruptors in the environment and their impact on human health. Environ Eng Manag J 11(9):1697–1706

Rajvanshi AK (1981) Large scale dew collection as a source of fresh water supply. Desalination 36(3):299–306. https://doi.org/10.1016/S0011-9164(00)88647-6

Rasheed T, Bilal M, Nabeel F, Adeel M, Iqbal HM (2019) Environmentally-related contaminants of high concern: potential sources and analytical modalities for detection, quantification, and treatment. Environ Int 122:52–66. https://doi.org/10.1016/j.envint.2018.11.038

Richardson SD, Kimura SY (2019) Water analysis: emerging contaminants and current issues. Anal Chem 92(1):473–505. https://doi.org/10.1021/acs.analchem.9b05269

Riva F, Zuccato E, Davoli E, Fattore E, Castiglioni S (2019) Risk assessment of a mixture of emerging contaminants in surface water in a highly urbanized area in Italy. J Hazard Mater 361:103–110. https://doi.org/10.1016/j.jhazmat.2018.07.099

Rosi-Marshall EJ, Snow D, Bartelt-Hunt SL, Paspalof A, Tank JL (2015) A review of ecological effects and environmental fate of illicit drugs in aquatic ecosystems. J Hazard Mater 282:18–25. https://doi.org/10.1016/j.jhazmat.2014.06.062

Rubio MA, Lissi E, Villena G (2002) Nitrite in rain and dew in Santiago city, Chile. Its possible impact on the early morning start of the photochemical smog. Atmos Environ 36:293–297. https://doi.org/10.1016/S1352-2310(01)00356-9

Rudel RA, Camann DE, Spengler JD, Korn LR, Brody JG (2003) Phthalates, alkylphenols, pesticides, polybrominated diphenyl ethers, and other endocrine disrupting compounds in indoor air and dust. Environ Sci Technol 37(20):4543–4553. https://doi.org/10.1021/es0264596

Saab OJ, Griesang F, Alves KA, Higashibara LR, Genta W (2017) Pesticides deposition in vineyards on different conditions of leaf wetness. Engenharia Agrícola 37(2):286–291. https://doi.org/10.1590/1809-4430-eng.agric.v37n2p286-291/2017

Sengelov G, Agerso Y, Halling Sorensen B, Baloda SB, Andersen JS, Jensen LB (2003) Bacterial antibiotic resistance levels in Danish farmland as a result of treatment with pig manure slurry. Environ Int 28:587–595. https://doi.org/10.1016/S0160-4120(02)00084-3

Sharan G, Beysens D, Milimouk-Melnytchouk I (2007) A study of dew water yields on Galvanized iron roofs in Kothara (North-West India). J Arid Environ 69(2):259–269. https://doi.org/10.1016/j.jaridenv.2006.09.004

Sharan G, Clus O, Singh S, Muselli M, Beysens D (2011) A very large dew and rain ridge collector in the Kutch area (Gujarat, India). J Hydrol 405(1–2):171–181. https://doi.org/10.1016/j.jhydrol.2011.05.019

Shohel M, Simol HA, Reid E, Reid JS, Salam A (2017) Dew water chemical composition and source characterization in the IGP outflow location (coastal Bhola, Bangladesh). Air Qual Atmos Health 10(8):981–990. https://doi.org/10.1007/s11869-017-0487-7

Singh SP, Khare P, Kumari MK, Srivastava SS (2006) Chemical characterization of dew at a regional representative site of North-Central India. Atmos Res 80:239–249. https://doi.org/10.1016/j.atmosres.2005.09.003

Slatyer RO (1967) Plant-water relationships. Academic, London

Sommer C, Bibby BM (2002) The influence of veterinary medicines on the decomposition of dung organic matter in soil. Eur J Soil Biol 38:155–159. https://doi.org/10.1016/S1164-5563(02)01138-X

Tomaszkiewicz M, Najm MA, Beysens D, Alameddine I, El-Fadel M (2015) Dew as a sustainable non-conventional water resource: a critical review. Environ Rev 23(4):425–442. https://doi.org/10.1139/er-2015-0035

Tomaszkiewicz M, Najm MA, Zurayk R, El-Fadel M (2017) Dew as an adaptation measure to meet water demand in agriculture and reforestation. Agric For Meteorol 232:411–421. https://doi.org/10.1016/j.agrformet.2016.09.009

Tran NH, Reinhard M, Khan E, Chen H, Nguyen VT, Li Y, Goh SG, Nguyen QB, Saeidi N, Gin KYH (2019) Emerging contaminants in wastewater, stormwater runoff, and surface water: application as chemical markers for diffuse sources. Sci Total Environ 676:252–267. https://doi.org/10.1016/j.scitotenv.2019.04.160

van Egmond HP (2004) Natural toxins: risks, regulations and the analytical situation in Europe. Anal Bioanal Chem 378:1152–1160. https://doi.org/10.1007/s00216-003-2373-4

Veleva L, Meraz E, Acosta M (2007) Zinc corrosion runoff process induced by humid tropical climate. Mater Corros 58(5). https://doi.org/10.1002/maco.200604007

Wang Z, Zheng Y, Zhao B, Zhang Y, Liu Z, Xu J, Chen Y, Yang Z, Wang F, Wang H, He J (2015) Human metabolic responses to chronic environmental polycyclic aromatic hydrocarbon exposure by a metabolomic approach. J Proteome Res 14(6):2583–2593. https://doi.org/10.1021/acs.jproteome.5b00134

Westergaard K, Muller AK, Christensen S, Bloem J, Sorensen SJ (2001) Effects of tylosin on the soil microbial community. J Soil Biochem 33:2061–2071. https://doi.org/10.1016/S0038-0717(01)00134-1

Wilkinson J, Hooda PS, Barker J, Barton S, Swinden J (2017) Occurrence, fate and transformation of emerging contaminants in water: an overarching review of the field. Environ Pollut 231:954–970. https://doi.org/10.1016/j.envpol.2017.08.032

Wombacher WD, Hornbuckle KC (2009) Synthetic musk fragrances in a conventional drinking water treatment plant with lime softening. J Environ Eng 135(11):1192–1198. https://doi.org/10.1061/(ASCE)EE.1943-7870.0000085

Xu Y, Yan B, Tang J (2015) The effect of climate change on variations in dew amount in a paddy ecosystem of the Sanjiang Plain, China. Adv Meteorol 2015(2):1–9. https://doi.org/10.1155/2015/793107

Yadav S, Kumar P (2014) Pollutant scavenging in dew water collected from an urban environment and related implications. Air Qual Atmos Health 7:559–566. https://doi.org/10.1007/s11869-014-0258-7

Yang L, Yan K, Zeng D, Lai X, Chen X, Fang Q, Guo H, Wu T, Zhang X (2017) Association of polycyclic aromatic hydrocarbons metabolites and risk of diabetes in coke oven workers. Environ Pollut 223:305–310. https://doi.org/10.1016/j.envpol.2017.01.027

# Chapter 4
# Particulate Matter Pollution and Global Agricultural Productivity

**Sujit Das, Debanjana Pal, and Abhijit Sarkar**

**Abstract** Following last few decades, global agriculture is facing tremendous challenges from diverse issues like increasing population, unrestrained urbanization, and uncontrolled environmental pollution and so on. Productivity of major crops has been declined severely leading to higher global hunger index as well as uncontrollable food crisis both in quantity and quality. Among the major issues, environmental pollution followed by the climate change play the most critical role. Particulate matter is one of the most underrated emerging contaminant which triggers adverse effects on plant and human health and efficacy. The main sources for particulate matter air pollution are both natural, which include forest fire, sea salt, dust (such as airborne soil), pollen, volcanic eruptions and particles formed from natural gaseous precursors, and anthropogenic, which include agricultural operations, industrial processes, combustion of wood and fossil fuels, construction and demolition activities, and entrainment of road dust into the air.

As an emerging contaminant, particulate matter pollution has been noticed and recognized globally for last couple of decades. Several reports showed that this pollution can directly and indirectly decrease the agricultural productivity by altering the physiological, i.e. inhibited photosynthetic rate and modified stomatal size and behaviour, and biochemical responses, i.e. increased ascorbic acid and phenolic compounds, etc. of crop plants worldwide. Now the effects of particulate matter pollution on agricultural crop can be mitigated by the diverse strategies including the adoption of precision nutrition management, zero tillage farming, the prevention of crop residue burning, improved water management, and the adoption of laser land levelling, etc. Though as a new domain, the number of research papers is limited till date. Present chapter mainly focuses on cataloguing of available literature and constructing road map which pave the way for future researchers.

**Keywords** Particulate matters · Air pollution · Biomass burning · climate change · Agricultural productivity

---

S. Das · D. Pal · A. Sarkar (✉)
Laboratory of Applied Stress Biology, Department of Botany, University of Gour Banga, Malda, West Bengal, India
e-mail: abhijitbhu@gmail.com



V. Kumar Singh et al. (eds.), *Sustainable Agriculture Reviews 50*, Sustainable Agriculture Reviews 50, https://doi.org/10.1007/978-3-030-63249-6_4

## Abbreviations

| | |
|---|---|
| RegCM3 | The three-dimensional hydrostatic regional climate model |
| CAMx | Comprehensive Air Quality Model with Extensions |
| CB-IV | Carbon Bond vs. 4 |
| ppmv | Parts per million by volume |
| IPCC | Intergovernmental Panel on Climate Change |
| $\mu gm^{-3}$ | Microgram per cubic meter |
| RSPM | Respirable Suspended Particulate matter |
| NAAQS | National Ambient Air Quality Standards |
| TSPMs | Total suspended particulate matters |
| $mgm^{-3}$ | Milligram per cubic meter |
| OSHA | Occupational Safety and Health Administration |
| ICP-AES | Indus Couple Plasma Atomic Emission Spectroscopy |
| ED-XRF | Energy Dispersive X-Ray Fluorescence |
| ICP-MS | Indus Couple Plasma Mass Spectroscopy |
| ICP-OES | Indus Couple Plasma Optical Emission Spectrometer |
| GC-MS | Gas Chromatography Mass Spectroscopy |
| DST & BT | Department of Science & Technology and Biotechnology |

## 4.1 Introduction

Agriculture and the environment are more complexly interconnected than normally occurred in the emissions and transmission to exposure and its effects (Hinz 2002). Various types of works such as sowing of seeds, augmentation, harvesting, breeding, animal husbandry, and horticulture are related with this agricultural sector (Jager 2005) and it is estimated that globally 450 million salaried men and women are involved in this particular sector. Specific work needs to be done in various fields, gardens, green-houses, animal husbandry facilities and agro-based industries to meet the demand for food, feed and fibre of mankind. But, the food crisis is the major concern for human beings worldwide (Sarkar and Agrawal 2010a, b; Agrawal et al. 2013). The production method, the level of mechanization and the labor requirement differ significantly in each work setting. Therefore, different agro-based industries such as feed mills, flour mills, cotton ginners, textile industries, etc. have different natures in the process (Arslan and Aybek 2012). Agriculture is crucial for the survival of human life. The activities related to agriculture have a tremendous impact on the environment such as deforestation, soil degradation, and clime change. Mainly due to emission of ammonia, toxic organic compounds, and particulate matters from agricultural practices, the environmental degradation especially the problems of the ambient air quality has emerged severely and affects the human beings (Erisman et al. 2008).

Particulate matter is an emerging contaminant which is a family of heterogeneous elements characterized by their size, chemical nature and suspended properties in the air. The nature of the particulate matter can be entirely minerals with some dissolved gases or black carbon in it. Particulates are the fine solid matter scattered through aeration (Forstner 1998) and demonstrates the characteristics of Brownian movement. Particulate matter pollution can be mostly primary or secondary and occurs naturally (pollen, spores, salt spray, volcano, mist/fog and soil erosion) and by anthropogenic activities (soot, fly ash, and cement dust) in an extensive range of particle sizes (Krupa 1997).

Particulate matter pollution is a concoction of acids, organic chemicals, smoke, metals and soil or dust and it can be called as aerodynamic equivalent diameter which is density sphere with a diameter of 1000 kg $m^{-3}$ with same constant velocity. Particulate matters are of different sizes and fractions related to various respiratory, cardiovascular and cerebrovascular diseases of human (Leepe et al. 2019; Fasola et al. 2020; Marchini et al. 2020). Particulate matter, smaller than 2.5 μm, is more responsible for long term effect of human health than the particulate matter smaller than 10 μm. It is estimated that the mortality rate is enhanced by 0.2–0.6% when subjected to short term exposure of particulate matter smaller than 10 μm. Due to prolonged exposure of particulate matter smaller than 2.5 μm, the mortality rate is raised of 6–13% (WHO 2013).

The concentration of particulate matter with an aerodynamic diameter less than 2.5 μm is increased by 18% between the years 2010–2016. $PM_{2.5}$ along with $PM_{2.5}$ poses greatest risk to health; as they can easily penetrate into lungs and the bloodstream. Globally, 4.1 million deaths of human beings have been occurred from heart disease and stroke, lungs cancer, chronic lungs disease, and respiratory infection in the year 2016 due to exposure of particulate matter smaller than 2.5 μm (WHO 2018).

Particulate matters can easily enter the respiratory tract which causes severe damage to the human respiratory system. The particulate matters are characterized by their concentration, size distribution, structure and chemical composition that are highly variable, both temporarily and spatially (Arslan and Aybek 2012). Maynard and Howard (1999) in their study have been reported that particulate matter smaller than 10 μm has short life span and more hazardous than the particulate matter with an aerodynamic diameter less than 2.5 μm.

Anthropogenic sources i.e. both stationary (industries) and mobile (vehicles) sources are the major contributor for the emission of huge amount of gaseous pollutants and particulate matters (Joshi and Swami 2007). Crop production is extremely dependent on environmental circumstances in which air quality plays main role (Agrawal et al. 2006) and these crop plants are susceptible to pollutant and used as indicator of air pollution (Muzika et al. 2004; Bytnerowitz et al. 2005; Petkovsek et al. 2008).

## 4.2 Particulate Matter Pollution and Air Quality

Several studies have been revealed that through changes in meteorology, climate change affects the concentration of particulate matter with an aerodynamic diameter less than 2.5 μm. Depending upon meteorological parameters, secondary particulate matter smaller than 2.5 μm is formed and dispersed and with the change of these parameter the formation and dispersion of particulate matter will also change (Racherla and Adams 2006; Tagaris et al. 2007, 2009; Tainio et al. 2013). Accordingly, the concentrations of particulate matter with an aerodynamic diameter less than 2.5 μm and climate change are correlated in different levels and it can be said that particulate matter cools the climate as an aerosol (IPCC 2007).

The three-dimensional hydrostatic regional climate model (RegCM3) has been created by Dickinson et al. (1989) and Giorgi and Bates (1989) and it is usually applied in climate change simulations (Giorgi et al. 1992; Hirakuchi and Giorgi 1995; Diffenbaugh et al. 2005; Gao et al. 2006a, b; Meleux et al. 2007a, b).

The simulations of air quality model has conducted by with version 4.40 of the Comprehensive Air Quality Model with Extensions (CAMx) which is the three-dimensional chemical transport model that continued consistent revision by science community over the past decade (Tainio et al. 2013) and it is applied for investigative study related to ozone and aerosols (Morris et al. 2005; Tesche et al. 2006). The model simulates important chemical and physical processes for understanding the conversion and distribution of ambient trace gas: (1) emission injection, (2) horizontal and vertical advocacy, (3) horizontal and vertical scattering, (4) dry and wet deposits, (5) chemical reactions in gas-phase and aqueous-phase, and (5) aerosol dynamics, thermodynamics and chemistry. The CAMx uses a mass conservative-and uninterrupted transport number in parallel processing. With the help of this model, it can evaluate integrated air from suburban continents and clarify large amounts of air pollution. The chemical process involved is Carbon Bond vs. 4 (CB-IV) which contains 117 reactions – 11 of which are photolytic and 67 species (37 state gases, 18 state parts and 12 radicals), isotropic thermodynamic equilibrium models (Nenes et al. 1998) are used for aerosol evaluation processes. Particulate matter pollution also occurs due to indoor activities. Indoor particulate matter exposure can be eliminated by spreading the use of particulate filters, reducing indoor combustion for heating and cooking, and stopping smoking. During peak traffic periods or badly air quality index day, particulate filters reduce household burning for heat and cook, and stop smoking. These general changes during peak traffic periods or time of unfortunate air quality possibly will profit human being with both short-range visible control and long-standing cardiovascular and respiratory problems (Anderson et al. 2012).

### 4.2.1 Modern Air Quality: Crisis and Concern

Air in total is a complex made up of several major and minor gaseous components. In standard ambient air, major gaseous components are – nitrogen with a concentration of 7, 80,840 parts per million by volume (ppmv) (almost 78.08% volume), oxygen with a concentration of 2,09,460 ppmv (almost 20.95% volume), argon with a concentration of 9340 ppmv and minor gaseous components are nitrous oxide with a concentration of 0.5 ppmv, xenon with a concentration of 0.009 ppmv etc. (Das et al. 2017). Owing to various anthropogenic as well as other natural causes, the atmospheric gaseous composition is changed across its limit. Different industrial productions, forest fire, dust storm, acid rain, and agricultural operation add hazardous gaseous air pollutants and various particulate matters. Due to the pollution of particulate matters, people, animals and plants are constantly exposed to particulate matters. Thus, addressing the issues of the pollution of particulate matters is very important. Agricultural operation, animal husbandry, and agro-industry are the source of indoor or outdoor air pollution, resulting in individual exposure to different concentration of dust from different source in fraction of different size of particulate matter.

Day by day our earth is getting warmer and more crowded. Because engines continue to emit pollutants, and half of the world does not have access to clean fuels or technologies, much of air we breathe is becoming dangerously polluted and now nine out of ten people breathe polluted air. It kills seven million people in a year. About 1/3 of death occurs from stroke, cancer and heart stroke are mainly due exposure of air pollutants which has the same effect of smoking tobacco to human (WHO 2016a).

It is difficult to avoid air pollution, no matter how developed the country is. It is all around us. Microscopic pollutants can remove our body's defences and penetrate deep into our breathing and blood circulation, eventually damaging our lungs, heart and brain. Air pollution is closely linked to major climate change. Fuel combustion is a major contributor to air pollution. Thus, 'efforts to alleviate one can successfully improve the other. Recently, the United Nations Intergovernmental Panel on Climate Change (IPCC) saw that if we could reduce global warming by 1.5 °C, coal-fired power should be depleted by 2050, if not, we would see a major climate crisis in just 20 years (WHO 2016b).

According to the Paris Agreement aimed at mitigating climate change, reducing air pollution alone could save the lives of about a million peoples worldwide by 2050. An economic benefit from air pollution control is significant in 15 countries. They emit greenhouse gases. The health impact of air is thought to consume more than 4% of their gross domestic product (WHO 2016c). Now both cities and villages of around the world are experiencing toxic pollutants in the air that exceed the average annual values proposed by World Health Organization's air quality guidelines.

A two-day discussion was held on 12–14 February 2019 at the headquarters of World Health Organization announcement and interference to reduce the Health of

Risk Reduction and Air Pollution. There was a presentation on the current review of evidence, methods and applications and had included significant time for discussion, identification of data gaps, research requirements and subsequent steps to seek expert advice in a consistent and harmonious manner to provide guidance material specific and practical advice forms. Most deaths from air pollution are caused by an infectious disease. One third of human deaths are caused by air pollution. Lung cancer responsible for 29% of deaths, stroke for 24%, heart disease for 25%, and lung disease for 43%. It has been studied that nine peoples out of ten in the world breathe contaminated air. Black carbon is able to enter deep into the lung and cardiovascular system (WHO 2018).

The levels of elevated outdoor particulate matters are directly associated with decreased lung functioning in children (Gehring et al. 2013). Elemental components of $PM_{2.5}$ produced from indoor sources may vary by particulate matter with an aerodynamic diameter less than 2.5 μm (Habre et al. 2014). Indoor particulate matter is a better predictor of lung function than outdoor particulate matter in children (Isiugo et al. 2019).

There are mostly the pollution of air is two different categories such as ambient or outdoor air pollution and household or indoor air pollution. The latter refers to the air pollution produced by household burning of fuels (coal, wood, and kerosene) through open fires or basic stove in inadequately spaces. Household and ambient air pollution are correlated because air from inside moves to the outside or from outside to inside. Due to household air pollution, about four million peoples are affected in the countries of Africa and Asia.

According to emission sources, air pollutants are two major categories such as primary pollutant which originate from the emission source and directly pollute the atmosphere; and secondary pollutants arise from the primary pollutants by chemical reaction of primary pollutants. For example, primary pollutants include –Sulphur dioxide, Oxides of Nitrogen, Oxides of carbon, particulate matters; and secondary pollutants include – Ozone and peroxy acetyl nitrite (Hickey et al. 2014). According to World Health Organization's guidelines, for particulate matter diameter less than 2.5 μm, the maximum safe level is 10 microgram per cubic meter (μg $m^{-3}$) or less. To encourage cities for reducing air pollution, but they are unable to meet the ideal safe levels, World Health Organization has set three intervening targets for cities such as 15 $\mu gm^{-3}$ (interim target 3); 25 μg $m^{-3}$ (interim target 2); 35 $\mu gm^{-3}$ (interim target 1). Many cities exceed the interim target 1 (WHO 2005).

Air pollution creates detrimental effect on children. Globally, 14% of 5–18 years children have faced asthma relating problems due to air pollution. More or less 5, 43,000 children (above 5 years old) have died every year from respiratory diseases related to air pollution. Air pollution is also related to childhood cancers. Fetal brain growth can be affected when subjected to exposure to air pollutants by the pregnant women. Air pollution is associated with damage in both of children and adult. While this particulate matter is affecting our health, pollutant in the air continue to cause long-standing environmental damage by continuing climate change which is a most important risk to health and welfare (Krupa 1997; Kunzli 2002)

### 4.2.2 Categorization of Major Particulate Matter Pollutants

Particulate pollutants are the concoction of liquids and solids particle present as a suspension into the atmosphere. In addition to natural sources (volcanic ashes, pollens of plants, yellow sand dusts, and sea salts), particulate matters are emitted from anthropogenic sources (soot, smog, fly ash, and chemical mist) (Lippmann 2020). Chemical analysis study shows that particulate matter smaller than 2.5 μm constitute with major ionic component such as sulphate, nitrate, ammonium, and organic carbon (Yao et al. 2002) which are existing in cement, fly ash (Cheng et al. 2008), explosive mixtures in coal mining, fertilizer and herbicide (Garba and Abubakar 2018), and civil construction. Huang et al. (2014) in their study in China depicted that organic carbons were rich component in particulate matter with an aerodynamic diameter less than 2.5 μm because it constituted with 30–50% of organic carbons in all city of China followed by sulphates (8–18%), nitrates (7–14%), ammoniums (5–10%), and chlorides (2–4%), respectively.

On the basis of size distribution, particulate matter can be of following types such as particulate matter with an aerodynamic diameter less than 10, 2.5, 10–2.5, and 1 μm. Aerodynamic diameter of particulate matter 10 μm or less is termed as Respirable Suspended particulate matter (RSPM). Particulate matter of 2.5 is recognized as fine fraction particle which are of 2.5 μm or less in aerodynamic diameter. But, particulate matter of 10–2.5 is identified as coarse fractions particle because its aerodynamic diameter is greater than 2.5 μm but equal to or less than 10 μm and particulate matter of 1 is generally recognized as the particles which diameter is a lesser than 1.0 μm (Arslan and Aybek 2012). Although, depending upon the nature of emission sources, particulate matter can fall into three different categories like coarse (natural emission sources), fine and ultrafine (anthropogenic emission sources) (EPA 2014). According to particle origin, there are various types of particulate matters such as Smoke particle, Dusts (heavy, settling, and suspended atmospheric dust), mists, fumes, cements dust, fly ashes, oil smokes, tobacco smokes, biological contaminant (Viruses and Bacteria), diesel exhausts, Coal dusts, and artificial nano-particle (EPA 2014).

### 4.2.3 Particulate Matter Pollution: Sources, Present Concentrations and Future Trends

#### 4.2.3.1 Source

There are different sources of particulate matters pollution. Particulate matters pollution may occur naturally and anthropogenically. Natural source of particulate matters pollution include soil dust and sea salt. Anthropogenic source of particulate matter pollution include traffic, domestic fuels burning, industry and other unstipulated sources of human being origin (Karagulian et al. 2015). Dust is considered by profuse materials in the earth's crust rock and the soils. These elements of particulate matters are connected with the re-suspension from the field or nude soil by local wind. It has been known that the dust of roads was included in the vehicular sources categories (Belis et al. 2013). Sea salt particle in the wind are found near the coastal regions and may be transported by the winds from coastal region, or road salts can come from the continental area (Liu et al. 2018). Secondary particles are formed from undetermined source originated from human beings. Primary particles include mechanically generated particle and primary carbonaceous particle. Primary particle also includes carbonaceous fly-ash particle formed from high temperature combustions of fossil fuel at coal based power plant (Karagulian et al. 2015).

### Other Sources of Particulate Matter Pollution

Particulate matter pollution occurs from different spheres. There are so many types of particulate matter pollution:

#### *Crop Residue Burning or Biomass Burning*

Biomass burning is the prime source of particulate matter pollution worldwide (Yang et al. 2008). In view of recent global atmospheric chemistry, it is the key source of air pollutants into the environment (Crutzen and Andreae 1990). At time of harvesting, huge crop residues (hay, sugarcane leaves and tops) are formed each year. Jain et al. (2014) reported that 8570, 141150, 37, 230, 120, 1460, 650, and 1210 kg of carbon monoxide, carbon dioxide, sulphur oxides, nitrogen oxides, ammonia, Non-metallic Volatile Organic Compounds (NMVOC), Non-metallic Hydrocarbons (NMHC), and particulate matters, respectively, are emitted from the burning of the remnants of the crop during the year 2008–2009.

#### *Combustion of Fossil Fuels*

The combustion sources of particulate matters include biomass burning in diesel engine, gasoline engine and coal combustion for energy use in different sectors (Park et al. 2018).

Globally, the pollution of air from different sectors (industries, thermal power plant and transportation) emerged severely by the burning of fossil fuel such as coal, diesel, petrol, and gasoline (USEPA 2017). In developed and developing countries of the world, fossil-fuel combustion contributed 85% of particulate matters pollution (Perera 2018).

*Industrial Smoke*

Different types of smokes are emitted from industries. These constitute mainly nitrogen oxides, sulphur oxides, and particulate matter and generally they are secondary pollutants in origin. Particulate matter with aerodynamic diameter less than 2.5 and 10 are generated from industrial smoke (Park et al. 2018).

*Agricultural Processes*

Agricultural processes include soil tillage, seed bed formation, and planting, the application of fertilizer, harvesting and postharvest processes that lead to the formation of particulate matters. These activities facilitate increased dust formation in air that may affect agricultural crop (Akbar-Khanzadeh et al. 2012).

*Wood Working Processes*

Wood working processes include sawing, shaping of wood, cutting, etc. The process is responsible for the formation of particulate matters in the atmosphere. Though, its function into the atmosphere is not so much significant, but it is also the important cause of atmospheric particulate matter related air pollution (Someshwar and Pinkerton 1992).

*Marble/Granite/Tiles Cutting Processes*

Recently it has become a tradition to make modern houses with marble, tiles, etc. Though it enhances the beautification of a house but during its shaping and cutting, enormous amount of dust is formed that increase the dust pollution through generation of particulate matter. Mainly, particulate matter with aerodynamic diameter smaller than 10 μm is generated during these activities that affect our atmosphere tremendously (Traverso et al. 2010).

*Household Activities*

There are so many activities of household which are responsible for smoke formation like from charcoal burning, cow dung cake burning, soot from the flame at the time of cooking, oil smoke from cooking etc. Particularly, these types of smoke generate particulate matter smaller than 10 μm which slowly degrade our atmosphere and cause particulate matter related air pollution (McCormack et al. 2008).

*Transportation or Automobile Smoke*

Transportation or automobiles are the serious cause of particulate matter pollution. Coarse particulate matter is generated from automobile smoke (Barrientos et al. 2016).

*Road Construction*

Bio-solids are used in the road/highways construction (Sharma et al. 2017) which is another cause of particulate matter pollution where dust is spread at final stage of road formation to remove its stickiness, which inhibits the particulate matter pollution (Muleski et al. 2005)

*Environmental Tobacco Smoke (ETS)*

Along with particulate matter, environmental tobacco smoke also causes atmospheric particulate matter pollution. The environmental tobacco smoke is related to other fossil fuel burning product which is mainly responsible for indoor particulate matter pollution (Repace and Lowrey 1980; Invernizzi et al. 2002, 2004; Curjuric et al. 2012).

#### 4.2.3.2 Present Condition of Particulate Emissions in India

With the help of satellite data, Greenpeace India (2019) has revealed that six Indian metropolitan cities or transport and industrial cluster are hotspots for air pollutants of nitrogen dioxide. Delhi, Bengaluru, Chennai, Hyderabad, Mumbai, and Kolkata have vehicle populations and diesel consumption. Nitrogen oxide is not emitted directly from any source, but it is formed from the reaction between air gases by the impact of daylight and elevated temperature.

After analysis, it was revealed that multiple Indian national capital cities are hotspots for rising levels of nitrogen oxides, dangerous pollutants that contribute to ozone formation. According to data collected from February, 2018 to May, 2019, coal mining and industrial group cities such as Sonbhadra-Singrauli in Madhya Pradesh and Uttar Pradesh, Korba in Chhattisgarh, Talchar in Orissa, Chandrapur in Maharashtra, Mundra in Gujarat and Durgapur in the West Bengal are all emitting $NO_x$ (Greenpeace India 2019).

Over the some years ago, several studies have identified that particulate matter smaller than 2.5 μm, nitrogen oxides, and ozone having significant effects on human health. It is estimated that air pollution (combination of ambient particulate matter less than 2.5 μm and ozone) killed 3.4 million worldwide in 2017 and more than 1.2 million in India alone. As per the available data of Greenpeace India, particulate matter with aerodynamic diameter less than 2.5 μm alone killed more than 6.7 lakh people in India in 2015. Citing a 2015 report by Indian Institute of Technology Kanpur, it was stated that 90% reduction of nitrogen oxides from power plants within a radius of 300 kilometer from Delhi can reduce nitrates by 45% which can effectively reduce the concentration of particulate matter smaller than 10 and 2.5 μm in Delhi. As per Greenpeace India's January, 2019 report, 241 cities are violating because these pollutants exceeded their standards (Greenpeace India 2019).

As air pollution becomes a global public health problem, in 2012 (WHO 2014) air pollution was declared as the principal health risk factor with seven million deaths worldwide. It was estimated that particulate matter related air pollution contributes for almost 6% of deaths worldwide (Lim et al. 2012). India is generally central economic country in South Asia. However, rapid urbanization, unrestrained industrialization, and unmanageable releases of solid wastes have led to a significant deterioration in urban air quality (Sarkar and Agrawal 2010a, b; Sarkar et al. 2012; Singh et al. 2014; Li et al. 2016). Therefore, 99.9% of the Indian population lived in vicinity where the particulate matter with aerodynamic diameter smaller than 2.5 μm exceeded World Health Organization's air quality guideline (10 $\mu g/m^3$) and also half of the population resided in the place where the National Ambient Air Quality Standards (NAAQS) of 40 $\mu g/m^3$ for particulate matter with aerodynamic diameter smaller than 2.5 μm has exceeded (Pant et al. 2019). The non-respirable part of the particulate matters is suspended particulate matters and RSPM and together forms Total suspended particulate matters (TSPMs). As per the available report of World Health Organization, Delhi is the most polluted city in the world (The Times of India 2014). It is noteworthy that the fine particle satellite systems made for the whole of India have revealed that people living in both urban and rural areas have exposed to dangerously high levels of particles. About 670 million people consisting 54.5% of the population live in areas that cannot assemble with the NAAQS for fine particulate matter (Dey et al. 2012). Various studies have shown that there is a compatible connection amid particle matter concentration and health compared to other air pollutants. Studies show statistically that there is a significant correlation between mortality and the concentration of surrounding particulate matter (Li et al. 2015). Thus, particulate matters and source and health effects need to be understood (Kaushik et al. 2006).

## 4.3 Impact of Particulate Matter Pollution on Agricultural Crops

Due to rapid urbanization and unplanned industrialization in the recent Anthropocene era, particulate matters pollution have been increased drastically in India. Particulate matters are the firm material, originated naturally and anthropogenically. The size fractionation of particulate matters and its undesirable impacts on human beings was well recognized in literature (Rai and Singh 2015).

The air pollutant has potential undesirable consequence on biochemical parameters of crop plants and reduces the whole enlargement and progress of plants. Plant adaptation with changes in environmental factors involves short-range physiological response and long-standing physiological, structural, and morphological changes. These changes help plants to reduce stress and make the most of internal and external organizations (Agrawal and Agrawal 1990; Rai et al. 2013; Rai and Panda 2014; Rai and Singh 2015).

### 4.3.1 Interaction of Particulate Matter with Crop Plants

Since plants are in constant contact with the air, they are primary receptors for both atmospheric gaseous and particulate pollutants. In terrestrial plant species, leaf surface acts as a natural sink for particulate matters. The foliar surface of trees has an effective contaminant-trap device and absorbed particulate matter in polluted environments (Maiti 1993; Samal and Santra 2002). The use of higher plants for the assessment of air quality is becoming widely popular day by day. Its main advantages are the accessibility of organic substance, effortlessness of species detection, sampling and treatment and ubiquity of some genres, making it possible to cover large areas. Lichens and algae are characterized by the distribution of unequal and patchy, and their sampling should be done by experts capable of distinguishing between species of similar colour (Maiti 1993).

Conventional agricultural activities of crop production and harvest also produce dust (Arslan and Aybek 2012). Different study have been reported that the emission of dust occurred from various agricultural practiced such as cutting, planting, weeding and mowing of cotton, nuts and wheat. During harvesting of wheat, rice, cotton, the emissions factor of particulate matter with aerodynamic diameter less than 10 is 1 (Graffney and Yu 2003). 13000 tons of particulate matter smaller than 10 are emitted per year at the time of land preparation and harvesting, respectively, in the San Joaquin Valley. Researchers consider this step important because they are looking for cost-effective ways to reduce emissions from agricultural operations and to determine the need for future research related to air quality (Arslan and Aybek 2012).

As air pollution is identified to be the consequence of industrialization and urbanization, agriculture was not considered as a major cause of air pollution, earlier. However, according to Arslan and Aybek (2012), the agricultural sector is the main source of particulate matters emissions. 5 and 25% of particulate matter with aerodynamic diameter less than 2.5 and 10 μm are emitted from the agricultural sector in Europe. Bogman et al. (2005) in their research have been depicted that a total yearly amount of 10100 ton of total suspended particulate matter and 200–3100 ton of particulate matter smaller than 10 μm, emitted from field farming in Flanders, Belgium. It has been assessed that the agricultural sectors emitted 35% of suspended particulate matters and 24% of particulate matter less than 10 μm.

The concentration of air pollutants becomes highest and exceeds their standards at certain times of the year, especially when conducting row crop farming, suggesting that row crop farming have to responsible for the emissions of particulate matter with aerodynamic diameter less than 10 μm (Madden et al. 2008). Globally, the stubble burning is very needful technique in farming for removing the harvest leftovers from the field, to prevent water logging and planting machines from freezing by straw, to obtain a smooth seed, and to save time before planting. Due to the hay burning in the fields of Haryana and Punjab, the emissions of smoke not only affects rural or agricultural environments, but also affects the surrounding urban areas, like Delhi, the most polluted cities in India. But, it was depicted that soil, vegetable burning, sulfate aerosols, vehicles and cooking are responsible for 38, 35, 20, 2 and

16% of particulate matter smaller than 2.5 μm emissions, respectively (Ravindra et al. 2019).

Particulate matter with aerodynamic diameter less than 10 μm consists mostly of soil-derived material. With intensive farming activities as well as correspondence on air quality violations, attention is being paid to allied crop farming as a potential major contributor to Particulate matter smaller than 10 μm. After research, it has been revealed that there are two mechanisms for reducing the concentration of particulate matter smaller than 10 μm – conservation tillage and standard tillage, conservation tillage is very crucial for the reduction of agricultural emissions of particulate matter with aerodynamic diameter less than 10 μm (Arslan and Aybek 2012). The emissions of particulate matter smaller than 10 μm reduced by about 85 and 52% by the conservation tillage on two separate cultivations (Madden et al. 2008). In this study it has been found that conservation tillage systems require zero or one operation, where conventional tillage needs six operations. In addition, work can be done on the land to conserve higher soil moisture content than in dry soil conditions. As less number of fields is needed in conventional tillage than standard tillage therefore, low levels of dust emitted in conservation tillage than standard tillage applications (Baker et al. 2005). Because of short-range exposure to particulate matter with aerodynamic diameter less than 10 μm, reduced lung function, cardiac anaemia, heart attack, and premature death (Krupa 1997; Kunzli 2002; Khaniabadi et al. 2017).

#### 4.3.1.1 Personal Respirable Quartz Concentration

It is also responsible for increasing the level of exposure to dust or particulate matter as well as quartz continuously emitted from agricultural operations. This directly affects farmers mainly tractor operators. The concentration of personal respirable quartz was increased by agriculture. For instance, it has been reported that the average concentration of respirable silica was 3.91 and 0.04 milligram per cubic meter (mg $m^{-3}$) for the planting of sweet potato and cotton (Archer et al. 2002). A large part of the change in quartz density was explained by agriculture and soil moisture (Swanepoel et al. 2010). Tractor operators were exposed to a quartz concentration of 2.0 and 0.05 mg $m^{-3}$ in an open cabin and in a closed cabin, respectively. Pull-type soil tillage equipment can produce large amounts of dust. It is not enough to scientifically determine the extent to which quartz exposure poses a risk to agriculture (Swanepoel et al. 2010).

The soil tilling processes with soil packer, rotary tiller, fertilizer, and, disc harrow produce inorganic dust (Arslan and Aybek 2012). The concentrations of particulate matter with aerodynamic diameter less than 10 μm have gravitationally greater than the Occupational Safety and Health Administration (OSHA) threshold (15000 μg $m^{-3}$) in rotary tilling (25770 μg $m^{-3}$), wheat harvesting (29300 μg $m^{-3}$), and hay making (24640 μg $m^{-3}$) (Arslan and Aybek 2012). In addition, the concentration of particulate matter with aerodynamic diameter smaller than 2.5 μm in these activities were 5888, 10560, 8470 μg $m^{-3}$ respectively, that is higher than the threshold limit value (5000 μg $m^{-3}$). The concentrations of particulate matter smaller than 10 μm were 3130 and 6026 μg $m^{-3}$ in wheat harvesting and hay making. It is an interesting fact that it was higher during hay making than the threshold limit value of particulate matter less than 2.5 μm. The concentration of particulate matter less than 2.5 and 10 μm has below than the threshold limit value in other field application (Arslan et al. 2010).

Another important area of interest in professional contact with dust is the agro-industry, where agricultural products are processed to be swallowed by humans, animals or plants. The concentration of responsible dust and very fine particles in the press, storage area, and the ginner of two cotton was higher than the threshold limit value of OSHA (1000 μg $m^{-3}$) (Arslan and Aybek 2012).

## 4.3.2 *Effect of Particulate Matter on Crop Plants*

### 4.3.2.1 Visible Symptoms

The deposition of particulate matter with aerodynamic diameter less than 1, 2.5, and 10 μm mainly occur on the foliar surface of the plant species and inhibit the regular respiration and photosynthetic activities of the plant species causing visible symptoms such as chlorosis and death of leaf tissue by the amalgamation of a thick crust formation and alkaline toxicity produced in damp weather. Dust coatings can also affect the normal application of pesticides and other agrochemicals used as sprays. Furthermore, the deposition of alkaline dust in the soil may increase the potential of hydrogen to adverse levels to the development of the crop (Rai 2016a, b).

### 4.3.2.2 Growth and Development Responses

Particulate matter less than 2.5 μm is composed of sulfate, ammonia, nitrates, mineral dust, black carbon, and water. It can have adverse effects on crop growth. The direct effect of Particulate matter less than 2.5 μm is acquired by deposition upon the foliar surface to reduce respiration and changes in leaf temperature (Hirano et al. 1995), which affect the yield of the crop. Ultimately, it can affect the environment on which crop development depends, for example, through the assimilation and spreading of solar radiation. Therefore, particulate matter with aerodynamic diameter less than 2.5 μm can reduce the amount of solar radiation available in plants (Liu et al. 2016), in that way reducing photosynthesis and limiting plant growth (Chameides et al. 1999).

Photosynthetic pigments are sufficiently responsive to air pollutants (Chauhan and Joshi 2010). The photosynthetic rate is measured with the help of portable photosynthetic system (Model LI-6200, LI-COR Inc.) (Sarkar et al. 2010; Sarkar and Agrawal 2012) but their sensitivity may be determining the response of plants to pollutants (Chauhan and Joshi 2010). Contaminants distort the synthesis of chlorophyll and increase its degradation. As a result, it is very popular that urban and industrial air pollution has become a serious threat to the agricultural ecosystem adjacent to cities and industrial areas (Agrawal et al. 2003). Chlorophyll pigment is crucial for photosynthetic activity, and reducing chlorophyll content has the capacity to indicate the reduced air pollution (Gilbert 1968; Giri et al. 2013). The continuous application of cement dust blocks stomata of leaf; accordingly interfere to the gases exchange (Lerman 1972). Carotenoids protect against photo-oxidation damage, therefore, there are serious consequences of reduced chlorophyll pigments (Xiao et al. 2011). Joshi and Swami (2007) further stated that there is a noteworthy reduction in carotenoid content of various plants grown in contaminated areas. Therefore, the physiological condition of plants can be better identified by its pigment content.

Suspended Particulate Matter affects plants in numerous ways based on the composition of the particles and is recognized for its direct or indirect effects on agricultural plants. Various gaseous and particulate pollutants emitting from brick kilns exhibit negative effects on adjacent plants. The relative density of different herbaceous plant species in the vicinity of the brick kiln decreases compared to the control area (Fatima 2011; Skinder et al. 2014)). Dust particles have local importance in road, industrial area, and brick klin (Lee and Dong 2011; Skinder et al. 2014). Excluding exposure to sunlight in the atmosphere, the uninterrupted effects of leaf dust can reduce the radioactivity of chloroplasts and stomatal conductance, and can also affect physical water loss control by stopping the stomata (Hirano et al. 1995; Wijayratne et al. 2009; Chaudhary and Rathore 2019).

#### 4.3.2.3 Physico-chemical Responses

Effect on Chlorophyll

Chlorophyll is a combination of xanthophylls, carotenoid, and various chlorophyll-a, b, c, d, e, which are found in plant cells. Chlorophyll-c is present in freshwater plants and chlorophyll-d is located only in sea red algae. Chlorophyll is very needful for assessing the impacts of air pollutants on plants and also shows the vital role in plant metabolism. Decreased chlorophyll content is directly related to plant growth (Wagh et al. 2006). Dust physically blocks stomata. Krajckova and Mejstrik (1984) studied that the stomatal diameter was 8–12 μm for different crops. The chemical effects of dust may be more important than any physical effects on the soil or directly on the surface of the plant. Xanthophylls, carotenoids and chlorophyll-b are the accessory pigments which are present mostly in plants, photosynthetic bacteria, algae and act as makers of photosynthesis (Joshi et al. 2009) (Table 4.1).

Dust can inhibit chlorophyll synthesis as the particles fix various metals and polycyclic hydrocarbons, so they can inhibit important enzymes needed for chlorophyll synthesis. Plants that have accumulated dust can attract light available for photosynthesis and block the stomatal pores for air expansion and creates stress on

**Table 4.1** Quantitative impacts of particulate matters (PM) on some important plants

| Sl. no. | Name of the plant | Common name | Country | Concentrations of particulate matter (Type) | Type of damage | Magnitude of damage (+ increased; – decreased) | References |
|---|---|---|---|---|---|---|---|
| 1 | *Rhamnus esquirolli* | Buckthorn | China | 150 μgm$^{-3}$ ($PM_{2.5}$) | Decreased photosynthetic rate | −9.90% | Li et al. (2019) |
| | | | | 250 μgm$^{-3}$ ($PM_{2.5}$) | Decreased stomatal conductance | −28.51% | |
| 2 | *Lindera kwangtungensis* | Spicebush | China | 150 μgm$^{-3}$ ($PM_{2.5}$) | Decreased Photosynthetic rate | −13.16% | Li et al. (2019) |
| | | | | 250μgm$^{-3}$ ($PM_{2.5}$) | Decreased stomatal conductance | −22.81% | |
| 3 | *Titicum asetivum cv PBW 343* | Wheat | India | 167.90 μgm$^{-3}$ (RSPM) | Increased carotenoid | +16.67 % | Joshi et al. (2009) |
| | | | | | Decreased chlorophyll a | −15.20 % | |
| | | | | | Decreased chlorophyll b | 18.80 % | |
| | | | | | Decreased total chlorophyll | −16.60 % | |
| | | | | | Decreased ascorbic acid | −35.38 % | |
| | | | | | Increased moisture content | +5.95 % | |
| 4 | *Brassica campestris* | Mustard | India | 160.67 μgm$^{-3}$ (RSPM) | Increased carotenoid | +22.94 % | Joshi et al. (2009) |
| | | | | | Decreased chlorophyll a | −6.92 % | |
| | | | | | Decreased chlorophyll b | −14.52 % | |
| | | | | | Decreased total chlorophyll | 10.63% | |
| | | | | | Decreased ascorbic acid | −8.46 % | |
| | | | | | Increased moisture content | +4.87 % | |
| 5 | *Clusia hilariana* | Scotch Attorney | Brazil. | 2.14 μgm$^{-3}$ (SPM) | Decreased chlorophyll a | −34.29 μg/cm$^2$ | Pereira et al. (2009) |
| | | | | | Decreased chlorophyll b | −27.30μg/cm$^2$ | |
| | | | | | Decreased carotenoid | −19.12 μg/cm$^2$ | |

| | | | | | | | |
|---|---|---|---|---|---|---|---|
| 6 | *Solanum lycopersicum* | Tomato | Italy | 22.40 μgm$^{-3}$ ($PM_{10}$) | Decreased chlorophyll a | −0.33mg/g | Daresta et al. (2015) |
| | | | | | Decreased chlorophyll b | −0.19 mg/g | |
| | | | | | Decreased total chlorcphyll | −0.50 mg/g | |
| | | | | | Increased carotenoid | +0.20 mg/g | |
| | | | | 19. 70 μgm$^{-3}$ ($PM_{10}$) | Decreased chlorophyll a | −0.33 mg/g | |
| | | | | | Decreased chlorophyll b | −0.13 mg/g | |
| | | | | | Decreased total chlorophyll | −0.46 mg/g | |
| | | | | | Increased carotenoid | +0.13 mg/g | |
| | | | | 24.30 μgm$^{-3}$ ($PM_{10}$) | Decreased chlorophyll a | −0.19 mg/g | |
| | | | | | Decreased chlorophyll b | −0.12 mg/g | |
| | | | | | Decreased total chlorophyll | −0.31 mg/g | |
| | | | | | Increased carotenoid | +0.16 mg/g | |
| | | | | 38.90 μgm$^{-3}$ ($PM_{10}$) | Decreased chlorophyll a | −0.47 mg/g | |
| | | | | | Decreased chlorophyll b | −0.18 mg/g | |
| | | | | | Decreased total chlorophyll | −0.65 mg/g | |
| | | | | | Increased carotenoid | +0.15 mg/g | |
| 7 | *Zea mays* | Maize | Czech Republic | Supplemental dose of fly ash particles | Increased stomatal length | 32 μm | Krajickova and Mejstrik (1984) |
| | | | | | Increased stomatal width | 12 μm | |
| 8 | *Phaseolus vulgaris* | Bean | Czech Republic | Supplemental dose of fly ash particles | Increased stomatal length | 14 μm | Krajickova and Mejstrik (1984) |
| | | | | | Increased stomatal width | 8 μm | |
| 9 | *Tectona grandis* | Teak | India | 122 μgm$^{-3}$ (RSPM) | Decreased chlorophyll a | −1.02 mg/g | Joshi et al. (2009) |
| | | | | | Decreased chlorophyll b | −0.59 mg/g | |
| | | | | | Decreased total chlorophyll | −1.61 mg/g | |
| | | | | | Increased carotenoid | +1.20 mg/g | |
| 10 | *Shorea robusta* | Sal | India | 122 μgm$^{-3}$ (RSPM) | Decreased chlorophyll a | −1.50 mg/g | Joshi et al. (2009) |
| | | | | | Decreased chlorophyll b | −0.80 mg/g | |
| | | | | | Decreased total chlorophyll | −2.30 mg/g | |
| | | | | | Increased carotenoid | +1.51 mg/g | |

plant metabolism (Borka 1980; Anthony 2001). Various studies depicted that both ascorbic acids and chlorophyll have been altered due to dust. Decrease of chlorophyll contents in the leaves caused due to alkaline conditions developed by dissolving the chemical present in the dusts particles into the cell fluids. Different studies have been reported that the plant species located in polluted site have lower total chlorophyll content than those located in control site (Mandal and Mukherji 2000; Samal and Santra 2002; Prajapati and Tripathi 2008).

### Effect on Ascorbic Acid

Ascorbic acids are the natural antioxidant and play the most vital role in pollution tolerance of the plant species (Rao and Dubey 1990; Keller and Schwager 1977). It also reacts with hydrogen peroxide that reply to various stress in plants (Joshi et al. 2009). The ascorbic acid contents of leaf increase to manage with these stresses while it retards leaf senescence (Garg and Kapoor 1972). According to Prajapati and Tripathi (2008), the level of ascorbic acid is low in winter pursued by summer and rainy season. Ascorbic acid is inversely proportional to total chlorophyll content. Ascorbic acid increases with the decrease in total chlorophyll content.

### Effect on Photosynthesis

The leaf pigment of the plants has declined due to the deposition of dust on leaves surface of plants as a result, photosynthetic activity become reduced (Sarma et al. 2017). Particulate matter with aerodynamic diameter smaller than 2.5 μm can easily intercept to trees and adhere by leaf surface roughness and trichomes (Li et al. 2019). Due to adsorption of particulate matter with aerodynamic diameter smaller than 2.5 μm, net photosynthetic rate and stomatal conductance decreased proportionally with the values of 9.90% and 28.51% of species *Rhamnus esquirolii* H. Lev. and that of value of 13.16% and 22.81% of species *Lindera kwangtungensis* (H.Liu) C.K. Allen, respectively (Li et al. 2019). The net photosynthetic rate varies with species which is caused by roughness of leaf surface (Terzaghi et al. 2013). The roughness of leaf surface showed the crucial role in the absorption of particulate matter with aerodynamic diameter smaller than 2.5 μm because of the different water potential of different leaves (Li et al. 2019) (Table 4.1).

### Effect on Respiration

Increased absorption of radiation by the dust can lift up the leaf temperature of plant and result in boosting photorespiration. Street dust is reported to enhance leaf temperatures by 2–4 °C (Armbrust 1986). Increasing temperature of leaf by 2–3 °C can lead better increase in photorespiration relative to photosynthesis.

Stomatal Responses

The stomata of leaves are blocked due to the deposition of the dust particles (Hirano et al. 1995). Then, the toxicity of particles of dust perturb the physiological activities such as the inhibition of plant growth, the rate of photosynthesis (Armbrust 1986), late blossoming and the hormonal discrepancy of plants (Farmer 1993). Particulate matter with aerodynamic diameter less than 2.5 μm can be easily absorbed through stomata and lenticels of the plants. When fine particles deposited on leaves, grooves and trichomes and providing a buffer time for plant leaves to close the stomata (Li et al. 2019).

Yield Responses

Yield responses are measured at the time of harvesting (Sarkar et al. 2015) and various type of damage like leaf injury, pre-mature senescence, and disturbed membrane permeability has been occurred due to deposition of particulate matter on leaves in sensitive plant species. As a result ultimately decreases photosynthetic activity and reduced growth and yield of that plant species (Atkinson et al. 1988). Regional air pollution can cause noteworthy yield losses in sensitive crops (Agrawal et al. 2003). Agrawal et al. (2003) in their study depicted that major yield loss in most important crop species occurred due to the effect of air pollutant. Several workers such as (Rajput and Agrawal 2005; Joshi et al. 2009) have been reported that photosynthetic pigments and yield in different crops reduced adversely due to the effect of dust and other air pollutants.

## 4.4 Sustainable Ways to Mitigate Particulate Matter Toxicity in Modern Agriculture

Agriculture and the environment are interconnected that means they release and collect contaminants like toxic organic compound and particulate matters from each other. Agriculture is source of diverse substances that can affect all essential components like soil, water, and air of the environment. However, agriculture is a recipient of lots of pollutants from various sources in the environment (Hinz 2002). Globally, India is the 3rd major emitter of greenhouse gas, where agricultural sector accounts for 18% of whole nationwide emission. Well-organized use of nitrogen fertilizers and pesticides, well management of water, zero land farming, and better residue management strategies provide effective solutions to reduce air pollution arising from the agriculture (Singh et al. 2020). This type of mitigation process of air pollution was developed, field tested, and extensively adopted by farmers across India, Bangladesh, Nepal and Pakistan.

Zero tillage farming knowledge plays very critical role in adopting sustainable intensity and suitable management strategies of rice-wheat graining systems. Fine particles can travel thousands of miles and also bear pathogens and destructive substances that cause acute and chronic respiratory problems of human beings. The strategy for mitigation of air pollution include the acceptance of proper nutrition measures, zero-tillage cultivation, and the hindrance of burning of crop residue, improved water management, and the adoption of laser land levelling (Singh et al. 2018; Vedachalam 2019).

## 4.5 Future Prospects

The source apportionment of particulate matter can be done by using various receptor models like Positive Matrix factorization, Chemical Mass Balance, Principal Component Analysis, Enrichment Factor, Multi Linear Regression, Aerosol Equilibrium, Time Series, and Aerosol Evolution (Ndamitso et al. 2016). Various instruments are used to detected particulate matter such as Atomic Absorption Spectrometer, Indus Couple Plasma Atomic Emission Spectroscopy (ICP-AES), Energy Dispersive X-Ray Fluoreacence (ED-XRF), Indus Couple Plasma Mass Spectroscopy (ICP-MS), Indus Couple Plasma Optical Emission Spectrometer (ICP-OES), Gas Chromatography Mass Spectroscopy (GC-MS), and Ion Chromatography (Ndamitso et al. 2016). In industrialized countries, the dust collectors are highly regulated for particulate matter emissions. To save environment, the industrial sector and vehicles need to install effective dust collection system to restrict particulate matter emissions. This system includes an inertial collector (cyclone collector), fabric filter collector (bag-house), wet scrubber, and electrostatic precipitator. Cyclone collectors are applied to remove large and coarse particles. These are applied in most cases as pre-cleaner to well-organized sectors. Fabric filter or bag-house is generally used in industries and forces the dust loaded air by bag-shaped fabric filter and collect the particulate matters on the outer surface of the bag. Sometimes, they allow the clean air to be exhausted into the atmosphere or in some cases recycled with ease. Common fabrics like polyester and fibreglass and common fabric coatings such as Teflon filter are also included in this system. Excess dust is cleaned from the bags and removed from the collector. But, wet scrubbers hold the polluted air by the scrubbing solution which is the assortment of water and other compounds that allowing the particulate to affix to the liquid molecules. However, electrostatic precipitators electrically charge the polluted air and then pass them by large electromagnetic plates and leaving the clean air to be exhausted or re-circulated (Mohapatra and Biswal 2014).

By adopting all these systems near the emission sources of particulates we have to reduce the pollution of particulate matter so that the effect of this pollution on agricultural plant will be reduced which will result in increased agricultural production and the economic condition of the country will prosper from this sector.

**Acknowledgement** Authors are eager to thank the editors for their kind invitation and suggestion. We also appreciatively acknowledge the Department of Science & Technology and Biotechnology (DST&BT), Government of West Bengal, for providing the financial support in the form of a research project (Memo No.: 207(Sanc.)-ST/P/S&T/5G-14/2018, dated: 20th February, 2019) to AS. The research project helped to develop the idea and outcome of this chapter.

## References

Agrawal M, Agrawal SB (1990) Effects of ozone exposure on enzymes and metabolites of nitrogen metabolism. Sci Hort 43:169–177. https://doi.org/10.1016/0304-4238(90)90048-J

Agrawal M, Singh B, Rajput M, Marshall F, Bell JNB (2003) Effects of air pollution on peri-urban agriculture: a case study. Environ Pollut 126:323–329. https://doi.org/10.1016/s0269-7491(03)00245-8

Agrawal M, Singh B, Agrawal SB, Bell JNB, Marshall F (2006) The effects of air pollution on yield and quality of mung bean grown in peri-urban areas of Varanasi. Water Air Soil Pollut 169:239–254. https://doi.org/10.1007/s11270-006-2237-6

Agrawal GK, Sarkar A, Righetti PG, Pedreschi R, Carpentier S, Wang T et al (2013) A decade of plant proteomics and mass spectrometry: translation of technical advancements to food security and safety issues. Mass Spec Rev 32(5):335–365. https://doi.org/10.1002/mas.21365

Akbar-Khanzadeh F, Ames A, Bisesi M, Milz S, Czajkowski K, Kumar A (2012) Particulate matter (PM) exposure assessment—Horizontal and vertical PM profiles in relation to agricultural activities and environmental factors in farm fields. J Occup Environ Hyg 9(8):502–516. https://doi.org/10.1080/15459624.2012.695216

Anderson JO, Thumdiyil JG, Stolbach A (2012) Clearing the air: a review of the effect of particulate matter air pollution on human health. J Med Toxicol 8:166–175. https://doi.org/10.1007/s13181-011-0203-1

Anthony P (2001) Dust from walking tracks: Impacts on rainforest leaves and epiphylls. https://www.wettropics.gov.au/site/user-assets/docs/DustyTracks.pdf. Data adopted on 4 Oct 2019

Archer JD, Cooper GS, Reist PC (2002) Exposure to respirable crystalline silica in eastern North Carolina farm workers. AIHA J 63:750–755. https://doi.org/10.1080/15428110208984765

Armbrust DV (1986) Effect of particulates (dust) on cotton growth, photosynthesis and respiration. Agron Am Soc Agron 6(78):1078–1081. https://doi.org/10.2134/agronj1986.00021962007800060027x

Arslan S, Aybek A (2012) Particulate matter exposure in Agriculture, chapter – 3. In: Haryanto B (ed) Air pollution – a comprehensive perspective. In Tech, Rijeka, pp 73–104. https://doi.org/10.5772/50084

Arslan S, Aybek A, Ekerbiçer H (2010) Measurement of personal $PM_{10}$, $PM_{2.5}$ and $PM_1$ exposures in tractor and combine operations and evaluation of health disturbances of operators. Int J Agric Sci 16:104–115. https://doi.org/10.1501/tarimbil_0000001127

Atkinson CJ, Robe SV, Winner WE (1988) The relationship between changes in photosynthesis and growth for radish plants fumigated with $SO_2$ and $NO_2$. New Phytol 110:173–184. https://doi.org/10.1111/j.1469-8137.1988.tb00250.x

Baker JB, Southard RJ, Mitchell JP (2005) Agricultural dust production in standard and conservation tillage systems in the San Joaquin Valley. J Environ Qual 34(4):1260–1269. https://doi.org/10.2134/jeq2003.0348

Barrientos EJ, Anderson JE, Maricq MM, Boehman AL (2016) Particulate matter indices using fuel smoke point for vehicle emissions with gasoline, ethanol blends, and butanol blends. Combust Flame 167:308–319. https://doi.org/10.1016/j.combustflame.2016.01.034

Belis C, Karagulian F, Bo L, Hopke PK (2013) Critical review and meta-analysis of ambient particulate matter source apportionment using receptor models in Europe. Atmos Environ 69:94–108. https://doi.org/10.1016/j.atmosenv.2012.11.009

Bogman P, Cornelis W, Rollé H, Gabriels D (2005) Prediction of TSP and PM10 emissions from agricultural operations in Flanders, Belgium. In 14th international conference "Transport and air pollution", Graz, Austria. http://citeseerx.ist.psu.edu/viewdoc/download?doi=10.1.1.530.1762&rep=rep1&type=pdf

Borka G (1980) The effects of cement dust pollution on growth and metabolism of Helianthus annuus. Environ Pollut A 22(1):75–79. https://www.cabdirect.org/cabdirect/abstract/19800705837edeeez

Bytnerowitz A, Badea O, Popescu F, Musselman R, Tanase M, Barbu I, Fraczek W, Gembasu N, Surdu A, Danescu F, Postelnicu D, Cenusa R, Vasile C (2005) Air pollution precipitation chemistry and forest health in the Retezat Mountains, Southern Carpathians, Romania. Environ Pollut 137:546–567. https://doi.org/10.1016/j.envpol.2005.01.040

Chameides WL, Yu H, Liu SC, Bergin M, Zhou X, Mearns L et al (1999) Case study of the effects of atmospheric aerosols and regional haze on agriculture: an opportunity to enhance crop yields in China through emission controls? Proc Natl Acad Sci USA 96(24):13626–13633. https://doi.org/10.1073/pnas.96.24.13626

Chaudhary IJ, Rathore D (2019) Dust pollution: Its removal and effect on foliage physiology of urban trees. Sustain Cities Soc 51:101696. https://doi.org/10.1016/j.scs.2019.101696

Chauhan A, Joshi PC (2010) Effect of ambient air pollutants on wheat and mustard crops growing in the vicinity of urban and industrial areas. N Y Sci J 3(2):52–60. https://www.researchgate.net/profile/Dr_Avnish_Chauhan/publication/260301296_Effect_Of_Ambient_Air_Pollutants_On_Wheat_and_Mustard_Crops_Growing_In_The_Vicinity_Of_Urban_and_Industrial_Areas/links/5444eecd0cf2a76a3ccdc3f9.pdf

Cheng CM, Taerakul P, Tu W, Zand B, Butalia T, Wolfe W, Walker H (2008) Surface runoff from full-scale coal combustion product pavements during accelerated loading. J Environ Eng 134(8):591–599. https://doi.org/10.1061/(ASCE)07339372(2008)134:8(591)

Crutzen PJ, Andreae MO (1990) Biomass burning in the tropics: Impact on atmospheric chemistry and biogeochemical cycles. Science 250(4988):1669–1678. https://doi.org/10.1126/science.250.4988.1669

Curjuric I, Imboden M, Nadif R, Kumar A, Schindler C, Haun M et al (2012) Different genes interact with particulate matter and tobacco smoke exposure in affecting lung function decline in the general population. PloS One 7(7):e40175. https://doi.org/10.1371/journal.pone.0040175

Daresta BE, Italiano F, de Gennaro G, Trotta M, Tutino M, Veronico P (2015) Atmospheric particulate matter (PM) effect on the growth of Solanum lycopersicum cv. Roma plants. Chemosphere 119:37–42. https://doi.org/10.1016/j.chemosphere.2014.05.054

Das S, Biswas A, Sarkar A (2017) Air pollution toxicity and global agricultural perspective: a brief review, Chapter 7. In: Padhy PK, Patra PK, Singh UK (eds) Emerging issues in environmental science: concerns and management. New Delhi Publishers, New Delhi, pp 75–89. isbn:978-93-85503-77-1

Dey S, Di Girolamo L, van Donkelaar A, Tripathi SN, Gupta T, Mohan M (2012) Variability of outdoor fine particulate ($PM_{2.5}$) concentration in the Indian Subcontinent: a remote sensing approach. Remote Sens Environ 127:153–161. https://doi.org/10.1016/j.rse.2012.08.021

Dickinson RE, Errico RM, Giorgi F, Bates G (1989) A regional climate model for the western United States. Clim Change 15:383–422. https://doi.org/10.1007/BF00240465

Diffenbaugh NS et al (2005) Fine-scale processes regulate the response of extreme events to global climate change. Proc Natl Acad Sci USA 102:15774–15778. https://doi.org/10.1073/pnas.0506042102

EPA (2014) Overview of particle air pollution ($PM_{2.5}$ and $PM_{10}$). https://www.epa.gov/sites/production/files/2014-05/documents/huff-particles.pdf

Erisman JW, Bleeker A, Hensen A, Vermeulen A (2008) Agricultural air quality in Europe and the future perspectives. Atmos Environ 42:3209–3217. https://doi.org/10.1016/j.atmosenv.2007.04.004

Farmer AM (1993) The effects of dust on vegetation—a review. Environ Pollut 79:63–75. https://doi.org/10.1016/0269-7491(93)90179-R

Fasola S, Maio S, Baldacci S, La Grutta S, Ferrante G, Forastiere F et al (2020) Effects of particulate matter on the incidence of respiratory diseases in the Pisan longitudinal study. Int J Environ Res Public Health 17(7):2540. https://doi.org/10.3390/ijerph17072540

Fatima I (2011). Impact of brick kiln emissions on the ambient air quality and vegetation: a case study of district Budgam. M.Phil dissertation, University of Kashmir. http://dspaces.uok.edu.in:8080/jspui/bitstream/1/1007/1/Irm%20Fatima%2C%202011%20%28M.Phil.%20Dissertation%29.pdf

Forstner U (1998) Pollution control technology in the private sector. In: Integrated pollution control. Springer, Berlin/Heidelberg, pp 49–80. https://doi.org/10.1007/978-3-642-80311-6_2

Gao XJ et al (2006a) Projected changes in mean and extreme precipitation over the Mediterranean region from a high resolution double nested RCM simulation. Geophys Res Lett 33:L03706. https://doi.org/10.1029/2005GL024954

Gao XJ et al (2006b) Projected changes in mean and extreme precipitation over the Mediterranean region from a high resolution double nested RCM simulation. Geophys Res Lett 33:L03706. https://doi.org/10.1029/2005GL024954

Garba ST, Abubakar MA (2018) Source and distribution of the heavy metals: Pb, Cd, Cu, Zn, Fe, Cr, and Mn in soils of Bauchi metropolis, Nigeria. Am J Eng Res 7(2):13–23. http://www.ajer.org/papers/Vol-7-issue-2/B07021323.pdf

Garg OP, Kapoor V (1972) Retardation of leaf senescence by ascorbic acid. J Exp Bot 23:699–703. https://doi.org/10.1093/jxb/23.3.699

Gehring U, Gruzieva O, Agius RM, Beelen R, Custovic A, Cyrys J (2013) Air pollution exposure and lung function in children: the ESCAPE project. Environ Health Perspect 121(11–12):1357. https://doi.org/10.1289/ehp.1306770

Gilbert OL (1968) Bryophytes as indicators of air pollution in the Tyne Valley. New Phytol 67:15–30. https://doi.org/10.1111/j.1469-8137.1968.tb05450.x

Giorgi F, Bates GT (1989) On the climatological skill of a regional model over complex terrain. Mon Weather Rev 117:2325–2347. https://doi.org/10.1175/1520-0493(1989)117<2325:TCSOAR>2.0.CO;2

Giorgi F, Marinucci MR, Visconti G (1992) A 2XCO$_2$ climate change scenario over Europe generated using a limited area model nested in a general circulation model 2. Climate change scenario. J Geophys Res Atmos 97(D9):10011–10028. https://doi.org/10.1029/92JD00614

Giri S, Shrivastava D, Deshmukh K, Dubey P (2013) Effect of air pollution on chlorophyll content of leaves. Curr Agric Res J 1(2):93–98. https://doi.org/10.12944/CARJ.1.2.04

Graffney P, Yu H (2003) Computing agricultural PM$_{10}$ fugitive dust emissions using process specific emission rates and GIS. U.S. EPA Annual Emission Inventory Conference San Diego, California. https://www3.epa.gov/ttn/chief/conference/ei12/fugdust/yu.pdf

Greenpeace, India, (4th July, 2019). Sixth Indian metros are hotspots of air pollutants nitrogen oxide. https://www.greenpeace.org/india/en/press/3715/new-satellite-data-shows-top-polluting-nox-hotspots-in-india-range-from-cities-to-industrial-clusters/. Data adopted on 4 Oct 2019

Habre R, Coull B, Moshier E, Godbold J, Grunin A, Nath A (2014) Sources of indoor air pollution in New York City residences of asthmatic children. J Expo Sci Environ Epidemiol 24(3):269. https://doi.org/10.1038/jes.2013.74

Hickey N, Boscarato I, Kaspar J (2014) Air pollution from mobile sources: formation and effects and abatement strategies. In: Current environmental issues and challenges. Springer, Dordrecht, pp 15–43. https://doi.org/10.1007/978-94-017-8777-2_2

Hinz T (2002) PM in and from agricultural- introduction and overview. Hinz T, Rönnpagel B, Linke S (eds.) Institutfür Technologie und Biosystemtechnik, Bundesforschungsanstaltfür,

Landwirtschaft (FAL) LandbauforschungVolkenrode–FAL J Agric Res Germany, pp 1–6. https://literatur.thuenen.de/digbib_extern/zi026867.pdf

Hirakuchi H, Giorgi F (1995) Multi-year present day and $2XCO_2$ simulations of monsoon-dominated climate over Eastern Asia and Japan with a regional climate model nested in a general circulation model. J Geophys Res 100:21105–21125. https://doi.org/10.1029/95JD01885

Hirano T, Kiyota M, Aiga I (1995) Physical effects of dust on leaf physiology of cucumber and kidney bean plants. Environ Pollut 89(3):255–261. https://doi.org/10.1016/0269-7491(94)00075-O

Huang RJ, Zhang Y, Bozzetti C, Ho KF, Cao JJ, Han Y, Daellenbach KR et al (2014) High secondary aerosol contribution to particulate pollution during haze events in China. Nature 514(7521):218–222. https://doi.org/10.1038/nature13774

Intergovernmental Panel on Climate Change (IPCC) (2007) Climate change 2007: the physical science basis. In: Solomon S et al (eds) Contribution of Working Group I to the fourth assessment report of the Intergovernmental Panel on Climate Change. Cambridge University Press, Cambridge

Invernizzi G, Ruprecht A, Mazza R (2002) Real-time measurement of indoor particulate matter originating from environmental tobacco smoke: a pilot study. Epidemiol Prev 26:30–34. https://moh-it.pure.elsevier.com/en/publications/real-time-measurement-of-indoor-particulate-matter-originating-fr

Invernizzi G, Ruprecht A, Mazza R, De Marco C, Boffi R (2004) Transfer of particulate matter pollution from smoking to non-smoking coaches: the explanation for the smoking ban on Italian trains. Tob Control 13(3):319–320. https://doi.org/10.1136/tc.2004.008433

Isiugo K, Jandarov R, Cox J, Ryan P, Newman N, Grinshpun SA, Reponen T (2019) Indoor particulate matter and lung function in children. Sci Total Environ 663:408–417. https://doi.org/10.1016/j.scitotenv.2019.01.309

Jager AC (2005) Exposure of poultry farm workers to ammonia, particulate matter and microorganisms in the Otchefstroom District, South Africa. PhD Thesis, North-West University, South Africa.

Jain N, Bhatia A, Pathak H (2014) Emission of air pollutants from crop residue burning in India. Aerosol Air Qual Res 14(1):422–430. https://doi.org/10.4209/aaqr.2013.01.0031

Joshi PC, Swami A (2007) Physiological responses of some tree species under roadside automobile pollution stress around city of Haridwar, India. Environmentalist 27:365–374. https://doi.org/10.1007/s10669-007-9049-0

Joshi N, Chauhan A, Joshi PC (2009) Impact of industrial air pollutants on some biochemical parameters and yield in wheat and mustard plants. Environmentalist 29:398–404. https://doi.org/10.1007/s10669-009-9218-4

Karagulian F, Belis CA, Dora CFC, Prüss-Ustün AM, Bonjour S, Adair-Rohani H, Amann M (2015) Contributions to cities' ambient particulate matter (PM): A systematic review of local source contributions at global level. Atmos Environ 120:475–483. https://doi.org/10.1016/j.atmosenv.2015.08.087

Kaushik CP, Ravindra K, Yadav K, Mehta S, Haritash AK (2006) Assessment of ambient air quality in urban centres of Haryana (India) in relation to different anthropogenic activities and health risks. Environ Monit Assess 122(1–3):27–40. https://doi.org/10.1007/s10661-005-9161-x

Keller T, Schwager H (1977) Air pollution and ascorbic acid. Eur J For Pathol 7:338–350. https://doi.org/10.1111/j.1439-0329.1977.tb00603.x

Khaniabadi YO, Goudarzi G, Daryanoosh SM, Borgini A, Tittarelli A, De Marco A (2017) Exposure to PM 10, NO 2, and O 3 and impacts on human health. Environ Sci Pollut Res 24(3):2781–2789. https://doi.org/10.1007/s11356-016-8038-6

Krajickova A, Mejstrik V (1984) The effect of fly-ash particles on the plugging of stomata. Environ Poll 36:83–93. https://doi.org/10.1016/0143-1471(84)90200-9

Krupa SV (1997) Air Pollution, People, and Plants: An Introduction. APS 19(4):397–403

Kunzli N (2002) The public health relevance of air pollution abatement. Eur Respir J 20:198–209. https://doi.org/10.1183/09031936.02.00401502

Lee BK, Dong TT (2011) Toxicity and source assignment of polycyclic aromatic hydrocarbons in road dust from urban residential and industrial areas in a typical industrial city in Korea. J Mater Cycles Waste Manag 13(1):34–42. https://doi.org/10.1007/s10163-010-0287-8

Leepe KA, Li M, Fang X, Hiyoshi A, Cao Y (2019) Acute effect of daily fine particulate matter pollution on cerebrovascular mortality in Shanghai, China: a population-based time series study. Environ Sci Poll Res 26(25):25491–25499. https://doi.org/10.1007/s11356-019-05689-8

Lerman S (1972) Cement-kiln dust and the bean plant (*Phaseolus vulgaris* L. Black Valentine Var.) indepth investigations into plant morphology, physiology and pathology. Ph.D. Thesis, University of California, Riverside. http://www.sidalc.net/cgi-bin/wxis.exe/?IsisScript=catalco.xis&method=post&formato=2&cantidad=1&expresion=mfn=055385

Li P, Xin J, Wang Y, Li G, Pan X, Wang S et al (2015) Association between particulate matter and its chemical constituents of urban air pollution and daily mortality or morbidity in Beijing City. Environ Sci Pollut Res 22(1):358–368. https://doi.org/10.1007/s11356-014-3301-1

Li G, Fang C, Wang S, Sun S (2016) The effect of economic growth, urbanization, and industrialization on fine particulate matter (PM2. 5) concentrations in China. Environ Sci 50(21):11452–11459. https://doi.org/10.1021/acs.est.6b02562

Li Y, Wang Y, Wang B, Wang Y, Yu W (2019) The Response of Plant Photosynthesis and Stomatal Conductance to fine Particulate Matter ($PM_{2.5}$) based on Leaf Factor Analyzing. J Plant Biol 62:120–128. https://doi.org/10.1007/s12374-018-0254-9

Lim SS, Vos T, Flaxman AD, Danaei G, Shibuya K, Adair-Rohani H et al (2012) A comparative risk assessment of burden of disease and injury attributable to 67 risk factors and risk factor clusters in 21 regions, 1990–2010: a systematic analysis for the Global Burden of Disease Study 2010. Lancet 380(9859):2224–2260. https://doi.org/10.1016/S0140-6736(12)61766-8

Lippmann M (2020) Ambient air particulate matter. In: Environmental toxicants: human exposures and their health effects. Wiley, Hoboken, pp 285–365. https://doi.org/10.1002/9781119438922

Liu JC, Mickley LJ, Sulprizio MP, Dominici F, Yue X, Ebisu K et al (2016) Particulate air pollution from wildfires in the Western US under climate change. Clim Chang 138(3–4):655–666. https://doi.org/10.1007/s10584-016-1762-6

Liu L, Yang X, Wang M, Long Y, Shen H, Nie Y et al (2018) Climate change, air quality and urban health: evidence from urban air quality surveillance system in 161 cities of China 2014. J Geosci Environ Protect 6(3):117–130. https://doi.org/10.4236/gep.2018.63011

Madden NM, Southard RJ, Mitchell JP (2008) Conservation tillage reduces PM10 emissions in dairy forage rotations. Atmos Environ 42(16):3795–3808. https://doi.org/10.1016/j.atmosenv.2007.12.058

Maiti SK (1993) Dust collection capacity of plants growing in coal mining areas. IJEP *13*(4):276–280. https://inis.iaea.org/search/search.aspx?orig_q=RN:24072213

Mandal M, Mukherji S (2000) Changes in chlorophyll context, chlorophyllase activity, Hill reaction, photosynthetic CO2 uptake, sugar and starch contents in five dicotyledonous plants exposed to automobile exhaust pollution. J Environ Biol 21:37–41. https://www.cabdirect.org/cabdirect/abstract/20013041584

Marchini T, Zirlik A, Wolf D (2020) Pathogenic role of air pollution particulate matter in cardiometabolic disease: evidence from mice and humans. Antioxid Redox Signal 33(4):263-279. https://doi.org/10.1089/ars.2020.8096

Maynard RL, Howard CV (1999) Particulate matter: properties and effects on health. Garland Publishing, New York. https://doi.org/10.1046/j.1365-2818.2000.00672.x

McCormack MC, Breysse PN, Hansel NN, Matsui EC, Tonorezos ES, Curtin-Brosnan J et al (2008) Common household activities are associated with elevated particulate matter concentrations in bedrooms of inner-city Baltimore pre-school children. Environ Res 106(2):148–155. https://doi.org/10.1016/j.envres.2007.08.012

Meleux F, Solmon F, Giorgi F (2007a) Increase in summer European ozone amounts due to climate change. Atmos Environ 41(35):7577–7587. https://doi.org/10.1016/j.atmosenv.2007.05.048

Meleux F, Solmon F, Giorgi F (2007b) Increase in summer European ozone amounts due to climate change. Atmos Environ 41(35):7577–7587. https://doi.org/10.1016/j.atmosenv.2007.05.048

Mohapatra K, Biswal SK (2014) Effect of Particulate matter (PM) on plants, climate, ecosystem and human health. Int J Adv Technol 2(4):2348–7550. https://www.semanticscholar.org/paper/EFFECT-OF-PARTICULATE-MATTER-(PM)-ON-PLANTS%2C-AND-Mohapatra-Biswal/24d4dfdd07cfbd70bcdb9ef92ec5cc52650cb28a

Morris R, Koo B, Yarwood G (2005) Evaluation of multisectional and two-section particulate matter photochemical grid models in the western United States. J Air Waste Manag Assoc 55(11):1683–1693. https://doi.org/10.1080/10473289.2005.10464764

Muleski GE, Cowherd C Jr, Kinsey JS (2005) Particulate emissions from construction activities. J Air Waste Manag Assoc 55(6):772–783. https://doi.org/10.1080/10473289.2005.10464669

Muzika RM, Guytte RP, Zielonka T, Liebhold AM (2004) The influence of $O_3$, $NO_2$ and $SO_2$ on growth of Picesabies and Fagussylvatica in the Carpathian Mountains. Environ Pollut 130:65–71. https://doi.org/10.1016/j.envpol.2003.10.021

Ndamitso MM, Abdulkadir A, Abulude FO (2016) Total atmospheric deposit source apportionment: a review. Environ Skep Crit 5(4):63–78. http://www.iaees.org/publications/journals/environsc/articles/2016-5(4)/total-atmospheric-deposit-source-apportionment-A-review.pdf

Nenes A, Pandis SN, Pilinis C (1998) ISORROPIA: a new thermodynamics equilibrium model for multiphase multicomponent inorganic aerosols. Aquat Geochem 4:123–152. https://doi.org/10.1023/A:1009604003981

Pant P, Lal RM, Guttikunda SK, Russell AG, Nagpure AS, Ramaswami A, Peltier RE (2019) Monitoring particulate matter in India: recent trends and future outlook. Air Qual Atmos Health 12(1):45–58. https://doi.org/10.1007/s11869-018-0629

Park M, Joo HS, Lee K, Jang M, Kim SD, Kim I et al (2018) Differential toxicities of fine particulate matters from various sources. Sci Rep 8(1):1–11. https://doi.org/10.1038/s41598-018-35398-0

Pereira EG, Oliva MA, Kuki KN, Cambraia J (2009) Photosynthetic changes and oxidative stress caused by iron ore dust deposition in the tropical CAM tree Clusia hilariana. Trees 23(2):277. https://doi.org/10.1007/s00468-008-0275-y

Perera F (2018) Pollution from fossil-fuel combustion is the leading environmental threat to global pediatric health and equity: solutions exist. Int J Environ Res Public Health 15(1):16. https://doi.org/10.3390/ijerph15010016

Petkovsek SAS, Batič F, Lasnik CR (2008) Norway spruce needles as bioindicator of air pollution in the area of influence of the Šoštanj Thermal Power Plant, Slovenia. Environ Pollut 151(2):287–291. https://doi.org/10.1016/j.envpol.2007.06.036

Prajapati SK, Tripathi BD (2008) Seasonal variation of leaf dust accumulation and pigment content in plant species exposed to urban particulates pollution. J Environ Qual 37(3):865–870. https://doi.org/10.2134/jeq2006.0511

Racherla PN, Adams PJ (2006) Sensitivity of global tropospheric ozone and fine particulate matter concentrations to climate change. J Geophys Res 1–11:D24103. https://doi.org/10.1029/2005JD006939

Rai PK (2016a) Biodiversity of roadside plants and their response to air pollution in an Indo-Burma hotspot region: implications for urban ecosystem restoration. J Asia Pac Biodivers 9(1):47–55. https://doi.org/10.1016/j.japb.2015.10.011

Rai PK (2016b) Impacts of particulate matter pollution on plants: implications for environmental biomonitoring. Ecotox Environ Safety 129:120–136. https://doi.org/10.1016/j.ecoenv.2016.03.012

Rai PK, Panda LL (2014) Dust capturing potential and air pollution tolerance index (APTI) of some road side tree vegetation in Aizawl, Mizoram, India: an Indo-Burma hot spot region. Air Qual Atmos Health 7(1):93–101. https://doi.org/10.1007/s11869-013-0217-8

Rai PK, Singh MM (2015) Lantana camara invasion in urban forests of an Indo–Burma hotspot region and its ecosustainable management implication through biomonitoring of particulate matter. J Asia Pac Biodivers 8(4):375–381. https://doi.org/10.1016/j.japb.2015.09.003

Rai PK, Panda LL, Chutia BM, Singh MM (2013) Comparative assessment of air pollution tolerance index (APTI) in the industrial (Rourkela) and non industrial area (Aizawl) of India: An ecomanagement approach. Afr J Environ Sci Technol 7(10) https://www.ajol.info/index.php/ajest/article/view/94959

Rajput M, Agrawal M (2005) Biomonitoring of air pollution in a seasonally dry tropical suburban area using wheat transplants. Environ Monit Assess 101:39–53. https://doi.org/10.1007/s10661-005-9129-x

Rao MV, Dubey PS (1990) Biochemical aspects (antioxidants) for development of tolerance in plants growing at different low levels of ambient air pollutants. Environ Pollut 64(1):55–66. https://doi.org/10.1016/0269-7491(90)90095-T

Ravindra K, Singh T, Mor S (2019) Emissions of air pollutants from primary crop residue burning in India and their mitigation strategies for cleaner emissions. J Clean Prod 208:261–273. https://doi.org/10.1016/j.jclepro.2018.10.031

Repace JL, Lowrey AH (1980) Indoor air pollution, tobacco smoke, and public health. J Sci 208:464–472. https://doi.org/10.1126/science.7367873

Samal AK, Santra SC (2002) Air quality of Kalyani Township (Nadia, West Bengal) and its impact on surrounding vegetation. Indian J Environ Health 44:71–76. https://pubmed.ncbi.nlm.nih.gov/12968728/

Sarkar A, Agrawal SB (2010a) Elevated ozone and two modern wheat cultivars: an assessment of dose dependent sensitivity with respect to growth, reproductive and yield parameters. Environ Exp Bot 69(3):328–337. https://doi.org/10.1016/j.envexpbot.2010.04.016

Sarkar A, Agrawal SB (2010b) Identification of ozone stress in Indian rice through foliar injury and differential protein profile. Environ Monit Assess 161(1-4):205–215. https://doi.org/10.1007/s10661-008-0738-z

Sarkar A, Agrawal SB (2012) Evaluating the response of two high yielding Indian rice cultivars against ambient and elevated levels of ozone by using open top chambers. J Environ Manage 95:S19–S24. https://doi.org/10.1016/j.jenvman.2011.06.049

Sarkar A, Rakwal R, Bhushan Agrawal S, Shibato J, Ogawa Y, Yoshida Y et al (2010) Investigating the impact of elevated levels of ozone on tropical wheat using integrated phenotypical, physiological, biochemical, and proteomics approaches. J Proteome Res 9(9):4565–4584. https://doi.org/10.1021/pr1002824

Sarkar A, Singh A, Agarawal SB (2012) Utilization of fly ash as soil amendments in agricultural fields on North-Eastern gangetic plains of India: potential benefits and risks assessments. Bull Nat Inst Ecol 23(1–2):9–20

Sarkar A, Singh AA, Agrawal SB, Ahmad A, Rai SP (2015) Cultivar specific variations in antioxidative defense system, genome and proteome of two tropical rice cultivars against ambient and elevated ozone. Ecotoxicol Environ Saf 115:101–111. https://doi.org/10.1016/j.ecoenv.2015.02.010

Sarma B, Chanda SK, Bhuyan M (2017) Impact of dust accumulation on three roadside plants and their adaptive responses at National Highway 37, Assam, India. Trop Plant Res 4(1):161–167. https://doi.org/10.22271/tpr.2017.v4.i1.023

Sharma B, Sarkar A, Singh P, Singh RP (2017) Agricultural utilization of biosolids: A review on potential effects on soil and plant grown. J Waste Manag 64:117–132. https://doi.org/10.1016/j.wasman.2017.03.002

Singh GK, Gupta K, Chaudhary S (2014) Solid waste management: its sources, collection, transportation and recycling. Int J Environ Sci Dev 5(4):347. https://doi.org/10.7763/IJESD.2014.V5.507

Singh S, Singh R, Mishra AK, Upadhyay S, Singh H, Raghubanshi AS (2018) Ecological perspectives of crop residue retention under the conservation agriculture systems. Trop Ecol 59(4):589–604. https://www.tropecol.com/pdf/open/PDF_59_4/5%20Singh%20et%20al.pdf

Singh V, Singh P, Singh R (2020) Sustainable strategies for rice-straw management from South Asian countries: a book review. J Clean Prod (in press). https://doi.org/10.1016/j.jclepro.2020.122439

Skinder BM, Sheikh AQ, Pandit AK, Ganai BA (2014) Brick kiln emissions and its environmental impact: a review. J Ecol 6(1):1–11. https://doi.org/10.5897/JENE2013.0423

Someshwar AV, Pinkerton J (1992) Wood processing industry. Air pollution engineering manual, Air and Waste Management Association. Van Nostrand Reinhold, New York, pp 835–849. https://p2infohouse.org/ref/33/32354.pdf

Swanepoel AJ, Rees D, Renton R, Swanepoel C, Kromhout H, Gardiner K (2010) Quartz exposure in agriculture: literature review and South African survey. Ann Occup Hyg 54(3):281–292. https://doi.org/10.1093/annhyg/meq003

Tagaris E, Manomaiphiboon K, Liao KJ, Leung LR, Woo JH, He S, Amar P, Russell AG (2007) Impacts of global climate change and emissions on regional ozone and fine particulate matter concentrations over the United States. J Geophys Res 1–11:D14312. https://doi.org/10.1029/2006JD008262

Tagaris E, Liao KJ, Delucia AJ, Deck L, Amar P, Russell AG (2009) Potential impact of climate change on air pollution-related human health effects. Environ SciTechnol 43(13):4979–4988. https://doi.org/10.1021/es803650w

Tainio M, Katarzyna JR, Reizer M, Warchałowski A, Trapp W, Skotak K (2013) Future climate and adverse health effects caused by fine particulate matter air pollution: case study for Poland. Reg Environ Change 13:705–715. https://doi.org/10.1007/s10113-012-0366-6

Terzaghi E, Wild E, Zacchello G, Cerabolini BEL, Jones KC, Di Guardo A (2013) Forest filter effect: role of leaves in capturing/releasing air particulate matter and its associated PAHs. Atmos Environ 74:378–384. https://doi.org/10.1016/j.atmosenv.2013.04.013

Tesche TW, Morris R, Tonnesen G, McNally D, Boylan J, Brewer P (2006) CMAQ/CAMx annual 2002 performance evaluation over the eastern US. Atmos Environ 40(26):4906–4919. https://doi.org/10.1016/j.atmosenv.2005.08.046

The Times of India (2014) Delhi has the worst air pollution in the world: WHO. https://timesofindia.indiatimes.com/home/environment/pollution/Delhi-has-the-worst-air-pollution-in-the-world-WHO/articleshow/34791079.cms

Traverso M, Rizzo G, Finkbeiner M (2010) Environmental performance of building materials: life cycle assessment of a typical Sicilian marble. Int J Life Cycle Assess 15(1):104. https://doi.org/10.1007/s11367-009-0135-z

US Environmental Protection Agency Greenhouse Gas Emissions. Available online: https://www3.epa.gov/climatechange/ghgemissions/

Vedachalam D (2019) A burning issue: How sustainable agriculture is helping mitigate air pollution in South Asia. CIMMYT, International Maize and Wheat Improvement Centre, News Report, June 4, 2019. https://www.cimmyt.org/multimedia/a-burning-issue/

Wagh ND, Shukla VP, Sarika BT, Ingle ST (2006) Biological monitoring of road-side plants exposed to vehicular pollution in Jalgaon city. J Environ Biol 27(2):419–421. http://www.jeb.co.in/journal_issues/200604_apr06_supp/paper_21.pdf

WHO (2005) Air quality guidelines for particulate matter, ozone, nitrogen dioxide and sulfur dioxide https://apps.who.int/iris/bitstream/handle/10665/69477/WHO_SDE_PHE_OEH_06.02_eng.pdf?sequence=1

WHO (2013) Research for universal health coverage. https://www.who.int/whr/2013/report/en/

WHO (2014) Burden of disease from household air pollution for 2012. https://www.who.int/phe/health_topics/outdoorair/databases/FINAL_HAP_AAP_BoD_24March2014.pdf?ua=1

WHO (2016a) Exposure to ambient air pollution from particulate matter for 2016. https://www.who.int/airpollution/ambient/AAP_exposure_Apr2018_final.pdf?ua=1

WHO (2016b) Global warming of 1.5 °C. https://report.ipcc.ch/sr15/pdf/sr15_spm_final.pdf

WHO (2016c) How air pollution is destroying our health. https://www.who.int/airpollution/news-and-events/how-air-pollution-is-destroying-our-health

WHO (2018) 9 out of 10 people worldwide breath polluted air, but more countries are taking action. https://www.who.int/news-room/detail/02-05-2018-9-out-of-10-people-worldwide-breathe-polluted-air-but-more-countries-are-taking-action

Wijayratne UC, Scoles-Sciulla SJ, Defalco LA (2009) Dust deposition effects on growth and physiology of the endangered Astragalus jaegerianus (Fabaceae). Madroño *56*(2):81–88. https://doi.org/10.3120/0024-9637-56.2.81

Xiao FG, Shen L, Ji HF (2011) On photoprotective mechanisms of carotenoids in light harvesting complex. Biochem Bioph Res CO 414(1):1–4. https://doi.org/10.1016/j.bbrc.2011.09.049
Yang S, He H, Lu S, Chen D, Zhu J (2008) Quantification of crop residue burning in the field and its influence on ambient air quality in Suqian, China. Atmos Environ 42(9):1961–1969. https://doi.org/10.1016/j.atmosenv.2007.12.007
Yao X, Chan CK, Fang M, Cadle S, Chan T, Mulawa P et al (2002) The water-soluble ionic composition of PM2. 5 in Shanghai and Beijing, China. Atmos Environ 36(26):4223–4234. https://doi.org/10.1016/S1352-2310(02)00342-4

# Chapter 5
# Dyes from Textile Industry Wastewater as Emerging Contaminants in Agricultural Fields

**Jatinder Singh, Parnika Gupta, and Anamika Das**

**Abstract** Industrialization has led to many devastating effects on the nature. Globally, textile industries are playing a vital role in generating high levels of toxic and recalcitrant compounds in the form of dyes in textile based industrial effluent. They are responsible for the disastrous effect on plants, humans and varied environmental components. Toxicity and carcinogenicity is totally dependent upon the quantity of various contaminants. It has been established by the researchers that dyes are recalcitrant and refractory pollutants that constitute a significant burden on the environment. So, it is need of the hour to understand the toxic effect on plants and further how mankind can be affected.

The chapter focuses on the occurrence, physicochemical behaviour, potential sources and significant routes of dyes from textile industry wastewater, and its toxic effects on plant's growth. Later the comparative accounts of different conventional vs. non-conventional remediation strategies are discussed briefly. This chapter has also presented the critical role played by immobilized oxidoreductase enzymes for management of dye contamination together with the effect of dyes released from textile wastewater on important crops as mostly agricultural plants are affected on a larger scale as they use dye wastewater for irrigation purpose which might be contaminated with many emerging contaminants. So the effluent must be treated before its release into the environment. Various physical, chemical and biological treatment processes are applied to eliminate the harmful components so as to transform or mineralize the contaminants into non-toxic forms.

**Keywords** Bioremediation · Biocatalysts · Contaminants · Dyes · Immobilization · Toxicity

---

J. Singh (✉) · P. Gupta
Department of Environmental Science and Technology, Central University of Punjab, Bathinda, India
e-mail: jatinderbiotech@gmail.com

A. Das
Department of Paramedical Sciences, Guru Kashi University, Bathinda, India


V. Kumar Singh et al. (eds.), *Sustainable Agriculture Reviews 50*, Sustainable Agriculture Reviews 50, https://doi.org/10.1007/978-3-030-63249-6_5

## Abbreviations

MoEF Ministry of Environment and Forests

## 5.1 Introduction

Rapid increment in industrialization and human population has prompted the arrival of undesirable substances in the nature causing contamination of habitat and toxicity to living creatures. W. H. Perkin found the primary engineered natural dye mauve (or aniline), in 1956. It is evaluated that as much as 10,000 tons of various manufactured dyes and shades are utilized in the textile business and more than 70,000 tons of engineered dyes are delivered each year around the world (Parshetti et al. 2006; Daneshvar et al. 2007; Aftab et al. 2011). Enterprises are the key players in the economy of numerous countries but at the same time are the significant polluters worldwide because of their possibly poisonous wastewater, which contains an assortment of natural and inorganic contaminants and which along these lines causes ecological contamination and toxicity in living creatures upon exposure (Bharagava and Mishra 2018). The usage of synthetic chemical dyes in numerous industrial processes like paper and pulp manufacturing, cloth dyeing, leather treatment, plastics and printing have increased manifolds over the last few years, thereby releasing the dye-containing industrial effluents into the aquatic and soil ecosystems (Aksu 2005). Dyes and dyestuffs are mostly utilized for cosmetics, foods, textiles, leather and pharmaceutical ventures. More than 10,000 financially accessible dyes exist and more than $7 \times 10^5$ tonnes of dyestuffs are produced every year.

Industries utilize enormous amounts of various engineered synthetic mixtures (predominantly dyes) for different purposes, including as a dyeing and finishing agent. Various manufactured dyes are utilized vigorously in various ventures and incorporated as azo, triphenylmethane, anthraquinone, phthalein, nitro, methane, and quinoline dyes (Khan et al. 2013; Saxena and Bharagava 2016). Therefore, these poisonous dyes are released alongside various industrial wastewaters into the regular natural ecosystem, including streams, lakes, ponds, and soil, and subsequently make ecological contamination. The wastewater-containing natural and inorganic toxins additionally reinforce the development of an assortment of pathogenic microscopic organisms, which likewise cause threat to the wellbeing of living creatures upon introduction (Saxena and Bharagava 2015; Mani and Bharagava 2016).

It is evaluated that roughly 200,000 tons of wastewater are created each year during washing, colouring, bleaching, and completing activities in the textile business. The wastewater unfavourably affects the water quality and chemistry, including colour, pH, biological oxygen demand, total organic carbon, chemical oxygen demand, total suspended solids, total nitrogen, total solids and an assortment of natural and inorganic contaminations (Senthilkumaar et al. 2006; Shengfang 2010; Mani and Bharagava 2016). A few enterprises are likewise producing a colossal measure of dye containing wastewater, for example, the material, paper, leather, and pharmaceutical enterprises (Saratale et al. 2006; Arulazhagan 2016). The textile industry releases wastewater with some profoundly environmentally hazardous substances including heavy metals and metalloids like As, Cr, Zn, Cd, Cu, Mn, and Co, which have mutagenic and cancer causing impacts (Kabra et al. 2011; Ambrosio et al. 2012).

The dye containing wastewater released from textile industries into the aquatic system causes decreased access of sunlight into water bodies, thereby diminishes the dissolved oxygen content leading to undesirable influences on the aquatic life by exerting negative effect on zooplanktons, phytoplanktons, amphibians and other life forms (Garg and Tripathi 2017). Dyes are of both natural and synthetic origin and are broadly utilized in the coloured material, synthesis, cosmetics, paper industries, foods, and pharmaceutical industries due to their ability to give simplicity of creation, solidness, in comparison to other naturalistic dyes. Thus, dyes are the compounds which are applied as shading specialists as a result of their ability of all time dyeing of the silk/filaments and in light of the fact that they are profoundly impervious to light, water, cleansers, oxidizing specialists, acids, soluble operators, and microbial activity (Saratale et al. 2011; Shah et al. 2013).

Further, as indicated by a yearly report by Union Ministry of Environment and Forests (MoEF), 13011 industrial units have delivered about 4.4 million tons of toxic dyes which squanders over more than 373 districts of the nation (Pointing 2001). Enterprises release around one ton of wastewater in day to day life through dyeing processes. As indicated by an estimate, by the end of the nineteenth century, around 10,000 engineered dyes have been found and utilized in industry for different purposes. Textile industries expend a huge volume of consumable water for various dyeing applications and, hence, release an immense amount of dye containing wastewater into nature, causing environmental contamination and threat to different life forms. Hence, it is basic to sufficiently get the wastewater treated before release to ensure the natural and general wellbeing. Physico-chemical methodologies are at present being applied for the treatment of wastewater by industrial units. These are compelling, yet environmentally dangerous as they use a colossal measure of poisonous synthetic substances and salts, which at that point end up in the environment and bestow a negative impact on the condition just as creating a colossal amount of secondary pollutants (Zhang et al. 2004; Pandey et al. 2007). Remediation processes has mostly exploited the biological methods which involved microbes and plants due to their non-toxic nature. However, bioremediation approaches are promising eco-friendly strategies using an assortment of microorganisms for natural clean-up and in this manner are seen as a minimal effort technique for the treatment of the industrial effluents (Forgacs et al. 2004; Saratale et al. 2006; Shah et al. 2013; Mani and Bharagava 2016).

This chapter focuses on the dye characterisation and the potential sources of these dyestuffs. Further in this chapter, the fate and transport of these dye laden effluents into the water bodies and eventually into the agricultural fields have been discussed, because those natural entities of water serve as a source of irrigation where those harmful emerging contaminants are also incorporated into the agricultural ecosystem. This is followed by the crop toxicity of those dyes and its toxicity mechanism. Lastly, the chapter throws light on the process of biodegradation and removal strategies of dyes. Along these lines, this chapter, thus, primarily centres on the harmful profile of dyes and bioremediation approaches for dye containing wastewater for maintaining natural integrity and homoeostasis.

## 5.2 Fate and Transport of Dye Contaminants in Agricultural Fields

The dyes employed are of two major types viz. natural and synthetic. The natural dyes are produced from renewable resources and are biodegradable, thereby less toxic and less allergenic as compared to synthetic dyes. Some natural dyes include marigold, safflower, weld, onion, morinda, quercetin, myrobolan, and turmeric dyes. Synthetic dyes are the organic dyes that are basically obtained from petroleum, coal tar derivatives, and every so often, a combination of mineral components extracted from benzene and its derivatives. Synthetic dyes and dyestuffs are widely used in many technological fields and industries such as food, pharmaceutical, leather, textile and due to this vast usage, the exact amount of dyes produced worldwide are unknown, but is roughly around 10,000 tons per year. The data representing the exact amount of dyes discharged as effluent is also not known but out of the produced dye, nearly 11% is released in effluents during the process of manufacture and application. Moreover, these synthetic textile dyes are so chemically stable and recalcitrant that traditional treatment technologies stand no chance. These highly stable reactive dyes, thus enter into the environment in the form of tinted wastewater, thereby making it difficult to treat the effluents released from these industries by implying common biological, chemical and physical methods due to its high colour intensity, heat, biological oxygen demand, chemical oxygen demand, pH and presence of several toxic metal ions (Senthilkumaar et al. 2006; Shengfang 2010; Mani and Bharagava 2016).

As depicted in the Fig. 5.1, the major source of water in agriculture fields is from the water bodies which are also the end points for effluents. The dye manufacturing industrial units and the sectors utilizing that dyestuff release the effluents into those water bodies. Apart from this, the dye particles which find their way into the atmosphere from these industries also combine with rain and join the water bodies (Choudhury 2017). These dyes and chemicals are then used as part of irrigation medium and proved to be highly toxic to soil microbial communities, plant growth and germination (Rehman et al. 2018). Dye effluents are exceedingly coloured, contain numerous toxic chemicals such as chlorine, formaldehyde, solvent, organic and inorganic compounds, aromatic amines, xenobiotics, pigments, alkali salts, and toxic heavy metals like lead, chromium, and mercury (Mishra and Bharagava 2016; Chowdhary et al. 2017; Yadav et al. 2017; Bharagava and Mishra 2018).

The solid wastes are cleared by numerous industries into water bodies such as ponds, lakes, rivers, streams, etc., where they show detrimental effects on water and soil ecology and bring about water and soil pollution as well as ecotoxicity in the

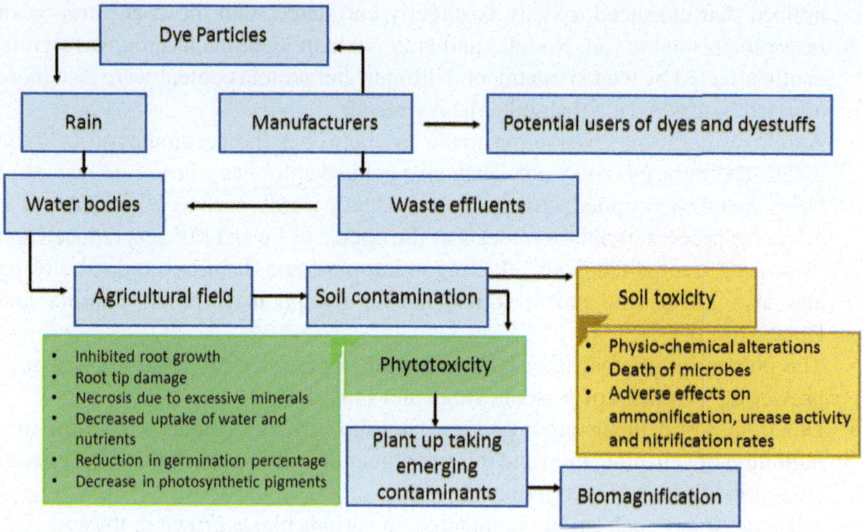


**Fig. 5.1** Fate and transport of dye contaminants in agricultural fields

environment (Kagalkar et al. 2010; Mani and Bharagava 2017). These contaminants in context of agriculture then become a part of food chain since plants uptake them. Moreover, they cause soil toxicity. These recalcitrant textile dyes significantly compromise photosynthesis, inhibit plant growth, are bioaccumulated, enter the food chain, thereby playing a crucial role in increasing mutagenicity, and induce carcinogenicity and toxicity (Aquino et al. 2014). In addition, dye effluents are also accountable for distressed biogeochemical (nutrient) cycling, which ensues in soil niches, and thus, creates soil contamination.

## 5.3 Toxic Effects and Mechanism of Toxicity

Textile as well as paper industry effluents contain organic and inorganic chemicals, balance of which may affect plant growth adversely. The effluents from textile industries comprise of numerous dyes, heavy metals, total dissolved solid and organic compounds such as plasticizers, cleaning solvents, etc. The wastewater also shows high biological and chemical oxygen demand due to numerous contaminants (Jaishree 2015). This is a matter of huge concern as the water from those water bodies is used for various purposes.

The components of effluent released from textile industry have been evaluated by various researchers for its impact on important plants. Such as:

- A study conducted by Athar and Ahmad (2002) on chickpea (*Cicer arietinum* L.) and green gram (*Vigna radiata* (L.) Wilczek) using heavy metals in the soil showed toxic effect on the dry matter yield of the legumes of test plants. They noticed that enhanced toxicity is directly correlated with the concentration of heavy metals in the soil. Nodules and grain yield of leguminous crop was significantly affected by the Cd treatment. Nitrogen and protein content were also more affected by Cd and Cu than other heavy metals.

- Another significant observation made by them was the accumulation of toxic metals and reduced crop yield. They also noticed chlorosis when combination of heavy metal was applied to the test plant. Heavy metals such as Pb is reported to affect the process of photosynthesis as the uptake of Fe and Mg gets reduced and these are essential elements affecting chloroplast and disturbs stomatal closing and also affects the enzymes responsible for photosynthesis (Sharma and Dubey 2005).
- The process of photosynthesis is decreased in presence of cadmium as it suppresses the synthesis of protein (Singh and Prasad 2014).
- Due to loss of cellular turgor pressure, mitotic activity is reduced resulting into inhibition of cell elongation and this phenomenon is the reason behind the growth reduction of plants (Gabbrielli et al. 1990).
- There are many such studies conducted on various plants grown in the soil contaminated with toxic materials by several researchers. The toxic metals accumulated in plants (Sulabha and Nandkar 2011) affect the nutritional quality. The toxicity is not limited to plants as via food chain further these affect human beings and domestic animals. Sometimes the domestic animals at several places consume water from nearby waterbodies which are being contaminated with textile wastewater. So, this is a matter of grave concern to look upon the disposal of textile wastewater. This water must be treated onsite up to the level of safe consumption.
- The foremost plant organ interacting with toxic metals are the roots and root tip is a crucial damageable site, leading to repressed root growth, stunted root system and declined yield from decreased water and nutrient uptake. The excessive nutrients in the effluents might also cause necrosis which eventually results in inhibition of root growth (Becker 2000). The toxic effects of dye on physiology and morphology of plants are illustrated in Fig. 5.2.

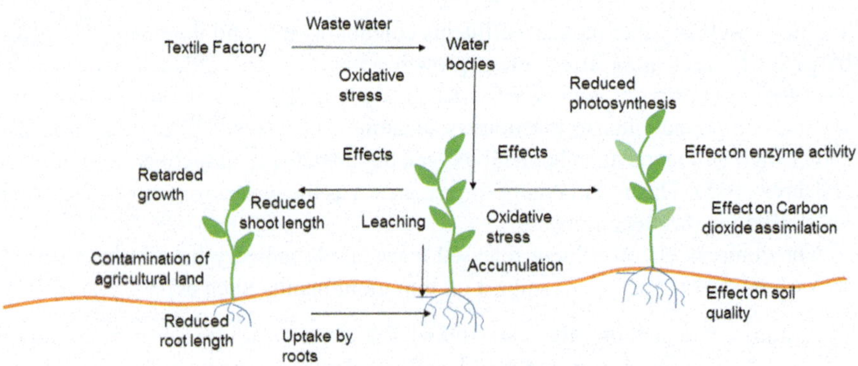


**Fig. 5.2** Toxic effects of dye on physiology and morphology of plants

- The industrial release reduces the germination percentage of ladyfinger (*Abelmoschus esculentus*) and kidney beans (*Phaseolus aureus*) (Mohammad and Khan 1985).
- Photosynthetic pigments such as chlorophyll 'a', 'b' and total chlorophyll are also decreased in certain plants subjected to untreated effluent water (Rehman and Bhatti 2009).
- Apart from this, dye pollutants are also noted to exert hazardous impacts on the soil. Certain dyes have the capabilities of altering the physical and chemical properties of the soil. Dyes display varying degree of stability in soils ranging from few days to several weeks depending upon the nature of chemical compounds. Irrigation with effluents from dye industries leads to considerable accumulation of total organic dyes in cultivated soil and accumulated dyes has also been reported to be transported in plant tissues (Zhou 2001).
- Phytotoxicity of dyes has been described by some researchers (Kalyani and Patil 2008; Ayed et al. 2011). There are direct evidences of dye uptake by plants (Uera et al. 2007; Muthunarayanan et al. 2011).
- Toxic nature of the dyestuffs eventually leads to the death of the soil microorganisms, thereby hampering the agricultural productivity. Certain azo-dyes adversely affect the growth of atmospheric $N_2$ fixing cyanobacterium *Anabaena* sp. (Hu and Wu 2001) and a substantial adverse consequence on ammonification, nitrification rates in the soil and urease activity was observed due to sulfonated azo-dye which lessens nitrogen use efficacy in crop production (Topac et al. 2009). These instances reveal considerable toxicity of dye pollutants to soil microorganisms and agriculturally important nutrient cycling processes.
- Oxidative stress, because of chromium in textile dyes, is another conundrum associated with recalcitrant character, posing a significant damage to the growth and development of plants, especially to $CO_2$ assimilation and photosynthesis (Copaciu et al. 2013).

## 5.4 Degradation Strategies of Dyes

### *5.4.1 Physical and Chemical Methods*

Physical and chemical methods used in dye degradation is summarized in Table 5.1. The methods are used individually and in combination too. The physical and chemical treatment methods are not popularly used and cannot be up-scaled to the field due to various limitations like cost, release of toxic and carcinogenic residues in the environment and production of chemical sludge (Gogate and Pandit 2004).

The physical and chemical methods both are accompanied by certain advantages and disadvantages. The chemical process of ozonation and Fenton's reagent both involve the method of oxidation reaction. The former shows no sludge formation as compared to the latter, which is effective for both soluble and insoluble dyes.

**Table 5.1** Physical and chemical dye treatment methods with their advantages and disadvantages

| S. No | Chemical/Physical methods | Method description | Advantages | Disadvantages |
|---|---|---|---|---|
| 1. | Ozonation | Oxidation reaction occurs in the presence of ozone gas | No production of sludge | Expensive with short life span |
| 2. | Fenton's reagent | $H_2O_2$–Fe(II) mainly use in oxidation reaction | Use for the effective treatment of soluble and insoluble dyes | Generates sludge |
| 3. | Sodium hypochlorite oxidation | Use NaClO at pH 7.0 for colour removal | It helps in the initiation and acceleration of cleavage of azo-bond | Produces carcinogenic by products |
| 4. | Photochemical | Reaction carried out in $H_2O_2$–UV | Zero sludge formation | Produces toxic by products |
| 5. | Electrochemical destruction | Reaction carried out in the presence of electricity | Release of non-hazardous compounds | Expensive |
| 6. | Adsorption | Bacterial, fungal, yeast and cellulose biomass, activated carbon, chitin and soil material | Use in the degradation of extensive variety of dyes | Expensive |
| 7. | Membrane filtration | Based on separation method | Efficient for every type of dye | Costly and need to replace membrane |
| 8. | Ion exchange | Based on ion exchange method | Regeneration of resins | Expensive and for specific dyes |
| 9. | Electrokinetic coagulation | Used ferrous sulphate and ferric chloride salts | Cost effective | Production of huge amount of sludge |
| 10. | Irradiation | Adequate amount of dissolved oxygen is required | Appropriate for all types of dyes | Lucrative |
| 11. | Sonication | Ultrasound waves are used | Complete and increase in rate of under spent dye bath conditions | High concentration of dye decreases its efficiency |
| 12. | Photocatalytic ($TiO_2$) | This method uses $TiO_2$ catalysts | $TiO_2$ supported on absorbents is more efficient | Generates vicious reactive radicals |

Source: Robinson et al. (2001)

Sodium hypochlorite oxidation method releases carcinogenic by-products which include organochlorines (Banat et al. 1999). Photochemical dye degradation method also does not produce sludge but do produce certain toxic by-products (Peralto-Zamora et al. 1999). Electrochemical destruction, membrane filtration, irradiation and adsorption have a wide scope but are expensive. On the contrary, ion exchange results in resin regeneration but is not suitable for all types (Pelegrini et al. 1999). Sonication completely decolourizes and increases the rate under spent dye bath conditions but efficiency decreases with increase in dye concentration. Lastly the photocatalytic method shows over 95% colour removal, but generates the harmful reactive radicals. Microbial remediation can best replace these physical and chemical methods for degradation of azo-dyes (Verma and Madamwar 2003).

### 5.4.2 *Biodegradation*

The textile dyes significantly compromise the water and soil ecology and in spite of this, the bioremediation of these emerging contaminants is feasible. Biodegradation is well-thought-out as one of the most effective and environment conscious mechanisms for the removal of emerging contaminants. Bioremediation can be defined as "*conversion or mineralization of harmful emerging contaminants by the enzymatic action of bacteria, plants, fungi and other biological agents*". Despite some disadvantages, these biodegradative techniques are quite positive and can be further improved by modern biotechnological methods. Bioremediation with the help of microorganisms provides the reduction, deduction or removal of harmful contaminants by microbial entities such as algae, bacteria, fungi and yeasts in wastewater, sludge, effluent or soil (Das and Dash 2017). One method is *in situ* bioremediation by using bioaugmentation or biostimulation comprising of the inoculation of exogenous microorganisms and introducing nutrients to favor the local microorganisms, respectively (Kasai 2011).

On the other end is the *ex situ* bioremediation by land farming and composting, hybrid crops, bioreactors, and genetically engineered organisms (Das and Dash 2014). Moreover, it is feasible to choose and separate wild microorganisms from the textile effluent, which have potential to degrade these chemical dyes (Kandelbauer and Guebitz 2005). Another prospect includes identification, isolation, cloning and transfer of genes encoding degradative enzymes which increase biodegradation capacity of native species. The engineered or hybrid strains (Kandelbauer and Guebitz 2005) are called super-degrading microorganisms (Pereira and Alves 2012).

#### 5.4.2.1 Microbial Degradation

The decolourization of dyes has long been a chief target of an extensive variety of dye effluents treatment processes. A huge amount of dye effluents discharged from various industrial units into the natural aquatic systems encompasses suspended solids and toxic chemicals with intense dark colour. As a result, this ultimately hinders the photosynthesis of aquatic plants and algae and affects other living organisms. The discharged dye containing effluents also contain a diversity of organic and inorganic pollutants like toxic metals, which results in serious water and soil pollution and toxicity in living beings (Asad et al. 2007; Aftab et al. 2011).

Biological methods for the decolourization of dye effluents is an excellent approach as compared with other physico-chemical treatment approaches because of their ease of application, environment friendly and inexpensive nature, and generation of non-toxic sludge with complete mineralization (Saratale et al. 2009a, b; Levin et al. 2010; Arulazhagan 2016). Biological approaches employ a variety of microbes such as bacteria, fungi, yeasts, actinomycetes, algae, and plants for the treatment of various dye effluents for environmental safety.

#### 5.4.2.2 Biodegradation by Bacteria

A bacterium is capable of degrading textile dyes under both aerobic and anaerobic conditions (Telke et al. 2015). As far as azo-dyes are concerned, bacterial degradation includes the reduction cleavage of azo-bonds (-N=N-) with the help of azo-reductase enzyme into aromatic amines with no colour and are potentially non-toxic to natural environment (Fig. 5.3) (Saratale et al. 2011).

The process of biodegradation includes the involvement of enzymes like laccase, tyrosinase, peroxidase, MG reductase and NADH-DCIP reductase (Tan et al. 2010). The bacterial biomass is excellent biosorbent material for the bioremediation of the textile dyes as they are the source of carbon and nitrogen (Roy et al. 2018). This involves complex mechanisms between living and dead cell biomass and textile dyes. The processes included are, for instance, adsorption, where there is an interaction between the textile dye molecules and the chemical groups located on the bacterial cell surface possessing electrical charges (Srinivasan and Viraraghavan 2010). The major drawback of bacterial biosorption is the adsorption capacity and the ultimate disposal of the biomass along with the pretreatment of the dyes and their effluents (Srinivasan and Viraraghavan 2010). However, there are systems with low operating cost and the likelihood of using hybrid adsorbent systems with greater efficiency for the purpose of treating textile dye effluents (Wawrzkiewicz et al. 2017).

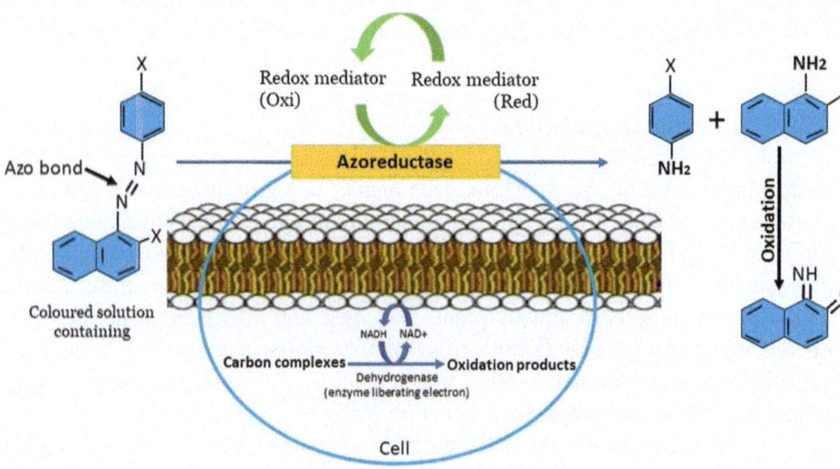


Fig. 5.3 Mechanism of azo-dye degradation by azo-reductase enzyme

The hybrid bacterial cultures or the co-cultures can provide higher biodegradation levels particularly since textile dye molecules can be attacked at different positions (Chandra 2016). For instance, the sulfated textile dye, HE4BD is degraded at a much elevated level by using the co-culture of *Proteus vulgaris* and *Micrococcus glutamicus*. The action of the oxidoreductive enzymes existing in co-culture leads to the formation of smaller molecular weight intermediates (Saratale et al. 2010). Likewise, the consortium of *Pseudomonas aeruginosa*, *Ochrobactrum* sp., and *Providencia vermicola* are reported to degrade red violet colour textile effluent of diverse composition with better results of biodegradation (Vijayalakshmidevi and Muthukumar 2015). The bacterial biodegradation of azo-dyes usually displays a higher rate of removal and mineralization of the toxic components. Along with this, the process is economically viable and produces very little sludge (Tomei et al. 2016). Moreover, the bacterial degradation process is faster than that performed by fungi but only with the absolute necessary and constant monitoring of the actual toxicity of the attained compounds and to govern various parameters involved such as oxygen levels and agitation, pH, temperature and concentration of the textile dye as well as the type and concentration of the substrate used (Saratale et al. 2011).

#### 5.4.2.3 Biodegradation by Extremophiles

The extremophiles are the organisms which can grow normally in conditions which are considered inhospitable, according to most eukaryotic organisms because of greater physical-chemical stress tolerance it encounters. The extremophiles and their extremozymes are amongst the most effective and striking bioremediation tools, especially in industrial sectors including textile industries which involve stressful conditions of temperature, pH, toxicity and salinity. In textile industries, dyeing process involves the usage of various salts which lay significant difficulties and complications for microbial communities. This is because high saline concentrations can cause plasmolysis, i.e. loss of water from the cytoplasm, thereby contracting in a way that plasma membrane slowly and eventually separates from the cell wall (Meng et al. 2012). Therefore, as an effective substitute, halotolerant microorganism can be utilized to biodegrade azo dyes in media with high salt concentration. The bacterial community comprising of *Sedimentibacter*, *Bacillus*, *Pseudomonas*, *Streptomyces* and *Clostridiales* exhibit excellent performance in the discolouration process under high concentrations of Reactive Brilliant Red X-3B dye and the whole discolouration happens in the presence of high concentrations of sodium chloride (Tan et al. 2009). Similarly, the bacterium, *Shewanella aquimarina* discolours the Acid Red 27 dye in NaCl medium (Meng et al. 2012). *Scheffersomyces spartinae* yeast can biodegrade aerobically and with high salinity, the Acid Scarlet 3R dye through azo-reduction, deamination and desulphurization. Similarly, another yeast *Picchia occidentalis* shows the same result in case of Acid Red B dye (Song et al. 2017). Another type of extremophiles includes thermophiles which can develop and sustain themselves at high temperatures. For instance, the species *Anoxybacillus pushchinoensis*, *Anoxybacillus flavithermus* and *Anoxybacillus kamchatkensis* demonstrated discolouration of Reactive Black 5 dye at high temperatures of 65 °C (Deive et al. 2010).

Certain organisms can sustain well in two or more than two extreme environments and are hence called polyextremophiles, for example, the bacterium *Bacillus pallidus* which degrades the textile effluents having pH between 9.3 and 10 and temperature of 60–65 °C (Paar et al. 2001). Thus, polyextremophiles and extremophiles can become a crucial part of research in bioremediation of textile effluents.

#### 5.4.2.4 Degradation by Fungi

A varied group of fungi can effectively degrade or decolourize a variety of complex dye substances into non-toxic metabolites (Fu and Viraraghavan 2001). Most fungal species can secrete large variety of extracellular ligninolytic enzymes such as manganese peroxidase, lignin peroxidase and laccase (Gomi et al. 2011) which are responsible for degradation of several dye effluents. *Pleurotus ostreatus*, *Penicillium* species, *Pichia* species and *Candida tropicalis* are proficient in decolourization of different dyes (Ali et al. 2009; Zhuo et al. 2011).

At present, white rot fungi are an exclusive group of fungal organisms, which has a strong capability for the degradation and/or decolourization of lignin at a large scale and has extracellular and non-specific enzymes that are implicated in the degradation of several persistent compounds and lignin (Weisburger 2002; Abedin 2008; Yang et al. 2009). Quite a large group of white rot fungi, for instance *Trametes versicolor*, *Irpex flavus*, *Pleurotus ostreatus*, *Pycnoporus sanguineus*, *Phellinus gilvus* and *Phanerochaete chrysosporium* can degrade various textile dyes such as azo, indigoid, and triphenylmethane dyes as well as heterocyclic dyes (Pajot et al. 2010; Khan et al. 2012). Recently, Singh et al. (2020) reported improved azo-dye degradation by co-culturing technique using two white rot fungal strains *Phanerochaete chrysosporium* and *Trametes versicolor.* Mycoremediation is, therefore, a natural, cost-effective and environmentally safe process.

#### 5.4.2.5 Degradation by Yeasts

Yeast is extensively reported in varied habitats such as soil, vegetation, and aquatic ecosystems. Yeast represents attractive features in comparison with bacteria and algae. Yeasts are reasonably priced and easily available source of biomass which can be cultivated faster than most filamentous fungi. The ability to endure adverse environmental conditions such as temperature extremes, minimal nutrient availability and pH variations as well as high pollutant concentrations is established in the yeast. Different classes of yeast species are involved in decolourization of a wide range of dyes effluents including, *Saccharomyces cerevisiae*, *Galactomyces geotrichum*, *Trichosporon beigelii*, etc. (Jadhav et al. 2008). Of late, *Candida palmioleophila* JKS4 isolated from activated sludge have been reported for degradation of several azo-dyes under aerobic condition (Waghmode et al. 2011; Khan et al. 2013).

*Kluyveromyces marxianus* IMB3 was also reported to have role in the decolourization of Remazol black-B into a less toxic form (Meehan et al. 2000).

#### 5.4.2.6 Degradation by Algae

Green algae are different assemblages of photosynthetic life forms found in both marine and freshwater environments and are documented almost everywhere on Earth. Actively growing green algae can decolourize textile effluents. A wide assortment of green algae such as *Spirogyra* species (Gupta et al. 2006), *Chlorella vulgaris*, *C. sorokiniana* (Khandare et al. 2011), *C. pyrenoidosa*, *Scenedesmus obliquus* and *Closterium lunula* (Yan and Pan 2004) have been accounted for the decolourization of a wide range of dyes. A number of species of *Chlorella* and *Oscillitoria* are proficient in decolourization of azo-dyes by breaking the azo-linkages into the aromatic amines or colourless intermediates, which are highly toxic and can be further degraded into the simpler non-toxic compounds like $CO_2$ or $H_2O$ (Karacakaya et al. 2009). In addition, the decolourization of colours relies upon the nature of colours and types of green algae utilized. Subsequently, in the adjustment of lakes, green algae can assume an immediate job in the decontamination of azo-colours, as opposed to just giving oxygen to bacterial development (Khandare et al. 2011). Further, cyanobacteria and diatoms are also reported in the decolourization of various dyes such as mono-azo and di-azo dyes. The cyanobacterium *Phormidium* can decolourize the indigo dye to a greater extent around (91%), but is unable to decolourize the sulfur black and Remazol Brilliant Blue R (RBBR) dyes (Caparkaya and Cavas 2008). Microalgae are also reported in the decolourization of textile dye effluents (Mostafa et al. 2009).

The physical or chemical methods of dye degradation are only effective if the volume of the effluent is small, therefore at small scale only membrane filtration method can be used. But use of these methods poses a constraint due to their cost. So, only lab-scale studies are possible with these methods and is incompetent at large-scale. In liquid state fermentation (LSF) on a continuous basis, microorganisms are incapable of dye removal from effluents because fermentation process requires few days. Though enzymes show very good dye degradation potential, they suffer from a lot of limitations. Free enzymes once used cannot be recovered. The cost associated with the large-scale production of enzymes is also very high. Due to these reasons the enzymatic degradation is not widely used in industries. To overcome these limitations, immobilization techniques have been developed.

#### 5.4.2.7 Immobilization

Researchers have exploited the enzymes playing pivotal role in remediation, solely and also along with nanotechnology. Immobilization is a technique wherein the compound binds with the bearer material. It allows the simpler separation of the compound from the reaction blend and brings down the expenses of the procedure.

Over the most recent couple of decades, a few techniques for immobilization on various support materials have been created. Because of the accessibility of various support materials, the technique of enzyme immobilization has expanded. In this connection, nanocarriers for enzyme immobilization have gained huge attention as it provides improved catalytic performance of enzyme. Basically they protect the loss of enzyme in harsher conditions. The catalytic activity of the immobilized enzymes will be altered with respect to the support matrix (Tischer and Wedekind 1999). Immobilization helps to retain the enzymatic activity for a series of cycles and allows repetitive and continuous use of enzymes (Cao 2005; Brena and Batista-Viera 2006).

Immobilization of enzymes over magnetic nanoparticles in contrast with original partners demonstrated more extensive pH and temperature range for the enzyme activity, therefore, improving thermal and storage stability of enzymes (Bilal et al. 2018). Phenolic azo dyes degradation was carried out with laccase enzyme immobilized over $Fe_3O_4/SiO_2$ nanoparticles (Wang et al. 2013). Sinirlioglu et al. (2013) prepared cross-linked laccase aggregates (CLEAs) of laccase enzyme and efficiently decolourized malachite green dye. Further, Kumar et al. (2014) prepared magnetic cross-linked laccase aggregates (M-CLEAs) for the degradation of recalcitrant dyes. Rapid degradation of methyl orange (azo-dye) was carried out using hollow cobalt nanoparticles (Sha et al. 2016). Zhang et al. (2019) used hydrolases immobilized magnetic nanoparticles to decolourize the molasses wastewater. Nadaroglu et al. (2019) immobilized laccase enzyme on chitosan coated magnetic nanoparticles and reported the high removal capacity of azo-dyes in wastewater. Iriarte-Mesa et al. (2019) immobilized laccase enzyme on superparamagnetic iron oxide nanoparticles and showed the enhance azo-dye reactivity as compared to free enzymes.

#### 5.4.2.8 Environmental Factors Affecting Dye Degradation

**Oxygen** It plays an important role in the physiology of the cell and high amount of $O_2$ will inhibit the process of dye reduction (Chang et al. 2004). Therefore, proper decolouration can be achieved by preventing the oxygen build-up.

**Temperature** Being the most influential parameter, it is directly proportional to the rate of colour removal, but only up to certain limit after which it decreases the colour removal (Chang and Kuo 2000).

**pH** Optimum pH for dye decolouration is between 6–10 as per some previous studies (Junnarkar et al. 2006).

## 5.5 Outlook and Concluding Notes

Rapid industrialization promoted the introduction of several emerging contaminants in the environment, which are responsible for pernicious effects on plants, humans and animals. Among all industries, textile industry is generating dye effluents in huge amount. The major issue is the lack of pre-treatment of these dye effluents before their release into the environment. Although, various physical, chemical and biological methods are involved in the treatment of dye contaminants, the latter one is more preferred by considering the disadvantages of physical and chemical methods. Despite the advantages of biological methods, recently nanotechnology overtakes it. Now-a-days, nanocarriers for enzyme immobilization have gained much attention. In addition, this chapter has also reviewed the effective role of immobilized oxidoreductase enzyme in dye. Following recommendations can be suggested for dye degradation and reducing the source of emerging contaminants in the agro-ecosystems:

- A more detailed and comprehensive investigation of the dye degradation potential of different methods should be carried out extensively.
- In case of biological methods, the biological entities must be properly exposed to effluents in order to establish their maximum response and maximum degradation potential.
- Different plant parts and tissues should be analysed separately to determine the effects of effluents comprehensively.
- Since the review has established that biological methods are advantageous over chemical and physical methods and nanocarriers even surpass biological treatments in its efficiency, it is suggested that large exploration of the nanotechnological methods must be carried out for the management of toxic and hazardous waste emanating from textile industry effluent.

## References

Abedin RMA (2008) Decolorization and biodegradation of crystal violet and malachite green by *Fusarium solani* (Martius) Saccardo. A comparative study on biosorption of dyes by the dead fungal biomass. Am Eurasian J Agric Environ Sci 1:17–31

Aftab U, Khan MR, Mahfooz M, Ali M, Aslam SH, Rehman A (2011) Decolourization and degradation of textile azo dyes by Corynebacterium sp. isolated from industrial effluent. Pak J Zool 43:1–8

Aksu Z (2005) Application of biosorption for the removal of organic pollutants: a review. Process Biochem 40:997–1026. https://doi.org/10.1016/j.procbio.2004.04.008

Ali H, Ahmed W, Haq T (2009) Decolorization and degradation of malachite green by *Aspergillus flavus* and *Alternaria solani*. Afr J Biotechnol 8:1574–1576

Ambrosio ST, Vilar JC, da Silva CAA Jr, Okada K, Nascimento RL, Campos Takaki GM (2012) A biosorption isotherm model for the removal of reactive azo dyes by inactivated mycelia of *Cunninghamella elegans* UCP542. Molecules 17:452–462. https://doi.org/10.3390/molecules17010452

Aquino JM, Rocha-Filho RC, Ruotolo LA, Bocchi N, Biaggio SR (2014) Electrochemical degradation of a real textile wastewater using β-$PbO_2$ and DSA® anodes. Chem Eng J 251:138–145. https://doi.org/10.1016/j.cej.2014.04.032

Arulazhagan P (2016) A study on microbial decolourization of reactive red M8B by *Bacillus subtilis* isolated from dye contaminated soil samples. Int J Curr Res Biol Med 1:1–13

Asad S, Amoozegar MA, Pourbabaee AA, Sarbolouki MN, Dastgheib SMM (2007) Decolorization of textile azo dyes by newly isolated halophilic and halotolerant bacteria. Bioresour Technol 98:2082–2088. https://doi.org/10.1016/j.biortech.2006.08.020

Athar R, Ahmad M (2002) Heavy metal toxicity in legume-microsymbiont system. J Plant Nutr 25:369–386. https://doi.org/10.1081/PLN-100108842

Ayed L, Mahdhi A, Cheref A, Bakhrouf A (2011) Decolorization and degradation of azo dye Methyl Red by an isolated *Sphingomonas paucimobilis*: biotoxicity and metabolites characterization. Desalination 274:272–277. https://doi.org/10.1016/j.desal.2011.02.024

Banat IM, McMullan G, Meehan C, Kirby N, Nigam P, Smyth WF, Marchant R (1999) Microbial decolorization of textile dyes present in textile industries effluent. In: Proceedings of the industrial waste technical conference, Indianapolis, USA, pp 1–16

Becker H (2000) Phytoremediation using plants to cleanup soil. Agri Res 23:1151–1157

Bharagava RN, Mishra S (2018) Hexavalent chromium reduction potential of *Cellulosimicrobium* sp. isolated from common effluent treatment plant of tannery industries. Ecotoxicol Environ Saf 147:102–109. https://doi.org/10.1016/j.ecoenv.2017.08.040

Bilal M, Zhao Y, Rasheed T, Iqbal HM (2018) Magnetic nanoparticles as versatile carriers for enzymes immobilization: a review. Int J Biol Macromol 120:2530–2544. https://doi.org/10.1016/j.ijbiomac.2018.09.025

Brena BM, Batista-Viera F (2006) Immobilization of enzymes: a literature survey. In: Guisan JM (ed) Immobilization of enzymes and cells. Humana Press, Totowa, pp 15–30. https://doi.org/10.1007/978-1-62703-550-7_2

Cao L (2005) Immobilised enzymes: science or art? Curr Opin Chem Biol 9:217–226. https://doi.org/10.1016/j.cbpa.2005.02.014

Caparkaya D, Cavas L (2008) Biosorption of methylene blue by a brown alga *Cystoseira barbatula* Kutzing. Acta Chim Slov 55:547–553

Chandra R (Ed) (2016) Environmental waste management. CRC Press. Chemosphere 69:89–98

Chang JS, Kuo TS (2000) Kinetics of bacterial decolorization of azo dye with *Escherichia coli* $NO_3$. Bioresour Technol 75:107–111. https://doi.org/10.1016/S0960-8524(00)00049-3

Chang JS, Chen BY, Lin YS (2004) Stimulation of bacterial decolorization of an azo dye by extracellular metabolites from *Escherichia coli* strain $NO_3$. Bioresour Technol 91:243–248. https://doi.org/10.1016/S0960-8524(03)00196-2

Choudhury AKR (2017) Sustainable chemical technologies for textile production. In: Sustainable fibres and textiles. Woodhead Publishing, Cambridge, MA, pp 267–322. https://doi.org/10.1016/B978-0-08-102041-8.00010-X

Chowdhary P, Yadav A, Kaithwas G, Bharagava RN (2017) Distillery wastewater: a major source of environmental pollution and it's biological treatment for environmental safety. In: Singh R, Kumar S (eds) Green technology and environmental sustainability. Springer, Cham., ISBN 978-3-319-50653-1. https://doi.org/10.1007/978-3-319-50654-8_18

Copaciu F, Opriş O, Coman V, Ristoiu D, Niinemets Ü, Copolovici L (2013) Diffuse water pollution by anthraquinone and azo dyes in environment importantly alters foliage volatiles, carotenoids and physiology in wheat (*Triticum aestivum*). Water Air Soil Pollut 224:1478. https://doi.org/10.1007/s11270-013-1478-4

Daneshvar N, Ayazloo M, Khataee AR, Pourhassan M (2007) Biological decolorization of dye solution containing Malachite Green by microalgae *Cosmarium* sp. Bioresour Technol 98:1176–1182. https://doi.org/10.1016/j.biortech.2006.05.025

Das S, Dash HR (2014) Microbial bioremediation: a potential tool for restoration of contaminated areas. In: Microbial biodegradation and bioremediation, pp 1–21. https://doi.org/10.1016/B978-0-12-800021-2.00001-7

Das S, Dash HR (eds) (2017) Handbook of metal-microbe interactions and bioremediation. CRC press, Boca Raton

Deive FJ, Domínguez A, Barrio T, Moscoso F, Morán P, Longo MA, Sanromán MA (2010) Decolorization of dye Reactive Black 5 by newly isolated thermophilic microorganisms from geothermal sites in Galicia (Spain). J Hazard Mater 182:735–742. https://doi.org/10.1016/j.jhazmat.2010.06.096

Forgacs E, Cserhati T, Oros G (2004) Removal of synthetic dyes wastewaters: a review. Environ Int 30:953–971. https://doi.org/10.1016/j.envint.2004.02.001

Fu Y, Viraraghavan T (2001) Removal of acid blue 29 from an aqueous solution by *Aspergillus niger*. Am Assoc Text Chem Color 1:36–40

Gabbrielli R, Pandolfini T, Vergano O, Palandic MR (1990) Comparison of two serpentine species with different nickel tolerance strategies. Plant Soil 122:271–277. https://doi.org/10.1007/BF02851985

Garg SK, Tripathi M (2017) Microbial strategies for discoloration and detoxifcation of azo dyes from textile effluents. Res J Microbiol 12:1–19. https://doi.org/10.3923/jm.2017.1.19

Gogate PR, Pandit AB (2004) A review of imperative technologies for wastewater treatment I: oxidation technologies at ambient conditions. Adv Environ Res 8:501–551. https://doi.org/10.1016/S1093-0191(03)00032-7

Gomi N, Yoshida S, Matsumoto K, Okudomi M, Konno H, Hisabori T, Sugano Y (2011) Degradation of the synthetic dye amaranth by the fungus *Bjerkandera adusta* Dec 1: inference of the degradation pathway from an analysis of decolorized products. Biodegradation 22:1239–1245. https://doi.org/10.1007/s10532-011-9478-9

Gupta VK, Rastogi A, Saini VK, Jain N (2006) Biosorption of copper (II) from aqueous solutions by *Spirogyra* species. J Colloid Interface Sci 296:59–63. https://doi.org/10.1016/j.jcis.2005.08.033

Hu T, Wu SC (2001) Assessment of the effect of azo dye RP2B on the growth of a nitrogen fixing cyanobacterium – *Anabaena* sp. Bioresour Technol 77:93–95. https://doi.org/10.1016/S0960-8524(00)00124-3

Iriarte-Mesa C, Díaz-Castañón S, Abradelo DG (2019) Facile immobilization of *Trametes versicolor* laccase on highly monodisperse superparamagnetic iron oxide nanoparticles. Colloids Surf B: Biointerfaces 181:470–479. https://doi.org/10.1016/j.colsurfb.2019.05.012

Jadhav SU, Kalme SD, Govindwar SP (2008) Biodegradation of methyl red by *Galactomyces geotrichum* MTCC 1360. Int Biodeterior Biodegrad 62:135–142. https://doi.org/10.1016/j.ibiod.2007.12.010

Jaishree KTI (2015) Effect of textile waste water on *Vigna unguiculata* var. RC-101 (cow pea) in a pot experiment with reference of heavy metal bioaccumulation. Int J Sci Res 4(1):1795–1800

Junnarkar N, Murty DS, Bhatt NS, Madamwar D (2006) Decolorization of diazo dye Direct Red 81 by a novel bacterial consortium. World J Microbiol Biotechnol 22:163–168. https://doi.org/10.1007/s11274-005-9014-3

Kabra AN, Khandare RV, Waghmode TR, Govindwar SP (2011) Differential fate of metabolism of a sulfonated azo dye remazol orange 3R by plants *Aster amellus* Linn., *Glandularia pulchella* (Sweet) Tronc. and their consortium. J Hazard Mater 190:424–431. https://doi.org/10.1016/j.jhazmat.2011.03.065

Kagalkar NA, Jagtap UB, Jadhav JP, Govindwar SP, Bapat SA (2010) Studies on phytoremediation potentially of *Typhonium flagelliforme* for the degradation of Brilliant blue R. Planta 232:271–285. https://doi.org/10.1007/s00425-010-1157-2

Kalyani DC, Patil PS (2008) Biodegradation of reactive textile dye Red BLI by an isolated bacterium *Pseudomonas* sp. SUK1. Bioresour Technol 99:4635–4641. https://doi.org/10.1016/j.biortech.2007.06.058

Kandelbauer A, Guebitz GM (2005) Bioremediation for the decolorization of textile dyes-a review. In: Environmental chemistry. Springer, Cham, pp 269–288. https://doi.org/10.1007/3-540-26531-7_26

Karacakaya P, Kilic NK, Duygu E, Donmez G (2009) Stimulation of reactive, dye removal by cyanobacteria in media containing triacontanol hormone. J Hazard Mater 172:1635–1639. https://doi.org/10.1016/j.jhazmat.2009.08.037

Kasai Y (2011) Molecular Technologies for Analysis of Petroleum Bioremediation. Microbial bioremediation of non-metals: current research. Caister Academic Press, Norfolk, pp 233–252

Khan TA, Dahiya S, Ali I (2012) Use of kaolinite as adsorbent: equilibrium, dynamics and thermodynamic studies on the adsorption of Rhodamine B from aqueous solution. Appl Clay Sci 69:58–66. https://doi.org/10.1016/j.clay.2012.09.001

Khan R, Bhawana P, Fulekar MH (2013) Microbial decolorization and degradation of synthetic dyes: a review. Rev Environ Sci Biotechnol 12:75–97. https://doi.org/10.1007/s11157-012-9287-6

Khandare RV, Kabra AN, Kurade MB, Govindwar SP (2011) Phytoremediation potential of *Portulaca grandiflora* Hook. (Moss-Rose) in degrading a sulfonated diazo reactive dye navy blue HE2R (reactive blue 172). Bioresour Technol 102:6774–6777. https://doi.org/10.1016/j.biortech.2011.03.094

Kumar VV, Sivanesan S, Cabana H (2014) Magnetic cross-linked laccase aggregates—bioremediation tool for decolorization of distinct classes of recalcitrant dyes. Sci Total Environ 487:830–839. https://doi.org/10.1016/j.scitotenv.2014.04.009

Levin L, Melignani E, Ramos AM (2010) Effect of nitrogen sources and vitamins on ligninolytic enzyme production by some white-rot fungi. Dye decolorization by selected culture filtrates. Bioresour Technol 101:4554–4563. https://doi.org/10.1016/j.biortech.2010.01.102

Mani S, Bharagava RN (2016) Exposure to crystal violet, its toxic, genotoxic and carcinogenic effects on environment and its degradation and detoxification for environmental safety. Rev Environ Contam Toxicol 237:71–104. https://doi.org/10.1007/978-3-319-23573-8_4

Mani S, Bharagava RN (2017) Isolation, screening and biochemical characterization of bacteria capable of crystal violet dye decolorization. Int J Appl Adv Sci Res 2:70–75

Meehan C, Banat IM, McMullan G, Nigam P, Smyth F, Marchant R (2000) Decolorization of ramazol black-B using a thermotolerant yeast, *Kluyveromyes marxianus* IMB3. Environ Int 26:75–79. https://doi.org/10.1016/s0160-4120(00)00084-2

Meng X, Liu G, Zhou J, Fu QS, Wang G (2012) Azo dye decolorization by *Shewanella aquimarina* under saline conditions. Bioresour Technol 114:95–101. https://doi.org/10.1016/j.biortech.2012.03.003

Mishra S, Bharagava RN (2016) Toxic and genotoxic effects of hexavalent chromium in environment and its bioremediation strategies. J Environ Sci Health C 34:1–32. https://doi.org/10.1080/10590501.2015.1096883

Mohammad A, Khan AU (1985) Effect of a textile factory effluent on soil and crop plants. Environ Pollut 37:131–148. https://doi.org/10.1016/0143-1471(85)90004-2

Mostafa M, Sheekhs E, Gharieb MM, Abou-El-Souod GW (2009) Biodegradation of dyes by some green algae and cyanobacteria. Int Biodeterior Biodegrad 63:699–704. https://doi.org/10.1016/j.ibiod.2009.04.010

Muthunarayanan V, Santhiya M, Swabna V (2011) Phytodegradation of textile dyes by Water Hyacinth (*Eichhornia crassipes*) from aqueous dye solutions. Int J Environ Sci 1:1702–1717

Nadaroglu H, Mosber G, Gungor AA, Adıguzel G, Adiguzel A (2019) Biodegradation of some azo dyes from wastewater with laccase from *Weissella viridescens* LB37 immobilized on magnetic chitosan nanoparticles. J Water Process Eng 31:100866. https://doi.org/10.1016/j.jwpe.2019.100866

Paar A, Costa S, Tzanov T, Gudelj M, Robra KH, Cavaco-Paulo A, Gübitz GM (2001) Thermo-alkali-stable catalases from newly isolated Bacillus sp. for the treatment and recycling of textile bleaching effluents. J Biotechnol 89:147–153. https://doi.org/10.1016/S0168-1656(01)00305-4

Pajot H, Delgado O, de Figueroa L, Fariña J (2010) Unraveling the decolourizing ability of yeast isolates from dye-polluted and virgin environments: an ecological and taxonomical overview. Antonie Van Leeuwenhoek 99:1–14. https://doi.org/10.1007/s10482-010-9495-4

Pandey A, Singh P, Iyengar L (2007) Bacterial decolorization and degradation of azo dyes: a review. Int Biodeterior Biodegrad 59:73–84. https://doi.org/10.1016/j.ibiod.2006.08.006

Parshetti G, Kalme S, Saratale G, Govindwar S (2006) Biodegradation of malachite green by *Kocuria rosea* MTCC 1532. Acta Chim Slov 4:492–498

Pelegrini R, Peralto-Zamora P, de Andrade AR, Reyers J, Duran N (1999) Electrochemically assisted photocatalytic degradation of reactive dyes. Appl Catal B Environ 22:83–90. https://doi.org/10.1016/S0926-3373(99)00037-5

Peralto-Zamora P, Kunz A, Gomez de Morales S, Pelegrini R, de Capos Moleiro P, Reyes J, Duran N (1999) Degradation of reactive dyes I. a comparative study of ozonation, enzymatic and photochemical processes. Chemosphere 38:835–852. https://doi.org/10.1016/S0045-6535(98)00227-6

Pereira L, Alves M (2012) Dyes—environmental impact and remediation. In: Environmental protection strategies for sustainable development. Springer, Dordrecht, pp 111–162. https://doi.org/10.1007/978-94-007-1591-2_4

Pointing S (2001) Feasibility of bioremediation by white-rot fungi. Appl Microbiol Biotechnol 57:20–33. https://doi.org/10.1007/s002530100745

Rehman A, Bhatti HN (2009) Textile effluents affected seed germination and early growth of some winter vegetable crops: a case study. Water Air Soil Pollut 198:155. https://doi.org/10.1007/s11270-008-9834-5

Rehman K, Shahzad T, Sahar A, Hussain S, Mahmood F, Siddique MH Rashid MI (2018) Effect of Reactive Black 5 azo dye on soil processes related to C and N cycling. Peer J 6:4802. https://doi.org/10.7717/peerj.4802

Robinson T, McMullan G, Marchant R, Nigam P (2001) Remediation of dyes in textile effluent: a critical review on current treatment technologies with a proposed alternative. Bioresour Technol 77:247–255. https://doi.org/10.1016/S0960-8524(00)00080-8

Roy U, Manna S, Sengupta S, Das P, Datta S, Mukhopadhyay A, Bhowal A (2018) Dye removal using microbial biosorbents. In: Green adsorbents for pollutant removal. Springer, Cham, pp 253–280. https://doi.org/10.1007/978-3-319-92162-4_8

Saratale GD, Kalme SD, Govindwar SP (2006) Decolorization of textile dyes by *Aspergillus ochraceus* (NCIM-1146). Indian J Biotechnol 5:407–410. http://nopr.niscair.res.in/handle/123456789/5603

Saratale RG, Saratale GD, Chang JS, Govindwar SP (2009a) Ecofriendly degradation of sulfonated diazo dye CI Reactive Green 19A using *Micrococcus glutamicus* NCIM-2168. Bioresour Technol 100:3897–3905. https://doi.org/10.1016/j.biortech.2009.03.051

Saratale RG, Saratale GD, Kalyani DC, Chang JS, Govindwar SP (2009b) Enhanced decolorization and biodegradation of textile azo dye Scarlet R by using developed microbial consortium-GR. Bioresour Technol 100:2493–2500. https://doi.org/10.1016/j.biortech.2008.12.013

Saratale RG, Saratale GD, Chang JS, Govindwar SP (2010) Decolorization and biodegradation of reactive dyes and dye wastewater by a developed bacterial consortium. Biodegradation 21(6):999-1015. https://doi.org/10.1007/s10532-010-9360-1

Saratale RG, Saratale GD, Chang JS, Govindwar SP (2011) Bacterial decolorization and degradation of azo dyes: a review. J Taiwan Inst Chem Eng 42:138–157. https://doi.org/10.1016/j.jtice.2010.06.006

Saxena G, Bharagava RN (2015) Persistent organic pollutants and bacterial communities present during the treatment of tannery wastewater. In: Environmental Waste Management (ed) Chandra R, 24 recent advances in environmental management. CRC Press/Taylor and Francis Group, Boca Raton, pp 217–247

Saxena G, Bharagava RN (2016) Ram Chandra: advances in biodegradation and bioremediation of industrial waste. Clean Techn Environ Policy 18:979–980

Senthilkumaar S, Kalaamani P, Subburaam CV (2006) Liquid phase adsorption of crystal violet onto activated carbons derived from male flowers of coconut tree. J Hazard Mater 136:800–808. https://doi.org/10.1016/j.jhazmat.2006.01.045

Sha Y, Mathew I, Cui Q, Clay M, Gao F, Zhang XJ, Gu Z (2016) Rapid degradation of azo dye methyl orange using hollow cobalt nanoparticles. Chemosphere 144:1530–1535. https://doi.org/10.1016/j.chemosphere.2015.10.040

Shah MP, Patel KA, Nair SS (2013) Microbiological removal of crystal violet dye by *Bacillus subtilis* ETL-2211. OA Biotechnology 2:9

Sharma P, Dubey RS (2005) Lead toxicity in plants. Braz J Plant Physiol 17:35–52. https://doi.org/10.1590/S1677-04202005000100004

Shengfang L (2010) Removal of crystal violet from aqueous solution by sorption into semiinterpenetrated networks hydrogels constituted of poly(acrylic acid-acrylamidemethacrylate) and amylase. Bioresour Technol 101:2197–2202. https://doi.org/10.1016/j.biortech.2009.11.044

Singh S, Prasad SM (2014) Growth, photosynthesis and oxidative responses of *Solanum melongena* L. seedlings to cadmium stress: mechanism of toxicity amelioration by kinetin. Sci Hortic 176:1–10. https://doi.org/10.1016/j.scienta.2014.06.022

Singh J, Das A, Yogalakshmi KN (2020) Enhanced laccase expression and azo dye decolourization during co-interaction of *Trametes versicolor* and *Phanerochaete chrysosporium*. SN Appl Sci 2:1–8. https://doi.org/10.1007/s42452-020-2832-y

Sinirlioglu ZA, Sinirlioglu D, Akbas F (2013) Preparation and characterization of stable cross-linked enzyme aggregates of novel laccase enzyme from *Shewanella putrefaciens* and using malachite green decolorization. Bioresour Technol 146:807–811. https://doi.org/10.1016/j.biortech.2013.08.032

Song L, Shao Y, Ning S, Tan L (2017) Performance of a newly isolated salt-tolerant yeast strain *Pichia occidentalis* G1 for degrading and detoxifying azo dyes. Bioresour Technol 233:21–29. https://doi.org/10.1016/j.biortech.2017.02.065

Srinivasan A, Viraraghavan T (2010) Decolorization of dye wastewaters by biosorbents: a review. J Environ Manag 91:1915–1929. https://doi.org/10.1016/j.jenvman.2010.05.003

Sulabha D, Nandkar PB (2011) The phytotoxicity of heavy metals on growth and metal uptake by spinach. Agric Biol Res 27:78–90

Tan L, Qu Y, Zhou J, Ma F, Li A (2009) Dynamics of microbial community for X-3B wastewater decolorization coping with high-salt and metal ions conditions. Bioresour Technol 100:3003–3009. https://doi.org/10.1016/j.biortech.2009.01.010

Tan CY, Li G, Lu XQ, Chen ZL (2010) Biosorption of basic Orange using dried *A. filiculoides*. Ecol Eng 36:1333–1340. https://doi.org/10.1016/j.ecoleng.2010.06.009

Telke AA, Kadam AA, Govindwar SP (2015) Bacterial enzymes and their role in decolorization of azo dyes. In: Microbial degradation of synthetic dyes in wastewaters. Springer, Cham, pp 149–168. https://doi.org/10.1007/978-3-319-10942-8_7

Tischer W, Wedekind F (1999) Immobilized enzymes: methods and applications. In: Biocatalysis-from discovery to application. Springer, Berlin/Heidelberg, pp 95–126. https://doi.org/10.1007/3-540-68116-7_4

Tomei MC, Angelucci DM, Daugulis AJ (2016) Sequential anaerobic-aerobic decolourization of a real textile wastewater in a two-phase partitioning bioreactor. Sci Total Environ 573:585–593. https://doi.org/10.1016/j.scitotenv.2016.08.140

Topac FO, Dindar E, Uçaroglu S, Başkaya HS (2009) Effect of a sulfonated azo dye and sulfanilic acid on nitrogen transformation processes in soil. J Hazard Mater 170:1006–1013. https://doi.org/10.1016/j.jhazmat.2009.05.080

Uera RB, Paz-Alberto AM, Sigua GC (2007) Phytoremediation potentials of selected tropical plants for ethidium bromide. Environ Sci Pollut Res 14:505–509. https://doi.org/10.1065/espr2007.02.391

Verma P, Madamwar D (2003) Decolourization of synthetic dyes by a newly isolated strain of *Serratia marcescens*. World J Microbiol Biotechnol 19:615–618. https://doi.org/10.1023/A:1025115801331

Vijayalakshmidevi SR, Muthukumar K (2015) Improved biodegradation of textile dye effluent by coculture. Ecotoxicol Environ Saf 114:23–30. https://doi.org/10.1016/j.ecoenv.2014.09.039

Waghmode TR, Kurade MB, Govindwar SP (2011) Time dependent degradation of mixture of structurally different azo and non azo dyes by using *Galactomyces geotrichum* MTCC 1360. Int Biodeterior Biodegrad 65:479–486. https://doi.org/10.1016/j.ibiod.2011.01.010

Wang H, Zhang W, Zhao J, Xu L, Zhou C, Chang L, Wang L (2013) Rapid decolorization of phenolic azo dyes by immobilized laccase with $Fe_3O_4/SiO_2$ nanoparticles as support. Ind Eng Chem Res 52:4401–4407. https://doi.org/10.1021/ie302627c
Wawrzkiewicz M, Bartczak P, Jesionowski T (2017) Enhanced removal of hazardous dye form aqueous solutions and real textile wastewater using bifunctional chitin/lignin biosorbent. Int J Biol Macromol 99:754–764. https://doi.org/10.1016/j.ijbiomac.2017.03.023
Weisburger JH (2002) Comments on the history and importance of aromatic and heterocyclic amines in public health. Mutat Res 22:506–507. https://doi.org/10.1016/S0027-5107(02)00147-1
Yadav A, Chowdhary P, Kaithwas G, Bharagava RN (2017) Toxic metals in the environment, their threats on ecosystem and bioremediation approaches. In: Das S, Singh HR (eds) Handbook of metal-microbe interaction and bioremediation. CRC Press/Taylor and Francis Group, Boca Raton, pp 128–141. https://doi.org/10.1201/978135153353-12
Yan H, Pan G (2004) Increase in biodegradation of dimethyl phthalate by *Closterium lunula* using inorganic carbon. Chemosphere 55:1281–1285. https://doi.org/10.1016/j.chemosphere.2003.12.019
Yang XQ, Zhao XX, Liu CY, Zheng Y, Qian SJ (2009) Decolorization of azo, triphenylmethane and anthraquinone dyes by a newly isolated *Trametes* sp. SQ01 and its laccase. Process Biochem 44:1185–1189. https://doi.org/10.1016/j.procbio.2009.06.015
Zhang F, Yediler A, Liang X, Kettrup A (2004) Effects of dye additives on the ozonation process and oxidation by products: a comparative study using hydrolyzed CI reactive red 120. Dyes Pigments 60:1–7. https://doi.org/10.1016/j.biortech.2019.02.049
Zhang Z, Li D, Zhang X (2019) Enzymatic decolorization of melanoidins from molasses wastewater by immobilized keratinase. Bioresour Technol 280:165–172. https://doi.org/10.1016/j.biortech.2019.02.049
Zhou Q (2001) Chemical pollution and transport of organic dyes in water soil crop system of the chinese coasts. Bull Environ Contam Toxicol 66:784–793. https://doi.org/10.1007/s001280077
Zhuo R, Ma L, Fan FF, Gong YM, Wan X, Jiang ML, Zhang XY, Yang Y (2011) Decolorization of different dyes by a newly isolated white-rot fungi strain *Ganoderma* sp. En3 and cloning and functional analysis of its laccase gene. J Hazard Mater 192:855–873. https://doi.org/10.1016/j.jhazmat.2011.05.106

# Part II
# Types of Emerging Contaminants in Agriculture

# Chapter 6
# Pharmaceuticals as Emerging Contaminant in Agriculture: Source, Transport, Ecological Risks and Removal Strategies

**Jaskiran Kaur**

**Abstract** Pharmaceuticals play a pivotal role in restoring the organic functions through curing and preventing diseases (both infectious and life-threatening) in humans and animals. But the release of an increasing quantity of such pharmaceuticals through urine and faeces into the water and soil create a serious problem for the resident flora and fauna in these environments. Even though present in low concentrations, they can persist in the soil for longer durations. There they impair soil fertility, induce toxicity in soil organisms; provoke antibiotic resistance in the soil bacteria; stunt the plant growth, and reduce the survival and reproduction of soil invertebrates. The sorption of pharmaceuticals involves the interplay of numerous factors involving soil types, soil properties, and the nature of the pharmaceutical compound itself.

The author reviews the possible sources of pharmaceuticals, their transport into agricultural soils, factors responsible for their sorption into the soil, and impacts exerted by them on soil components. At last, I explore the innovative technologies employed to embark upon pharmaceuticals' eco-toxicological impacts. The sorption of pharmaceuticals into the soil is likely to be observed at acidic pH as well as higher distribution coefficient and hydrophobicity of pharmaceuticals. At a higher dose, pharmaceuticals results in inhibition of soil microbial growth, reduced microbial activity, decreased substrate utilization potential in microbes, induction of antibiotic resistance in soil microbiota, retarded shoot, and root growth and impaired reproduction in soil invertebrates. The principal techniques adopted for their removal are bioremediation and phytoremediation which removes more than 95% of pharmaceuticals in the soil.

J. Kaur (✉)
Department of Biotechnology, Doaba College, Jalandhar, Punjab, India
e-mail: jaskiran_kaur49@yahoo.com



V. Kumar Singh et al. (eds.), *Sustainable Agriculture Reviews 50*, Sustainable Agriculture Reviews 50, https://doi.org/10.1007/978-3-030-63249-6_6

**Keywords** Antibiotic resistance · Eco-toxicological impacts · Pharmaceuticals · Distribution coefficient · Soil microbiota · Soil microbial activity · Substrate utilization · Soil invertebrates · Bioremediation · Phytoremediation

## Abbreviations

| | |
|---|---|
| BCFs | Bio-concentration factors |
| CFU | Colony forming unit |
| CIP | Ciprofloxacin |
| GDP | Gross Domestic Product |
| $K_d$ | Distribution coefficient |
| MBR | Membrane bioreactor |
| NER | Non-extractable residues |
| SMM | Sulfamonomethoxine |
| TC | Tetracycline |
| UV | Ultra-violet |

## 6.1 Introduction

Over decennium, growth of agriculture remains central to the rise of human civilization. The world economy unquestionably more or less depends upon agriculture. As per reports, around 1% of total Gross Domestic Product (GDP) of the world has been contributed by the agriculture sector (Zavatta 2014). In the Indian scenario, participation of agriculture to GDP was found to be nearly 17% (Arun 2017). Besides influencing the nation's economy, agriculture has also registered an imperative position in terms of generating employment to two thirds of the Indian population. The progress of the agriculture sector which is generally reflected in increased crop production, however, involves the intensive use of available water resources. According to Carpenter et al. (2011), agriculture withdraws 76% of the world's freshwater resources. As a result of consumption of excess amounts of water by the agriculture sector, an exceedingly high pressure is inserted on the water thereby creating a situation of water crisis. To cope up with the issue of water scarcity, a certain alternative sources of water is the prerequisite of the present time.

Wastewater reuse for irrigation is considered as a feasible option to secure the water resources for coming generations without compromising the peak demand of water in the agricultural sector; especially an essential need in areas receiving scarce rainfall. In addition, it also minimizes the impacts associated with effluent discharge in the freshwater ecosystems and constitutes high nutrient content, hence reducing the dependence on costly chemical fertilizers (Toze 2006; Rusan et al. 2007; Agrafioti and Diamadopoulos 2012). The application of wastewater for irrigating variety of crops ranging from staple, fiber, forest and horticultural crops has been experimented by various researchers so far (Al-Jamal et al. 2002; Antolin et al. 2005; Carter et al. 2005; Libutti et al. 2018). However, apart from the aforesaid advantages, the wastewater reuse option also has a major share of risks to agriculture.

One of the major risks is the entry of emerging contaminants in agriculture. The term emerging contaminant is meant by all those compounds (either natural or man-made) that are not usually recognized in the environment but have an inherent ability to cause detrimental effects in humans and entire ecology (Sauve and Desrosiers 2014; Mohamed and Paleologos 2017). Emerging contaminants includes a widespread category of compounds such as pesticides, human and veterinary pharmaceuticals, paints, flame retardants, preservatives, cosmetics, antiperspirants and man-made compounds such as engineered nanomaterials and ultra-violet (UV) filters which are indispensable part of our society (Fatta-Kassinos et al. 2011; Barroso et al. 2019). They are mainly present in water bodies as well as in treated and untreated wastewater resulting from municipal and industrial activities (Barnes et al. 2004; Watkinson et al. 2007; Gavrilescu et al. 2015). Figure 6.1 illustrates some of the most significant emerging contaminants and routes of their entry into the agricultural fields.

Degradation of environmental quality from emerging contaminants is now known as a major concern in several nations. In the present-day, awareness has been focused at pharmaceuticals, that are although biologically active but still not quantified properly due to absence of authenticated analytical methods (Fatta-Kassinos et al. 2011). Though, the wastewater treatment reduces the toxicity of effluent to a larger extent but certain pharmaceuticals are still reported which tend to release continuously into the environment (Thiele-Bruhn 2003; Carter et al. 2014). Such wastewater when used for irrigation ultimately translocated from the agricultural

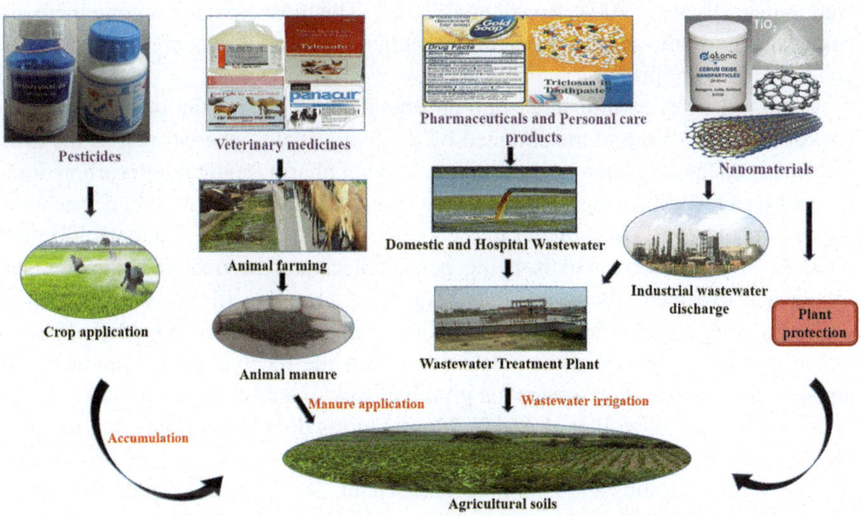


**Fig. 6.1** Different routes of entry of different emerging contaminants into the agricultural systems

soils into the plant tissues. It is well known from the previous literature that the pharmaceuticals concentrations of up to several kilograms per hectare reached the soils (Thiele-Bruhn 2003; Kaczala and Blum 2016). As a result, concerns have been raised with regard to the biological potency of pharmaceuticals towards the soil dwelling fauna and flora. Some studies reported the accumulation of antibiotics in the soil invertebrates such as earthworms (Kinney et al. 2008; Berge and Vulliet 2015; Carter et al. 2016). Colinas et al. (1994) reported inhibition of the soil bacteria, protozoa, nematodes and microarthropods due to the occurrence of antibiotics, oxytetracycline and penicillin in the soil. Conversely, Hoper et al. (2002) observed an increase in microbial growth and activity in presence of pharmaceuticals. The current chapter stressed on the various categories of pharmaceuticals screened in the soil environment and the sources of their uptake in agroecosystem. Furthermore, more precise information about transformation pathways of pharmaceuticals in the soil environment; assessment of their potential impacts on soil properties, soil organisms and plants; and application of diverse treatment processes for their elimination and/or mineralization is explained in an elaborative manner.

## 6.2 Pharmaceuticals as Emerging Contaminants

Pharmaceutical is a drug or medicine that is manufactured for use in medical treatment. It includes a range of drugs that are prescribed by the doctors as well as those drugs which are sold in the markets without any prescription. Their prevalent uses include human and animal medicine, where they are employed for prophylactic, therapeutic purposes as well as growth promoters (Halling-Sørensen et al. 1998). More than 4000 pharmaceuticals that fall into different classes have been discovered to date (Beausse 2004; Boxall et al. 2012). The most common among them are analgesics and anti-inflammatory drugs, anticonvulsants, antidepressants, human, and veterinary antibiotics.

When a drug is administered into an organism (either human or animal), only part of it is absorbed and metabolized by the body tissues whereas rest of the drug follows three fates: (1) one is the mineralization of pharmaceutical into carbon dioxide and water, (2) the drug with lipophilic characteristics is not readily degradable, hence part of it will be withheld in the sludge, and (3) drug is metabolized into a more hydrophilic form which, being persistent cannot be treated completely by wastewater treatment plants, and thus, remain intact in the effluent, discharge from such treatment plants (Carlsson et al. 2006). Such pharmaceutical loaded effluents or wastewaters will then ultimately release into the receiving water bodies. The aquatic environment thus serves as a principal sink of a mixture of pharmaceuticals on a global scale (Boxall et al. 2012). Many reports documented the occurrence of pharmaceuticals in river water, groundwater, surface water, sea water, wastewater, sewage flows, biosolids and landfills (Barnes et al. 2004; Xia et al. 2005; Batt et al. 2006; Kim et al. 2007; Focazio et al. 2008; Sim et al. 2010; Zuccato et al. 2010; Fatta-Kassinos et al. 2011; Lim et al. 2013; Li 2014; Liu et al. 2016). Table 6.1

**Table 6.1** Studies on occurrence of pharmaceuticals in different environments

| Pharmaceuticals class | Examples | Occurrence | Range in concentration | Reference |
|---|---|---|---|---|
| Analgesics and anti-inflammatory | Acetaminophen | Ground water sample collected from landfills | 0.009 μg/L | Barnes et al. (2004) |
| | | Effluent | 9.5 ng/L | Kim et al. (2007) |
| | | Surface water | 33 ng/L | Kim et al. (2007) |
| | | Ground water | 0.38 μg/L | Barnes et al. (2008) |
| | | Untreated drinking water | 0.16 μg/L | Focazio et al. (2008) |
| | | Domestic wastewater and surface water | 5.8–58.7 μg/L | Agunbiade and Moodley (2014) |
| | | River water | 18.8 ng/L | Riva et al. (2015) |
| | | Wastewater from sewage treatment plant | 340–690 ng/L | Subedi et al. (2017) |
| | | Wastewater treatment plant | 11.3 μg/L | Kanama et al. (2018) |
| | | River transect | 327.7 ng/L | Cantwell et al. (2018) |
| | Codeine | Ground water sample collected from landfills | 0.1–0.24 μg/L | Barnes et al. (2004) |
| | | Wastewater from sewage treatment plant | 25–82 ng/L | Subedi et al. (2017) |
| | Diclofenac | Effluent | 40 ng/L | Kim et al. (2007) |
| | | Surface water | 3 ng/L | Kim et al. (2007) |
| | | River water | 121 ng/L | Riva et al. (2015) |
| | | Domestic wastewater and surface water | 1.1–15.6 μg/L | Agunbiade and Moodley (2014) |
| | | Landfill leachate | 4.8–19.3 μg/L | Sui et al. (2017) |
| | | Wastewater treatment plant | 0.07–0.7 μg/L | Kanama et al. (2018) |

(continued)

**Table 6.1** (continued)

| Pharmaceuticals class | Examples | Occurrence | Range in concentration | Reference |
|---|---|---|---|---|
| | Hydrocodone | Effluent | 41 ng/L | Kim et al. (2007) |
| | | Surface water | 1.6 ng/L | Kim et al. (2007) |
| | Ibuprofen | River water | 4.4–78.5 ng/L | Calamari et al. (2003) |
| | | Ground water sample collected from landfills | 0.018 μg/L | Barnes et al. (2004) |
| | | River water | 20 ng/L | Zuccato et al. (2005) |
| | | Effluent | 65 ng/L | Kim et al. (2007) |
| | | Surface water | 28 ng/L | Kim et al. (2007) |
| | | Untreated drinking water | 0.27 μg/L | Focazio et al. (2008) |
| | | Ground water | 3.11 μg/L | Barnes et al. (2008) |
| | | Domestic wastewater and surface water | 0.8–18.9 μg/L | Agunbiade and Moodley (2014) |
| | | River water | 79.5 ng/L | Riva et al. (2015) |
| | | Wastewater from sewage treatment plant | 630–980 ng/L | Subedi et al. (2017) |
| | | Wastewater treatment plant | 13.6 μg/L | Kanama et al. (2018) |
| | Naproxen | Effluent | 128 ng/L | Kim et al. (2007) |
| | | Surface water | 11 ng/L | Kim et al. (2007) |
| | | Sewage treatment plant effluent | 33.9 μg/L | Kone et al. (2013) |
| | | River water | 62.4 ng/L | Riva et al. (2015) |
| | Salicylic acid | Sewage treatment plant effluent | 59.6 μg/L | Kone et al. (2013) |
| | | Surface water | 17 μg/L | Kone et al. (2013) |
| Antiasthmatic | Salbutamol | River water | 2.5 ng/L | Zuccato et al. (2005) |

| | | | | |
|---|---|---|---|---|
| Anticonvulsants | Carbamazepine | River water | 175.3 ng/L | Zuccato et al. (2005) |
| | | Effluent | 226 ng/L | Kim et al. (2007) |
| | | Surface water | 25 ng/L | Kim et al. (2007) |
| | | Sewage treatment plant effluent | 3.287 μg/L | Kone et al. (2013) |
| | | Surface water | 0.7 μg/L | Kone et al. (2013) |
| | | River water | 86 ng/L | Riva et al. (2015) |
| | | Wastewater from sewage treatment plant | 480–580 ng/L | Subedi et al. (2017) |
| | | Landfill leachate | 2.1–6.2 μg/L | Sui et al. (2017) |
| | | River transect | 542.6 ng/L | Cantwell et al. (2018) |
| | | Ground water | 162 ng/L | Bexfield et al. (2019) |
| | Dilantin | Effluent | 44 ng/L | Kim et al. (2007) |
| | | Surface water | 4.3 ng/L | Kim et al. (2007) |
| Antidepressants | Desmethylvenlafaxine | Surface water | 1.472 μg/L | Kone et al. (2013) |
| | Fluoxetine | Ground water | 0.056 μg/L | Barnes et al. (2008) |
| | | Sewage treatment plant effluents | 0.799 μg/L | Kone et al. (2013) |
| | | Surface water | 0.141 μg/L | Kone et al. (2013) |
| | Venlafaxine | Sewage treatment plant effluent | 0.808 μg/L | Kone et al. (2013) |
| Antihistamine | Diphenhydramine | Untreated drinking water | 0.023 μg/L | Focazio et al. (2008) |
| | Ranitidine | River water | 0.86–38.5 ng/L | Calamari et al. (2003) |
| | | River water | 38.5 ng/L | Zuccato et al. (2005) |
| | | River transect | 1002.1 ng/L | Cantwell et al. (2018) |
| Antiparasitic | Amino fenbendazole | Livestock Wastewater treatment plant | 0.554–94.1 μg/L | Sim et al. (2013) |
| | Fenbendazole | Livestock Wastewater treatment plant | 3.85–241 μg/L | Sim et al. (2013) |
| | | Municipal Wastewater treatment plant | 0.018 μg/L | Sim et al. (2013) |
| | | River water | 0.006–1.3 μg/L | Sim et al. (2013) |
| | Fenbendazole sulfone | Livestock Wastewater treatment plant | 0.283–93.9 μg/L | Sim et al. (2013) |
| | p-hydroxyfenbendazol | Livestock Wastewater treatment plant | 0.693–92.7 μg/L | Sim et al. (2013) |

(continued)

**Table 6.1** (continued)

| Pharmaceuticals class | Examples | Occurrence | Range in concentration | Reference |
|---|---|---|---|---|
| Antirheumatic | Sulfasalazine | Urban wastewater effluents | 10 ng/L | Watkinson et al. (2007) |
| | | Creeks and river waters | 202–321 ng/L | Fisher and Scott (2008) |
| | | Urban wastewater effluents | 0.15 μg/L | Watkinson et al. (2009) |
| | | Surface waters | 0.03 μg/L | Watkinson et al. (2009) |
| Anxiolytic | Meprobamate | Effluent | 6 ng/L | Kim et al. (2007) |
| | | Ground water | 164 ng/L | Bexfield et al. (2019) |
| β-blocker | Atenolol | River water | 3.4–241 ng/L | Calamari et al. (2003) |
| | | River water | 241 ng/L | Zuccato et al. (2005) |
| | | Sewage treatment plant effluents | 1.68 μg/L | Kone et al. (2013) |
| | | Domestic wastewater and surface water | 0.9–39.1 μg/L | Agunbiade and Moodley (2014) |
| | | River water | 166 ng/L | Riva et al. (2015) |
| | | Wastewater from sewage treatment plant | 590–1500 ng/L | Subedi et al. (2017) |
| | | River transect | 1074.3 ng/L | Cantwell et al. (2018) |
| | | Wastewater treatment plant | 0.33–3.2 μg/L | Kanama et al. (2018) |
| | | Ground water | 8.7 ng/L | Bexfield et al. (2019) |
| Calcium channel blockers | Dehydronifedipine | Untreated drinking water | 0.019 μg/L | Focazio et al. (2008) |
| | | Ground water | 0.022 μg/L | Barnes et al. (2008) |
| Cardiovascular drug | Enalapril | River water | 3.6 ng/L | Riva et al. (2015) |
| Contrast media | Iopromide | Effluent | 2630 ng/L | Kim et al. (2007) |
| | | Surface water | 134 ng/L | Kim et al. (2007) |
| Diuretic | Furosemide | River water | 1.72–254.7 ng/L | Calamari et al. (2003) |
| | | River water | 254.7 ng/L | Zuccato et al. (2005) |
| | | River water | 27 ng/L | Riva et al. (2015) |
| | | River transect | 1234.8 ng/L | Cantwell et al. (2018) |

| | | | | |
|---|---|---|---|---|
| | Hydrochlorothiazide | River water | 255.8 ng/L | Zuccato et al. (2005) |
| | | River water | 314 ng/L | Riva et al. (2015) |
| Feed additive | Ractopamine | Groundwater | 50 ng/L | Bartelt-Hunt et al. (2011) |
| Hemorrheologic agents | Pentoxifylline | Effluent | 2.9 ng/L | Kim et al. (2007) |
| | | Surface water | 1.6 ng/L | Kim et al. (2007) |
| Lipid-regulator | Atorvastatin | River water | 0.8 ng/L | Riva et al. (2015) |
| | Bezafibrate | River water | 0.79–57.1 ng/L | Calamari et al. (2003) |
| | | River water | 57.2 ng/L | Zuccato et al. (2005) |
| | | Domestic wastewater and surface water | 0.81–8.7 μg/L | Agunbiade and Moodley (2014) |
| | | River water | 28.2 ng/L | Riva et al. (2015) |
| | Gemfibrozil | Effluent | 11.2 ng/L | Kim et al. (2007) |
| | | Surface water | 6.6 ng/L | Kim et al. (2007) |
| | | Sewage treatment plant effluents | 2.09 μg/L | Kone et al. (2013) |
| | | Surface waters | 0.58 μg/L | Kone et al. (2013) |
| | | Landfill leachate | 2–4.4 μg/L | Sui et al. (2017) |
| | | River transect | 1440.4 ng/L | Cantwell et al. (2018) |
| Tranquilizers | Diazepam | River water | 0.13–1.16 ng/L | Calamari et al. (2003) |
| | | Wastewater from sewage treatment plant | 9.5–36 ng/L | Subedi et al. (2017) |
| Human and veterinary antibiotics | Amoxicillin | Urban wastewater effluents | 30 ng/L | Watkinson et al. (2007) |
| | | Urban wastewater effluents | 0.05 μg/L | Watkinson et al. (2009) |
| | | Surface waters | 0.2 μg/L | Watkinson et al. (2009) |
| | | Hospital effluents | 0.9 μg/L | Watkinson et al. (2009) |
| | | River waters | <2.08 ng/L | Zuccato et al. (2010) |
| | | River water | 16.7 ng/L | Riva et al. (2015) |
| | Azithromycin | Untreated drinking water | 0.029 μg/L | Focazio et al. (2008) |

(continued)

**Table 6.1** (continued)

| Pharmaceuticals class | Examples | Occurrence | Range in concentration | Reference |
|---|---|---|---|---|
| | Carbodox | Ground water sample collected from landfills | 0.1 μg/L | Barnes et al. (2004) |
| | Cefaclor | Urban wastewater effluents | 60 ng/L | Watkinson et al. (2007) |
| | | Urban wastewater effluents | 1.8 μg/L | Watkinson et al. (2009) |
| | | Surface waters | 0.2 μg/L | Watkinson et al. (2009) |
| | Cephalexin | Surface waters | 15–27 ng/L | Costanzo et al. (2005) |
| | | Urban wastewater effluents | 0.25 μg/L | Watkinson et al. (2009) |
| | | Surface waters | 0.1 μg/L | Watkinson et al. (2009) |
| | | Hospital effluents | 10 μg/L | Watkinson et al. (2009) |
| | Chloramphenicol | Domestic wastewater and surface water | 0.5–10.7 μg/L | Agunbiade and Moodley (2014) |
| | | River sample | n.d.-3.9 ng/L | Chen and Zhou (2014) |
| | | River sediment | n.d.-0.7 μg/kg | Chen and Zhou (2014) |
| | Chlortetracycline | Surface water | n.d.- 150 ng/L | Lindsey et al. (2001) |
| | | Ground water sample collected from landfills | 0.05 μg/L | Barnes et al. (2004) |
| | | Surface water | n.d.- 100 ng/L | Kolpin et al. (2004) |
| | | Urban wastewater effluents | 5 ng/L | Watkinson et al. (2007) |
| | | Urban wastewater effluents | 0.25 μg/L | Watkinson et al. (2009) |
| | | Surface waters | 0.6 μg/L | Watkinson et al. (2009) |
| | | River sample | n.d.- 46.7 ng/L | Chen and Zhou (2014) |
| | | River sediment | n.d.- 6.3 μg/kg | Chen and Zhou (2014) |

| | | | | |
|---|---|---|---|---|
| | Ciprofloxacin | River water | 14.3–26.1 ng/L | Calamari et al. (2003) |
| | | Ground water sample collected from landfills | 0.02 μg/L | Barnes et al. (2004) |
| | | Surface water | n.d.- 30 ng/L | Kolpin et al. (2004) |
| | | River water | 14.4 ng/L | Zuccato et al. (2005) |
| | | Surface waters | 23–42 ng/L | Costanzo et al. (2005) |
| | | Urban wastewater effluents | 720 ng/L | Watkinson et al. (2007) |
| | | Untreated drinking water | 0.03 μg/L | Focazio et al. (2008) |
| | | Surface waters | 1.3 μg/L | Watkinson et al. (2009) |
| | | Hospital effluents | 15 μg/L | Watkinson et al. (2009) |
| | | River waters | 1.3–16 ng/L | Zuccato et al. (2010) |
| | | Domestic wastewater and surface water | 16.9 μg/L | Agunbiade and Moodley (2014) |
| | | River sample | n.d.- 34.2 ng/L | Chen and Zhou (2014) |
| | | River water | 6.7 ng/L | Riva et al. (2015) |
| | | River water | 509 ng/L | Ngumba et al. (2016) |
| | | Wastewater treatment plant effluent | 67 ng/L | Ngumba et al. (2016) |
| | | Wastewater treatment plant | 0.08–1.4 μg/L | Kanama et al. (2018) |
| | Clarithromycin | River water | 0.4–20.3 ng/L | Calamari et al. (2003) |
| | | River water | 8.3 ng/L | Zuccato et al. (2005) |
| | | River waters | 0.8–2.1 ng/L | Zuccato et al. (2010) |
| | | River water | 149 ng/L | Riva et al. (2015) |
| | Clindamycin | Urban wastewater effluents | 5 ng/L | Watkinson et al. (2007) |
| | | Urban wastewater effluents | 0.07 μg/L | Watkinson et al. (2009) |
| | | Surface waters | 0.01 μg/L | Watkinson et al. (2009) |
| | | Hospital effluents | 0.09 μg/L | Watkinson et al. (2009) |
| | | Wastewater from sewage treatment plant | 25 ng/L | Subedi et al. (2017) |
| | Cloxacillin | Urban wastewater effluents | 0.7 μg/L | Watkinson et al. (2009) |
| | Demeclocycline | Surface water | 120–325 ng/L | Yang and Carlson (2003) |

(continued)

Table 6.1 (continued)

| Pharmaceuticals class | Examples | Occurrence | Range in concentration | Reference |
|---|---|---|---|---|
| | Doxycycline | Urban wastewater effluents | 40 ng/L | Watkinson et al. (2007) |
| | | Urban wastewater effluents | 0.15 μg/L | Watkinson et al. (2009) |
| | | Surface waters | 0.4 μg/L | Watkinson et al. (2009) |
| | | Hospital effluents | 0.2 μg/L | Watkinson et al. (2009) |
| | Enrofloxacin | Surface water | n.d.- 10 ng/L | Kolpin et al. (2004) |
| | | Urban wastewater effluents | 10 ng/L | Watkinson et al. (2007) |
| | | Untreated drinking water | 0.04 μg/L | Focazio et al. (2008) |
| | | Urban wastewater effluents | 0.05 μg/L | Watkinson et al. (2009) |
| | | Surface waters | 0.3 μg/L | Watkinson et al. (2009) |
| | | Hospital effluents | 0.1 μg/L | Watkinson et al. (2009) |
| | | River sample | n.d.- 14.6 ng/L | Chen and Zhou (2014) |
| | | River sediment | n.d.- 8.9 μg/kg | Chen and Zhou (2014) |
| | Erythromycin | River water | 1.4–15.9 ng/L | Calamari et al. (2003) |
| | | River water | 4.5 ng/L | Zuccato et al. (2005) |
| | | Effluent | 130 ng/L | Kim et al. (2007) |
| | | Surface water | 3.4 ng/L | Kim et al. (2007) |
| | | River waters | 0.7–4.6 ng/L | Zuccato et al. (2010) |
| | | Groundwater | 72–2380 ng/L | Bartelt-Hunt et al. (2011) |
| | | Domestic wastewater and surface water | 0.58–22.5 μg/L | Agunbiade and Moodley (2014) |
| | | River sample | 0.4–6.9 ng/L | Chen and Zhou (2014) |
| | | River sediment | 1.5–24.6 μg/kg | Chen and Zhou (2014) |
| | Erythromycin-$H_2O$ | Surface water | n.d.- 220 ng/L | Kolpin et al. (2004) |
| | | River water | 9–41 ng/L | Managaki et al. (2007) |
| | | River waters | 1.6–5.3 ng/L | Zuccato et al. (2010) |
| | Levofloxacin | River water | 30.7 ng/L | Riva et al. (2015) |

| | | | | |
|---|---|---|---|---|
| | Lincomycin | Surface water | n.d.- 10 ng/L | Kolpin et al. (2004) |
| | | River water | 24.4 ng/L | Zuccato et al. (2005) |
| | | Urban wastewater effluents | 60 ng/L | Watkinson et al. (2007) |
| | | Ground water | 0.32 μg/L | Barnes et al. (2008) |
| | | Urban wastewater effluents | 0.3 μg/L | Watkinson et al. (2009) |
| | | Surface waters | 0.05 μg/L | Watkinson et al. (2009) |
| | | Hospital effluents | 1.7 μg/L | Watkinson et al. (2009) |
| | | River waters | 3.7–7.4 ng/L | Zuccato et al. (2010) |
| | | Creek waters | 0.006 μg/L | Bernot et al. (2013) |
| | | Wastewater from sewage treatment plant | 130–430 ng/L | Subedi et al. (2017) |
| | Monensin | Creeks and river waters | 3 ng/L | Fisher and Scott (2008) |
| | | Urban wastewater effluents | 0.02 μg/L | Watkinson et al. (2009) |
| | | Surface waters | 0.15 μg/L | Watkinson et al. (2009) |
| | | Groundwater | 180–2350 ng/L | Bartelt-Hunt et al. (2011) |
| | | Carcass burial leachate | 191–11,980 ng/L | Yuan et al. (2013) |
| | Nalidixic acid | Creeks and river waters | 9.6 ng/L | Fisher and Scott (2008) |
| | | Hospital effluents | 0.04 μg/L | Watkinson et al. (2009) |
| | | Surface waters | 0.75 μg/L | Watkinson et al. (2009) |
| | | Urban wastewater effluents | 0.45 μg/L | Watkinson et al. (2009) |
| | | Domestic wastewater and surface water | 1.7–30.8 μg/L | Agunbiade and Moodley (2014) |
| | Norfloxacin | Surface waters | 30–80 ng/L | Costanzo et al. (2005) |
| | | Urban wastewater effluents | 40 ng/L | Watkinson et al. (2007) |
| | | Hospital effluents | 0.2 μg/L | Watkinson et al. (2009) |
| | | Surface waters | 1.15 μg/L | Watkinson et al. (2009) |
| | | Urban wastewater effluents | 0.25 μg/L | Watkinson et al. (2009) |
| | | River sample | n.d.- 0.2 ng/L | Chen and Zhou (2014) |
| | | Wastewater treatment plant | 0.03–0.35 μg/L | Kanama et al. (2018) |

(continued)

**Table 6.1** (continued)

| Pharmaceuticals class | Examples | Occurrence | Range in concentration | Reference |
|---|---|---|---|---|
| | Ofloxacin | River waters | 0.65–18.06 ng/L | Zuccato et al. (2010) |
| | | River sample | n.d.- 28.5 ng/L | Chen and Zhou (2014) |
| | | River sediment | n.d.- 12.4 μg/kg | Chen and Zhou (2014) |
| | Oleandomycin | Urban wastewater effluents | 0.15 μg/L | Watkinson et al. (2009) |
| | | Surface waters | 0.02 μg/L | Watkinson et al. (2009) |
| | | Hospital effluents | 0.04 μg/L | Watkinson et al. (2009) |
| | | River waters | <0.31 ng/L | Zuccato et al. (2010) |
| | Oxytetracycline | Surface water | n.d.- 340 ng/L | Lindsey et al. (2001) |
| | | Surface water | 60–150 ng/L | Yang and Carlson (2003) |
| | | River water | 0.19–19.2 ng/L | Calamari et al. (2003) |
| | | Urban wastewater effluents | 20 ng/L | Watkinson et al. (2007) |
| | | Creeks and river waters | 109–423 ng/L | Fisher and Scott (2008) |
| | | Urban wastewater effluents | 0.007 μg/L | Watkinson et al. (2009) |
| | | Surface waters | 0.1 μg/L | Watkinson et al. (2009) |
| | | River waters | <1.1–1.8 ng/L | Zuccato et al. (2010) |
| | | River sample | n.d.- 219.8 ng/L | Chen and Zhou (2014) |
| | | River sediment | 0.6–18.6 μg/kg | Chen and Zhou (2014) |
| | Penicillin G | Surface waters | 0.25 μg/L | Watkinson et al. (2009) |
| | | Urban wastewater effluents | 0.3 μg/L | Watkinson et al. (2009) |
| | Penicillin V | Urban wastewater effluents | 80 ng/L | Watkinson et al. (2007) |
| | | Hospital effluents | 0.01 μg/L | Watkinson et al. (2009) |
| | | Surface waters | 0.01 μg/L | Watkinson et al. (2009) |
| | | Urban wastewater effluents | 2 μg/L | Watkinson et al. (2009) |

| | | | | |
|---|---|---|---|---|
| | Roxithromycin | Urban wastewater effluents | 100 ng/L | Watkinson et al. (2007) |
| | | Urban wastewater effluents | 0.5 μg/L | Watkinson et al. (2009) |
| | | Surface waters | 0.35 μg/L | Watkinson et al. (2009) |
| | | Hospital effluents | 0.4 μg/L | Watkinson et al. (2009) |
| | | River sample | 0.2–2.2 ng/L | Chen and Zhou (2014) |
| | | River sediment | 0.3–4.1 μg/kg | Chen and Zhou (2014) |
| | Salinomycin | Surface waters | 0.15 μg/L | Watkinson et al. (2009) |
| | Sarafloxacin | Untreated drinking water | 0.02 μg/L | Focazio et al. (2008) |
| | Spiramycin | River water | 74.2 ng/L | Zuccato et al. (2005) |
| | | River waters | <1.4–2.3 ng/L | Zuccato et al. (2010) |
| | Sulfadimethoxine | Surface water | n.d.- 15,000 ng/L | Lindsey et al. (2001) |
| | | Surface water | n.d.- 10 ng/L | Kolpin et al. (2004) |
| | | Confined animal feeding operation waste lagoons | 2.03 μg/L | Batt et al. (2006) |
| | | Private wells for drinking water purpose | 0.04–0.06 μg/L | Batt et al. (2006) |
| | Sulfamethazine | Surface water | n.d.- 220 ng/L | Lindsey et al. (2001) |
| | | Private wells for drinking water purpose | 0.07–0.22 μg/L | Batt et al. (2006) |
| | | Confined animal feeding operation waste lagoons | 45.3 μg/L | Batt et al. (2006) |
| | | River water | 18,500–19,200 ng/L | Managaki et al. (2007) |
| | | Ground water | 0.36 μg/L | Barnes et al. (2008) |
| | | River sample | 19.9–389.4 ng/L | Chen and Zhou (2014) |
| | | River sediment | 0.2–2.7 μg/kg | Chen and Zhou (2014) |

(continued)

Table 6.1 (continued)

| Pharmaceuticals class | Examples | Occurrence | Range in concentration | Reference |
|---|---|---|---|---|
| | Sulfamethoxazole | Ground water | 220 ng/L | Lindsey et al. (2001) |
| | | Surface water | n.d.- 1000 ng/L | Lindsey et al. (2001) |
| | | Surface water | n.d.- 160 ng/L | Yang and Carlson (2003) |
| | | Surface water | n.d.- 70 ng/L | Kolpin et al. (2004) |
| | | Effluent | 136 ng/L | Kim et al. (2007) |
| | | Surface water | 20 ng/L | Kim et al. (2007) |
| | | Urban wastewater effluents | 320 ng/L | Watkinson et al. (2007) |
| | | River water | 20–174 ng/L | Managaki et al. (2007) |
| | | Ground water | 1.11 μg/L | Barnes et al. (2008) |
| | | Urban wastewater effluents | 0.2 μg/L | Watkinson et al. (2009) |
| | | Surface waters | 2 μg/L | Watkinson et al. (2009) |
| | | Hospital effluents | 0.3 μg/L | Watkinson et al. (2009) |
| | | River waters | 1.8–2.3 ng/L | Zuccato et al. (2010) |
| | | Sewage treatment plant effluents | 3.278 μg/L | Kone et al. (2013) |
| | | Surface waters | 0.286 μg/L | Kone et al. (2013) |
| | | River sample | 2.2–764.9 ng/L | Chen and Zhou (2014) |
| | | River sediment | 0.05–0.6 μg/ kg | Chen and Zhou (2014) |
| | | River water | 13,765 ng/L | Ngumba et al. (2016) |
| | | Wastewater treatment plant effluent | 336 ng/L | Ngumba et al. (2016) |
| | | Wastewater from sewage treatment plant | 25–260 ng/L | Subedi et al. (2017) |
| | | River transect | 616.6 ng/L | Cantwell et al. (2018) |
| | | Ground water | 120 ng/L | Bexfield et al. (2019) |
| | Sulfathiazole | Surface water | n.d.- 80 ng/L | Lindsey et al. (2001) |
| | | Urban wastewater effluents | 5 ng/L | Watkinson et al. (2007) |
| | | Urban wastewater effluents | 0.6 μg/L | Watkinson et al. (2009) |
| | | Surface waters | 0.04 μg/L | Watkinson et al. (2009) |
| | | River sample | n.d.- 121.1 ng/L | Chen and Zhou (2014) |
| | | River sediment | n.d.- 0.6 μg/kg | Chen and Zhou (2014) |

| | | | | |
|---|---|---|---|---|
| | Tetracycline | Surface water | n.d.- 110 ng/L | Lindsey et al. (2001) |
| | | Surface water | 60–160 ng/L | Yang and Carlson (2003) |
| | | Surface water | n.d.- 300 ng/L | Kolpin et al. (2004) |
| | | Urban wastewater effluents | 30 ng/L | Watkinson et al. (2007) |
| | | Urban wastewater effluents | 0.002 μg/L | Watkinson et al. (2009) |
| | | Hospital effluents | 0.04 μg/L | Watkinson et al. (2009) |
| | | Surface waters | 0.08 μg/L | Watkinson et al. (2009) |
| | | River sample | n.d.- 54.3 ng/L | Chen and Zhou (2014) |
| | | River sediment | n.d.- 21.7 μg/kg | Chen and Zhou (2014) |
| | | Wastewater treatment plant | 0.5–1.4 μg/L | Kanama et al. (2018) |
| | Tilmicosin | River waters | <0.7–8.9 ng/L | Zuccato et al. (2010) |
| | Trimethoprim | Surface water | n.d.- 80 ng/L | Kolpin et al. (2004) |
| | | Urban wastewater effluents | 70 ng/L | Watkinson et al. (2007) |
| | | Effluent | 58 ng/L | Kim et al. (2007) |
| | | Surface water | 4 ng/L | Kim et al. (2007) |
| | | River water | 7–44 ng/L | Managaki et al. (2007) |
| | | Hospital effluents | 19 μg/L | Watkinson et al. (2009) |
| | | Surface waters | 0.15 μg/L | Watkinson et al. (2009) |
| | | Urban wastewater effluents | 0.25 μg/L | Watkinson et al. (2009) |
| | | Sewage treatment plant effluents | 3.5 μg/L | Kone et al. (2013) |
| | | River water | 2650 ng/L | Ngumba et al. (2016) |
| | | Wastewater treatment plant effluent | 66 ng/L | Ngumba et al. (2016) |
| | | River transect | 350 ng/L | Cantwell et al. (2018) |
| | | Ground water | 14.9 ng/L | Bexfield et al. (2019) |

(continued)

Table 6.1 (continued)

| Pharmaceuticals class | Examples | Occurrence | Range in concentration | Reference |
|---|---|---|---|---|
| | Tylosin | River water | 0.29–2.7 ng/L | Calamari et al. (2003) |
| | | Surface water | 0–20 ng/L | Kolpin et al. (2004) |
| | | Urban wastewater effluents | 65 ng/L | Watkinson et al. (2007) |
| | | Urban wastewater effluents | 3.4 μg/L | Watkinson et al. (2009) |
| | | Surface waters | 0.06 μg/L | Watkinson et al. (2009) |
| | | River waters | <0.7 ng/ L | Zuccato et al. (2010) |
| | | Domestic wastewater and surface water | 0.21–21.9 μg/L | Agunbiade and Moodley (2014) |
| | Vancomycin | River waters | 0.5–11.6 ng/L | Zuccato et al. (2010) |
| | Virginiamycin | Ground water sample collected from landfills | 0.1 μg/L | Barnes et al. (2004) |

summarizes an extensive list of the occurrence of pharmaceuticals belonging to different classes in an array of the aquatic environment.

The most studied pharmaceuticals are acetaminophen, ibuprofen, carbamazepine, atenolol, chlortetracycline, tetracycline, oxytetracycline, erythromycin, ciprofloxacin, enrofloxacin, sulfamethoxazole, sulfamethazine, trimethoprim, tylosin, and lincomycin among others. The concentrations of commonly occurring pharmaceuticals in wastewater and aquatic environment vary in the range from ng/L to μg/L. It is worthwhile to mention that the concentration of some pharmaceuticals e.g. acetaminophen may even reach up to the level 58.7 μg/L as reported by Agunbiade and Moodley (2014) in domestic wastewater and surface waters. Moreover, as they are continuously released into the soil environment, the soil flora and fauna are most probably in contact with these compounds during their entire life. The transport mechanism of different pharmaceuticals in the soil environment is elaborated in the next section.

## 6.3 Transport of Pharmaceuticals into the Soil Environment

The pharmaceuticals, either by irrigation with wastewater and contaminated water or through land application of sewage sludge (biosolids) and animal slurries as a fertilizer, may enter into the soil environment (Christian et al. 2003; Carter et al. 2005; Wang et al. 2014; Petrie et al. 2015). A summary of some of the pharmaceuticals found in manure or soil environments is mentioned in Table 6.2. A number of studies have confirmed that the accumulation of the antibiotic in the soil is due to the fertilization of soil with cattle, poultry, swine and chicken manure (Christian et al. 2003; Brambilla et al. 2007; Awad et al. 2014; Wang et al. 2014). The most commonly used antibiotics such as tetracyclines, chlortetracycline, oxytetracycline, sulfamethazine, and sulfamethoxazole could be found in chicken farm manure at concentrations of 8901 μg/kg, 65,696 μg/kg, 50,923 μg/kg, 2179 μg/kg and 753 μg/kg, respectively (Wang et al. 2014). Sulphonamide antibiotics are known to be highly mobile in the terrestrial environment, and therefore, they tend to be transported into the surrounding water bodies via run-off. Karci and Balcıoglu (2009) found sulfonamide antibiotics in cattle and poultry manure amended soils at concentrations as high as 0.4 mg/kg whereas the concentration of one of the sulfonamide (sulfachloropyridazine) in manure itself was found to be around 35.53 mg/kg.

The fate of pharmaceuticals in agriculture is reliant on the sorption behavior of a particular pharmaceutical compound in the soil which in turn is controlled by the compound's physicochemical properties as well as soil properties (Boxall 2004; Morais et al. 2013). The pharmaceuticals belonging to various classes are different in terms of their molecular structure and physicochemical properties. In particular, certain pharmaceuticals exhibit greater hydrophobicity, for example, Carbamazepine and Bezafibrate, resulting in their retention within the matrix of soil (Tolls 2001; Thiele-Bruhn 2003; Fatta-Kassinos et al. 2011). On the other hand, Petrie et al. (2015) reviewed that antibiotics such as norfloxacin, ciprofloxacin, and ofloxacin

**Table 6.2** Presence of pharmaceuticals in soil, manure and animal slurries

| Pharmaceuticals | Sample | Concentrations reported | References |
|---|---|---|---|
| Carbamazepine | Soil fertilized with dewatered municipal biosolids | 30 μg/kg | Gottschall et al. (2012) |
| | Soil | 0.04 and 0.26 μg/kg | Azzouz and Ballesteros (2012) |
| Chlortetracycline | Liquid manure | 46,000 μg/kg | Martinez-Carballo et al. (2007) |
| | Manure of pig farm | 98–349.2 μg/kg | Wang et al. (2014) |
| | Soil of pig farm | 12.8–21.6 μg/kg | Wang et al. (2014) |
| | Manure of chicken farm | 65,696 μg/kg | Wang et al. (2014) |
| | Soil of chicken farm | 4331 μg/kg | Wang et al. (2014) |
| | Soil closed to swine manure composting facility | 0.13–0.8 μg/L | Awad et al. (2014) |
| Chlortetracycline+Oxytetracycline | Pig manure | 100,000 μg/L | Alvarez et al. (2010) |
| Ciprofloxacin | Manure of chicken farm | 9224 μg/kg | Wang et al. (2014) |
| | Soil of chicken farm | 505 μg/kg | Wang et al. (2014) |
| Diclofenac | Soil | 0.009 and 0.09 μg/kg | Azzouz and Ballesteros (2012) |
| Enrofloxacin | Soil fertilized with cattle and poultry manure | 20–50 μg/kg | Karci and Balcıoglu (2009) |
| | Manure of chicken farm | 8725 μg/kg | Wang et al. (2014) |
| | Soil of chicken farm | 197 μg/kg | Wang et al. (2014) |
| Ibuprofen | Soil | 0.09 and 0.19 μg/kg | Azzouz and Ballesteros (2012) |
| Ketoprofen | Soil | 0.2 and 0.46 μg/kg | Azzouz and Ballesteros (2012) |
| Naproxen | Soil | 0.15 and 0.2 μg/kg | Azzouz and Ballesteros (2012) |

| | | | |
|---|---|---|---|
| Oxytetracycline | Calf manure | 3100 μg/L | Arikan et al. (2006) |
| | Weaning pig slurry | 24,000 μg/kg 4-epi-OXY | Brambilla et al. (2007) |
| | Soil fertilized with weaning pig slurry | 171 μg/kg 4-epi-OXY | Brambilla et al. (2007) |
| | Liquid manure | 29,000 μg/kg | Martinez-Carballo et al. (2007) |
| | Cattle and poultry manure | 200 μg/kg | Karci and Balcıoglu (2009) |
| | Soil fertilized with cattle and poultry manure | 100 μg/kg | Karci and Balcıoglu (2009) |
| | Manure of pig farm | 394–559 μg/kg | Wang et al. (2014) |
| | Soil of pig farm | 28.9–36.8 μg/kg | Wang et al. (2014) |
| | Manure of cattle farm | 515.5 μg/kg | Wang et al. (2014) |
| | Soil of cattle farm | 34.8 μg/kg | Wang et al. (2014) |
| | Manure of chicken farm | 50,923 μg/kg | Wang et al. (2014) |
| | Soil of chicken farm | 3212 μg/kg | Wang et al. (2014) |
| | Soil closed to swine manure composting facility | 0.5–0.7 μg/L | Awad et al. (2014) |
| Sulfachloropyridazine | Cattle and poultry manure | 35,530 μg/kg | Karci and Balcıoglu (2009) |
| | Soil | ~3000–9600 μg/kg | Accinelli et al. (2007) |
| Sulfadiazine | Liquid manure | 91,000 μg/kg | Martinez-Carballo et al. (2007) |
| | Manure of chicken farm | 870 μg/kg | Wang et al. (2014) |
| | Soil of chicken farm | 21.8 μg/kg | Wang et al. (2014) |

(continued)

**Table 6.2** (continued)

| Pharmaceuticals | Sample | Concentrations reported | References |
|---|---|---|---|
| Sulfamethazine | Cattle liquid manure | < 100 μg/kg | Christian et al. (2003) |
| | Swine liquid manure | 1100 μg/kg | Christian et al. (2003) |
| | Soil treated with swine manure | 15 μg/kg | Christian et al. (2003) |
| | Soil | ~ 2000–12,000 μg/kg | Accinelli et al. (2007) |
| | Manure of pig farm | 256.2–426.6 μg/kg | Wang et al. (2014) |
| | Soil of pig farm | 14.4–18.8 μg/kg | Wang et al. (2014) |
| | Manure of chicken farm | 2179 μg/kg | Wang et al. (2014) |
| | Soil of chicken farm | 105.9 μg/kg | Wang et al. (2014) |
| | Soil closed to swine manure composting facility | 0.29–1.1 μg/L | Awad et al. (2014) |
| Sulfamethoxazole | Fresh poultry manure | 3760 μg/kg | Karci and Balcıoglu (2009) |
| | Manure of chicken farm | 753 μg/kg | Wang et al. (2014) |
| | Soil of chicken farm | 20.9 μg/kg | Wang et al. (2014) |
| | Manure of pig farm | 19.6 μg/kg | Wang et al. (2014) |
| | Soil of pig farm | 0.33 μg/kg | Wang et al. (2014) |
| | Soil closed to swine manure composting facility | 0.61–1.1 μg/L | Awad et al. (2014) |
| Sulfathiazole | Manure of pig farm | 14.6 μg/kg | Wang et al. (2014) |
| | Soil closed to swine manure composting facility | 0.08–0.3 μg/L | Awad et al. (2014) |

| | | | |
|---|---|---|---|
| Tetracycline | Liquid manure | 23,000 µg/kg | Martinez-Carballo et al. (2007) |
| | Cattle manure | 1650 µg/kg | Kim et al. (2011) |
| | Pig manure | 16,500 µg/kg | Kim et al. (2011) |
| | Poultry manure | 15,600 µg/kg | Kim et al. (2011) |
| | Manure of chicken farm | 8901 µg/kg | Wang et al. (2014) |
| | Soil of chicken farm | 753 µg/kg | Wang et al. (2014) |
| | Soil closed to swine manure composting facility | 30.1–177.6 µg/L | Awad et al. (2014) |
| Thiamphenicol | Pig waste slurry | 80,000 and 160,000 µg/L | Lallai et al. (2002) |
| Trimethoprim | Liquid manure | 17,000 µg/kg | Martinez-Carballo et al. (2007) |

have more likeability to transport to the aqueous medium because of their hydrophilic mobility. In the case of charge compounds, hydrophobicity alone is not the deciding factor for sorption. In such cases, the cation exchange or anion exchange is the possible mechanism operating during sorption. For instance, the persistence of cationic pharmaceuticals, fluoxetine in the soil is the result of the formation of a cationic complex with clay mineral plus its capacity to be adsorbed onto organic matter (Droge and Goss 2013).

Likewise, soil pH is also one of the deciding factors behind antibiotics sorption into the soil. The sorption capacity of antibiotics is strongly controlled by medium pH. For example, the tetracycline adsorption on montmorillonite and kaolinite as a solution pH increases from 6 to 8. The efficient adsorption of antibiotics is expected to occur with the soil having acidic pH and clay content (Sassman and Lee 2005). Awad et al. (2014) reported a high concentration of tetracycline in sandy loam soils which are having lower pH values of 5.5 and 6.05 and cation exchange capacity of 10.1 and 8.9 cmol/kg, respectively.

The potential of adsorption of pharmaceuticals into the soil matrix is estimated by a parameter distribution coefficient which is usually represented by $K_d$. It represents the adsorption capacity of pharmaceuticals between the sorbent (soil or sewage sludge) and the aqueous phase (pore water) (Fatta-Kassinos et al. 2011). The higher $K_d$ values suggest the greater liability of pharmaceuticals to accumulate in the soil matrix and decreased mobility in soil (Wang and Wang 2015). $K_d$ values generally vary depending upon the properties of different classes of pharmaceuticals. Carter et al. (2016) observed $K_d$ values for diclofenac, carbamazepine, fluoxetine, and orlistat in the range from 5.63 to 18.37 L/kg, 1.34 to 4.45 L/kg, 55.48 to 71.44 L/kg and 28.99 to 110.01 L/kg, respectively. The $K_d$ values for the sulphonamide class of pharmaceuticals are much lower that indicates their greater mobility in soil (Accinelli et al. 2007). The effects of pharmaceuticals on different soil parameters have been described in the next section.

## 6.4 Effect on Soil Parameters

The soil's physical, biochemical and microbiological parameters i.e. pH, soil organic carbon, organic matter, radioactivity, water content, total soil respiration, etc. altered with time which appears to be associated with nature of the pharmaceutical compounds and soil types. Carter et al. (2016) while studying the radioactivity in different soil types; silty sand, clayey loam, and loamy sand observed a decrease in radioactivity after 1 day irrespective of the soil types. In all these soil types though carbamazepine exhibit persistent characteristics but other pharmaceuticals such as diclofenac and orlistat, the radioactivity decreases due to mineralization. Both of these antibiotics can form non-extractable residues with soil and the degree of non-extractable residues formation is dependent upon the soil type. With time, the non-extractable residues fractions increased, attaining a maximum 97.4% of the total radioactivity in silty sand and 49.9% in clayey loam for diclofenac and orlistat exposures, respectively. Change in soil pH had also been seen as a result of pharmaceutical uptake in soil. It has been seen that over 21 days incubation of pharmaceuticals, diclofenac, fluoxetine, and orlistat with the silty sand soil leads to a change in soil pH to 6.6–8.8 from 6.8. Another parameter i.e. organic matter content in the soil, however, remains unaffected even at a high concentration of pharmaceuticals (Al-Farsi et al. 2018).

One of the most significant parameters i.e. the total soil respiration varied with the type of pharmaceutical applied. Soil respiration is an indicator of the carbon stock and is monitored by measuring the carbon dioxide produced through the biological activity of soil organisms (soil microbes, plant roots, and soil fauna). Xu et al. (2016) reported the inhibition of soil basal respiration when incubated with 10 and 100 mg/kg of sulfadiazine for 24 h and 7 days. But on the 28th day, an increase in basal respiration has been found. This is because of the development of resistance against antibiotics by the soil microbes with time (Deng et al. 2009). The soil respiration, however, is not affected so much during incubation of soil with tetracycline, chlortetracycline, sulfapyridine, and tylosin (Thiele-Bruhn and Beck 2005; Liu et al. 2009).

Zielezny et al. (2006) also found no change in respiratory activity when orthic luvisol soil was incubated with sulfadiazine and chlortetracycline at concentrations: 1, 10, and 50 mg/kg for 48 days at a temperature of 20 °C. In contrast, Vaclavik et al. (2004) reported a 1.3–1.7 times increase in soil respiration after pre-treatment with chlortetracycline and sulfachlorpyridazine at initial concentrations of 60 and 600 mg/kg.

## 6.5 Effect on Soil Microbiota

As an indispensable community of soil, the microorganisms play a vital role in maintaining soil quality. Several processes such as nutrient and carbon cycling, energy flow, contaminants degradation, and repression of soil associated diseases are controlled by microorganisms. The study of diversification and functioning of soil microbiota is hence, the pre-requisite in assessing the impact of pharmaceuticals. Table 6.3 throws light on the possible impacts of pharmaceuticals on soil microbiota. Numerous reports documented the possible influence of pharmaceuticals on the structural diversity and functioning of bacteria in agricultural soils (Thiele-Bruhn and Beck 2005; Kong et al. 2006; Zielezny et al. 2006; Cermak et al. 2008; Unger et al. 2013; Xu et al. 2016).

Kong et al. (2006) assessed the functional diversity, evenness, and substrate utilization of the soil microbial community. Oxytetracycline addition in the soil could result in a 63% reduction in functional diversity of soil microbial community at 43 μM and a 41% decrease in functional evenness at 109 μM. The substrate utilization potential of microbes decreased abruptly with an increase in concentration of oxytetracycline. At 1 μM of oxytetracycline, substrate groups (i.e. carbohydrates, amino acids, amines, carboxylic acids) utilization was suppressed by 50% whereas a 65% decrease in polymer utilization was observed. Some antibiotics can induce dose-related effects on soil microbial activity.

**Table 6.3** Studies pertaining to the impacts of pharmaceuticals on soil fauna and soil processes

| Pharmaceuticals | Organism/ Plant/ parameter studied | Example of sample studied | Pharmaceutical dosage | Impacts | References |
|---|---|---|---|---|---|
| Amoxicillin | Bacteria | *Arthrobacter* sp., *Pseudomonas* sp., *Pseudomonas marginalis*, *Pseudomonas lini*, *Lactobacillus reuterii* | 10 mg/kg | Increase in colony forming unit (CFU) and amoxicillin resistant CFU count | Binh et al. (2007) |
| | Vegetable | *Raphanus raphanistrum subsp. sativius* | 1 and 5 mg/ L | Sulfamethoxazole was accumulated in *Raphanus raphanistrum subsp. sativius* roots at 0.81 ± 1.41 and 0.91 ± 1.26 mg/kg for 1 and 5 mg/ L concentrations, respectively. Around 1.73 ± 0.63 mg/kg sulfamethoxazole was translocated from the soil irrigated with treated wastewater with 5 mg/ L of pharmaceuticals | Al-Farsi et al. (2018) |
| Carbamazepine | Earthworm | *Eisenia fetida* | 26 μg/kg | Increase in internal pH of earthworm | Carter et al. (2016) |
| Chlortetracycline | Sewage sludge bacteria | *Nitrosomonas europaea* | $EC_{50}$–0.4 mg/L | Growth inhibition and reduction in nitrification rate of bacteria | Halling-Sørensen (2001) |
| | Acid phosphatase activity | – | 1–300 mg/kg | Decline in soil acid phosphatase activity during 22 days incubation | Liu et al. (2009) |
| | Vegetables of pig farm | Wheat, turnips, cabbage, and leeks | 0.1- 0.53 μg/kg | – | Wang et al. (2014) |
| | Vegetables of chicken farm | Wheat, turnips, cabbage, and leeks | 29.7 μg/kg | – | Wang et al. (2014) |
| Ciprofloxacin | Nitrogen cycling | – | 100 mg/kg | A slight decrease in nitrogen-cycling genes (chiA, amoA, nifH, nirK, nirS, and narG) was observed | Ma et al. (2014) |
| | Vegetables of chicken farm | Wheat, turnips, cabbage, and leeks | 15.7 μg/kg | – | Wang et al. (2014) |

| | | | | | |
|---|---|---|---|---|---|
| Diclofenac | Earthworm | *Eisenia fetida* | 25 μg/kg | Increase in internal pH of earthworm | Carter et al. (2016) |
| | Urease activity | – | 1 and 10 mg/kg | Increase in urease activity with 10 mg/kg of antibiotic until 30th day | Cycon et al. (2016) |
| Enrofloxacin | Crop and vegetables | *Triticum aestivum* L., *Brassica campestris* L., *Cyphomandra betacea* | >5–700 mg/kg | Elongation of roots and shoots are affected | Jin et al. (2009) |
| | Vegetables of chicken farm | Wheat, turnips, cabbage, and leeks | 337 μg/kg | – | Wang et al. (2014) |
| Fluoxetine | Earthworm | *Eisenia fetida* | 28 μg/kg | Decrease in internal pH of earthworm | Carter et al. (2016) |
| Ibuprofen | Urease activity | – | 1 and 10 mg/kg | Stimulation of urease activity during the 90 days experiment | Cycon et al. (2016) |
| Ivermectin | Earthworm | *Eisenia fetida* | 0.5–10 mg/kg of soil | Earthworm survival and reproduction were affected | Rombke et al. (2010) |
| Lincomycin | Bacteria | *Streptomyces lincolnensis* DSM 40355 | 50 and 1000 μg/mL | Fewer colonies at 1000 μg/ mL; larger decrease in the number of putative actinomycetes in the Srbsko soil in comparison to Trebon soil at both the concentrations | Cermak et al. (2008) |
| Metronidazole | Sewage sludge bacteria | *Nitrosomonas europaea* | NOEC-100 mg/L | Growth inhibition of bacteria | Halling-Sørensen (2001) |
| Orlistat | Earthworm | *Eisenia fetida* | 44 μg/kg | Increase in internal pH of earthworm | Carter et al. (2016) |

(continued)

**Table 6.3** (continued)

| Pharmaceuticals | Organism/ Plant/ parameter studied | Example of sample studied | Pharmaceutical dosage | Impacts | References |
|---|---|---|---|---|---|
| Oxytetracycline | Soil fauna | *Folsomia fimetaria, Enchytraeus crypticus, Aporrectodea caliginosa* | 0–5000 mg/kg | Low toxic effect on soil fauna at 3000 mg/ kg | Baguer et al. (2000) |
| | Sewage sludge bacteria | *Nitrosomonas europaea* | $EC_{50}$–1.2 mg/L | Growth inhibition and reduction in nitrification rate of bacteria | Halling-Sørensen (2001) |
| | Plant sample | Corn seeds | < 20 ng/g 4-epi-OXY | – | Brambilla et al. (2007) |
| | Plant sample | *Medicago sativa* L. | > 0.002 mM | 61% and 85% inhibition in shoot and root growth, respectively. | Kong et al. (2007) |
| | Vegetables of pig farm | Wheat, turnips, cabbage, and leeks | 0.3–0.9 μg/kg | – | Wang et al. (2014) |
| | Vegetables of cattle farm | Wheat, turnips, cabbage, and leeks | 0.9 μg/kg | – | Wang et al. (2014) |
| | Vegetables of chicken farm | Wheat, turnips, cabbage, and leeks | 50.1 μg/kg | – | Wang et al. (2014) |
| | Dehydrogenase activity | – | 1–30 mg/kg | Stimulation of dehydrogenase activity after 14 days and decreased sharply on 120th day | Ma et al. (2016) |
| | Microbial biomass carbon ($C_{mic}$) | – | 1–30 mg/kg | Increase in $C_{mic}$ by 1.37–2.08 times after 42 days incubation with antibiotic as compared to control soil (without antibiotic addition) | Ma et al. (2016) |
| | Nitrification process | – | 1–30 mg/kg | 3.01–10.9 times increase after 28 days incubation with antibiotic | Ma et al. (2016) |
| Penicillin G | Sewage sludge bacteria | *Nitrosomonas europaea* | $EC_{50}$–84.6 mg/L | Growth inhibition of bacteria | Halling-Sørensen (2001) |
| Streptomycin | Sewage sludge bacteria | *Nitrosomonas europaea* | $EC_{50}$–0.47 mg/L | Growth inhibition and reduction in nitrification rate of bacteria | Halling-Sørensen (2001) |

| | | | | | |
|---|---|---|---|---|---|
| Sulfadiazine | Sewage sludge bacteria | *Nitrosomonas europaea* | NOEC- 60 mg/L | Growth inhibition and increased nitrification rate of bacteria | Halling-Sørensen (2001) |
| | Crop and vegetables | *Triticum aestivum* L., *Brassica campestris* L., *Cyphomandra betacea* | 5–92 mg/kg | Inhibition of root and shoot elongation | Jin et al. (2009) |
| | Vegetables of chicken farm | Wheat, turnips, cabbage, and leeks | 0.53 µg/kg | – | Wang et al. (2014) |
| | Bacteria | *Pseudomonas*, *Lysobacter*, *Hydrogenophaga* and *Adhaeribacter* | 10 and 100 mg/kg | Decrease in bacterial abundance which results in decrease in soil quality | Ding et al. (2014) |
| | Dehydrogenase activity | – | 10–100 mg/kg | Increase in dehydrogenase activity in treatment SD10 (sulfadiazine concentration- 10 mg/ kg) at end of incubation period | Xu et al. (2016) |
| Sulfamonomethoxine | Crop and vegetables | *Triticum aestivum* L., *Brassica campestris* L., *Cyphomandra betacea* | 10–121 mg/kg | Root and shoot elongation are affected | Jin et al. (2009) |
| | Nitrogen cycling | – | 100 mg/kg | A decrease in nitrogen-cycling genes (chiA, amoA, nifH, nirK, nirS, and narG) was observed | Ma et al. (2014) |
| Sulfadimethoxine | Weed plant | *Lythrum salicaria* | 0.005 and 50 mg/L | Organ-dependent response; roots, cotyledons and cotyledon petioles were prone to antibiotic toxicity, increased growth of internodes and leaves length at the lower drug concentrations | Migliore et al. (2010) |
| | Plant | *Salix fragilis* | 0.01–10 mg/L | Increase in willow tolerance to the antibiotic with the exposure duration | Michelini et al. (2014) |

(continued)

**Table 6.3** (continued)

| Pharmaceuticals | Organism/ Plant/ parameter studied | Example of sample studied | Pharmaceutical dosage | Impacts | References |
|---|---|---|---|---|---|
| Sulfamethazine | Acid phosphatase activity | – | 1–100 mg/kg | Decline in acid phosphatase activity during 22 days incubation period | Liu et al. (2009) |
| | Vegetables of pig farm | Wheat, turnips, cabbage, and leeks | 1.3–1.4 μg/kg | – | Wang et al. (2014) |
| | Vegetables of chicken farm | Wheat, turnips, cabbage, and leeks | 6.4 μg/kg | – | Wang et al. (2014) |
| Sulfamethoxazole | Acid phosphatase activity | | 1–100 mg/kg | Decline in acid phosphatase activity during 22 days incubation period | Liu et al. (2009) |
| | Vegetables of chicken farm | Wheat, turnips, cabbage, and leeks | 1.8 μg/kg | – | Wang et al. (2014) |
| | Vegetable | *Raphanus raphanistrum subsp. sativius* | 1 and 5 mg/ L | Sulfamethoxazole was accumulated in *Raphanus raphanistrum subsp. sativius* roots at 1.06 ± 0.63 and 1.89 ± 0.68 mg/kg for 1 and 5 mg/ L concentrations, respectively. Around 6.04 ± 3.5 mg/kg sulfamethoxazole was translocated from the soil irrigated with treated wastewater with 5 mg/ L of pharmaceuticals | Al-Farsi et al. (2018) |

| | | | | | |
|---|---|---|---|---|---|
| Tetracycline | Sewage sludge bacteria | *Nitrosomonas europaea* | $EC_{50}$–2.2 mg/L | Growth inhibition of bacteria | Halling-Sørensen (2001) |
| | Acid phosphatase activity | – | 1–300 mg/kg | Decline in acid phosphatase activity during 22 days incubation period | Liu et al. (2009) |
| | Vegetables of chicken farm | Wheat, turnips, cabbage, and leeks | 4.3 μg/kg | – | Wang et al. (2014) |
| | Nitrogen cycling | – | 100 mg/kg | A slight decrease in nitrogen-cycling genes (chiA, amoA, nifH, nirK, nirS, and narG) was observed | Ma et al. (2014) |
| | Bacteria | Firmicutes and Chloroflexi | 1–10 μg/kg | Inhibition of biological activity | Zhao et al. (2019) |
| Trimethoprim | Acid phosphatase activity | – | 1–300 mg/kg | Decline in acid phosphatase activity during 22 days incubation period | Liu et al. (2009) |
| Tylosin | Soil fauna | *Folsomia fimetaria, Enchytraeus crypticus, Aporrectodea caliginosa* | 0–5000 mg/kg | Low toxic effect on soil fauna at 3000 mg/ kg | Baguer et al. (2000) |
| | Sewage sludge bacteria | *Nitrosomonas europaea* | $EC_{50}$–54.7 mg/L | Growth inhibition and increased nitrification rate of bacteria | Halling-Sørensen (2001) |
| | Bacteria, fungi, protozoa | – | 100 μg/mL | Community structure in terms of colony morphology and sole carbon source utilization was altered | Westergaard et al. (2001) |
| | Acid phosphatase activity | – | 1–300 mg/kg | Decline in acid phosphatase activity during 22 days incubation period | Liu et al. (2009) |

*NOEC* No observed effect concentrations, $E_{50}$ Effect concentration to 50% inhibition

Thiele-Bruhn and Beck (2005) observed a 10% inhibition in soil microbial activity when tetracycline and sulfapyridine were added to the soil at concentrations of 0.003–7.35 μg/g. Furthermore, these dose-related effects are dependent upon the incubation time. The incubation of tetracycline and sulfapyridine with sandy Cambisol soil results in a significant inhibitory effect in 24 h as compared to the 4 h incubation period. Selective pressure is exerted by tetracycline and sulfapyridine on soil microbial communities which results in dose-related shifts to soil fungi from soil bacteria and this shift was more pronounced during 14 day.

Xu et al. (2016) assessed the single and combined effects of different concentrations of antibiotics; sulfadiazine (10 mg/kg and 100 mg/kg and metal); copper (20 mg/kg and 200 mg/kg) on community structure and microbial metabolic activity. The study concluded that the sulfadiazine or copper showed a reduction in microbial metabolic activity at high concentrations during the 28 days incubation period, while at low concentrations, the insignificant inhibitory effect was seen. However, in a study by Accinelli et al. (2007), it has been found that the pharmaceuticals sulfachlorpyridazine and sulfamethazine at a concentration up to 100 μg/g, exert a negligible effect on the soil microbes.

Aside from the growth inhibition and interference with the community structure as well as the functioning of soil microbes, pharmaceuticals also impact the nitrogen cycling, enzymatic activities and induce antibiotic resistance in soil organisms (Thiele-Bruhn 2003; Liu et al. 2009; Ma et al. 2014). The studies exploring the effect of antibiotic exposure on soil bacteria enzyme activity and soil processes such as nitrogen cycle are given in Table 6.3.

The prolonged exposure of soil bacteria towards pharmaceuticals increases the risk of antibiotic resistance development in the soil bacterial species (Halling-Sørensen et al. 1998; Ma et al. 2014; Atashgahi et al. 2018). According to the study by Ma et al. (2014), the short time exposure of about 7 days to antibiotics; sulfamonomethoxine (SMM), tetracycline (TC), and ciprofloxacin (CIP) results in the increase in a number of the SMM-, TC- and CIP-resistant bacteria. More specifically, SMM-, CIP- and TC- amended soils revealed 24.4% higher SMM-resistant bacteria, 69.44% higher CIP-resistant bacteria, and 416.67% higher TC-resistant bacteria at the 7th day of incubation when compared to control soils (no addition of antibiotics). The food crops grown in soil having such antibiotic-resistant bacteria pose serious health risks to humans and animals via the food chain. The transfer of such resistance through food chain interferes with the success of pharmacotherapies for curing animals and humans (Thiele-Bruhn 2003; Ribeiro et al. 2018).

## 6.6 Effect on Soil Invertebrates

The soil fauna is another explicable component of soil. Different species of earthworms, springtails, enchytraeids, predatory mites, and mollusks are some of the widely occurring fauna in the soil. Their role in mineralization of organic matter, supporting and regulating the primary productivity; and in soil structure development and maintenance motivated the researchers to explore soil fauna reaction towards the entry of antibiotics into the soil. Table 6.3 summarizes the various reports representing the effects of pharmaceuticals on soil invertebrates. A study by Carter et al. (2016) demonstrated the uptake of diclofenac, fluoxetine, orlistat, and carbamazepine into the earthworm *Eisenia fetida*. The uptake rate differs based on soil type. For treatment with fluoxetine, the uptake rate in all soils was reported to be around 0.96–2.35 ml/g/d while the fastest depuration rate was exhibited by carbamazepine (16–0.24/d). Pore water-based bio-concentration factors (BCFs) were highest for orlistat (between 30.5 and 115.9). This is followed by carbamazepine, diclofenac, and fluoxetine having BCFs ranging between 1.1 and 1.6, 7.0 and 69.6, and 14.1 and 20.4, respectively.

The rationale behind studying pore water concentration of respective pharmaceuticals is based upon the fact that the exposures having its high value can exhibit the highest internal concentrations. However, alone pore water concentration is not considered as the deciding factor behind the pharmaceuticals uptake in earthworms. Additionally, the pharmaceuticals also change the internal pH of earthworms as compared to control earthworms (without antibiotic exposure). Conversely, Kinney et al. (2008) conclude that the pharmaceuticals (carbamazepine, trimethoprim, and diphenhydramine) present in earthworm tissues were less than those reported in the corresponding soil samples. It has been seen that on other soil fauna such as springtails and enchytraeids, no effect of pharmaceuticals was found at environmentally relevant concentrations (Baguer et al. 2000). But, the antihelmintic antibiotic i.e. ivermectin affects the reproduction and survival of springtails (Rombke et al. 2010). In addition to impacts on various soil components, pharmaceuticals also affect the plant and related components which have been elaborated in the next section.

## 6.7 Effect on Plants

The extensive application of pharmaceuticals in the soil through biosolids and manure may target the ion channels and enzymes in addition to the inhibition of essential elements that are needed during plant growth (Williams and Cook 2007). The reports on the effect of pharmaceuticals on various crops viz., vegetable, staple, and horticulture crops are mentioned in Table 6.3. The uptake and effect of pharmaceuticals are mainly dependent upon the pharmaceutical type and the plant species. This fact has been explored in a study by Carter et al. (2014) who reported uptake of 52 μg/g and 33 μg/g of carbamazepine in the *Raphanus sativus* and *Lolium perenne*, respectively. Other tested pharmaceuticals, diclofenac, fluoxetine, propranolol, sulfamethazine, however, were below the limit of quantification. Being hydrophilic, carbamazepine instead of attaching to soil particles tends to reside in the aqueous phase. The high amount of carbamazepine in pore water facilitates a central role in its uptake in plants. Further, the difference in carbamazepine uptake between different crop species is attributable to certain factors viz., transpiration rates, degree of root growth, leaf material's shape and size and lipid content which come into play during sorption.

Kumar et al. (2005) reported the accumulation of antibiotic chlortetracycline in the above-ground tissue of cabbage, onion, and corn whereas tylosin uptake in these plants was negligible. Though tylosin being applied at high concentrations in soils, the molecular weight of tylosin is quite large, thereby making its uptake in plants difficult.

The performance of different tissues and organs of the plant towards pharmaceuticals varies depending upon the dose concentration and the time of exposure. Most leaf material exhibited greater pharmaceutical accumulation and hence became more toxic when compared to roots (Carter et al. 2014). However, some studies documented the root retention of hydrophobic neutral pharmaceuticals (Herklotz et al. 2010; Wu et al. 2010). In a study by Al-Farsi et al. (2018), negative effects were foreseen in plants having high concentration (5 mg/L) of spiked pharmaceuticals. Such plants are more prone to attack by pests (aphids and whiteflies) due to the lowest level of resistance and also lose the mature leaves. In the worst cases, few of them may even die.

## 6.8 Removal of Pharmaceuticals from the Agricultural Systems

For long-lasting protection of soil from pharmaceuticals' toxicity, one of the best possible solutions is to remove them from the aqueous phase of the wastewater treatment plant. Biological treatment such as membrane bioreactor, sorption, photolytic degradation, advanced oxidation technologies, bioremediation, and phytoremediation of contaminants are some of the possible practices to treat the xenobiotic pharmaceuticals.

The potentiality of membrane bioreactor in removing pharmaceuticals from wastewater has been investigated during the past few years. Kim et al. (2007) found the removal rate of 99% for the acetaminophen and ibuprofen by membrane bioreactor, but the process proved to be inefficient in removing antibiotics such as erythromycin, naproxen, diclofenac, trimethoprim, and carbamazepine. Tadkaew et al. (2011) reported 95%, 96.7%, 91.9%, and 96.9% removal of paracetamol, ibuprofen, sulfamethoxazole, and atenolol, respectively, by the laboratory scale membrane bioreactor system.

Advanced oxidation technologies, because of their less sophisticated mode of operation and environment-friendly nature have been regarded as an outstanding technology for removing pharmaceuticals from aqueous samples. In this technology, highly reactive radicals, such as sulfate radicals and hydroxyl radicals are generated under ambient conditions that attack and decompose these toxic compounds and even mineralize them into $CO_2$ and $H_2O$. So far, various advanced oxidation technologies, for instance, $TiO_2$/UV, $H_2O_2$/UV, UV/$O_3$, peroxydisulfate/UV systems, and Fenton processes have been used (Tan et al. 2013; Wang and Xu 2012). Amidst these, $H_2O_2$/UV is the most popular among the scientific community attributable to its low cost. Kim et al. (2009) reported more than 90% removal of 23 pharmaceuticals within 30 min time when UV/$H_2O_2$ treatment was used (at a UV dose of 691 mJ/cm). The hydroxyl radical tends to react with the existing organic as well as inorganic water matrices with similar degradation rate constants. However, in another study by Xiao et al. (2019) sulfate radical was used for degradation due to its redox potential of 2.5–3.1 V, high selectivity, and appreciable lifetime. More recently, degradation of antibiotic-clonidine was carried out through hydroxyl radical in UV/$H_2O_2$ and sulfate radicals in UV/peroxydisulfate (Xiao et al. 2020).

Zhang et al. (2020) constructed asymmetric double Z-scheme $BiFeO_3$/$CuBi_2O_4$/$BaTiO_3$ photocatalyst for degradation of norfloxacin. It has been seen that with the increase in light illumination time, the extent of degradation increases. In particular, norfloxacin removal of around 7.8%, 11.0%, 45.5% and 63.5% was obtained within 60 min light illumination for $BaTiO_3$, $CuBi_2O_4$, $BiFeO_3$ and $BiFeO_3$/$CuBi_2O_4$/$BaTiO_3$ photocatalytic systems, respectively. The effective dose of photocatalyst required during degradation of norfloxacin was 1 mg/L. At higher catalyst doses, hindrance in light transmission was observed (Huang et al. 2019).

Due to the high cost of equipment of membrane bioreactor and advanced oxidation, it becomes more appropriate to shift focus towards certain cost-effective approaches for the treatment of pharmaceutical rich wastewater. The application of different remediation approaches is therefore expected to be a well efficient method for reducing the toxicity of pharmaceuticals. Microbes-based bioremediation and phytoremediation are two of the most effective approaches which employ the action of microbes and plants, respectively, for the elimination of pharmaceuticals from the soil. A lot of researchers worldwide isolated pharmaceuticals degrading bacteria such as *Bacillus cereus*, *Labrys portucalensis*, *Enterobacter hormaechei*, *Ochrobactrum* sp., *Gordonia* sp., *Streptomyces* sp., *Gemmatimonas* sp., *Peredibacter* sp., *Ilumatobacter* sp., *Ohtaekwangia* sp., *Legionella* sp., *Hyphomicrobium* sp., *Escherichia* sp. and *Candida* sp. from bovine feces, sediments, activated sludge, wastewater, pig manure and antibiotics contaminated soils (Erickson et al. 2014; Amorim et al. 2014; Aissaoui et al. 2017; Bessa et al. 2017; Zhang et al. 2017; Mulla et al. 2018; Wen et al. 2018; Thelusmond et al. 2018). In particular, pharmaceuticals contaminated soil inhabitant bacteria, *Ochrobactrum* sp. has shown 97% degradation of 100 mg/L of erythromycin A after 72 h of incubation at a temperature of 32 °C and pH of 6.5 (Zhang et al. 2017). A study by Mulla et al. (2018) compared the degradation of sulfamethoxazole by *Ochrobactrum* sp., *Labrys* sp., and *Gordonia* sp. that were isolated from pig manure, wastewater and activated sludge of wastewater treatment plant, respectively. The highest degradation of 62.2% was exhibited by *Labrys* sp. followed by *Gordonia* sp. (51.4%) and *Ochrobactrum* sp. (45.2%) within the incubation period of 288 h. The degradation efficiency can be improved after the addition of growth substrates such as glucose, ammonium, and sodium acetate. Aissaoui et al. (2017) demonstrated that the bacterium *Enterobacter hormaechei* isolated from activated sludge sample degraded

52.8% of diclofenac after 48 h of incubation. On the other hand, under the co-metabolic system (in the presence of 50 mg/L of glucose), 82% elimination of diclofenac was registered.

Utilizing the approach of phytoremediation, successful removal of tetracycline, chlortetracycline, aspirin, sulfamethazine, and sulfadimethoxine can be attained (Lee et al. 2009; Kang et al. 2013; Michelini et al. 2014; Gahlawat and Gauba 2016; Li et al. 2020). A study carried out by Kang et al. (2013) have reported that the vegetable crops (cabbage, carrot, garlic, green bell pepper, lettuce, onion, potato, radish, spinach, sweet corn, and tomato) can uptake five pharmaceuticals namely, chlortetracycline, monensin, sulfamethazine, tylosin, and virginiamycin when grown in clay loam and sandy loam soils that were fertilized with four manures i.e. raw turkey manure, composted turkey manure, raw hog manure, and composted hog manure. Li et al. (2020) found that the ornamental hyperaccumulators, *Mirabilis jalapa* L. and *Tagetes patula* L. have the potential to dissipate 99% of tetracycline when grown in alkaline soils co-contaminated with tetracycline antibiotics and cadmium.

## 6.9 Conclusions and Future Research Recommendations

Pharmaceuticals, nevertheless one of the commendable breakthroughs in improving the life expectancy of humans and animals through the prevention of diseases but its presence beyond optimum levels is detrimental to the integrity of the soil. The irrigation of fields with industrial and hospital wastewater or land application of biosolids as manure provides a route for the entry of such xenobiotics in the soils. The trace amount of pharmaceuticals in the soils is a matter of concern due to their effect on soil indigenous communities. Reduction in soil fertility, stunted plant growth, toxicity to the soil organisms (and subsequent biomagnifications by the toxic soil organisms), development of resistant bacteria and interference with the soil processes including nitrogen cycle, basal respiration rate, and microbial activity are some of the possible risks faced from time to time due to exposure of agricultural lands to pharmaceuticals. Physico-chemical and biological treatment methods could be useful to remove or treat one or more classes of pharmaceuticals but the effectiveness of a particular method primarily depends upon many factors. This requires a thorough understanding of how a particular pharmaceutical bound with the soil components and its potential to undergo degradation. Certain environmental condition such as soil water content, pharmaceutical's physicochemical characteristics and soil properties also need the utmost attention to interpret the success rate of a particular treatment method. Moreover, the soil environment is exposed to a concoction of pharmaceuticals at a particular time but limited studies are available in the literature related to this prospect, hence necessitating elaborative research.

**Acknowledgements** Author is highly obliged to Doaba College, Jalandhar, Punjab, for providing infrastructure for this work. Author also humbly acknowledges the series editors as well as reviewers for providing valuable comments and suggestions that helps a lot in improving the chapter.

## References

Accinelli C, Koskinen WC, Becker JM, Sadowsky MJ (2007) Environmental fate of two sulfonamide antimicrobial agents in soil. J Agric Food Chem 55(7):2677–2682. https://doi.org/10.1021/jf063709j

Agrafioti E, Diamadopoulos E (2012) A strategic plan for reuse of treated municipal wastewater for crop irrigation on the Island of Crete. Agric Water Manag 105:57–64. https://doi.org/10.1016/j.agwat.2012.01.002

Agunbiade FO, Moodley B (2014) Pharmaceuticals as emerging organic contaminants in Umgeni River water system, KwaZulu-Natal, South Africa. Environ Monit Asess 186(11):7273–7291. https://doi.org/10.1007/s10661-014-3926-z

Aissaoui S, Ouled-Haddar H, Sifour M, Harrouche K, Sghaier H (2017) Metabolic and co-metabolic transformation of diclofenac by *Enterobacter hormaechei* D15 isolated from activated sludge. Curr Microbiol 74(3):381–388. https://doi.org/10.1007/s00284-016-1190-x

Al-Farsi R, Ahmed M, Al-Busaidi A, Choudri BS (2018) Assessing the presence of pharmaceuticals in soil and plants irrigated with treated wastewater in Oman. Int J Recycl Org Waste Agric 7(2):165–172. https://doi.org/10.1007/s40093-018-0202-1

Al-Jamal MS, Sammis TW, Mexal JG, Picchioni GA, Zachritz WH (2002) A growth-irrigation scheduling model for wastewater use in forest production. Agric Water Manag 56(1):57–79. https://doi.org/10.1016/S0378-3774(02)00003-3

Alvarez JA, Otero L, Lema JM, Omil F (2010) The effect and fate of antibiotics during the anaerobic digestion of pig manure. Bioresour Technol 101(22):8581–8586. https://doi.org/10.1016/j.biortech.2010.06.075

Amorim CL, Moreira IS, Maia AS, Tiritan ME, Castro PM (2014) Biodegradation of ofloxacin, norfloxacin, and ciprofloxacin as single and mixed substrates by *Labrys portucalensis* F11. Appl Microbiol Biotechnol 98(7):3181–3190. https://doi.org/10.1007/s00253-013-5333-8

Antolin MC, Pascual I, Garcia C, Polo A, Sanchez-Diaz M (2005) Growth, yield and solute content of barley in soils treated with sewage sludge under semiarid Mediterranean conditions. Field Crops Res 94(2–3):224–237. https://doi.org/10.1016/j.fcr.2005.01.009

Arikan OA, Sikora LJ, Mulbry W, Khan SU, Rice C, Foster GD (2006) The fate and effect of oxytetracycline during the anaerobic digestion of manure from therapeutically treated calves. Process Biochem 41(7):1637–1643. https://doi.org/10.1016/j.procbio.2006.03.010

Arun B (2017) Indian agriculture-status, importance and role in Indian Economy. J Stud Manag Plann 3(13):212–213

Atashgahi S, Sanchez-Andrea I, Heipieper HJ, van der Meer JR, Stams AJ, Smidt H (2018) Prospects for harnessing biocide resistance for bioremediation and detoxification. Science 360(6390):743–746

Awad YM, Kim SC, El-Azeem SAA, Kim KH, Kim KR, Kim K, Jeon C, Lee SS, Ok YS (2014) Veterinary antibiotics contamination in water, sediment, and soil near a swine manure composting facility. Environ Earth Sci 71(3):1433–1440. https://doi.org/10.1007/s12665-013-2548-z

Azzouz A, Ballesteros E (2012) Combined microwave-assisted extraction and continuous solid-phase extraction prior to gas chromatography–mass spectrometry determination of pharmaceuticals, personal care products and hormones in soils, sediments and sludge. Sci Total Environ 419:208–215. https://doi.org/10.1016/j.scitotenv.2011.12.058

Baguer AJ, Jensen J, Krogh PH (2000) Effects of the antibiotics oxytetracycline and tylosin on soil fauna. Chemosphere 40(7):751–757. https://doi.org/10.1016/S0045-6535(99)00449-X

Barnes KK, Christenson SC, Kolpin DW, Focazio MJ, Furlong ET, Zaugg SD, Meyer MT, Barber LB (2004) Pharmaceuticals and other organic waste water contaminants within a leachate plume downgradient of a municipal landfill. Ground Water Monit Remidiat 24(2):119–126. https://doi.org/10.1111/j.1745-6592.2004.tb00720.x

Barnes KK, Kolpin DW, Furlong ET, Zaugg SD, Meyer MT, Barber LB (2008) A national reconnaissance of pharmaceuticals and other organic wastewater contaminants in the United States-I) groundwater. Sci Total Environ 402(2–3):192–200. https://doi.org/10.1016/j.scitotenv.2008.04.028

Barroso PJ, Santos JL, Martín J, Aparicio I, Alonso E (2019) Emerging contaminants in the atmosphere: analysis, occurrence and future challenges. Crit Rev Environ Sci Technol 49(2):104–171. https://doi.org/10.1080/106443389.2018.1540761

Bartelt-Hunt S, Snow DD, Damon-Powell T, Miesbach D (2011) Occurrence of steroid hormones and antibiotics in shallow groundwater impacted by livestock waste control facilities. J Contaminant Hydrol 123(3–4):94–103. https://doi.org/10.1016/j.jconhyd.2010.12.010

Batt AL, Snow DD, Aga DS (2006) Occurrence of sulfonamide antimicrobials in private water wells in Washington County, Idaho, USA. Chemosphere 64(11):1963–1971. https://doi.org/10.1016/j.chemosphere.2006.01.029

Beausse J (2004) Selected drugs in solid matrices: a review of environmental determination, occurrence and properties of principal substances. TrAC Trends Anal Chem 23(10–11):753–761. https://doi.org/10.1016/j.trac.2004.08.005

Berge A, Vulliet E (2015) Development of a method for the analysis of hormones and pharmaceuticals in earthworms by quick, easy, cheap, effective, rugged and safe (QuEChERS) extraction followed by liquid chromatography-tandem mass spectrometry (LC-MS/MS). Anal Bioanal Chem 407(26):7995–8008. https://doi.org/10.1007/s00216-015-8972-z

Bernot MJ, Smit L, Frey J (2013) Human and veterinary pharmaceutical abundance and transport in a rural Central Indiana stream influenced by confined animal feeding operations (CAFOs). Sci Total Environ 445:219–230. https://doi.org/10.1016/j.scitotenv.2012.12.039

Bessa VS, Moreira IS, Tiritan ME, Castro PML (2017) Enrichment of bacterial strains for the biodegradation of diclofenac and carbamazepine from activated sludge. Int Biodeterior Biodegrad 120:135–142. https://doi.org/10.1016/j.ibiod.2017.02.008

Bexfield LM, Toccalino PL, Belitz K, Foreman WT, Furlong ET (2019) Hormones and pharmaceuticals in groundwater used as a source of drinking water across the United States. Environ Sci Technol 53(6):2950–2960. https://doi.org/10.1021/acs.est.8b05592

Binh CTT, Heuer H, Gomes NCM, Kotzerke A, Fulle M, Wilke BM, Schloter M, Smalla K (2007) Short-term effects of amoxicillin on bacterial communities in manured soil. FEMS Microbiol Ecol 62(3):290–302. https://doi.org/10.1111/j.1574-6941.2007.00393.x

Boxall AB (2004) The environmental side effects of medication. EMBO Rep 5(12):1110–1116. https://doi.org/10.1038/sj.embor.7400307

Boxall AB, Rudd MA, Brooks BW, Caldwell DJ, Choi K, Hickmann S, Innes E, Ostapyk K, Staveley JP, Verslycke T, Ankley GT, Beazley KF, Belanger SE, Berninger JP, Carriquiriborde P, Coors A, DeLeo PC, Dyer SD, Ericson JF, Gagné F, Giesy JP, Gouin T, Hallstrom L, Karlsson MV, Larsson DGJ, Lazorchak JM, Mastrocco F, McLaughlin A, McMaster ME, Meyerhoff RD, Moore R, Parrott JL, Snape JR, Murray-Smith R, Servos MR, Sibley PK, Straub JO, Szabo ND, Topp E, Tetreault GR, Trudeau VL, Kraak GVD (2012) Pharmaceuticals and personal care products in the environment: what are the big questions? Environ Health Perspec 120(9):1221–1229. https://doi.org/10.1289/ehp.1104477

Brambilla G, Patrizii M, De Filippis SP, Bonazzi G, Mantovi P, Barchi D, Migliore L (2007) Oxytetracycline as environmental contaminant in arable lands. Anal Chim Acta 586(1–2):326–329. https://doi.org/10.1016/j.aca.2006.11.019

Calamari D, Zuccato E, Castiglioni S, Bagnati R, Fanelli R (2003) Strategic survey of therapeutic drugs in the rivers Po and Lambro in northern Italy. Environ Sci Technol 37(7):1241–1248. https://doi.org/10.1021/es020158e

Cantwell MG, Katz DR, Sullivan JC, Shapley D, Lipscomb J, Epstein J, Juhl AR, Knudson C, O'Mullan GD (2018) Spatial patterns of pharmaceuticals and wastewater tracers in the Hudson River Estuary. Water Res 137:335–343. https://doi.org/10.1016/j.watres.2017.12.044

Carlsson C, Johansson AK, Alvan G, Bergman K, Kuhler T (2006) Are pharmaceuticals potent environmental pollutants? Part I: environmental risk assessments of selected active pharmaceutical ingredients. Sci Total Environ 364(1–3):67–87. https://doi.org/10.1016/j.scitotenv.2005.06.035

Carpenter SR, Stanley EH, Vander Zanden MJ (2011) State of the world's freshwater ecosystems: physical, chemical, and biological changes. Annu Rev Environ Resour 36:75–99. https://doi.org/10.1146/annurev-environ-021810-094524

Carter CT, Grieve CM, Poss JA, Suarez DL (2005) Production and ion uptake of Celosia argentea irrigated with saline wastewaters. Sci Hortic 106(3):381–394. https://doi.org/10.1016/j.scienta.2005.04.007

Carter LJ, Harris E, Williams M, Ryan JJ, Kookana RS, Boxall AB (2014) Fate and uptake of pharmaceuticals in soil–plant systems. J Agric Food Chem 62(4):816–825. https://doi.org/10.1021/jf404282y

Carter LJ, Ryan JJ, Boxall AB (2016) Effects of soil properties on the uptake of pharmaceuticals into earthworms. Environ Pollut 213:922–931. https://doi.org/10.1016/j.envpol.2016.03.044

Cermak L, Kopecky J, Novotna J, Omelka M, Parkhomenko N, Plhackova K, Sagova-Mareckova M (2008) Bacterial communities of two contrasting soils reacted differently to lincomycin treatment. Appl Soil Ecol 40(2):348–358. https://doi.org/10.1016/j.apsoil.2008.06.001

Chen K, Zhou JL (2014) Occurrence and behavior of antibiotics in water and sediments from the Huangpu River, Shanghai, China. Chemosphere 95:604–612. https://doi.org/10.1016/j.chemosphere.2013.09.119

Christian T, Schneider RJ, Farber HA, Skutlarek D, Meyer MT, Goldbach HE (2003) Determination of antibiotic residues in manure, soil, and surface waters. Acta Hydrochim Hydrobiol 31(1):36–44. https://doi.org/10.1002/aheh.200390014

Colinas C, Ingham E, Molina R (1994) Population responses of target and non-target forest soil organisms to selected biocides. Soil Biol Biochem 26(1):41–47. https://doi.org/10.1016/0038-0717(94)90193-7

Costanzo SD, Murby J, Bates J (2005) Ecosystem response to antibiotics entering the aquatic environment. Mar Pollut Bull 51(1–4):218–223. https://doi.org/10.1016/j.marpolbul.2004.10.038

Cycon M, Borymski S, Zołnierczyk B, Piotrowska-Seget Z (2016) Variable effects of non-steroidal anti-inflammatory drugs (NSAIDs) on selected biochemical processes mediated by soil microorganisms. Front Microbiol 7:1–19. https://doi.org/10.3389/fmicb.2016.01969

Deng H, Li XF, Cheng WD, Zhu YG (2009) Resistance and resilience of Cu-polluted soil after Cu perturbation, tested by a wide range of soil microbial parameters. FEMS Microbiol Ecol 70(2):293–304. https://doi.org/10.1111/j.1574-6941.2009.00741.x

Ding GC, Radl V, Schloter-Hai B, Jechalke S, Heuer H, Smalla K, Schloter M (2014) Dynamics of soil bacterial communities in response to repeated application of manure containing sulfadiazine. PLoS One 9:1–10. https://doi.org/10.1371/journal.pone.0092958

Droge ST, Goss KU (2013) Development and evaluation of a new sorption model for organic cations in soil: contributions from organic matter and clay minerals. Environ Sci Technol 47(24):14233–14241. https://doi.org/10.1021/es4031886

Erickson BD, Elkins CA, Mullis LB, Heinze TM, Wagner RD, Cerniglia CE (2014) A metallo-β-lactamase is responsible for the degradation of ceftiofur by the bovine intestinal bacterium *Bacillus cereus* P41. Vet Microbiol 172(3–4):499–504. https://doi.org/10.1016/j.vetmic.2014.05.032

Fatta-Kassinos D, Meric S, Nikolaou A (2011) Pharmaceutical residues in environmental waters and wastewater: current state of knowledge and future research. Anal Bioanal Chem 399(1):251–275. https://doi.org/10.1007/s00216-010-4300-9

Fisher PM, Scott R (2008) Evaluating and controlling pharmaceutical emissions from dairy farms: a critical first step in developing a preventative management approach. J Cleaner Product 16(14):1437–1446. https://doi.org/10.1016/j.jclepro.2008.04.024

Focazio, MJ, Kolpin DW, Barnes KK, Furlong ET, Meyer MT, Zaugg SD, Barber LB, Thurman ME (2008) A national reconnaissance for pharmaceuticals and other organic wastewater contaminants in the United States—II) untreated drinking water sources. Sci Total Environ 402(2–3):201–216. https://doi.org/10.1016/j.scitotenv.2008.02.021

Gahlawat S, Gauba P (2016) Phytoremediation of aspirin and tetracycline by *Brassica juncea*. Int J Phytoremediation 18(9):929–935. https://doi.org/10.1080/15226514.2015.1131230

Gavrilescu M, Demnerova K, Aamand J, Agathos S, Fava F (2015) Emerging pollutants in the environment: present and future challenges in biomonitoring, ecological risks and bioremediation. New Biotechnol 32(1):147–156. https://doi.org/10.1016/j.nbt.2014.01.001

Gottschall N, Topp E, Metcalfe C, Edwards M, Payne M, Kleywegt S, Russell P, Lapen DR (2012) Pharmaceutical and personal care products in groundwater, subsurface drainage, soil, and wheat grain, following a high single application of municipal biosolids to a field. Chemosphere 87(2):194–203. https://doi.org/10.1016/j.chemosphere.2011.12.018

Halling-Sørensen B (2001) Inhibition of aerobic growth and nitrification of bacteria in sewage sludge by antibacterial agents. Arch Environ Contam Toxicol 40(4):451–460. https://doi.org/10.1007/s002440010197

Halling-Sørensen B, Nielsen SN, Lanzky PF, Ingerslev F, Lützhøft HH, Jørgensen S (1998) Occurrence, fate and effects of pharmaceutical substances in the environment-a review. Chemosphere 36(2):357–393. https://doi.org/10.1016/S0045-6535(97)00354-8

Herklotz PA, Gurung P, Heuvel BV, Kinney CA (2010) Uptake of human pharmaceuticals by plants grown under hydroponic conditions. Chemosphere 78(11):1416–1421. https://doi.org/10.1016/j.chemosphere.2009.12.048

Hoper H, Kues J, Nau H, Hamscher G (2002) Eintrag und Verbleib von Tierarzneimittelwirkstoffen in Böden. Bodenschutz 4(2):141–148

Huang D, Li J, Zeng G, Xue W, Chen S, Li Z, Deng R, Yang Y, Cheng M (2019) Facile construction of hierarchical flower-like Z-scheme AgBr/$Bi_2WO_6$ photocatalysts for effective removal of tetracycline: degradation pathways and mechanism. Chem Eng J 375:121991. https://doi.org/10.1016/j.cej.2019.121991

Jin C, Chen Q, Sun R, Zhou Q, Liu J (2009) Eco-toxic effects of sulfadiazine sodium, sulfamonomethoxine sodium and enrofloxacin on wheat, Chinese cabbage and tomato. Ecotoxicology 18(7):878–885. https://doi.org/10.1007/s10646-009-0349-7

Kaczala F, Blum SE (2016) The occurrence of veterinary pharmaceuticals in the environment: a review. Curr Anal Chem 12(3):169–182. https://doi.org/10.2174/1573411012666151009193108

Kanama KM, Daso AP, Mpenyana-Monyatsi L, Coetzee MA (2018) Assessment of pharmaceuticals, personal care products, and hormones in wastewater treatment plants receiving inflows from health facilities in North West Province, South Africa. J Toxicol 2018:1–15. https://doi.org/10.1155/2018/3751930

Kang DH, Gupta S, Rosen C, Fritz V, Singh A, Chander Y, Murray H, Rohwer C (2013) Antibiotic uptake by vegetable crops from manure-applied soils. J Agric Food Chem 61(42):9992–10001. https://doi.org/10.1021/jf404045m

Karci A, Balcıoglu IA (2009) Investigation of the tetracycline, sulfonamide, and fluoroquinolone antimicrobial compounds in animal manure and agricultural soils in Turkey. Sci Total Environ 407(16):4652–4664. https://doi.org/10.1016/j.scitotenv.2009.04.047

Kim SD, Cho J, Kim IS, Vanderford BJ, Snyder SA (2007) Occurrence and removal of pharmaceuticals and endocrine disruptors in South Korean surface, drinking, and waste waters. Water Res 41(5):1013–1021. https://doi.org/10.1016/j.watres.2006.06.034

Kim I, Yamashita N, Tanaka H (2009) Photodegradation of pharmaceuticals and personal care products during UV and UV/$H_2O_2$ treatments. Chemosphere 77(4):518–525. https://doi.org/10.1016/j.chemosphere.2009.07.041

Kim KR, Owens G, Kwon SI, So KH, Lee DB, Ok YS (2011) Occurrence and environmental fate of veterinary antibiotics in the terrestrial environment. Water Air Soil Pollut 214(1–4):163–174. https://doi.org/10.1007/s11270-010-0412-2

Kinney CA, Furlong ET, Kolpin DW, Burkhardt MR, Zaugg SD, Werner SL, Bossio JP, Benotti MJ (2008) Bioaccumulation of pharmaceuticals and other anthropogenic waste indicators in earthworms from agricultural soil amended with biosolid or swine manure. Environ Sci Technol 42(6):1863–1870. https://doi.org/10.1021/es702304c

Kolpin DW, Skopec M, Meyer MT, Furlong ET, Zaugg SD (2004) Urban contribution of pharmaceuticals and other organic wastewater contaminants to streams during differing flow conditions. Sci Total Environ 328(1–3):119–130. https://doi.org/10.1016/j.scitotenv.2004.01.015

Kone M, Cologgi DL, Lu W, Smith DW, Ulrich AC (2013) Pharmaceuticals in Canadian sewage treatment plant effluents and surface waters: occurrence and environmental risk assessment. Environ Technol Rev 2(1):17–27. https://doi.org/10.1080/21622515.2013.865793

Kong WD, Zhu YG, Fu BJ, Marschner P, He JZ (2006) The veterinary antibiotic oxytetracycline and Cu influence functional diversity of the soil microbial community. Environ Pollut 143(1):129–137. https://doi.org/10.1016/j.envpol.2005.11.003

Kong WD, Zhu YG, Liang YC, Zhang J, Smith FA, Yang M (2007) Uptake of oxytetracycline and its phytotoxicity to alfalfa (*Medicago sativa* L.). Environ Pollut 147(1):187–193. https://doi.org/10.1016/j.envpol.2006.08.016

Kumar K, Gupta SC, Baidoo SK, Chander Y, Rosen CJ (2005) Antibiotic uptake by plants from soil fertilized with animal manure. J Environ Qual 34(6):2082–2085. https://doi.org/10.2134/jeq2005.0026

Lallai A, Mura G, Onnis N (2002) The effects of certain antibiotics on biogas production in the anaerobic digestion of pig waste slurry. Bioresour Technol 82(2):205–208. https://doi.org/10.1016/S0960-8524(01)00162-6

Lee HY, Lim JE, Kwon OK, Kim SC, Yang JE, Ok YS (2009) Removal of antibiotics in soil and water: a literature review. J Agric Life Sci 20:45–54

Li WC (2014) Occurrence, sources, and fate of pharmaceuticals in aquatic environment and soil. Environ Pollut 187:193–201. https://doi.org/10.1016/j.envpol.2014.01.015

Li X, Zhu W, Meng G, Guo R, Wang Y (2020) Phytoremediation of alkaline soils co-contaminated with cadmium and tetracycline antibiotics using the ornamental hyperaccumulators *Mirabilis jalapa* L. and *Tagetes patula* L. Environ Sci Pollut Res 27:14175–14183. https://doi.org/10.1007/s11356-020-07975-2

Libutti A, Gatta G, Gagliardi A, Vergine P, Pollice A, Beneduce L, Disciglio G, Tarantino E (2018) Agro-industrial wastewater reuse for irrigation of a vegetable crop succession under Mediterranean conditions. Agric Water Manag 196:1–14. https://doi.org/10.1016/j.agwat.2017.10.015

Lim SJ, Seo CK, Kim TH, Myung SW (2013) Occurrence and ecological hazard assessment of selected veterinary medicines in livestock wastewater treatment plants. J Environ Sci Health Part B 48(8):658–670. https://doi.org/10.1080/03601234.2013.778604

Lindsey ME, Meyer M, Thurman EM (2001) Analysis of trace levels of sulfonamide and tetracycline antimicrobials in groundwater and surface water using solid-phase extraction and liquid chromatography/mass spectrometry. Anal Chem 73(19):4640–4646. https://doi.org/10.1021/ac010514w

Liu F, Ying GG, Tao R, Zhao JL, Yang JF, Zhao LF (2009) Effects of six selected antibiotics on plant growth and soil microbial and enzymatic activities. Environ Poll 157(5):1636–1642. https://doi.org/10.1016/j.envpol.2008.12.021

Liu X, Zhang H, Li L, Fu C, Tu C, Huang Y, Wu L, Tang J, Luo Y, Christie P (2016) Levels, distributions and sources of veterinary antibiotics in the sediments of the Bohai Sea in China and surrounding estuaries. Mar Pollut Bull 109(1):597–602. https://doi.org/10.1016/j.marpolbul.2016.05.033

Ma J, Lin H, Sun W, Wang Q, Yu Q, Zhao Y, Fu J (2014) Soil microbial systems respond differentially to tetracycline, sulfamonomethoxine, and ciprofloxacin entering soil under pot experimental conditions alone and in combination. Environ Sci Pollut Res 21(12):7436–7448. https://doi.org/10.1007/s11356-014-2685-2

Ma T, Pan X, Liu W, Christie P, Luo Y, Wu L (2016) Effects of different concentrations and application frequencies of oxytetracycline on soil enzyme activities and microbial community diversity. Eur J Soil Biol 76:53–60. https://doi.org/10.1016/j.ejsobi.2016.07.004

Managaki S, Murata A, Takada H, Tuyen BC, Chiem NH (2007) Distribution of macrolides, sulfonamides, and trimethoprim in tropical waters: ubiquitous occurrence of veterinary antibiotics in the Mekong Delta. Environ Sci Technol 41(23):8004–8010. https://doi.org/10.1021/es0709021

Martinez-Carballo E, Gonzalez-Barreiro C, Scharf S, Gans O (2007) Environmental monitoring study of selected veterinary antibiotics in animal manure and soils in Austria. Environ Poll 48(2):570–579. https://doi.org/10.1016/j.envpol.2006.11.035

Michelini L, Gallina G, Capolongo F, Ghisi R (2014) Accumulation and response of willow plants exposed to environmental relevant sulfonamide concentrations. Int J Phytoremediation 16(9):947–961. https://doi.org/10.1080/15226514.2013.810576

Migliore L, Rotini A, Cerioli NL, Cozzolino S, Fiori M (2010) Phytotoxic antibiotic sulfadimethoxine elicits a complex hormetic response in the weed *Lythrum salicaria* L. Dose Response 8(4):414–427

Mohamed AMO, Paleologos EK (2017) Fundamentals of geoenvironmental engineering: understanding soil, water, and pollutant interaction and transport, 1st edn. Butterworth-Heinemann, Oxford, pp 327–358

Morais SA, Delerue-Matos C, Gabarrell X, Blanquez P (2013) Multimedia fate modeling and comparative impact on freshwater ecosystems of pharmaceuticals from biosolids-amended soils. Chemosphere 93(2):252–262. https://doi.org/10.1016/j.chemosphere.2013.04.074

Mulla SI, Hu A, Sun Q, Li J, Suanon F, Ashfaq M, Yu CP (2018) Biodegradation of sulfamethoxazole in bacteria from three different origins. J Environ Manag 206:93–102. https://doi.org/10.1016/j.jenvman.2017.10.029

Ngumba E, Gachanja A, Tuhkanen T (2016) Occurrence of selected antibiotics and antiretroviral drugs in Nairobi River Basin, Kenya. Sci Total Environ 539:206–213. https://doi.org/10.1016/j.scitotenv.2015.08.139

Petrie B, Barden R, Kasprzyk-Hordern B (2015) A review on emerging contaminants in wastewaters and the environment: current knowledge, understudied areas and recommendations for future monitoring. Water Res 72:3–27. https://doi.org/10.1016/j.watres.2014.08.053

Ribeiro AR, Sures B, Schmidt TC (2018) Cephalosporin antibiotics in the aquatic environment: a critical review of occurrence, fate, ecotoxicity and removal technologies. Environ Pollut 241:1153–1166. https://doi.org/10.1016/j.envpol.2018.06.040

Riva F, Zuccato E, Castiglioni S (2015) Prioritization and analysis of pharmaceuticals for human use contaminating the aquatic ecosystem in Italy. J Pharm Biomed Anal 106:71–78. https://doi.org/10.1016/j.jpba.2014.10.003

Rombke J, Krogh KA, Moser T, Scheffczyk A, Liebig M (2010) Effects of the veterinary pharmaceutical ivermectin on soil invertebrates in laboratory tests. Arch Environ Contam Toxicol 58(2):332–340. https://doi.org/10.1007/s00244-009-9414-8

Rusan MJM, Hinnawi S, Rousan L (2007) Long term effect of wastewater irrigation of forage crops on soil and plant quality parameters. Desalination 215(1–3):143–152. https://doi.org/10.1016/j.desal.2006.10.032

Sassman SA, Lee LS (2005) Sorption of three tetracyclines by several soils: assessing the role of pH and cation exchange. Environ Sci Technol 39(19):7452–7459. https://doi.org/10.1021/es0480217

Sauve S, Desrosiers M (2014) A review of what is an emerging contaminant. Chem Cent J 8(1):1–7. https://doi.org/10.1186/1752-153X-8-15

Sim WJ, Lee JW, Oh JE (2010) Occurrence and fate of pharmaceuticals in wastewater treatment plants and rivers in Korea. Environ Pollut 158(5):1938–1947. https://doi.org/10.1016/j.envpol.2009.10.036

Sim WJ, Kim HY, Choi SD, Kwon JH, Oh JE (2013) Evaluation of pharmaceuticals and personal care products with emphasis on anthelmintics in human sanitary waste, sewage, hospital wastewater, livestock wastewater and receiving water. J Hazard Mat 248:219–227. https://doi.org/10.1016/j.jhazmat.2013.01.007

Subedi B, Balakrishna K, Joshua DI, Kannan K (2017) Mass loading and removal of pharmaceuticals and personal care products including psychoactives, antihypertensives, and antibiotics in two sewage treatment plants in southern India. Chemosphere 167:429–437. https://doi.org/10.1016/j.chemosphere.2016.10.026

Sui Q, Zhao W, Cao X, Lu S, Qiu Z, Gu X, Yu G (2017) Pharmaceuticals and personal care products in the leachates from a typical landfill reservoir of municipal solid waste in Shanghai, China: occurrence and removal by a full-scale membrane bioreactor. J Hazard Mat 323:99–108. https://doi.org/10.1016/j.jhazmat.2016.03.047

Tadkaew N, Hai FI, McDonald JA, Khan SJ, Nghiem LD (2011) Removal of trace organics by MBR treatment: the role of molecular properties. Water Res 45(8):2439–2451. https://doi.org/10.1016/j.watres.2011.01.023

Tan C, Gao N, Deng Y, Zhang Y, Sui M, Deng J, Zhou S (2013) Degradation of antipyrine by UV, UV/$H_2O_2$ and UV/PS. J Hazard Mater 260:1008–1016. https://doi.org/10.1016/j.jhazmat.2013.06.060

Thelusmond JR, Kawka E, Strathmann TJ, Cupples AM (2018) Diclofenac, carbamazepine and triclocarban biodegradation in agricultural soils and the microorganisms and metabolic pathways affected. Sci Total Environ 640:1393–1410. https://doi.org/10.1016/j.scitotenv.2018.05.403

Thiele-Bruhn S (2003) Pharmaceutical antibiotic compounds in soils- a review. J Plant Nutr Soil Sc 166(2):145–167. https://doi.org/10.1002/jpln.200390023

Thiele-Bruhn S, Beck IC (2005) Effects of sulfonamide and tetracycline antibiotics on soil microbial activity and microbial biomass. Chemosphere 59(4):457–465. https://doi.org/10.1016/j.chemosphere.2005.01.023

Tolls J (2001) Sorption of veterinary pharmaceuticals in soils: a review. Environ Sci Technol 35(17):3397–3406. https://doi.org/10.1021/es0003021

Toze S (2006) Reuse of effluent water-benefits and risks. Agric Water Manag 80(1–3):147–159. https://doi.org/10.1016/j.agwat.2005.07.010

Unger IM, Goyne KW, Kennedy AC, Kremer RJ, McLain JE, Williams CF (2013) Antibiotic effects on microbial community characteristics in soils under conservation management practices. Soil Sci Soc Am J 77(1):100–112. https://doi.org/10.2136/sssaj2012.0099

Vaclavik E, Halling-Sørensen B, Ingerslev F (2004) Evaluation of manometric respiration tests to assess the effects of veterinary antibiotics in soil. Chemosphere 56(7):667–676. https://doi.org/10.1016/j.chemosphere.2004.02.018

Wang S, Wang H (2015) Adsorption behavior of antibiotic in soil environment: a critical review. Front Environ Sci Eng 9(4):565–574. https://doi.org/10.1007/s11783-015-0801-2

Wang JL, Xu LJ (2012) Advanced oxidation processes for wastewater treatment: formation of hydroxyl radical and application. Crit Rev Environ Sci Technol 42(3):251–325. https://doi.org/10.1080/10643389.2010.507698

Wang N, Guo X, Xu J, Kong X, Gao S, Shan Z (2014) Pollution characteristics and environmental risk assessment of typical veterinary antibiotics in livestock farms in Southeastern China. J Environ Sci and Health Part B 49(7):468–479.https://doi.org/10.1080/03601234.2014.896660

Watkinson AJ, Murby EJ, Costanzo SD (2007) Removal of antibiotics in conventional and advanced wastewater treatment: implications for environmental discharge and wastewater recycling. Water Res 41(18):4164–4176. https://doi.org/10.1016/j.watres.2007.04.005

Watkinson AJ, Murby EJ, Kolpin DW, Costanzo SD (2009) The occurrence of antibiotics in an urban watershed: from wastewater to drinking water. Sci Total Environ 407(8):2711–2723. https://doi.org/10.1016/j.scitotenv.2008.11.059

Wen X, Wang Y, Zou Y, Ma B, Wu Y (2018) No evidential correlation between veterinary antibiotic degradation ability and resistance genes in microorganisms during the biodegradation of doxycycline. Ecotox Ennviron Saf 147:759–766. https://doi.org/10.1016/j.ecoenv.2017.09.025

Westergaard K, Muller AK, Christensen S, Bloem J, Sorensen SJ (2001) Effects of tylosin as a disturbance on the soil microbial community. Soil Biol Biochem 33(15):2061–2071. https://doi.org/10.1016/S0038-0717(01)00134-1

Williams RT, Cook JC (2007) Exposure to pharmaceuticals present in the environment. Ther Innov Regul Sci 41(2):133–141. https://doi.org/10.1177/009286150704100202

Wu C, Spongberg AL, Witter JD, Fang M, Czajkowski KP (2010) Uptake of pharmaceutical and personal care products by soybean plants from soils applied with biosolids and irrigated with contaminated water. Environ Sci Technol 44(16):6157–6161. https://doi.org/10.1021/es1011115

Xia K, Bhandari A, Das K, Pillar G (2005) Occurrence and fate of pharmaceuticals and personal care products (PPCPs) in biosolids. J Environ Qual 34(1):91–104. https://doi.org/10.2134/jeq2005.0091

Xiao R, Liu K, Bai L, Minakata D, Seo Y, Goktaş RK, Dionysisou DD, Tang CJ, Wei Z, Spinney R (2019) Inactivation of pathogenic microorganisms by sulfate radical: present and future. Chem Eng J 371:222–232. https://doi.org/10.1016/j.cej.2019.03.296

Xiao R, He L, Luo Z, Spinney R, Wei Z, Dionysiou DD, Zhao F (2020) An experimental and theoretical study on the degradation of clonidine by hydroxyl and sulfate radicals. Sci Total Environ 710:136333. https://doi.org/10.1016/j.scitotenv.2019.136333

Xu Y, Yu W, Ma Q, Wang J, Zhou H, Jiang C (2016) The combined effect of sulfadiazine and copper on soil microbial activity and community structure. Ecotoxicol Environ Saf 134:43–52. https://doi.org/10.1016/j.ecoenv.2016.06.041

Yang S, Carlson K (2003) Evolution of antibiotic occurrence in a river through pristine, urban and agricultural landscapes. Water Res 37(19):4645–4656. https://doi.org/10.1016/S0043-1354(03)00399-3

Yuan Q, Snow DD, Bartelt-Hunt SL (2013) Potential water quality impacts originating from land burial of cattle carcasses. Sci Total Environ 456:246–253. https://doi.org/10.1016/j.scitotenv.2013.03.083

Zavatta G (2014) Agriculture remains central to the world economy. 60% of the population depends on agriculture for survival. Expo 2015, Milano

Zhang W, Qiu L, Gong A, Yuan X (2017) Isolation and characterization of a high-efficiency erythromycin A-degrading *Ochrobactrum* sp. strain. Mar Pollut Bull 114(2):896–902. https://doi.org/10.1016/j.marpolbul.2016.10.076

Zhang X, Wang X, Chai J, Xue S, Wang R, Jiang L, Wang J, Zhang Z, Dionysiou DD (2020) Construction of novel symmetric double Z-scheme $BiFeO_3/CuBi_2O_4/BaTiO_3$ photocatalyst with enhanced solar-light-driven photocatalytic performance for degradation of norfloxacin. Appl Catal B: Environ 119017. https://doi.org/10.1016/j.apcatb.2020.119017

Zhao F, Yang L, Chen L, Xiang Q, Li S, Sun L, Yu X, Fang L (2019) Soil contamination with antibiotics in a typical peri-urban area in eastern China: seasonal variation, risk assessment, and microbial responses. J Environ Sci 79:200–212. https://doi.org/10.1016/j.jes.2018.11.024

Zielezny Y, Groeneweg J, Vereecken H, Tappe W (2006) Impact of sulfadiazine and chlorotetracycline on soil bacterial community structure and respiratory activity. Soil Biol Biochem 38(8):2372–2380. https://doi.org/10.1016/j.soilbio.2006.01.031

Zuccato E, Castiglioni S, Fanelli R (2005) Identification of the pharmaceuticals for human use contaminating the Italian aquatic environment. J Hazard Mat 122(3):205–209. https://doi.org/10.1016/j.jhazmat.2005.03.001

Zuccato E, Castiglioni S, Bagnati R, Melis M, Fanelli R (2010) Source, occurrence and fate of antibiotics in the Italian aquatic environment. J Hazard Mat 179(1–3):1042–1048. https://doi.org/10.1016/j.jhazmat.2010.03.110

# Chapter 7
# Antibiotics and Antibiotic Resistance Genes in Agroecosystems as Emerging Contaminants

**Vipin Kumar Singh, Rishikesh Singh, Ajay Kumar, Rahul Bhadouria, Pardeep Singh, and Kin Israel Notarte**

**Abstract** Rapid rise in pharmaceutical industries during past decades in order to cure challenging human diseases has released considerable amounts of antibiotics in natural ecosystem including soil and aquatic ecosystem. The presence of antibiotics beyond the acceptable limits in agro-ecosystems entering through different sources like wastewater irrigation and manure application has witnessed multiple negative consequences on environmental homoeostasis. Moreover, the injudicious application of antibiotics for treatment of human diseases, improvement in crop yield and enhancement in productivity of livestock based meat production have triggered the resistance development in exposed microorganisms dwelling in agricul-

V. K. Singh (✉)
Department of Botany, Centre of Advanced Study in Botany, Institute of Science, Banaras Hindu University, Varanasi, India
e-mail: vipinks85@gmail.com

R. Singh
Institute of Environment and Sustainable Development, Banaras Hindu University, Varanasi, India
e-mail: rishikesh.iesd@gmail.com

A. Kumar
Agriculture Research Organization (ARO), Volcani Center, Rishon LeZion, Israel
e-mail: ajaykumar_bhu@yahoo.com

R. Bhadouria
Department of Botany, University of Delhi, New Delhi, India
e-mail: rahulbhadouriya2@gmail.com

P. Singh
Department of Environmental Science, PGDAV College, University of Delhi, New Delhi, India
e-mail: psingh.rs.apc@itbhu.ac.in

K. I. Notarte
Faculty of Medicine and Surgery, University of Santo Tomas, Manila, Philippines
e-mail: kinotarte@gmail.com



V. Kumar Singh et al. (eds.), *Sustainable Agriculture Reviews 50*, Sustainable Agriculture Reviews 50, https://doi.org/10.1007/978-3-030-63249-6_7

tural soils, putting severe environmental threat to humans and other components of the food chain. The resistance to exposed antibiotics conferred through antibiotic resistance genes is well documented and agro-ecosystem contamination with antibiotic resistance genes as a rising risk is registered globally. So far, numbers of antibiotics and antibiotic resistance genes have been identified as emerging contaminants in soil ecosystem, suggesting the deployment of suitable strategies falling in the categories of physical, chemical and biological strategies to decontaminate the agricultural soils affected with antibiotics and associated resistance genes.

The present chapter entails contamination of soil with different antibiotics and genes responsible for antibiotic resistance, changes in soil microbiological characteristics and important soil processes including biogeochemical cycling of vital nutrients. In addition, possible sources of contamination including manure application, effluents from wastewater treatment plants, discharges from hospitals, households and pharmaceutical industries are also provided. Some of the currently used quantification techniques like high performance liquid chromatography (HPLC), liquid chromatography-mass spectroscopy (LC-MS), quantitative-polymerase chain reaction (q-PCR) and next generation sequencing (NGS) are presented in the article in order to elucidate the fate and transport. Furthermore, impact on human health leading to considerable changes in gut microflora and microbial resistance development against commonly used antibiotics and spread of antibiotic resistance genes in environment along with possible sustainable strategies falling in the category of microbial degradation and phytoremediation have been discussed briefly to deal with such emerging contaminants.

**Keywords** Animal manure · Antimicrobial therapy · Fate and transport · Pharmaceutical wastes · Resistance emergence · Residual limit · Sewage sludge · Toxicity

## Abbreviations

| | |
|---|---|
| ELISA | Enzyme-linked immunosorbent assay |
| HPLC | High performance liquid chromatography |
| LC-MS | Liquid chromatography-mass spectroscopy |
| q-PCR | Quantitative polymerase chain reaction |
| UPLC | Ultra-performance liquid chromatography |

## 7.1 Introduction

Antibiotics, "the wonder drugs" have been largely used in human health care, aquaculture, swine industries (Cheng et al. 2019) and increased meat production from livestock. Presently, more than 260 antibiotics belonging to nearly 20 classes are recognized (Everage et al. 2014). Some of the commonly used antibiotics for treatment of human diseases are lincomycin, cephalothin and erythromycin (Ray et al. 2001), and concentrations of antibiotics belonging to class penicillins, tetracyclines, sulfonamides and macrolides are frequently assessed in environmental matrices (Boxall et al. 2003; Giger et al. 2003). Worldwide, the utilization of antibiotics has been estimated as $10^5$ to $2 \times 10^5$ tons annually (Ahmad et al. 2014) and is recently being considered as emerging contaminant (Muhammad et al. 2020). The large scale and indiscriminate application of human and veterinary antibiotics have contaminated both aquatic and terrestrial ecosystem putting considerable risks to human health and environment. Moreover, treated discharges from wastewater and landfill leachate treatment facilities all around the world, and application of farm manures, introducing substantial contents of antibiotics has emerged as important sources of antibiotics contamination. Therefore, the contamination with antibiotics has been regarded as one of the major global concern due to hazardous impacts on human health and natural ecosystem. The minute content of antibiotics has been found to be associated with modulation in endocrine functions, metabolic activities, developmental processes (Zhang et al. 2013) and microbiota in humans with the resultant of disease susceptibility (Keeney et al. 2014). Further, because of inherent antimicrobial attributes, long term exposure may lead to toxicity, thereby affecting microbiological characteristics of aquatic and terrestrial environment. In addition, introduction of diverse antibiotics into soil system is evidenced to induce resistance development in microorganisms via induction of resistant genes mainly harbored in bacterial systems and negative impact on annelid *Eisenia fetida* (Li et al. 2016). The negative influence of antibiotics on plant root development, singly or in combination has recently been demonstrated by Timmerer et al. (2020).

Antibiotics may enter into soil ecosystem through different pathways including, sewage sludge wastewater from pharmaceutical industries and utilization of farmyard (Timmerer et al. 2020), poultry or other livestock manure in agricultural soils containing antibiotic residues. Very large amount of the antibiotics used in animal feed (Shen et al. 2014) is excreted unabsorbed in form of feaces and urine. As many of the nations around globe are currently giving preferences for the application of organic fertilizers for enhancement of crop productivity, the employment of animal manure is the chief source of soil contamination with antibiotics (Tasho and Cho 2016).

The introduction of different human and veterinary antibiotics in a given soil or water environment is one of the chief factors responsible for the presence of resistant microbes and antibiotic resistance genes (García et al. 2020) rendered by selection pressure posing severe risks to natural ecosystem and overall human health. Furthermore, the excess utilization of antimicrobial substances like antibiotics has resulted into presence of resistant genes in different components of environment fostering the evolution of strains of microorganisms tolerant to numbers of commonly employed antibiotics eventually threatening human health. The antibiotics, even at concentrations lower than required for growth inhibition may provoke the emergence resistant bacterial species (Qiao et al. 2018). So far very large numbers of resistance genes for antibiotics like sulfonamides and tetracyclins have been reported in treated effluents released from sewage treatment facilities globally. Most importantly, the antibiotic resistant bacteria and fungi containing resistance genes is reported to cause higher than 2.8 million contagions and mortality of $3.5 \times 10^4$ humans annually in US only (CDC 2019).

In general, transfer of genes fostering resistance and gene mutation are regarded as the principal mechanisms underlying the introduction of antibiotic resistance genes in bacterial systems. The important route for transfer of antibiotic resistance genes in microbial systems is horizontal gene transfer (Salyers et al. 2004) mediated by transformation, conjugation and transduction. This process involves interspecific as well as intraspecific transference of resistant genes within bacteria. Transfer of resistance genes mediated by transposon is presented by Lu and Lu (2019). Transfer of antibiotic resistance genes from animals to humans via meat, feed and waste products is also presented (Founou et al. 2016). With the rapid division of microbial communities the copy number of particular resistant gene is also multiplied rapidly in a given environment. Most notably, the number of resistant genes in microorganisms is determined by the numbers of antibiotics they are exposed. For instance, 38 genes conferring resistance to tetracycline is explored and represented in bacteria (Roberts 2005; Grabert et al. 2018). Surprisingly, the infections caused by resistant microorganisms could be an important factor for thousands of death by 2050 (O'neill 2014). The antibiotic resistance genes have been identified in different biotic and abiotic environmental components including aquatic and terrestrial ones. The identification of antibiotic resistance genes in drinking water (Zhang et al. 2020a, b, c, d), sediments (Lu and Lu 2020), rivers (Koniuszewska et al. 2020), soil (Lu and Lu 2019), water treatment plants (Xu et al. 2020), manures (Lin et al. 2020), and wild animals (Francesco et al. 2020) is reported globally as major concern to natural ecosystem and the presence is influenced by factors like antibiotics, microbiological community, environmental conditions and chemical pollutants. In the present chapter, we have discussed about different antibiotics and antibiotic resistance genes present in soil, methods of identification, effect on environmental components, transport and fate together with the possible strategies for successful elimination.

## 7.2 Antibiotics

Antibiotic abuse has arisen as a global threat to human health and environmental integrity. Antibiotics have long been used as antimicrobial therapy to control human and animal diseases throughout the globe, apart from employment as supplement in feed to augment the livestock production for increased meat products. Moreover, there is rise in the application of common antibiotics than previous decades. The incomplete metabolism and minute absorption of antibiotics introduced into humans and animals body release very large amount of antibiotics in the natural ecosystem in form of excreta as well as urine (Chen et al. 2017). The inefficiency of sludge, wastewater, water and drinking water treatment plants (Arshad and Zafar 2020) have also contributed much in the release of antibiotic residues in the environment, thereby negatively affecting aquatic and terrestrial ecosystems. Interestingly, concentrations of antibiotics even lower than effective for antimicrobial activity has led to the emergence of antibiotic resistant bacteria often difficult to treat.

Most of the antibiotics being employed for combating various diseases are natural products of microbial origin. Although classifying them is a major challenge, they can be classified according to the chemical structure, origin, mechanism of action, biosynthetic pathway, physicochemical characteristics and biological effects they cause (Béahdy 1974; Calderón and Sabundayo 2007; O'Rourke et al. 2020). Depending on chemical structure, antibiotics can be classified as beta lactams, oxazolidones, sulfonamides, quinolones, aminoglycosides, glycopeptides, macrolides, and tetracyclins. The expansion in available antibiotics has further classified a particular group of antibiotics in first, second, third and fourth generation (King et al. 2000), putting a challenge in classification. On the basis of action mechanism (Fig. 7.1), antibiotics can be grouped as inhibitors of DNA synthesis, RNA synthesis, protein synthesis, cell wall biosynthesis, plasmamembrane biosynthesis and fatty acid biosynthesis (O'Rourke et al. 2020). Classification of antibiotics as antibacterial (Pulingam et al. 2020), antifungal (Salehi et al. 2020), antiviral (Naruse et al. 1990), anticandidal (de Oliveira et al. 2020), and antiprotozoal agent (Maarouf et al. 1997) is also proposed. In following section, we have discussed the sources, fate and transport of antibiotics as well as decontamination approaches currently being practiced for successful elimination from contaminated environmental components.

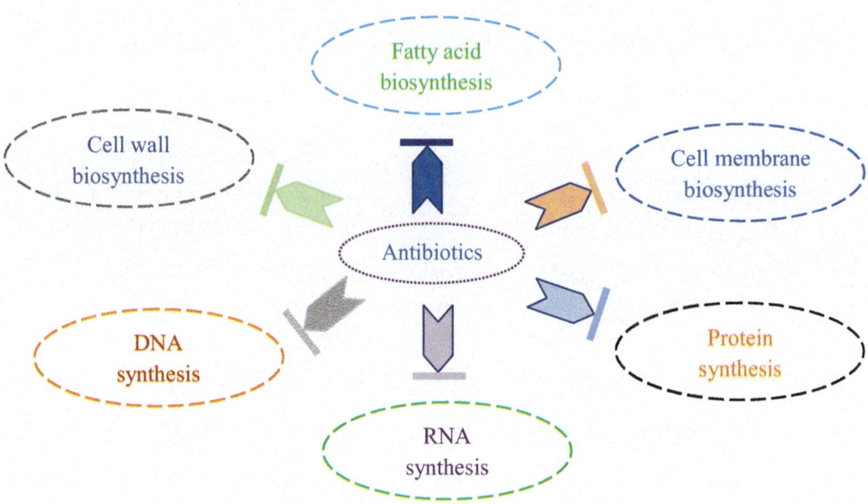


**Fig. 7.1** Important inhibitory mechanisms of action of antibiotics. (Source: O'Rourke et al. 2020)

### 7.2.1 *Sources of Antibiotics Contamination in Soil*

Contamination with antibiotics from different sources has potential to negatively influence the natural characteristics of soil ecosystem. The contamination of soil has increased multiple folds because of rapid rise in employment of antibiotics (Xu et al. 2003). Important direct and indirect sources of antibiotics (Fig. 7.2) with characteristic soil contaminating potential include human and livestock excreta (He et al. 2016; Wang et al. 2020a, b, c), pharmaceutical wastes (Liang et al. 2020a, b), livestock manures (Lin et al. 2020), runoff from soil treated with organic fertilizers, and effluents from sewage sludge, municipal solid waste, water and wastewater treatment facilities (Ngigi et al. 2020) posing harmful impact to associated ecosystem (Fig. 7.3). Most notably, different sources vary considerably in terms of composition, characteristics and antibiotics quantity. The identification of newer possible sources using modern tools of remote sensing could be quite helpful in fostering the strategies essentially needed for the contaminant management.

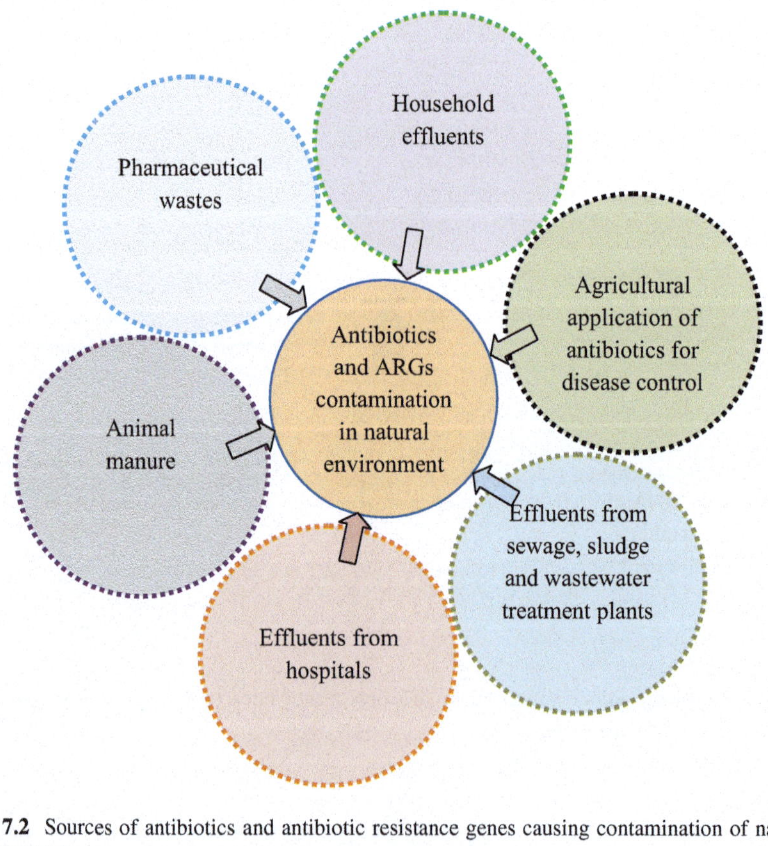


**Fig. 7.2** Sources of antibiotics and antibiotic resistance genes causing contamination of natural environment

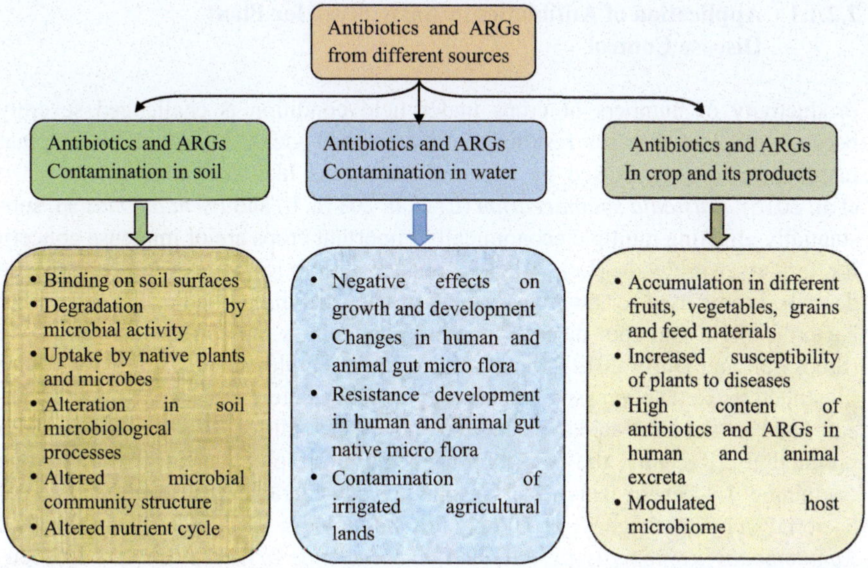


**Fig. 7.3** Impacts associated with antibiotics and antibiotic resistance genes (ARGs) contamination in natural environment

Further, leaching of antibiotics may contaminate the nearby water and groundwater resources. The content of antibiotics in soil, and crop produce may range from microgram to gram per kilogram (Boxall et al. 2002). The presence of tetracyclin antibiotics in animal based manure is reported by Sengeløv et al. (2003). The report on presence of aminoglycosides and phenicol based antibiotics in wastes generated from pharmaceutical operation plants and treatment facilities meant for management of wastewater in Tunisia is represented by Tahrani et al. (2016). The concentration of neomycin, kanamycin B and chloramphenicol in wastewater effluent was determined as 16.4, 7.5, and 3.0 ng $ml^{-1}$, respectively, indicating insufficient elimination of target antibiotics even after treatment. The concentrations of antibiotics lincomycin and tiamulin equivalent to 597 and 115 ng $l^{-1}$, respectively in run-off from soil amended with manure are registered recently (Barrios et al. 2020). Further, the amount of antibiotics lost from manure applied soil was time dependent. Increased timing between manure application and rainfall attributed to the declined level of antibiotics in run off. The occurrence of different antibiotics present in sewage sludge is mentioned by Ajibola et al. (2020). The highest concentration of antibiotics in sewage sludge was determined for oxytetracycline and ciprofloxacin as 4689 ng $g^{-1}$ and 1201 ng $g^{-1}$, respectively based on dry weight basis, providing information pertaining to the antibiotic menace from treatment facilities. In addition to regular monitoring and assessment of antibiotic residues in environmental components, the precise measurement and information regarding fate and transport, therefore, is essentially required to mitigate the negative influences on humans, environment and soil microbiological attributes.

#### 7.2.1.1 Application of Antibiotics in Agriculture for Plant Disease Control

Productivity of numbers of crops under field condition is challenged severely because of plant diseases resulting into decreased yield. Among different plant pathogens, diseases caused by bacterial pathogens like *Xanthomonas* (Lorenzo et al. 2016), *Ralstonia solanacearum* (Li et al. 2011a, b) and *Pseudomonas* sp. substantially affecting multiple economically important crops are of immense concern for farmers worldwide (Lei 2019). Globally, antibiotics as crop protectants are described to be used for the management of bacterial diseases in plants mainly for bacterial blight and spot diseases and streptomycin is more frequently used one (Stockwell and Duffy 2012). More strikingly, the employment of antibiotics were illustrated to exert suppressive impact on pathogens before the development of diseases and are demonstrated to be utilized under the conditions of raised risks to crop productivity. Recently, the contribution of antibiotic potential of propanolamine containing 1,4-benzoxazin-3-ones displaying considerable efficacy in control of bacterial phytopathogens including *Pseudomonas syringae* pv *actinidiae*, *Xanthomonas axonopodis* pv *citri* and *Xanthomonas oryzae* pv *oryzae* is documented by Rao et al. (2020). The utilization of test compound exhibited superior $EC_{50}$ as compared to commercially employed antibiotics and effectiveness in controlling rice blast disease. The mechanism of action revealed destructive effect on cell wall with the resultant cellular death. Antibiotics identified as antifungal peptides may also serve as important foundation for the control of plant diseases induced by fungal pathogen infections (Zhang et al. 2020a, b, c, d). The important antifungal activities are attributed to the interferences in biosynthesis of cell wall and plasma membrane. Antifungal peptides responsible for growth inhibition in different fungi such as *Botrytis cinerea*, *Aspergillus flavus* (Moreno et al. 2003; Muhialdin et al. 2020; Tóth et al. 2020), *Alternaria mali*, *Botrytis cinerea*, *Cladosporium cucumerinum*, *Colletotrichum lagenarium*, *Didimella bryoniae*, and *Magnaporthe grisea* (Kim et al. 2007). The participation of polyene macrolide in management of postharvest losses of grapes is described by Kim et al. (2020). The filtrate extract derived compound showed considerable antifungal action over *Botrytis cinerea* and reduced disease severity substantially.

Although, most of the antibiotics used in agriculture for management of plant diseases are demonstrated to have negative consequences on soil and water environment as well as human health through induction of resistance development not only in pathogenic microorganisms but also in beneficial ones, in addition to the spread of antibiotic resistance genes in natural environment. This problem could be considerably minimized by cutting down the doses via development of newer antibiotics, however cost and effectiveness should also be taken into consideration.

#### 7.2.1.2 Soil Contamination with Antibiotics

Soil has been regarded as the chief sink for diverse antibiotics. Soil quality deterioration with emerging contaminants like antibiotics (Zhu et al. 2020a, b) because of no regulatory restriction posed (Thiele-Bruhn 2003) has been a globally rising environmental concern due to toxicity exerted in soils (Fan et al., 2020), aquatic organisms (Escobar-Huerfano et al. 2020), plants and humans (Jin et al. 2009; Williams-Nguyen et al. 2016). Different sources like poultry manures, sewage sludge (Picó and Andreu 2007) and water released from treatment facilities meant for treating municipal wastes (Pan and Chu 2017a, b), pharmaceutical wastes, sewage sludge, and wastewater (Fig. 7.2) containing substantial amount of antibiotics after application into soil systems are described to contaminate the agricultural soils, food chains and, modulate the soil microbiological communities, thereby associated processes governing the cycling of important minerals and nutrients. Most of the antibiotics detected in soil are of polar nature and contain groups with characteristic ionization. The content of residual antibiotics in soil of cultivated system ranging from 450 to 900 μg $kg^{-1}$, 13 to 67 μg $kg^{-1}$ and 6 to 52 μg $kg^{-1}$ for tetracyclins, macrolides and fluoroquinolones, respectively is documented by different workers (Schüller 1998; Winckler and Grafe 2000). The maximum detectable antibiotics content in manure supplemented soil was illustrated to possess nearly 200 μg $kg^{-1}$, 7 μg $kg^{-1}$, and 11 μg $kg^{-1}$of tetracyclins, chlortetracyclins and sulfadimidine, respectively (Hamscher et al. 2001; Höper et al. 2002). In China, the content of antibiotics residue in farm soil and vegetable growing farms has been reported to reach 5.17 mg $kg^{-1}$ (Kang et al. 2013; Shen et al. 2014). The content of antibiotics in soil ranging from ng $g^{-1}$ to μg $g^{-1}$ is presented by different workers (Li et al. 2011a, b; Pan et al. 2014; Li et al. 2015). Concentration of different antibiotics including veterinary ones in agricultural soils upto 1331 μg $kg^{-1}$ for flumequine (Ho et al. 2014), in suburban area with average value of antibiotics detected as 43.64 μg $kg^{-1}$ (Yang et al. 2016) and in urban surface soil reaching maximally upto 94.6 μg $kg^{-1}$ norfloxacin (Gao et al. 2015) is illustrated by various researchers. The presence of norfloxacin as dominating antibiotic in soil beside river reaching a concentration equivalent to 27.21 μg $kg^{-1}$ is described recently (Zhu et al. 2020a, b) and a gradient in antibiotics allocation along river soil was also noticed. Recently, the antibiotics residue in paddy cultivated field is presented by Uddin et al. (2020). The contamination of soil was ascribed to human overexploitation and the farms associated with livestock and poultry production. The important factors controlling antibiotic concentration were pH, organic materials, amount of sand and ion exchange potential. In conclusion, regular monitoring and risk assessment of human as well as veterinary antibiotics must be taken into account for the proper management of affected soil so as to avoid toxic impact on natural ecosystem and human health. List of some important sources causing the contamination of agro-ecosystem with antibiotics is presented in Table 7.1.

**Table 7.1** List of different sources responsible for introduction of antibiotics and antibiotic resistance genes (ARGs) in agro-ecosystem

| S.N. | Source | Country | Antibiotics/class | ARGs/mobile genetic elements | Method of detection/ quantification | Reference |
|---|---|---|---|---|---|---|
| 1. | Sewage sludge | Nigeria | Sulfonamides, fluoroquinolones, macrolides, tetracyclines, β-lactams, trimethoprim, thiamphenicol | – | Liquid chromatography (LC)-tandem mass spectrometry (MS) | Ajibola et al. (2020) |
| 2. | Manure | Canada | Sulfamethazine, chlortetracycline, tylosin | – | LC–MS/MS | Aust et al. (2008) |
| 3. | Swine manure slurry | United States | Chlortetracycline, lincomycin, tiamulin, penicillin G | Tetracycline resistance genes like *tet(A)*, *tet(B)*, *tet(C)*, *tet(D)*, *tet(E)*, *tet(O)*, *tet(Q)*, *tet(W)*, *tet(X))* and lincomycin resistance genes *erm(A)*, *erm(B)*, *erm(C)*, *erm(F)*, and *erm*(G) | HPLC-MS, q-PCR | Barrios et al. (2020) |
| 4. | Raw wastewater | Cameroon | Enrofloxacin, oxytetracycline, sulfamethoxazole, sulfadimidine, oxytetracycline, trimethoprim, ciprofloxacin | *catIII*, *cat(Q)*, *flo(R)*, $bla_{rm3}$, *sul3*, *tet(C)*, *flo(R)*, $bla_{CARB-3}$, $bla_{OXA-347}$, $bla_{OXA-5}$ $bla_{Rm3}$ | LC-MS and high-throughput (HT) sequencing | Bougnom et al. (2020) |
| 5. | Municipal wastewater solids | United States | – | *Erm(B)*, *qnr(A)*, *sul(1)*, *tet(A)*, *tet(W)*, *tet(X)*, and class 1 Integrons | q-PCR | Burch et al. (2017) |
| 6. | Sewage sludge | China | – | *Orp(J)*, *flo(R)*, *aad(E)*, *pik(R2)*, *erm(B)*, *erm(D)*, *mex(F)* | High-throughput sequencing and high-throughput q-PCR | Chen et al. (2016) |
| 7. | Municipal wastewater solids | United States | – | *tet(A)*, *tet(L)*, *tet(O)*, *tet(W)*, *tet(X)* | q-PCR | Diehl and LaPara (2010) |

| | | | | | | |
|---|---|---|---|---|---|---|
| 8. | Mariculture sites | China | – | *Bac(A)*, *mex(F)*, *mex(B)*, *sul(1)* and *sul(2)* | q-PCR and high-throughput sequencing | Gao et al. (2018) |
| 9. | Wastewaters, sewage sludges, surface waters | Switzerland | Ciprofloxacin and norfloxacin, clarithromycin, erythromycin-$H_2O$ | – | LC-fluorescence detection (FLD) | Giger et al. (2003) |
| 10. | Animal manure | China | – | *Vat(E-01)*, *erm(F)*, *qac(Edelta l-01)*, *erm(B)*, *tet(M-02)* | HT-qPCR | Han et al. (2018) |
| 11. | Swine wastes | China | Eight sulfonamides and diaminopyrimidines (sulfaguanidine, sulfadiazine, sulfamethazine, sulfamonomethoxine, sulfachlorpyridazine, sulfamethoxazole, sulfaquinoxaline, trimethoprim), five tetracyclines (oxytetracycline, tetracycline, chlortetracycline, methacycline, doxycycline), one chloramphenicols (florfenicol), three macrolides (erythromycin-$H_2O$, leucomycin and tylosin) and one lincosamides (lincomycin) | *Sul(1)*, *sul(2) sul(3)*, *tet(A)*, *tet(G)*, *tet(H)*, *tet(M)*, *tet(O)*, *tet(Q)*, *tet(S)*, *tet(W)*, *tet(B/P)*, *tet(T)*, *tet(X)*, *cml(A)*, *flo(R)*, *fex(A)*, *fex(B)*, *cfr*, *erm(B)*, *erm(C)*, *erm(E)* | q-PCR | He et al. (2016) |
| 12. | Dairy manure | Germany | Oxytetracycline and monensin | – | Enzyme-linked immunosorbent assay (ELISA) | Hilaire et al. (2020) |
| 13. | Broiler manure | Malaysia | Doxycycline, enrofloxacin, flumequine, norfloxacin, trimethoprim, tylosin | – | LC-MS/MS | Ho et al. (2014) |

(continued)

**Table 7.1** (continued)

| S.N. | Source | Country | Antibiotics/class | ARGs/mobile genetic elements | Method of detection/ quantification | Reference |
|---|---|---|---|---|---|---|
| 14. | Activated sludge | China | – | Tetracycline resistant genes *tet(A)*, *tet(B)*, *tet(C)*, *tet(G)*, *tet(K)* *tetP(A)*, *tet(M)*, *tet(O)*, *tet(S)*, *tet(W)*, *tet(X)* | q-PCR and next-generation sequencing | Huang et al. (2014) |
| 15. | Pharmaceutical wastewater | Pakistan | Ofloxacin, ciprofloxacin, levofloxacin, oxytetracycline, doxycycline | – | HPLC | Hussain et al. (2016) |
| 16. | Animal manure | China | – | Resistant genes against macrolide-lincosamide-streptogramin, multidrug, tetracycline, aminoglycoside, β-lactams, vancomycin, fluoroquinolone, quinolone, florfenicol, chloramphenicol, and amphenicol sulfonamide and *aad(A-1-01)*, *mex(F)* | HT q-PCR | Li et al. (2020a, b) |
| 17. | Pharmaceutical wastewater | China | Sulfamethoxazole | – | UPLC with a UV detector | Liang et al. (2020a, b) |
| 18. | Swine manure | China | Sulfadiazine, sulfamethazine, sulfamonomethoxine, sulfamethoxypyridazine | *Sul1*, *sul2*, *sul3*, and *sul(A)* and mobile genetic element *IntI1* | q-PCR | Lin et al. (2020) |

| | | | | | | |
|---|---|---|---|---|---|---|
| 19. | Agricultural watershed | United States | – | Macrolide *erm(B)*, *erm(F)*, tetracycline *tet(A)*, *tet(M)*, *tet(O)*, *tet(W)*, sulfonamide *sul1*, *sul2*, aminoglycoside *aad(A2)*, *vga(A)*, *vga(B)* | q-PCR | Neher et al. (2020) |
| 20. | Hospital wastewater, treatment plant, and surface water | Kenya | A total of 17 belonging to class sulfonamides, β-lactams, macrolides and aminoglycosides | – | LC/MS/MS | Ngigi et al. (2020) |

#### 7.2.1.3 Methods for Determination of Antibiotics Content

Precise determination of antibiotics and degradation products, if any, in different environmental samples such as manures, human and animal excreta, soils, effluents from wastewater, and sewage sludge treatment plants is very much important to control the further contamination spread of resistant microbial forms as well as antibiotic resistance genes. Prior to determination, the selection of suitable methods effective in extracting the antibiotics is also very much important. The extraction can be facilitated through utilization of different solvents, buffers and centrifugation to concentrate the samples for quantification of antibiotics in given samples. Further, a particular extraction procedures as well as determination techniques could not be equally effective for precise quantification of antibiotics from diverse environmental matrices. Some of the important techniques for determination of antibiotics include ultraviolet visible spectroscopy, high performance liquid chromatography (HPLC), HPLC coupled with mass spectroscopy (MS), liquid chromatography (LC)-MS, LC-electrospray ionization (ESI), tandem mass spectrometry and HPLC-MS/MS (Tang et al. 2015; He et al. 2016; Ajibola et al. 2020; Barrios et al. 2020; Neher et al. 2020; Sun et al. 2020; Wang et al. 2020a, b, c; Zhang et al. 2020a, b, c, d). However, the limit of detection (LOD) and limit of quantification (LOQ) for different analytical instruments varies considerably, therefore selection of suitable extraction techniques, instrument used and operating conditions including external standard should be given due preference for reliable accuracy in measurement of antibiotics and degradation products in environmental samples.

#### 7.2.1.4 Safety Limits of Antibiotics for Soil and Other Components

Safety limits to any antibiotic meant for agricultural applications must be applied strictly to avoid any associated harmful impacts on soil natural attributes, human health and other environmental components. The indiscriminate utilization of antibiotics directly to protect the crop productivity losses rendered by devastating plant diseases together with the application of livestock based manure, and treated water emanating from wastewater, sewage sludge and pharmaceuticals wastes in agricultural field has necessitated the enforcement of strict regulatory measures to set allowable limit or maximum residual limit in order to mitigate the environmental consequences of antibiotic abuse (Aitken et al. 2016). The soil applied antibiotics may be taken by plants (Kang et al. 2013; Pan and Chu 2017a, b; Li et al. 2020a, b) and variably accumulated in different important parts like roots (Yan et al. 2020), vegetables (Hu et al. 2010; Hussain et al. 2016), fruits and other parts important for feed application. Unfortunately, no regulations have been set for allowable concentration of antibiotics either in soil or in soil water (Thiele-Bruhn 2003). Nevertheless, model based evaluations on pre-directed concentrations are determined. The antibiotic concentration for animal excreta, dung and soil, and groundwater is determined as 10 μg $kg^{-1}$, 100 μg $kg^{-1}$, and 0.10 μg $L^{-1}$, respectively. The maximum residual limit of antibiotic residue in food as set by CODEX Alimentarius for Veterinary

Medicine Residues in Food (1996) is zero, 0.004 μg $ml^{-1}$, 0.03 μg $ml^{-1}$ and 0.05 μg $ml^{-1}$ for chloramphenicol, ampicillin and benzylpenicillin, dicloxacillin, and erythromycin, respectively. Similarly, European Union (EU 2009) has defined maximum residual limit of antibiotics ofloxacin and streptomycin as 30 ng $ml^{-1}$ and 200 ng $ml^{-1}$, respectively. Hence, the consumption of fruits and other plant derived products may likely cause the entry of substantial amount of antibiotics in humans and animals. Moreover, the antibiotics applied directly to fruits for controlling post-harvest losses could lead to antibiotics access to humans eventually raising concern over food safety, thus implying the development of regulatory laws (Ramírez et al. 2003) to limit growing antibiotic menace in soils.

### 7.2.2 Fate of Antibiotics in Contaminated Soil

The antibiotics once introduced into agro-ecosystem are amenable to considerable changes governed by soil natural characteristics. The antibiotics entered into soil could have three fates (i) adsorption, (ii) degradation and, (iii) leaching. The binding of antibiotics to soil matrices is influenced substantially by physico-chemical characteristics. Tetracyclines and macrolines are demonstrated to have greater affinity for given soil in comparison to other antibiotics encountered in soil environment (Rabølle and Spliid 2000). The study conducted by Pan and Chu (2016) described variations in adsorption attributes of selected antibiotics and recorded highest affinity of tetracyclines and least of sulfonamides for soil having clay loam nature. The sorption of antibiotics is regulated chiefly through functional group of basic structural motif (Kulshrestha et al. 2004). Moreover, antibiotics within a class can interact differentially with minerals existing in soil matrices leading to significant changes in adsorption pattern. Besides these, organic material content, soil organic nitrogen, pH, ion exchange potential, and clay materials influence the sorption of antibiotics in soil (Ahmed 2017). Higher the value of organic carbon, nitrogen, clay particles, and cation exchange potential higher is the adsorption of antibiotics as recorded for doxycycline (Álvarez-Esmorís et al. 2020). Partitioning coefficient is another important attribute of antibiotics modulating the adsorption of antibiotics onto soil. In general, the antibiotics belonging to class tetracyclines have high partitioning coefficient implying the retention of compound in soil. In contrast, the antibiotics falling within the group of sulfonamides display minimum partitioning coefficient advocating higher mobility in soil system (Pan and Chu 2017a, b). The strong adsorption of antibiotics on soil is helpful in alleviating the toxicity to human and other environmental components.

The antibiotics introduced into soil may undergo various modifications or degradation facilitated by biological or non-biological processes (Fig. 7.4). The microbiological activities of diverse bacteria and fungi found in soil are the important drivers determining the fate of numerous antibiotics. The degradation by microbial

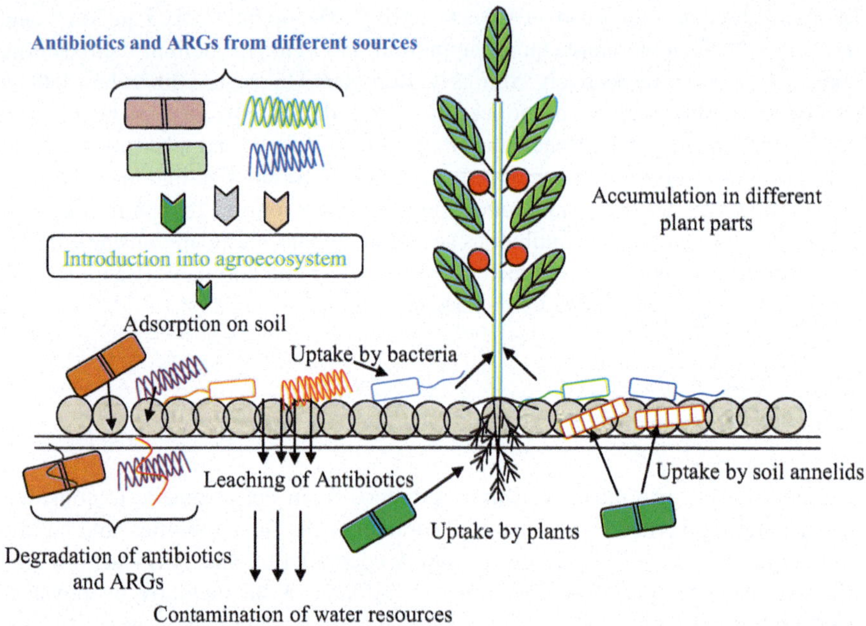


**Fig. 7.4** Fate and transport of antibiotics and antibiotic resistance genes (ARGs) in agro-ecosystem

enzymes has the potential to eliminate antibiotics therefore the toxicity from a given environmental matrix (Yang et al. 2012; Pan and Chu 2016). Most notably the biological degradation potential of microbes is regulated by environmental variables like pH, availability of nutrients, moisture, oxygen, microbial adaptation features, microbial communities and chemical characteristics of compounds (Selvam and Wong 2017; Reis et al. 2020). The higher loss in available antibiotics for non-sterilized soil as compared to sterilized soil pointing towards the active contribution of microbial population in degradation is illustrated by Pan and Chu (2016). Nevertheless, there is no consistency in antibiotics degradation behavior of non-sterilized and sterilized soils indicting complex nature of biotic and abiotic components of different soils.

Soil introduced antibiotics have potential to transport and leach into surrounding water and groundwater resources causing contamination and associated environmental impacts. Rainfall leading to surface run-off and downward percolation of water containing high concentration of antibiotics is an important factor facilitating the contamination of surface water resources like river and groundwater. However, much information regarding the leaching of antibiotics entered into soil is very much scarce. Some studies under column conditions have been conducted to decipher the leaching phenomenon (Kay et al. 2005a, b). Rainfall simulation based investigations in soil column experiment to reveal the leaching behavior of five different classes of antibiotics in natural agro-ecosystem is presented by Pan and Chu

(2017a, b) and precipitation for longer duration under natural field condition was ascribed to increased percolation of antibiotic residues. The increased rainfall enriching the soil antibiotics at manure applied locations is also presented (Zhao et al. 2020). The leaching and sequestration characteristics of antibiotics, hence is considered to be controlled by different attributes like chemistry of antibiotics, pH, adsorption affinity, ionization tendency, soil characteristics, partition coefficient, miscibility in aqueous environment, soil organic material content and slope (Kay et al. 2004; Thiele-Bruhn et al. 2004; Jones et al. 2005; Aust et al. 2008; Zhao et al. 2020).

### 7.2.3 Decontamination Strategies Employed for Antibiotics Removal

The presence of antibiotics in soil may modulate the microbially driven biological process (Grenni et al. 2018; Kovalakova et al. 2020) eventually affecting the soil characteristics and crop productivity. Therefore, application of sustainable methods for elimination of antibiotics from contaminated agro-ecosystem is imperative for maintaining the productivity of economically important crops. The elimination of antibiotics may be performed before and after the introduction of antibiotics in a given environmental component. The treatment of wastes emanating from pharmaceutical industries, hospitals, farmyards and even from treatment facilities requires the employment of efficient processes minimizing the concentration of antibiotics to a safer level. Nevertheless, most of the treatment plants are not successful in removing the target antibiotics absolutely (Liang et al. 2019; Manasfi et al. 2020). In contrast, biodegradation of target antibiotics of recalcitrant nature in soil may be sought as an effective strategy to detoxify the contaminants (Kumar et al. 2019; Liang et al. 2020a, b; Manasfi et al. 2020). The mineralization and biotransformation may determine the extent of detoxification and elimination of hazardous antibiotics (Girardi et al. 2011). Improvement in degradation by soil amendment could be employed to further facilitate the antibiotics degradation (Liang et al. 2020a, b). In this connection, the isolation of important bacterial and fungal strains resistant to higher concentrations of antibiotics (Sodhi et al. 2020) and application of microbial enzymes like laccase effective against large numbers of antibiotics (Ding et al. 2016) even at low temperature (Tian et al. 2020) is much preferable.

The extensive information about the antibiotic degradation pathway could provide an important insight into the exploitation of candidate microorganisms for managing the menace of antibiotics residues into soil environment (Reis et al. 2020). Phytoremediation is another approach for cleansing of agro-ecosystem badly affected with antibiotics residue (Li et al. 2020a, b). The plants with efficient uptake potential together with the rapid multiplication rate under natural environmental condition could help solve the problem to a greater extent (Kumar et al. 2005; Hilaire et al. 2020; Panja et al. 2020). The phytoremediation characteristics, however, is expected to be influenced by nature of antibiotics, plants used, soil microbes and pollutants other than antibiotics. Therefore, consideration of all such factors must be taken into consideration for improving the phytoremediation potential by chosen plants and thwarting of the eco-toxicological effects of antibiotics.

## 7.3 Antibiotic Resistance Genes in Soil/Agroecosystems

Antibiotics have long been used for the cure of innumerable diseases not only of humans but also of livestocks. The indiscriminate exploitation of antibiotics has contaminated the agricultural soil via the application of animal based manures and treated wastewater from different sources causing the expression of antibiotic resistance genes in microbial forms. Antibiotic resistance genes are considered as an emerging contaminant in various environmental matrices (Ben et al. 2019; Zarei-Baygi et al. 2019). The microbes possessing antibiotic resistance genes could put great risks to human health through spread of diseases. Moreover, the mobility of antibiotic resistance genes to native microorganism and its access to human may create plethora of burdens to human health and immunity. To circumvent the risks of rapid dissemination of antibiotic resistance genes in various components of environment, the active participation of personnel from varied discipline including environment, medicine, agriculture and microbiology is essentially required in order to safeguard the environment and human health (Bougnom et al. 2020).

### *7.3.1 Sources of Antibiotic Resistance Genes*

Antibiotic resistance genes may enter into soil ecosystem via different routes and important sources of antibiotic resistance genes include treated wastewaters and sludge, human and animal excreta, household and hospital discharges and farmlands (Fig. 7.2). The identification of potential sources leading to contamination with antibiotic resistance genes is very much important to protect the natural environment and biodiversity of native microbes inhabiting in agro-ecosystem. The introduction of antibiotic resistance genes in soil through liberation of fermentation drug containing antibiotic penicillin is presented currently (Wang et al. 2020a, b, c). The *in vitro* study reflected the modulation in microbial community structure as the primary driving force responsible for abundance of microbes harboring antibiotic resistance genes. Wastewater treatment plants as an important overlooked source leading to dissemination of antibiotic resistance genes is presented by Zhang et al. (2018) as many of the currently deployed treatment facilities are not able to absolutely eliminate the hazardous antibiotic resistance genes. Moreover, the direct application of antibiotics into agro-ecosystem as largely been practiced for controlling bacterial diseases of economically important crops may facilitate the increased occurrence of resistant bacteria as well as antibiotic resistance genes (Scarano et al.

2014; Gao et al. 2018). Mariculture is being practiced for production of marine organisms to fulfill the ever rising demand of food to rapidly increasing human population. The application of substantial concentrations of antibiotics is commonly evidenced in aquaculture. Such production site contributing in release and dissemination of antibiotic resistance genes is represented by Gao et al. (2018). The increasing advocacy for organic agriculture in many parts of the world, has exacerbated the exploitation of organic fertilizers derived from animals. Approximately 30–90% of the antibiotics taken by different livestocks are excreted unabsorbed and largely remain in unmodified form in excreta (Sarmah et al. 2006). The addition of manure as a source of organic materials into agro-ecosystem leading to soil contamination with antibiotic resistance genes is elaborated by different researchers (Heuer et al. 2011; Tang et al. 2015; Pu et al. 2020), suggesting the view of manure not only as a reservoir of diverse antibiotics but also of antibiotic resistance bacteria conferring increased availability of antibiotic resistance genes. The identification of different sources carrying antibiotic resistance genes, resistance to particular antibiotics and processes influencing their dissemination under natural soil environment is an important prerequisite for the management of antibiotic resistance genes in agro-ecosystem. List of some important sources causing the contamination of agro-ecosystem with antibiotic resistance genes is presented in Table 7.1.

#### 7.3.1.1 Resistance Genes Identified Against the Antibiotics

Identification of resistance genes against antibiotics is vital step to prevent the spread in agro-ecosystem and thereby human health and environment. So far, numbers of different resistance genes conferring resistance to human and veterinary antibiotics because of overexploitation are reported from urban and cultivated soil environment across the globe (Knapp et al. 2010; Heuer et al. 2011; Chen et al. 2016; Han et al. 2018; Li et al. 2020a, b). The presence of antibiotic resistance genes in sediments and paddy cultivated field is explored by Awad et al. (2015) indicating the abundance of antibiotic resistance genes in sediment samples against selected antibiotics belonging to sulfonamide class. The antibiotic resistance genes for different selected antibiotics class were documented in soil with rice cultivation. The availability of resistance genes against different antibiotic groups in dairy agro-ecosystem as revealed through metagenomics technology is described by Pitta et al. (2016). Large numbers of the resistances genes were found to be associated with transporters for multiple drug molecules, different classes of antibiotics like vancomycin, tatracyclin, bacitracin in descending order and proteins facilitating drug transport exterior of the cellular environment. Most strikingly, most of the resistance genes were documented to be linked with microbes falling in the phylum Proteobacteria. Irrigation of agricultural fields with wastewater causing predominance of antibiotic resistance genes conferring resistance to antibiotics amphenicols, β-lactams, and tetracycline is elaborated currently by Bougnom et al. (2020). The investigation on occurrence of antibiotic resistance genes in forest and grassland procured soil is demonstrated by Willms et al. (2020). The identified antibiotic resistance genes included *mefA*, *sul2*, *aac(6′)-lb*, and $bla_{IMP\text{-}12}$ providing resistance to antibiotics belonging to classes macrolides, sulfonamide, aminoglycosides and beta lactams, respectively. The dissemination of resistance genes in cultivated fields were registered to be induced by amplicons. Identification of cultivated ecosystems harboring the pool of different antibiotic resistance genes are therefore necessary for safeguarding the crop products, native soil microbial characteristics and human health. In addition, retrieving detailed information pertaining to the contribution of environmental parameters influencing the dissemination of antibiotic resistance genes and mobile genes are of immense significance to protect the agro-ecosystem.

#### 7.3.1.2 Methods for Identification of Antibiotic Resistance Genes

Both cultivation and cultivation independent approaches have been reported for the identification of antibiotic resistance genes in natural as well as engineered ecosystem. The quantitative polymerase chain reaction (q-PCR), high throughput q-PCR and next generation sequencing (NGS) is one of the most widely employed techniques for the identification of various antibiotic resistance genes from diverse environmental components (Huang et al. 2014; Burch et al. 2017; Liu et al. 2019; Lu et al. 2019; Yuan et al. 2020; Zhang et al. 2020a, b, c, d). Since, the identification of particular antibiotic resistance genes is relied on available primer sequences, the recognition of novel resistance genes is a major challenge globally (Schmieder and Edwards 2012). The deployment of culture independent metagenomic approaches may prove quite advantageous for rapid identification not only of previously known antibiotic resistance genes but also for newly emerging ones, hence sufficiently eliminates the limitations of conventional PCR based identification techniques. Metagenomics based approaches have been found successful in detection of antibiotic resistance genes existing in wastewater treatment facilities (Paiva et al. 2017). Interestingly, the metagenomics approach could foster help in comparison of different antibiotic resistance genes from various environmental samples and efficiency of specific treatment technology in elimination of antibiotic resistance genes as well. Apart from antibiotic resistance genes, identification of mobile genetic elements responsible for resistance transfer in appropriate hosts is also possible with metagenomics dependent technologies (Wang et al. 2013). Besides metagenomics, the utilization of small sequences of antibiotic resistance genes has also been proposed for easy identification; however the technique being very much prone to errors is not much advocated (Pärnänen et al. 2016). The evolution of rapid identification techniques for antibiotic resistance genes may be helpful as early warning tool to restrict the further progression of antibiotic resistance in a given environmental system.

### 7.3.2 Mechanism of Antibiotic Resistance Gene Transfer in Other Organisms

Horizontal gene transfer is regarded as one of the important processes conferring dissemination of antibiotic resistance genes in a given environment (Li et al. 2020a, b). The horizontal gene transfer process is assisted by processes like conjugation, transduction and transformation. The participation of mobile genetic elements, insertion sequences, viruses attacking on bacteria as well as integrons helping in antibiotic resistance genes transfer together with the active contribution of plasmids is well illustrated (Gogarten and Townsend 2005; Bennett 2008; Stalder et al. 2012; Rizzo et al. 2013; Ghaly et al. 2020). The involvement of conjugative transposable elements from *Bacteroides* rendering spread of antibiotic resistance genes is investigated by Whittle et al. (2002). Antibiotics even at very low environmentally relevant concentrations are known to provoke the antibiotic resistance genes transfer assisted by conjugative transposons falling within the CTnDOT/ERL family, suggesting the active contribution of small contents of antibiotics in dissemination of antibiotic resistance genes. Contribution of insertion sequences associated with plasmids harboring resistance genes against carbapenem fostering transfer of antibiotic resistance genes is illustrated (Iacono et al. 2008).

Plasmid associated transfer of antibiotic resistance genes is known to occur via recombination between chromosomal and plasmid segments (Manson et al. 2010). In addition, transfer of antibiotic resistance genes through plasmids was more efficient routes for transfer as compared to resistance genes located on chromosomal regions (Liu et al. 2019). Furthermore, majority of the plasmid linked antibiotic resistance genes displayed increased transcription again advocating increased risk of resistance gene spread. Thanks to the fact not all antibiotic resistance genes existing in bacterial systems are linked with mobile genetic elements putting severe restriction on fast spread. In addition to horizontal gene transfer, the spread of antibiotic resistance genes via mutation is also illustrated (Zhu et al. 2020a, b). The detailed mechanistic information regarding the transfer of resistome (pool of antibiotic resistance genes) and factors influencing dissemination is essentially needed for risk assessment and preventing transfer in pathogenic microorganisms as well as in humans via contaminated food chain.

### 7.3.3 Fate of Antibiotic Resistance Genes

Comprehensive knowledge on fate of antibiotic resistance genes in soils receiving the treated wastewater and sludge is important to maintain the natural attributes of agro-ecosystem. Nevertheless, the facts regarding the persistence and fate of antibiotic resistance genes in environment are not fully elucidated. The extracellular antibiotic resistance genes under soil environment are susceptible to adsorption on soil matrices, degradation and acquisition (Fig. 7.4) by different microbial forms

(Chee-Sanford et al. 2009). The antibiotic resistance genes existing intra-cellularly may be transferred to closely or distantly related organisms from different environmental components (Yuan et al. 2020) via horizontal gene transfers. The binding of bacterial species onto the soil surfaces in constructed wetlands has been proposed as a means of elimination of antibiotic resistance genes responsible for tetracycline resistance (Liu et al. 2013). Similarly, the role of microplastics in controlling the fate of antibiotic resistance genes through adsorption is recently proposed (Lu et al. 2020a, b). The adsorption of antibiotic resistance genes was influenced by characteristics of microplastics. The adsorption of antibiotic resistance genes onto soil components may protect the DNA from degradation rendered by enzymatic activities (Romanowski et al. 1993), causing persistence of resistance genes. The soil mixed residual solids derived from treatment facilities was analyzed for decline in antibiotic resistance genes as well as class I integrons (Burch et al. 2014). The study revealed very low reduction in content of five different antibiotic resistance genes fostering resistance against tetracyclins, sulfonamides and erythromycin together with the integrase gene. The half-life ranged from 13 days to 81 days for selected antibiotic resistance genes and integrase gene. The increased persistence in agro-ecosystem rises the chances of dissemination to soil microorganisms, food chain components and finally human. Thus, the fate of antibiotic resistance genes introduced into agro-ecosystem is affected by different factors including amount of organic fertilizer utilized, pH, organic material content, ion exchange characteristics, nutritional status, source organism, cropping, harvesting and time gap between application of treated wastewater/bioslids and rainfall (Chen et al. 2016; Li et al. 2018; Qiao et al. 2018; Barrios et al. 2020). The in depth information could be helpful in dealing with the continuously rising global risks of antibiotic resistance genes in various environmental matrices and human health.

### 7.3.4 Measures to Control the Spread of Antibiotic Resistance Genes

The widespread dissemination of antibiotic resistance genes in different environmental compartment is a global concern necessitating for the development and deployment of suitable control measures for mitigating the negative consequences. The treatment strategies could be simply categorized before and after the dissemination of antibiotic resistance genes. Elimination of antibiotic resistance genes present in different sources like wastewater, sewage sludge, biosolids, pharmaceutical wastes, and other water resources used for irrigation purposes may be important strategy for removal of antibiotic resistance genes. Once after the antibiotic resistance genes are introduced into the agro-ecosystem, further progression into pathogenic microbes and food chain (Zhang et al. 2019) poses severe challenges in treating such emerging contaminants. The contribution of aerobic and anaerobic digestion, pasteurization, pyrolysis, alkali stabilization together with the air drying beds and constructed wetlands has been reported to differentially sequester or inactivate antibiotic resistance genes from wastewater and solids thereof (Diehl and LaPara 2010; Ma et al. 2011; Liu et al. 2013; Yuan et al. 2015; Burch et al. 2017; Kimbell et al. 2018). In contrast, there are also reports suggesting increase in copy number of some target antibiotic resistance genes like *erm*(B), *erm*(F), and *tet*(O) during anaerobic digestion process (Diehl and LaPara 2010; Ma et al. 2011) and conditioning of wastes with sulfur based compounds (Lu et al. 2020a, b). Further process efficiency may be improved by additional pretreatment steps and enhancement in temperature. The study of Burch et al. (2017) demonstrated the promising potential of alkali stabilization, anaerobic digestion, and pasteurization in sequestering six antibiotic resistance genes along with the integrase genes to a level reaching equivalent to control samples. Nevertheless, majority of the treatment methodologies are ineffective in reducing the concentrations of antibiotic resistance genes, implying dissemination in natural environment and eventually human exposure via different agricultural products. The treatment of antibiotic resistance genes present in biosolids and treated water as well as wastewater treatment plants is influenced by samples characteristics like pH, presence of metals, nutrients, operational conditions like treatment time, energy input, temperature etc., and most strikingly existing microbial communities; therefore, a single treatment procedure may not necessarily equally effective in decontaminating antibiotic resistance genes from all types of solids and wastewater. Further improvement in existing technologies and evolution of novel techniques could be a breakthrough for elimination of environmentally hazardous antibiotic resistance genes.

The indiscriminate application of antibiotics in agro-ecosystems via different sources has given rise to emergence of resistance development not only in native agro-ecosystem's microorganisms, but also in those surviving in human and animal's gut. The selection pressure imposed by antibiotics has favored the multiplication and survival of pathogenic microbes possessing antibiotic resistance genes, thereby putting great risks to human health and agriculture ecosystem. Furthermore, the food chain contamination with antibiotics and resistance genes has negatively influenced the inherent functioning of natural environment to a greater extent.

## 7.4 Conclusion and Future Recommendations

The extensive application of antibiotics in cure of human and animal diseases, aquaculture, and agriculture together with animal growth promotion for increased meat products has led to contamination of agro-ecosystem. The antibiotics even at very low concentrations are markedly associated with the changes in microbiological characteristics of different ecosystem including agro-ecosystem, thereby affecting the cycling of important minerals and nutrients. The enrichment of antibiotics has facilitated the selection of genes conferring resistance against numbers of antibiotics. The risks imposed by antibiotics and antibiotic resistance genes to human health and environment need to be managed properly through environmentally friendly tools in order to protect the natural environment and restricting the spread of antibiotic resistance genes. The thorough investigations on nature of antibiotics, antibiotic resistances genes, factors governing the mobility, together with the precise and accurate determination of human as well as veterinary antibiotics, and role of environmental factors, therefore is essentially needed to cope up with rapidly rising danger. In this connection following future studies should be considered for management of affected agro-ecosystems.

- Research and development for considerable improvement in extraction procedures and analytical techniques for precise quantification of antibiotics, degradation products and associated resistance genes.
- Assessment of antibiotics and degradation products accumulation in different environmental matrices largely unknown so far.
- Enforcement of stringent regulatory laws for application of antibiotics in agro-ecosystems.
- Exploration of impact of antibiotics and antibiotic resistance genes on soil annelids and arthropods could help us substantially in tackling the continuously rising menace of antibiotics and antibiotic resistance genes in nature.

## References

Ahmad M, Vithanage M, Kim K, Cho JS, Lee YH, Joo YK, Lee SS, Ok YS (2014) Inhibitory effect of veterinary antibiotics on denitrification in groundwater: a microcosm approach. Sci World J. https://doi.org/10.1155/2014/879831

Ahmed MJ (2017) Adsorption of quinolone, tetracycline, and penicillin antibiotics from aqueous solution using activated carbons. Environ Toxicol Pharmacol 50:1–10. https://doi.org/10.1016/j.etap.2017.01.004

Aitken SL, Dilworth TJ, Heil EL, Nailor MD (2016) Agricultural applications for antimicrobials. A danger to human health: an official position statement of the Society of Infectious Diseases Pharmacists. Pharmacotherapy 36:422–432. https://doi.org/10.1002/phar.1737

Ajibola AS, Tisler S, Zwiener C (2020) Simultaneous determination of multiclass antibiotics in sewage sludge based on QuEChERS extraction and liquid chromatography-tandem mass spectrometry. Anal Methods 12:576–586. https://doi.org/10.1039/C9AY02188D

Álvarez-Esmorís C, Conde-Cid M, Fernández-Calviño D, Fernández-Sanjurjo MJ, Núñez-Delgado A, Álvarez-Rodríguez E, Arias-Estévez M (2020) Adsorption-desorption of doxycycline in agricultural soils: batch and stirred-flow-chamber experiments. Environ Res:109565. https://doi.org/10.1016/j.envres.2020.109565

Arshad M, Zafar R (2020) Chapter 9: Antibiotics, AMRs, and ARGs: fate in the environment. In: Hashmi MZ (ed) Antibiotics and antimicrobial resistance genes in the environment. Elsevier, pp 138–154. https://doi.org/10.1016/B978-0-12-818882-8.00009-7

Aust MO, Godlinski F, Travis GR, Hao X, McAllister TA, Leinweber P, Thiele-Bruhn S (2008) Distribution of sulfamethazine, chlortetracycline and tylosin in manure and soil of Canadian feedlots after subtherapeutic use in cattle. Environ Pollut 156:1243–1251. https://doi.org/10.1016/j.envpol.2008.03.011

Awad YM, Kim KR, Kim SC, Kim K, Lee SR, Lee SS, Ok YS (2015) Monitoring antibiotic residues and corresponding antibiotic resistance genes in an agro-ecosystem. J Chem:974843. https://doi.org/10.1155/2015/974843

Barrios RE, Khuntia HK, Bartelt-Hunt SL, Gilley JE, Schmidt A, Snow DD, Li X (2020) Fate and transport of antibiotics and antibiotic resistance genes in runoff and soil as affected by the timing of swine manure slurry application. Sci Total Environ:136505. https://doi.org/10.1016/j.scitotenv.2020.136505

Béahdy J (1974) Recent developments of antibiotic research and classification of antibiotics according to chemical structure. Adv Appl Microbiol 18:309–406. https://doi.org/10.1016/S0065-2164(08)70573-2

Ben Y, Fu C, Hu M, Liu L, Wong MH, Zheng C (2019) Human health risk assessment of antibiotic resistance associated with antibiotic residues in the environment: a review. Environ Res 169:483–493. https://doi.org/10.1016/j.envres.2018.11.040

Bennett PM (2008) Plasmid encoded antibiotic resistance: acquisition and transfer of antibiotic resistance genes in bacteria. Br J Pharmacol 153:S347–S357. https://doi.org/10.1038/sj.bjp.0707607

Bougnom BP, Thiele-Bruhn S, Ricci V, Zongo C, Piddock LJV (2020) Raw wastewater irrigation for urban agriculture in three African cities increases the abundance of transferable antibiotic resistance genes in soil, including those encoding extended spectrum β-lactamases (ESBLs). Sci Total Environ 698:134201. https://doi.org/10.1016/j.scitotenv.2019.134201

Boxall AB, Blackwell P, Cavallo R, Kay P, Tolls J (2002) The sorption and transport of a sulphonamide antibiotic in soil systems. Toxicol Lett 131:19–28. https://doi.org/10.1016/s0378-4274(02)00063-2

Boxall AB, Kolpin DW, Halling-Sørensen B, Tolls J (2003) Peer reviewed: are veterinary medicines causing environmental risks? Environ Sci Technol 37:286A–294A. https://doi.org/10.1021/es032519b

Burch TR, Sadowsky MJ, LaPara TM (2014) Fate of antibiotic resistance genes and class 1 integrons in soil microcosms following the application of treated residual municipal wastewater solids. Environ Sci Technol 48:5620–5627. https://doi.org/10.1021/es501098g

Burch TR, Sadowsky MJ, LaPara TM (2017) Effect of different treatment technologies on the fate of antibiotic resistance genes and class 1 integrons when residual municipal wastewater solids are applied to soil. Environ Sci Technol 51:14225–14232. https://doi.org/10.1021/acs.est.7b04760

Calderón CB, Sabundayo BP (2007) Antimicrobial classifications: drugs for bugs. In: Schwalbe R, Steele-Moore L, Goodwin AC (eds) Antimicrobial susceptibility testing protocols. CRC Press, Taylor and Frances Group. isbn:978-0-8247-4100-6

Centers for Disease Control and Prevention (CDC) (2019) Antibiotic resistance threats in the United States 2019. US Department of Health and Human Services, CDC, Atlanta, p 148

Chee-Sanford JC, Mackie RI, Koike S, Krapac IG, Lin YF, Yannarell AC, Maxwell S, Aminov RI (2009) Fate and transport of antibiotic residues and antibiotic resistance genes following land application of manure waste. J Environ Qual 38:1086–1108. https://doi.org/10.2134/jeq2008.0128

Chen Q, An X, Li H, Su J, Ma Y, Zhu YG (2016) Long-term field application of sewage sludge increases the abundance of antibiotic resistance genes in soil. Environ Int 92:1–10. https://doi.org/10.1016/j.envint.2016.03.026

Chen Y, Zhou JL, Cheng L, Zheng YY, Xu J (2017) Sediment and salinity effects on the bioaccumulation of sulfamethoxazole in zebrafish (*Danio rerio*). Chemosphere 180:467–475. https://doi.org/10.1016/j.chemosphere.2017.04.055

Cheng D, Ngo HH, Guo W, Chang SW, Nguyen DD, Liu Y, Wei Q, Wei D (2019) A critical review on antibiotics and hormones in swine wastewater: water pollution problems and control approaches. J Hazard Mater 121682. https://doi.org/10.1016/j.jhazmat.2019.121682

CODEX Codex Alimentarius (1996) Vol. 3. FAO/OMS. Comisión del Codex Alimentarius, Rome, 60.

de Oliveira JA, Williams DE, Andersen RJ, Sarragiotto MH, Baldoqui DC (2020) Pumilacidins AE from sediment-derived bacterium *Bacillus* sp. 4040 an their antimicrobial activity evaluation. J Braz Chem Soc 31:357–363. https://doi.org/10.21577/0103-5053.20190188

Diehl DL, LaPara TM (2010) Effect of temperature on the fate of genes encoding tetracycline resistance and the integrase of class 1 integrons within anaerobic and aerobic digesters treating municipal wastewater solids. Environ Sci Technol 44:9128–9133. https://doi.org/10.1021/es102765a

Ding H, Wu Y, Zou B, Lou Q, Zhang W, Zhong J, Lu L, Dai G (2016) Simultaneous removal and degradation characteristics of sulfonamide, tetracycline, and quinolone antibiotics by laccase-mediated oxidation coupled with soil adsorption. J Hazard Mater 307:350–358. https://doi.org/10.1016/j.jhazmat.2015.12.062

Escobar-Huerfano F, Gómez-Oliván LM, Luja-Mondragón M, SanJuan-Reyes N, Islas-Flores H, Hernández-Navarro MD (2020) Embryotoxic and teratogenic profile of tretracycline at environmentally relevant concentrations on *Cyprinus carpio*. Chemosphere 240:124969. https://doi.org/10.1016/j.chemosphere.2019.124969

EU consolidated version of Commission Regulation No 37/2010 of December 2009 on Pharmacologically active substances and their classification regarding maximum residue limits in foodstuffs of animal origin.

Everage TJ, Boopathy R, Nathaniel R, LaFleur G, Doucet J (2014) A survey of antibiotic-resistant bacteria in a sewage treatment plant in Thibodaux, Louisiana, USA. Int Biodeterior Biodegradation 95:2–10. https://doi.org/10.1016/j.ibiod.2014.05.028

Fan NS, Bai YH, Chen QQ, Shen YY, Huang BC, Jin RC (2020) Deciphering the toxic effects of antibiotics on denitrification: process performance, microbial community and antibiotic resistance genes. J Environ Manag 262:110375. https://doi.org/10.1016/j.jenvman.2020.110375

Founou LL, Founou RC, Essack SY (2016) Antibiotic resistance in the food chain: a developing country-perspective. Front Microbiol 7:1881. https://doi.org/10.3389/fmicb.2016.01881

Francesco AD, Renzi M, Borel N, Marti H, Salvatore D (2020) Detection of tetracycline-resistance genes in European hedgehogs (Erinaceus europaeus) and crested porcupines (*Hystrix cristata*). J Wildl Dis 56:219–223. https://doi.org/10.7589/2019-03-068

Gao L, Shi Y, Li W, Liu J, Cai Y (2015) Occurrence and distribution of antibiotics in urban soil in Beijing and Shanghai, China. Environ Sci Pollut Res 22:11360–11371. https://doi.org/10.1007/s11356-015-4230-3

Gao Q, Li Y, Qi Z, Yue Y, Min M, Peng S, Shi Z, Gao Y (2018) Diverse and abundant antibiotic resistance genes from mariculture sites of China's coastline. Sci Total Environ 630:117–125. https://doi.org/10.1016/j.scitotenv.2018.02.122

García J, García-Galán MJ, Day JW, Boopathy R, White JR, Wallace S, Hunter RG (2020) A review of emerging organic contaminants (EOCs), antibiotic resistant bacteria (ARB), and antibiotic resistance genes (ARGs) in the environment: increasing removal with wetlands and reducing environmental impacts. Bioresour Technol 123228. https://doi.org/10.1016/j.biortech.2020.123228

Ghaly TM, Geoghegan JL, Tetu SG, Gillings MR (2020) The peril and promise of integrons: beyond antibiotic resistance. Trends Microbiol. https://doi.org/10.1016/j.tim.2019.12.002

Giger W, Alder AC, Golet EM, Kohler HPE, McArdell CS, Molnar E, Siegrist H, Suter MJF (2003) Occurrence and fate of antibiotics as trace contaminants in wastewaters, sewage sludges, and surface waters. CHIMIA Int J Chem 57(9):485–491. https://doi.org/10.2533/000942903777679064

Girardi C, Greve J, Lamshöft M, Fetzer I, Miltner A, Schäffer A, Kästner M (2011) Biodegradation of ciprofloxacin in water and soil and its effects on the microbial communities. J Hazard Mater 198:22–30. https://doi.org/10.1016/j.jhazmat.2011.10.004

Gogarten JP, Townsend JP (2005) Horizontal gene transfer, genome innovation and evolution. Nat Rev Microbiol 3:679–687. https://doi.org/10.1038/nrmicro1204

Grabert R, Boopathy R, Nathaniel R, LaFleur G (2018) Effect of tetracycline on ammonia and carbon removal by the facultative bacteria in the anaerobic digester of a sewage treatment plant. Bioresour Technol 267:265–270. https://doi.org/10.1016/j.biortech.2018.07.061

Grenni P, Ancona V, Caracciolo AB (2018) Ecological effects of antibiotics on natural ecosystems: a review. Microchem J 136:25–39. https://doi.org/10.1016/j.microc.2017.02.006

Hamscher G, Sczesny S, Höper H, Nau H. Tierarzneimittel als persistente organische Kontaminanten von Böden. 10 Jahre BodenDauerbeobachtung in Niedersachsen, 2001, 10

Han XM, Hu HW, Chen QL, Yang LY, Li HL, Zhu YG, Li XZ, Ma YB (2018) Antibiotic resistance genes and associated bacterial communities in agricultural soils amended with different sources of animal manures. Soil Biol Biochem 126:91–102. https://doi.org/10.1016/j.soilbio.2018.08.018

He LY, Ying GG, Liu YS, Su HC, Chen J, Liu SS, Zhao JL (2016) Discharge of swine wastes risks water quality and food safety: antibiotics and antibiotic resistance genes from swine sources to the receiving environments. Environ Int 92–93:210–219. https://doi.org/10.1016/j.envint.2016.03.023

Heuer H, Schmitt H, Smalla K (2011) Antibiotic resistance gene spread due to manure application on agricultural fields. Curr Opin Microbiol 14:236–243. https://doi.org/10.1016/j.mib.2011.04.009

Hilaire SS, Bellows B, Brady JA, Muir JP (2020) Oxytetracycline and monensin uptake by Tifton 85 bermudagrass from dairy manure-applied soil. Agronomy 10:468. https://doi.org/10.3390/agronomy10040468

Ho YB, Zakaria MP, Latif PA, Saari N (2014) Occurrence of veterinary antibiotics and progesterone in broiler manure and agricultural soil in Malaysia. Sci Total Environ 488:261–267. https://doi.org/10.1016/j.scitotenv.2014.04.109

Höper H, Kues J, Nau H, Hamscher G (2002) Eintrag und verbleib von tierarzneimittelwirkstoffen in Böden. Bodenschutz 4:141–148

Hu X, Zhou Q, Luo Y (2010) Occurrence and source analysis of typical veterinary antibiotics in manure, soil, vegetables and groundwater from organic vegetable bases, northern China. Environ Pollut 158:2992–2998. https://doi.org/10.1016/j.envpol.2010.05.023

Huang K, Tang J, Zhang XX, Xu K, Ren H (2014) A comprehensive insight into tetracycline resistant bacteria and antibiotic resistance genes in activated sludge using next-generation sequencing. Int J Mol Sci 15:10083–10100. https://doi.org/10.3390/ijms150610083

Hussain S, Naeem M, Chaudhry MN, Iqbal MA (2016) Accumulation of residual antibiotics in the vegetables irrigated by pharmaceutical wastewater. Expo Health 8:107–115. https://doi.org/10.1007/s12403-015-0186-2

Iacono M, Villa L, Fortini D, Bordoni R, Imperi F, Bonnal RJ, Sicheritz-Ponten T, De Bellis G, Visca P, Cassone A, Carattoli A (2008) Whole-genome pyrosequencing of an epidemic multidrug-resistant *Acinetobacter baumannii* strain belonging to the European clone II group. Antimicrob Agents Chemother 52:2616–2625. DOI: doi:https://doi.org/10.1128/AAC.01643-07

Jin C, Chen Q, Sun R, Zhou Q, Liu J (2009) Eco-toxic effects of sulfadiazine sodium, sulfamonomethoxine sodium and enrofloxacin on wheat, Chinese cabbage and tomato. Ecotoxicology 18:878–885. https://doi.org/10.1007/s10646-009-0349-7

Jones AD, Bruland GL, Agrawal SG, Vasudevan D (2005) Factors influencing the sorption of oxytetracycline to soils. Environ Toxicol Chem 24:761–770. https://doi.org/10.1897/04-037R.1

Kang DH, Gupta S, Rosen C, Fritz V, Singh A, Chander Y, Murray H, Rohwer C (2013) Antibiotic uptake by vegetable crops from manure-applied soils. J Agric Food Chem 61:9992–10001. https://doi.org/10.1021/jf404045m

Kay P, Blackwell PA, Boxall AB (2004) Fate of veterinary antibiotics in a macroporous tile drained clay soil. Environ Toxicol Chem 23:1136–1144

Kay P, Blackwell PA, Boxall AB (2005a) Column studies to investigate the fate of veterinary antibiotics in clay soils following slurry application to agricultural land. Chemosphere 60:497–507. https://doi.org/10.1016/j.chemosphere.2005.01.028

Kay P, Blackwell PA, Boxall AB (2005b) A lysimeter experiment to investigate the leaching of veterinary antibiotics through a clay soil and comparison with field data. Environ Pollut 134:333–341. https://doi.org/10.1016/j.envpol.2004.07.021

Keeney KM, Yurist-Doutsch S, Arrieta MC, Finlay BB (2014) Effects of antibiotics on human microbiota and subsequent disease. Annu Rev Microbiol 68:217–235. https://doi.org/10.1146/annurev-micro-091313-103456

Kim YS, Kim HM, Chang C, Hwang IC, Oh H, Ahn JS, Kim KD, Hwang BK, Kim BS (2007) Biological evaluation of neopeptins isolated from a *Streptomyces* strain. Pest Manage Sci 63:1208–1214. https://doi.org/10.1002/ps.1450

Kim JD, Kang JE, Kim BS (2020) Postharvest disease control efficacy of the polyene macrolide lucensomycin produced by *Streptomyces plumbeus* strain CA5 against gray mold on grapes. Postharvest Biol Technol 162:111115. https://doi.org/10.1016/j.postharvbio.2019.111115

Kimbell LK, Kappell AD, McNamara PJ (2018) Effect of pyrolysis on the removal of antibiotic resistance genes and class I integrons from municipal wastewater biosolids. Environ Sci Water Res Technol 4:1807–1818. https://doi.org/10.1039/c8ew00141c

King DE, Malone R, Lilley SH (2000) New classification and update on the quinolone antibiotics. Am Fam Physician 61:2741–2748

Knapp CW, Dolfing J, Ehlert PA, Graham DW (2010) Evidence of increasing antibiotic resistance gene abundances in archived soils since 1940. Environ Sci Technol 44:580–587. https://doi.org/10.1021/es901221x

Koniuszewska I, Korzeniewska E, Harnisz M, Kiedrzyńska E, Kiedrzyński M, Czatzkowska M, Jarosiewicz P, Zalewski M (2020) The occurrence of antibiotic-resistance genes in the Pilica River, Poland. Ecohydrol Hydrobiol 20:1–11. https://doi.org/10.1016/j.ecohyd.2019.09.002

Kovalakova P, Cizmas L, McDonald TJ, Marsalek B, Feng M, Sharma VK (2020) Occurrence and toxicity of antibiotics in the aquatic environment: a review. Chemosphere 126351. https://doi.org/10.1016/j.chemosphere.2020.126351

Kulshrestha P, Giese RF, Aga DS (2004) Investigating the molecular interactions of oxytetracycline in clay and organic matter: insights on factors affecting its mobility in soil. Environ Sci Technol 38:4097–4105. https://doi.org/10.1021/es034856q

Kumar K, Gupta SC, Baidoo SK, Chander Y, Rosen CJ (2005) Antibiotic uptake by plants from soil fertilized with animal manure. J Environ Qual 34:2082–2085. https://doi.org/10.2134/jeq2005.0026

Kumar M, Jaiswal S, Sodhi KK, Shree P, Singh DK, Agrawal PK, Shukla P (2019) Antibiotics bioremediation: perspectives on its ecotoxicity and resistance. Environ Int 124:448–461. https://doi.org/10.1016/j.envint.2018.12.065

Lei L (2019) A viral alternative to antibiotics. Nat Plants 6:4. https://doi.org/10.1038/s41477-019-0584-8

Li R, Zhang Y, Lee CC, Liu L, Huang Y (2011a) Hydrophilic interaction chromatography separation mechanisms of tetracyclines on amino-bonded silica column. J Separ Sci 34:1508–1516. https://doi.org/10.1002/jssc.201100130

Li Z, Wu S, Bai X, Liu Y, Lu J, Liu Y, Xiao B, Lu X, Fan L (2011b) Genome sequence of the tobacco bacterial wilt pathogen *Ralstonia solanacearum*. J Bacteriol 193:6088–6089. https://doi.org/10.1128/JB.06009-11

Li C, Chen J, Wang J, Ma Z, Han P, Luan Y, Lu A (2015) Occurrence of antibiotics in soils and manures from greenhouse vegetable production bases of Beijing, China and an associated risk assessment. Sci Total Environ 521:101–107. https://doi.org/10.1016/j.scitotenv.2015.03.070

Li Y, Tang H, Hu Y, Wang X, Ai X, Tang L, Matthew C, Cavanagh J, Qiu J (2016) Enrofloxacin at environmentally relevant concentrations enhances uptake and toxicity of cadmium in the earthworm *Eisenia fetida* in farm soils. J Hazard Mater 308:312–320. https://doi.org/10.1016/j.jhazmat.2016.01.057

Li S, Shi W, Liu W, Li H, Zhang W, Hu J, Ke Y, Sun W, Ni J (2018) A duodecennial national synthesis of antibiotics in China's major rivers and seas (2005–2016). Sci Total Environ 615:906–917. https://doi.org/10.1016/j.scitotenv.2017.09.328

Li S, Yao Q, Liu J, Wei D, Zhou B, Zhu P, Cui XA, Jin J, Liu X, Wang G (2020a) Profiles of antibiotic resistome with animal manure application in black soils of northeast China. J Hazard Mater 384:121216. https://doi.org/10.1016/j.jhazmat.2019.121216

Li X, Zhu W, Meng G, Guo R, Wang Y (2020b) Phytoremediation of alkaline soils co-contaminated with cadmium and tetracycline antibiotics using the ornamental hyperaccumulators *Mirabilis jalapa* L. and *Tagetes patula* L. Environ Sci Pollut Res 27:14175–14183. https://doi.org/10.1007/s11356-020-07975-2

Liang B, Ma J, Cai W, Li Z, Liu W, Qi M, Zhao Y, Ma X, Deng Y, Wang A, Zhou J (2019) Response of chloramphenicol-reducing biocathode resistome to continuous electrical stimulation. Water Res 148:398–406. https://doi.org/10.1016/j.watres.2018.10.073

Liang DH, Hu Y, Cheng J, Chen Y (2020a) Simultaneous sulfamethoxazole biodegradation and nitrogen conversion in low C/N ratio pharmaceutical wastewater by *Achromobacter* sp. JL9. Sci Total Environ 703:135586. https://doi.org/10.1016/j.scitotenv.2019.135586

Liang J, Tang S, Gong J, Zeng G, Tang W, Song B, Zhang P, Yang Z, Luo Y (2020b) Responses of enzymatic activity and microbial communities to biochar/compost amendment in sulfamethoxazole polluted wetland soil. J Hazard Mater 385:121533. https://doi.org/10.1016/j.jhazmat.2019.121533

Lin H, Sun W, Yu Q, Ma J (2020) Acidic conditions enhance the removal of sulfonamide antibiotics and antibiotic resistance determinants in swine manure. Environ Pollut:114439. https://doi.org/10.1016/j.envpol.2020.114439

Liu L, Liu C, Zheng J, Huang X, Wang Z, Liu Y, Zhu G (2013) Elimination of veterinary antibiotics and antibiotic resistance genes from swine wastewater in the vertical flow constructed wetlands. Chemosphere 91(8):1088–1093. https://doi.org/10.1016/j.chemosphere.2013.01.007

Liu Z, Klümper U, Liu Y, Yang Y, Wei Q, Lin JG, Gu JD, Li M (2019) Metagenomic and metatranscriptomic analyses reveal activity and hosts of antibiotic resistance genes in activated sludge. Environ Int 129:208–220. https://doi.org/10.1016/j.envint.2019.05.036

Lorenzo FD, Palmigiano A, Silipo A, Desaki Y, Garozzo D, Lanzetta R, Shibuya N, Molinaro A (2016) The structure of the lipooligosaccharide from *Xanthomonas oryzae* pv. Oryzae: the causal agent of the bacterial leaf blight in rice. Carbohydr Res 427:38–43. https://doi.org/10.1016/j.carres.2016.03.026

Lu XM, Lu PZ (2019) Distribution of antibiotic resistance genes in soil amended using *Azolla imbricata* and its driving mechanisms. Sci Total Environ 692:422–431. https://doi.org/10.1016/j.scitotenv.2019.07.285

Lu XM, Lu PZ (2020) Seasonal variations in antibiotic resistance genes in estuarine sediments and the driving mechanisms. J Hazard Mater 383:121164. https://doi.org/10.1016/j.jhazmat.2019.121164

Lu J, Zhang Y, Wu J, Wang J, Zhang C, Lin Y (2019) Occurrence and spatial distribution of antibiotic resistance genes in the Bohai Sea and Yellow Sea areas, China. Environ Pollut 252:450–460. https://doi.org/10.1016/j.envpol.2019.05.143

Lu XM, Lu PZ, Liu XP (2020a) Fate and abundance of antibiotic resistance genes on microplastics in facility vegetable soil. Sci Total Environ 709:136276. https://doi.org/10.1016/j.scitotenv.2019.136276

Lu Y, Xiao Y, Zheng G, Lu J, Zhou L (2020b) Conditioning with zero-valent iron or $Fe^{2+}$ activated peroxydisulfate at an acidic initial sludge pH removed intracellular antibiotic resistance genes but increased extracellular antibiotic resistance genes in sewage sludge. J Hazard Mater 386:121982. https://doi.org/10.1016/j.jhazmat.2019.121982

Ma Y, Wilson CA, Novak JT, Riffat R, Aynur S, Murthy S, Pruden A (2011) Effect of various sludge digestion conditions on sulfonamide, macrolide, and tetracycline resistance genes and class I integrons. Environ Sci Technol 45:7855–7861. https://doi.org/10.1021/es200827t

Maarouf M, Lawrence F, Brown S, Robert-Gero M (1997) Biochemical alterations in paromomycin-treated *Leishmania donovani* promastigotes. Parasitol Res 83:198–202. https://doi.org/10.1007/s004360050232

Manasfi R, Chiron S, Montemurro N, Perez S, Brienza M (2020) Biodegradation of fluoroquinolone antibiotics and the climbazole fungicide by *Trichoderma* species. Environ Sci Pollut Res:1–11. https://doi.org/10.1007/s11356-020-08442-8

Manson JM, Hancock LE, Gilmore MS (2010) Mechanism of chromosomal transfer of Enterococcus faecalis pathogenicity island, capsule, antimicrobial resistance, and other traits. Proc Natl Acad Sci 107:12269–12274. https://doi.org/10.1073/pnas.1000139107

Moreno AB, del Pozo ÁM, Borja M, Segundo BS (2003) Activity of the antifungal protein from *Aspergillus giganteus* against *Botrytis cinerea*. Phytopathology 93:1344–1353. DOI: doi:https://doi.org/10.1094/PHYTO.2003.93.11.1344

Muhammad J, Khan S, Su JQ, Hesham AEL, Ditta A, Nawab J, Ali A (2020) Antibiotics in poultry manure and their associated health issues: a systematic review. J Soil Sediments 20:486–497. https://doi.org/10.1007/s11368-019-02360-0

Muhialdin BJ, Algboory HL, Kadum H, Mohammed NK, Saari N, Hassan Z, Hussin ASM (2020) Antifungal activity determination for the peptides generated by *Lactobacillus plantarum* TE10 against *Aspergillus flavus* in maize seeds. Food Control 109:106898. https://doi.org/10.1016/j.foodcont.2019.106898

Naruse N, Tenmyo O, Kobaru S, Kamei H, Miyaki T, Konishi M, Oki T (1990) Pumilacidin, a complex of new antiviral antibiotics. J Antibiot 43:267–280. https://doi.org/10.7164/antibiotics.43.267

Neher TP, Ma L, Moorman TB, Howe AC, Soupir ML (2020) Catchment-scale export of antibiotic resistance genes and bacteria from an agricultural watershed in central Iowa. PLoS One 15:e0227136. https://doi.org/10.1371/journal.pone.0227136

Ngigi AN, Magu MM, Muendo BM (2020) Occurrence of antibiotics residues in hospital wastewater, wastewater treatment plant, and in surface water in Nairobi County, Kenya. Environ Monit Assess 192:18. https://doi.org/10.1007/s10661-019-7952-8

O'neill J (2014) Review on antimicrobial resistance: tackling a crisis for the health and wealth of nations. Review on Antimicrobial Resistance, London. https://amrreview.org/Publications.html

O'Rourke A, Beyhan S, Choi Y, Morales P, Chan AP, Espinoza JL, Dupont CL, Meyer KJ, Spoering A, Lewis K, Nierman WC (2020) Mechanism-of-action classification of antibiotics by global transcriptome profiling. Antimicrob Agents Chemother. https://doi.org/10.1128/AAC.01207-19

Paiva MC, Reis MP, Costa PS, Dias MF, Bleicher L, Scholte LL, Nardi RM, Nascimento AM (2017) Identification of new bacteria harboring qnrS and aac (6′)-Ib/cr and mutations possibly involved in fluoroquinolone resistance in raw sewage and activated sludge samples from a full-scale WWTP. Water Res 110:27–37. https://doi.org/10.1016/j.watres.2016.11.056

Pan M, Chu LM (2016) Adsorption and degradation of five selected antibiotics in agricultural soil. Sci Total Environ 545:48–56. https://doi.org/10.1016/j.scitotenv.2015.12.040

Pan M, Chu LM (2017a) Fate of antibiotics in soil and their uptake by edible crops. Sci Total Environ 599:500–512. https://doi.org/10.1016/j.scitotenv.2017.04.214

Pan M, Chu LM (2017b) Leaching behavior of veterinary antibiotics in animal manure-applied soils. Sci Total Environ 579:466–473. https://doi.org/10.1016/j.scitotenv.2016.11.072

Pan M, Wong CK, Chu LM (2014) Distribution of antibiotics in wastewater-irrigated soils and their accumulation in vegetable crops in the Pearl River Delta, southern China. J Agric Food Chem 62:11062–11069. https://doi.org/10.1021/jf503850v

Panja S, Sarkar D, Datta R (2020) Removal of antibiotics and nutrients by vetiver grass (*Chrysopogon zizanioides*) from secondary wastewater effluent. Int J Phytoremediation:1–10. https://doi.org/10.1080/15226514.2019.1710813

Pärnänen K, Karkman A, Tamminen M, Lyra C, Hultman J, Paulin L, Virta M (2016) Evaluating the mobility potential of antibiotic resistance genes in environmental resistomes without metagenomics. Sci Rep 6:35790. https://doi.org/10.1038/srep35790

Picó Y, Andreu V (2007) Fluoroquinolones in soil-risks and challenges. Anal Bioanal Chem 387:1287–1299. https://doi.org/10.1007/s00216-006-0843-1

Pitta DW, Dou Z, Kumar S, Indugu N, Toth JD, Vecchiarelli B, Bhukya B (2016) Metagenomic evidence of the prevalence and distribution patterns of antimicrobial resistance genes in dairy agro-ecosystem s. Foodborne Pathog Dis 13:296–302. https://doi.org/10.1089/fpd.2015.2092

Pu Q, Zhao LX, Li YT, Su JQ (2020) Manure fertilization increase antibiotic resistance in soils from typical greenhouse vegetable production bases, China. J Hazard Mater 391:122267. https://doi.org/10.1016/j.jhazmat.2020.122267

Pulingam T, Thong KL, Appaturi JN, Nordin NI, Dinshaw IJ, Lai CW, Leo BF (2020) Synergistic antibacterial actions of graphene oxide and antibiotics towards bacteria and the toxicological effects of graphene oxide on human epidermal keratinocytes. Eur J Pharm Sci 142:105087. https://doi.org/10.1016/j.ejps.2019.105087

Qiao M, Ying GG, Singer AC, Zhu YG (2018) Review of antibiotic resistance in China and its environment. Environ Int 110:160–172. https://doi.org/10.1016/j.envint.2017.10.016

Rabølle M, Spliid NH (2000) Sorption and mobility of metronidazole, olaquindox, oxytetracycline and tylosin in soil. Chemosphere 40:715–722. https://doi.org/10.1016/s0045-6535(99)00442-7

Ramírez A, Gutiérrez R, Díaz G, González C, Pérez N, Vega S, Noa M (2003) High-performance thin-layer chromatography–bioautography for multiple antibiotic residues in cow's milk. J Chromatogr B 784:315–322. https://doi.org/10.1016/S1570-0232(02)00819-X

Rao J, Liu L, Zeng D, Wang M, Xiang M, Yang S (2020) Antibiotic activities of propanolamine containing 1, 4-benzoxazin-3-ones against phytopathogenic bacteria. RSC Adv 10:682–688. https://doi.org/10.1039/C9RA09639F

Ray RS, Mehrotra S, Shankar U, Babu GS, Joshi PC, Hans RK (2001) Evaluation of UV-induced superoxide radical generation potential of some common antibiotics. Drug Chem Toxicol 24:191–200. https://doi.org/10.1081/dct-100102610

Reis AC, Kolvenbach BA, Nunes OC, Corvini PF (2020) Biodegradation of antibiotics: the new resistance determinants – part I. New Biotechnol 54:34–51. https://doi.org/10.1016/j.nbt.2019.08.002

Rizzo L, Manaia C, Merlin C, Schwartz T, Dagot C, Ploy MC, Michael I, Fatta-Kassinos D (2013) Urban wastewater treatment plants as hotspots for antibiotic resistant bacteria and genes spread into the environment: a review. Sci Total Environ 447:345–360. https://doi.org/10.1016/j.scitotenv.2013.01.032

Roberts MC (2005) Update on acquired tetracycline resistance genes. FEMS Microbiol Lett 245:195–203. https://doi.org/10.1016/j.femsle.2005.02.034

Romanowski G, Lorenz MG, Wackernagel W (1993) Plasmid DNA in a groundwater aquifer microcosm-adsorption, DNAase resistance and natural genetic transformation of *Bacillus subtilis*. Mol Ecol 2:171–181. https://doi.org/10.1111/j.1365-294X.1993.tb00106.x

Salehi Z, Fatahi N, Izadi A, Taran M, Badali H, Hashemi SJ, Rezaie S, Daie R, Bara A, Ghaffari M, Aala F (2020) Comparison of in vitro antifungal activity of novel triazoles with old antifungal agents against dermatophyte species caused *Tinea pedis*. J Mycol Méd:100935. https://doi.org/10.1016/j.mycmed.2020.100935

Salyers AA, Gupta A, Wang Y (2004) Human intestinal bacteria as reservoirs for antibiotic resistance genes. Trends Microbiol 12:412–416. https://doi.org/10.1016/j.tim.2004.07.004

Sarmah AK, Meyer MT, Boxall AB (2006) A global perspective on the use, sales, exposure pathways, occurrence, fate and effects of veterinary antibiotics (VAs) in the environment. Chemosphere 65:725–759. https://doi.org/10.1016/j.chemosphere.2006.03.026

Scarano C, Spanu C, Ziino G, Pedonese F, Dalmasso A, Spanu V, Virdis S, De Santis EP (2014) Antibiotic resistance of *Vibrio* species isolated from *Sparus aurata* reared in Italian mariculture. New Microbiol 37:329–337

Schmieder R, Edwards R (2012) Insights into antibiotic resistance through metagenomic approaches. Future Microbiol 7(1):73–89. https://doi.org/10.2217/fmb.11.135

Schüller S (1998) Anwendung antibiotisch wirksamer Substanzen beim Tier und Beurteilung der Umweltsicherheit entsprechender Produkte. 3. Statuskolloquium ökotoxikologischer Forschungen in der Euregio Bodensee

Selvam A, Wong JWC (2017) Chapter 12: Degradation of antibiotics in livestock manure during composting. In: JWC W, Tyagi RD, Pandey A (eds) Current developments in biotechnologyandbioengineering:solidwastemanagement.https://doi.org/10.1016/B978-0-444-63664-5.00012-5

Sengeløv G, Agersø Y, Halling-Sørensen B, Baloda SB, Andersen JS, Jensen LB (2003) Bacterial antibiotic resistance levels in Danish farmland as a result of treatment with pig manure slurry. Environ Int 28:587–595. https://doi.org/10.1016/S0160-4120(02)00084-3

Shen DS, Tao XQ, Shentu JL, Wang MZ (2014) Residues of veterinary antibiotics in pig feeds and manures in Zhejiang Province. Adv Mater Res 1010-1012:301–304. https://doi.org/10.4028/www.scientific.net/AMR.1010-1012.301

Sodhi KK, Kumar M, Singh DK (2020) Potential application in amoxicillin removal of *Alcaligenes* sp. MMA and enzymatic studies through molecular docking. Arch Microbiol:1–7. https://doi.org/10.1007/s00203-020-01868-1

Stalder T, Barraud O, Casellas M, Dagot C, Ploy MC (2012) Integron involvement in environmental spread of antibiotic resistance. Front Microbiol 3:119. https://doi.org/10.3389/fmicb.2012.00119

Stockwell VO, Duffy B (2012) Use of antibiotics in plant agriculture. Rev Sci Tech OIE 31(1). https://doi.org/10.20506/rst.31.1.2104

Sun J, Li N, Yang P, Zhang Y, Yuan Y, Lu X, Zhang H (2020) Simultaneous antibiotic degradation, nitrogen removal and power generation in a microalgae-bacteria powered biofuel cell designed for aquaculture wastewater treatment and energy recovery. Int J Hydrogen Energ 45:10871–10881. https://doi.org/10.1016/j.ijhydene.2020.02.029

Tahrani L, Van Loco J, Ben Mansour H, Reyns T (2016) Occurrence of antibiotics in pharmaceutical industrial wastewater, wastewater treatment plant and sea waters in Tunisia. J Water Health 14:208–213. DOI: doi:https://doi.org/10.2166/wh.2015.224

Tang X, Lou C, Wang S, Lu Y, Liu M, Hashmi MZ, Liang X, Li Z, Liao Y, Qin W, Fan F (2015) Effects of long-term manure applications on the occurrence of antibiotics and antibiotic resistance genes (ARGs) in paddy soils: evidence from four field experiments in south of China. Soil Biol Biochem 90:179–187. https://doi.org/10.1016/j.soilbio.2015.07.027

Tasho RP, Cho JY (2016) Veterinary antibiotics in animal waste, its distribution in soil and uptake by plants: a review. Sci Total Environ 563:366–376. https://doi.org/10.1016/j.scitotenv.2016.04.140

Thiele-Bruhn S (2003) Pharmaceutical antibiotic compounds in soils – a review. J Plant Nutr Soil Sci 166:145–167. https://doi.org/10.1002/jpln.200390023

Thiele-Bruhn S, Seibicke T, Schulten HR, Leinweber P (2004) Sorption of sulfonamide pharmaceutical antibiotics on whole soils and particle-size fractions. J Environ Qual 33:1331–1342. https://doi.org/10.2134/jeq2004.1331

Tian Q, Dou X, Huang L, Wang L, Meng D, Zhai L, Shen Y, You C, Guan Z, Liao X (2020) Characterization of a robust cold-adapted and thermostable laccase from *Pycnoporus* sp. SYBC-L10 with a strong ability for the degradation of tetracycline and oxytetracycline by laccase-mediated oxidation. J Hazard Mater 382:121084. https://doi.org/10.1016/j.jhazmat.2019.121084

Timmerer U, Lehmann L, Schnug E, Bloem E (2020) Toxic effects of single antibiotics and antibiotics in combination on germination and growth of *Sinapis alba* L. Plan Theory 9:107. https://doi.org/10.3390/plants9010107

Tóth L, Boros É, Poór P, Ördög A, Kele Z, Váradi G, Holzknecht J, Bratschun-Khan D, Nagy I, Tóth GK, Rákhely G (2020) The potential use of the *Penicillium chrysogenum* antifungal protein PAF, the designed variant PAFopt and its γ-core peptide Pγopt in plant protection. Microb Biotechnol. https://doi.org/10.1111/1751-7915.13559

Uddin M, Chen J, Qiao X, Tian R, Zhu M (2020) Insight into dynamics and bioavailability of antibiotics in paddy soils by in situ soil moisture sampler. Sci Total Environ 703:135562. https://doi.org/10.1016/j.scitotenv.2019.135562

Wang Z, Zhang XX, Huang K, Miao Y, Shi P, Liu B, Long C, Li A (2013) Metagenomic profiling of antibiotic resistance genes and mobile genetic elements in a tannery wastewater treatment plant. PLoS One 8:10. https://doi.org/10.1371/journal.pone.0076079

Wang B, Yan J, Li G, Zhang J, Zhang L, Li Z, Chen H (2020a) Risk of penicillin fermentation dreg: increase of antibiotic resistance genes after soil discharge. Environ Pollut 113956. https://doi.org/10.1016/j.envpol.2020.113956

Wang M, Liu H, Dai X (2020b) Dosage effects of lincomycin mycelial residues on lincomycin resistance genes and soil microbial communities. Environ Pollut 256:113392. https://doi.org/10.1016/j.envpol.2019.113392

Wang Q, Duan YJ, Wang SP, Wang LT, Hou ZL, Cui YX, Hou J, Das R, Mao DQ, Luo Y (2020c) Occurrence and distribution of clinical and veterinary antibiotics in the faeces of a Chinese population. J Hazard Mater 383:121129. https://doi.org/10.1016/j.jhazmat.2019.121129

Whittle G, Shoemaker NB, Salyers AA (2002) The role of *Bacteroides* conjugative transposons in the dissemination of antibiotic resistance genes. Cell Mol Life Sci 59:2044–2054. https://doi.org/10.1007/s000180200004

Williams-Nguyen J, Sallach JB, Bartelt-Hunt S, Boxall AB, Durso LM, McLain JE, Singer RS, Snow DD, Zilles JL (2016) Antibiotics and antibiotic resistance in agro-ecosystems: state of the science. J Environ Qual 45:394–406. https://doi.org/10.2134/jeq2015.07.0336

Willms IM, Yuan J, Penone C, Goldmann K, Vogt J, Wubet T, Schöning I, Schrumpf M, Buscot F, Nacke H (2020) Distribution of medically relevant antibiotic resistance genes and mobile genetic elements in soils of temperate forests and grasslands varying in land use. Genes 11:150. https://doi.org/10.3390/genes11020150

Winckler C, Grafe A (2000) Stoffeintrag Durch Tierarzneimittel und Pharmakologisch Wirksame Fuutterzusatzstoffe unter Besonderer Berücksichtigung von Tetrazyklinen. Berlin: UBA-Texte 44/00, 145

Xu Z, Zhang Q, Fang HH (2003) Applications of porous resin sorbents in industrial wastewater treatment and resource recovery. Crit Rev Environ Sci Technol 33:363–389. https://doi.org/10.1080/10643380390249512

Xu L, Zhou Z, Zhu L, Han Y, Lin Z, Feng W, Liu Y, Shuai X, Chen H (2020) Antibiotic resistance genes and microcystins in a drinking water treatment plant. Environ Pollut 258:113718. https://doi.org/10.1016/j.envpol.2019.113718

Yan Y, Pengmao Y, Xu X, Zhang L, Wang G, Jin Q, Chen L (2020) Migration of antibiotic ciprofloxacin during phytoremediation of contaminated water and identification of transformation products. Aquat Toxicol 219:105374. https://doi.org/10.1016/j.aquatox.2019.105374

Yang JF, Ying GG, Liu S, Zhou LJ, Zhao JL, Tao R, Peng PA (2012) Biological degradation and microbial function effect of norfloxacin in a soil under different conditions. J Environ Sci Health Part B 47:288–295. https://doi.org/10.1080/03601234.2012.638886

Yang Y, Owino AA, Gao Y, Yan X, Xu C, Wang J (2016) Occurrence, composition and risk assessment of antibiotics in soils from Kenya, Africa. Ecotoxicology 25:1194–1201. https://doi.org/10.1007/s10646-016-1673-3

Yuan QB, Guo MT, Yang J (2015) Fate of antibiotic resistant bacteria and genes during wastewater chlorination: implication for antibiotic resistance control. PLoS One 10(3). https://doi.org/10.1371/journal.pone.0119403

Yuan W, Tian T, Yang Q, Riaz L (2020) Transfer potentials of antibiotic resistance genes in *Escherichia* spp. strains from different sources. Chemosphere 246:125736. https://doi.org/10.1016/j.chemosphere.2019.125736

Zarei-Baygi A, Harb M, Wang P, Stadler LB, Smith AL (2019) Evaluating antibiotic resistance gene correlations with antibiotic exposure conditions in anaerobic membrane bioreactors. Environ Sci Technol 53:3599–3609. https://doi.org/10.1021/acs.est.9b00798

Zhang H, Liu P, Feng Y, Yang F (2013) Fate of antibiotics during wastewater treatment and antibiotic distribution in the effluent-receiving waters of the yellow sea, northern China. Marine Pollut Bull 73:282–290. https://doi.org/10.1016/j.marpolbul.2013.05.007

Zhang Y, Li A, Dai T, Li F, Xie H, Chen L, Wen D (2018) Cell-free DNA: a neglected source for antibiotic resistance genes spreading from WWTPs. Environ Sci Technol 52:248–257. https://doi.org/10.1021/acs.est.7b04283

Zhang YJ, Hu HW, Chen Q, Singh BK, Yan H, Chen D, He JZ (2019) Transfer of antibiotic resistance from manure-amended soils to vegetable microbiomes. Environ Int 130:104912. https://doi.org/10.1016/j.envint.2019.104912

Zhang D, Lu Y, Chen H, Wu C, Zhang H, Chen L, Chen X (2020a) Antifungal peptides produced by actinomycetes and their biological activities against plant diseases. J Antibiot:1–18. https://doi.org/10.1038/s41429-020-0287-4

Zhang H, Chen S, Zhang Q, Long Z, Yu Y, Fang H (2020b) Fungicides enhanced the abundance of antibiotic resistance genes in greenhouse soil. Environ Pollut:113877. https://doi.org/10.1016/j.envpol.2019.113877

Zhang J, Buhe C, Yu D, Zhong H, Wei Y (2020c) Ammonia stress reduces antibiotic efflux but enriches horizontal gene transfer of antibiotic resistance genes in anaerobic digestion. Bioresour Technol 295:122191. https://doi.org/10.1016/j.biortech.2019.122191

Zhang K, Xin R, Zhao Z, Ma Y, Zhang Y, Niu Z (2020d) Antibiotic resistance genes in drinking water of China: occurrence, distribution and influencing factors. Ecotoxicol Environ Saf 188:109837. https://doi.org/10.1016/j.ecoenv.2019.109837

Zhao F, Chen L, Yang L, Sun L, Li S, Li M, Feng Q (2020) Effects of land use and rainfall on sequestration of veterinary antibiotics in soils at the hillslope scale. Environ Pollut 260:114112. https://doi.org/10.1016/j.envpol.2020.114112

Zhu D, Giles M, Daniell T, Neilson R, Yang XR (2020a) Does reduced usage of antibiotics in livestock production mitigate the spread of antibiotic resistance in soil, earthworm guts, and the phyllosphere? Environ Int 136:105359. https://doi.org/10.1016/j.envint.2019.105359

Zhu Y, Miao J, Wen H, Li T, Zhao Z, Guo X, Li H, Zhang G (2020b) The occurrence and spatial distribution of typical antibiotics in soils along the Fenhe River in Shanxi Province. J Soils Sediments 20:889–899. https://doi.org/10.1007/s11368-019-02402-7

# Chapter 8
# Soil Contamination by Polycyclic Aromatic Hydrocarbons in the Agroecosystems

**Rajni Yadav, Abhishek Kumar, Diksha Tokas, and Anand Narain Singh**

**Abstract** To maximize better yield of the crops by any means in the recent agricultural practices, cultivators are applying today different kind of organic and inorganic chemical compounds in the form of fertilizers and pesticides with high rates of application. These chemical compounds often get deposited into soils and increase their concentrations beyond a certain threshold. Whenever the concentration of some chemical compounds increases beyond a threshold level then corresponding compounds may start interfering with the ecological processes, such soils are referred as 'contaminated soil'. However, agricultural soil contamination not only interferes with the physiology and metabolism of crop plants, but also results in various health disorders to human beings. Among several contaminants of the soils, polycyclic aromatic hydrocarbons are highly hazardous and ubiquitously present in environment. Thus, soil contamination by polycyclic aromatic hydrocarbons needs to be carefully monitored and managed to minimize the impacts on global ecosystems and human health.

Based on a literature survey, present chapter provides the baseline information about the source, ecological impact and health risks due to soil contamination by polycyclic aromatic hydrocarbons in the agro-ecosystems. Although these compounds are generated from both natural and anthropogenic processes, the contribution from later is very high. Further, pyrogenic sources such as those generated via industrialization and urbanization processes have substantially increased the accumulation of these compounds in the soils. This increased concentration considerably alters the structure and functions of agro-ecosystems and results in compromised ecosystem services. Also, the normal growth and metabolism of crops is affected due to alterations in the physiology like chlorophyll concentration, protein synthesis and development of tissues. Since polycyclic aromatic hydrocarbons are potent carcinogens, they can risk the human health whenever inhaled, touched or consumed some affected crops. Farmers and other people associated with agriculture are at higher risks of health disorders and cancer. The residents of urban areas are at high risk of health disorders as compared to rural areas. Therefore, considering the

R. Yadav · A. Kumar · D. Tokas · A. N. Singh (✉)
Soil Ecosystem and Restoration Ecology Lab, Department of Botany, Panjab University, Chandigarh, India
e-mail: dranand1212@gmail.com



V. Kumar Singh et al. (eds.), *Sustainable Agriculture Reviews 50*, Sustainable Agriculture Reviews 50, https://doi.org/10.1007/978-3-030-63249-6_8

negative impacts of polycyclic aromatic hydrocarbons, an immediate attention is required from the policymakers to frame out a policy for checking the emission and management of these harmful chemical compounds.

**Keywords** Anthropogenic activities · Carcinogenic · Persistent organic pollutants · Soil contamination

## 8.1 Introduction

Many man-made activities including agriculture, development and industrialization release a substantial amount of harmful chemical compounds into the atmosphere. These compounds then cascade into major ecological processes such as hydrological cycle, bio-geo-chemical cycles and food-webs. Consequently, the ecosystem structure and function are altered leading to compromised ecosystem services and human health risks. Polycyclic aromatic hydrocarbons (PAHs) are actually a major category of organic pollutants, which have drastic impacts on ecosystems and human health (Abbas et al. 2018). These compounds are widespread in the environment (Rong et al. 2007; Zhang et al. 2015) and their hazardous nature has stimulated the interests of researchers across the globe. Furthermore, increasing concentrations of PAHs in the environment primarily due to industrialization, agriculture and urbanization is hindering our goal to sustainability. Therefore, it becomes imperative to assess the sources and impacts of PAHs on ecosystem functioning and human health. This will help to policymakers to develop effective policies for regulating emissions and mitigating their harmful effects.

Although the soil contamination by polycyclic aromatic hydrocarbons (PAHs) is a global concern, developing countries are more affected by the impacts of such soil contamination due to higher industrial and developmental activities. Many previous studies agreed the persistence of PAHs in environment (Alshaarawy et al. 2016; Cao et al. 2016; White et al. 2016; Hu et al. 2017) and suggested occurrence of PAHs in different ecosystem with varying concentrations (Rong et al. 2007; Agarwal et al. 2009; Farooq et al. 2011; Zhang et al. 2015). For instance, the annual concentration of PAHs is estimated to be more than 50 tons in the Chenab river, which is shared by India and Pakistan (Farooq et al. 2011). Similarly, its concentration was 22.1–1256.9 ng $g^{-1}$ in some parts of China (Rong et al. 2007; Zhang et al. 2015) whereas it was 830–3880 μg $kg^{-1}$ in agricultural soils of Delhi, India (Agarwal et al. 2009). These PAHs either remains in the air or move into soil and groundwater with rain. Further, it has been observed that urban areas have higher atmospheric concentrations of toxic PAHs than rural areas and some studies have shown that due to air pollution, urban areas soil are more contaminated as compared to the soils from rural areas (Fazeli et al. 2018; Eghbal et al. 2019). Not only in atmosphere, but also in soils the PAHs concentration is comparatively higher in the soil near industrial activities or urban areas than in the soil from rural regions. The individual compounds evaluation indicated that they are derived from two types of sources:

(i) background diffuse pollution, and (ii) anthropogenic depositions originating mainly from the combustion of fossil fuels. The content of PAHs in soil is controlled by physico-chemical characteristics of the soil, mainly SOM level, moisture content, and soil texture at regional or national level (Maliszewska-Kordybach et al. 2008).

Polycyclic aromatic hydrocarbons are generally generated from both natural and man-made activities. Natural sources include forest fires and volcanic eruptions, whereas, man-made activities such as urbanization, industrialization, vehicular emission, burning of fossil fuels, improper disposal of wastes and excessive use of pesticides in agriculture are major sources of PAHs emissions (Jiang et al. 2016; Franco et al. 2017). Also, incomplete combustion of fossil fuels and lignocellulosic biomass are considered as the systematic sources of PAHs in environment. On the other hand, agricultural soils sustain PAHs contamination from atmospheric deposition, incomplete combustion of crop debris, fertilizers or pesticides and other organic chemicals that are used to increase the yield of agricultural products (Maliszewska-Kordybach et al. 2008). Since these compounds are unsusceptible to the degradation and present in surrounding for a longer duration, these are called as persistent organic pollutants (Flowers et al. 2002; Alshaarawy et al. 2016; Cao et al. 2016; White et al. 2016; Hu et al. 2017).

According to Office of Solid Waste and Emergency Response (OSWER 2015), evaluating risk on human health, water resources, air and soil pollution may originate from various sources such as local or long-range atmospheric deposition, sludge disposal from sewage treatment plants. Actually, polycyclic aromatic hydrocarbons (PAHs) are sparingly soluble in water and soil organic matter act as a good adsorbent for PAHs, hence, soil can be ideal indicator for their presence and may acts as sink for their deposition (Tang et al. 2005; Yin et al. 2008; Agarwal et al. 2009; Liu et al. 2016). Due to their regular presence in the atmosphere because of direct deposition, crop residue burning, from organic substances and also from biowastes such as fertilizers/pesticides, agricultural land is under pressure of soil contamination. However, in mechanized modern agricultural systems, due to applications of agro-chemicals, soils have become the significant recipient of a number of organic chemicals and inorganic amendments. Therefore, the accumulation of different kinds of heavy metals and pesticides in the soil medium hampering soil biological fertility and further all those harmful elements transferring via agricultural crops to the human food-chain.

When polycyclic aromatic hydrocarbons are transferred into the soil via rainfall, these compounds affect the entire soil ecosystem by modifying the soil's physical properties such as bulk density, soil pH and soil electrical conductivity. The altered soil structure and conditions directly affect the composition and abundance of soil biota. Further, these compounds can enter to the complex food web, and therefore, may result in bio-magnification and bio-accumulation in different organisms. Thus, bio-accumulation of pesticides and PAHs in the soil has the potential to restrict the soil functions, causing toxicity to plants and contaminate the food chain. After cascading into ecosystems, they eventually pass into human body through various modes such as food, crop, inhalation, and direct touch or contact. Substantial studies have shown that some of the polycyclic aromatic hydrocarbons (PAHs) have carcinogenic properties, they can promote epigenetic effects and induce cancer by genotoxic effects via affecting the proliferative capacity of mutated cells (Abbas et al. 2018; Skrbic et al. 2019). Presence of PAHs in the soil creates a risk to the human health; through food chain contamination and accidental ingestion, also toxic activities exhibited by them towards various biological elements present in the soil environment such as microorganisms and plants (Maliszewska-Kordybach et al. 2008). This has become a serious challenge nowadays in the changing environmental conditions. To determine the risk of presence of PAHs in agricultural soils on human beings and terrestrial biota, a quantitative ecological and health risk assessment is much needed.

This chapter first provides a baseline overview of polycyclic aromatic hydrocarbons (PAHs) and specifically focus on various sources, ecological impacts, effects on soil-biota and soil biological fertility and human health risks. Therefore, a brief overview has been given on the sources of emerging contaminants, particularly PAHs, their ecological impacts and health risks to the humans into various sub-sections.

## 8.2 Polycyclic Aromatic Hydrocarbons (PAHs): An Overview

Polycyclic aromatic hydrocarbons (PAHs) are chemically characterized by two or more fused benzene rings, are a large group of organic compounds. Polycyclic (poly-nuclear) aromatic hydrocarbons are a group of complex organic chemicals made up of carbon and hydrogen with a fused ring structure which contains at least two benzene rings. These compounds broadly distributed in the air, soil, and water, and are known as first atmospheric pollutants and recognized as suspected carcinogens. The molecular weight of PAHs is directly proportional to the carcinogenicity. For example, carcinogenicity of PAHs increases with the increase in molecular weight. Benzo[a]pyrene (B[a]P) is a well-known PAHs which have five rings. Due to low vapour pressure, few PAHs are ambient in air, as associated with particles and gas phase. Phenanthrene is a lighter PAHs and present particularly in a gaseous phase. B[a]P is a heavier PAHs that absolutely absorbed on to particles (Ravindra et al. 2008).

### *8.2.1 Formation of Polycyclic Aromatic Hydrocarbons*

Polycyclic aromatic hydrocarbons may be synthesized under oxygen-deficient conditions from saturated hydrocarbons. Two mechanisms that can define PAHs formation are petrogenesis and pyro-synthesis. As the temperature goes beyond 500 °C, the bonds between C–C and C–H break which resulted in the formation of free radicals. These free radicals combine to acetylene which further condenses as thermally

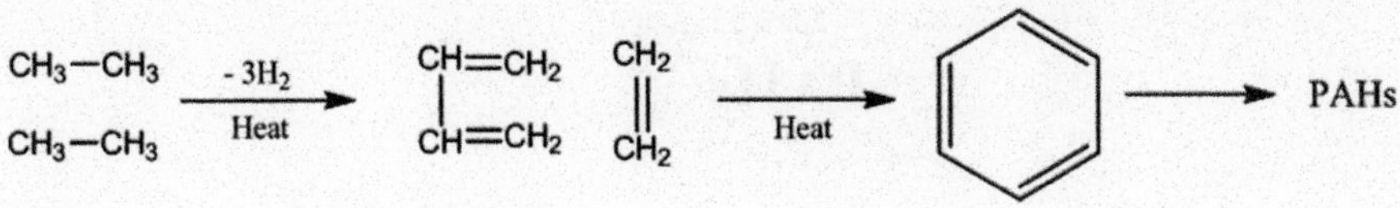


**Fig. 8.1** Pyro-synthesis of polycyclic aromatic hydrocarbons (PAHs) starting with ethane. In this process, ethane is first converted into ethene through dehydrogenation, subsequently transformed into Buta-1,3-diene. These molecules may combine to form benzene and PAHs

resistant aromatic ring structures. The formation of PAHs starting with ethane has been illustrated in the Fig. 8.1. The formation of PAHs from hydrocarbons by pyro-synthesis varies from paraffins to aromatics through olefins and cyclo-olefins (Manahan 1994).

There are three possible mechanisms for the formation of PAHs through combustion, viz. (i) slow Diels–Alder condensation, (ii) rapid radical reactions, and (iii) ionic reaction mechanism (Haynes 1991). These compounds are generally produced by four generalized pathways:

1. It occurs at low temperature lower than 70 °C, relatively rapid (days to years), conversion of organic matter into biomolecules after initial deposition in sediments.
2. Long-term, modest temperature in the range of 100–300 °C and slow formation of petrogenic PAHs.
3. High temperature greater than 500 °C, rapid, incomplete combustion of organic biomass under oxygen-starved conditions which form pyrogenic PAHs.
4. The biosynthesis via plants and animals of PAHs or relatively simple mixtures.

### 8.2.2 *Sources of Polycyclic Aromatic Hydrocarbons*

The distribution and degree of alkylation of PAHs depend on system temperature. In the environment, different distributional profiles of various sources can be used for identifying various sources of PAHs. Major sources of PAHs in the environment are diagenic, biogenic, petrogenic and pyrogenic. PAHs enter via different routes in the environment and are commonly found as a blend of two or more soot compounds. Though PAHs arise from natural sources such as forest fires and volcanic eruptions, and anthropogenic sources such as vehicles internal combustion, sewage sludge, burning of fossil fuels, pesticides, and fertilizer formulations, etc., the latter contributes major proportions to the environmental PAHs. Figures 8.2 represent the formation of PAHs from natural and anthropogenic sources and various categories of their formation, respectively.

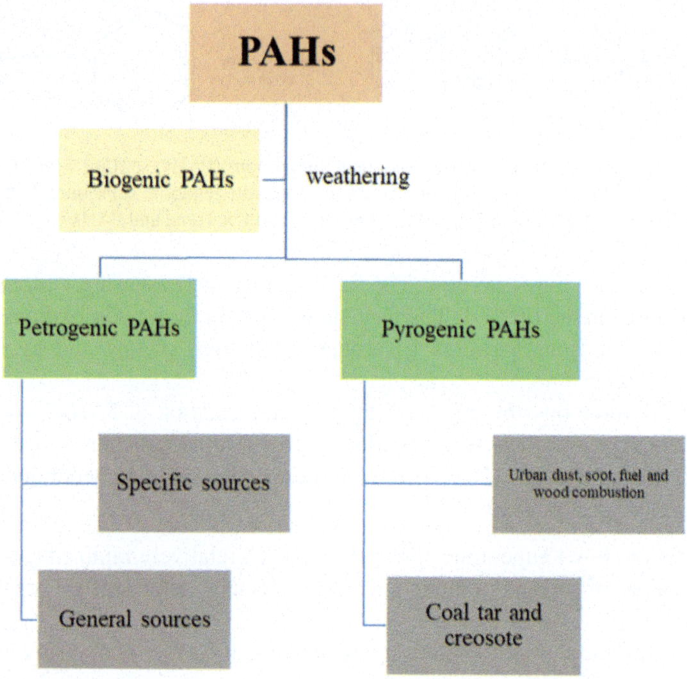


**Fig. 8.2** Generalized representation of different categories of polycyclic aromatic hydrocarbons based on their compositions

## 8.3 Emerging Sources of Agricultural Contamination

In India, the history of agriculture can be traced from the ancient time period. Nowadays, India ranks second after China world-widely in agricultural outputs. Most of the Indian population depends on agricultural practices for their livelihood and economic growth. Therefore, agriculture plays a vital role in India's economy (Srivastava et al. 2016). However, on the other hand, human's profession for livelihood in the world, agriculture is one of the top anthropogenic activities amongst others. No doubt, through this profession, human development has been successfully achieved in a substantial manner and got victory over the food crisis that saved a number of famished human populations across the world. Practitioner always has a mind-set of gaining maximum yield by using different pesticides, fertilizers and other chemicals which are responsible for high yield of crops (Singh et al. 2019). But in modern agricultural systems, due to applications of inorganic and organic fertilizers as well as different kind of pesticides, soils have become recipient of a number of organic chemicals and inorganic amendments; therefore, accumulation of different kind of heavy metals and pesticides in the soil medium hampering soil biological fertility and further all those harmful elements are transferring via

**Table 8.1** Various natural and anthropogenic sources of polycyclic aromatic hydrocarbons (PAHs) in the agro-ecosystems via petrogenic and pyrogenic processes

| PAHs sources | |
|---|---|
| Natural | Anthropogenic |
| Petroleum/coal | Pesticides/fertilizers formulation |
| Vegetative decay | Municipal discharges/sewage sludge |
| Minerals | PAHs contaminated media |
| Plant synthesis | Road dust |
| Fires and volcanic eruption | Vehicles/jet aircraft/marine seeps |
| Atmospheric inputs | Incineration |
| – | Wood burning |
| – | Industrial discharges |
| – | Cigarette smoke |
| – | Urban run off |

Source: Abdel-Shafy and Mansour (2016)

agricultural crops to the human and ultimately leading contamination of food chain affecting human health and ecosystem (Srivastava et al. 2016; Singh et al. 2019).

Agricultural contamination is introduced via natural processes as well as anthropogenic processes. Forest fires and volcanic eruptions are common sources of natural processes of contamination. In the environment, the majority of contaminants emitted from fossil fuel combustion sources such as automobiles, asphalt production, coking manufacturing plants, and manufacturing or production units that use fossil fuels are more common sources of anthropogenic processes of soil contamination which dramatically increased the quantity of PAHs and other poly-aromatic organic pollutants (Soltani et al. 2015) (Table 8.1).

### *8.3.1 Diagenic Polycyclic Aromatic Hydrocarbons*

Diagenic polycyclic aromatic hydrocarbons are generated by natural processes occurring while organic matter is deposited in soils and/or sediments. The process starts after deposition of organic matter in nature (sediments or soils) and the process occurred at low temperatures in the absence of oxygen. It is also termed as "aromatization" reactions producing a variety of aromatic biomarkers. Under the above process, many other components of PAHs are produced. For example, perylene, a five-ringed PAH, is commonly spotted in sediments of water bodies such as lakes, rivers, and oceans at a depth where oxygen is limited.

Soil organic matter is derived from diatoms and other plant materials under oceans sediments, during diagenesis are considered as the major source of perylene (Louda and Baker 1984; Venkatesan 1988). Other PAHs which are also produced via diagenesis such as Retene (l-methyl-7-isopropylphenanthrene), smaller $C_2$ through $C_4$ alkyl phenanthrenes, occur at very low concentrations near-shore sediments. Phenanthrenes and chrysenes are derived from the dehydrogenation of diterpenoid, triterpenoid, abietic acids and pimaric acids precursors profuse in the pine resin, wood ash and terrestrial plants (Tan and Heit 1981).

### 8.3.2 Fossil Fuels Related Polycyclic Aromatic Hydrocarbons

Petrogenic PAHs are produced from petroleum reservoirs (0.2% to more than 7% PAHs) and coal-beds over geological time. Petrogenic PAHs are produced from biological organic matter from plankton converted into petroleum and coal. These are formed at high temperatures (higher than diagenesis) and high pressure under the deeply buried layers of sediments. Aromatic rings fused in petrogenic PAHs varies from three to five rings, and up to ten aromatic rings under certain structures (Berkowitz 1988). The nature and dependence of processes on organic matter, subsurface migrations, burial conditions of temperature, pressure and biodegradation converting organic matter into fossil fuels, encompasses semi-random chemical processes.

Composition of PAHs is greatly varied, for example, fossil fuels like crude oil and coal have mainly two to six or more than six ringed PAHs. With an increasing molecular weight of alkylated structures linked with the two to four ringed compounds, the abundance of aromatic hydrocarbons in petroleum is usually decreased remarkably. A complex variety of parent materials are responsible for the formation of different types of PAHs from fossil fuels like un-substituted and alkylated PAHs. The phenanthrene homologous series of PAHs include a series of alkylated homologues of phenanthrene with several alkyl substitutions. Within the petrogenic PAHs, the comparative abundance of the alkylated PAHs surpasses the abundance of parent i.e. un-substituted compound or $C_0$-phenanthrene. So that main feature of petrogenic alkylated PAHs increases the parent PAHs. Chrysenes, Dibenzothiophenes, Phenanthrenes, Fluorenes are formed by petrogenic processes.

### 8.3.3 Pyrogenic Polycyclic Aromatic Hydrocarbons

Pyrogenic polycyclic aromatic hydrocarbons (PAHs) are formed by both processes (natural and anthropogenic) during the high-temperature processes and complete combustion. The main source of these PAHs is the burning of fossil fuels in closed combustion engines such as diesel, gasoline, used motor oils and wood-burning such as campfires, stoves. Such types of PAHs are also formed during processing of coals at high temperature in gasification processes. The residues of the coal gas process, known as coal-tars, are the rich source pyrogenic PAHs (Emsbo-Mattingly et al. 2001).

Pyrogenic PAHs are characterized by greater abundance of 4, 5 and 6+ ringed structures as compared to other PAHs present in most of the petroleum types. Soot particles and PAHs are formed simultaneously during combustion processes (Richter and Howard 2000), where high molecular weight PAHs acts as the molecular precursors for soot formation. Moreover, PAHs formation by combustion is independent of the fuel sources (Jenkins et al. 1996), whereas the configurations of PAHs are dependent on oxygen and temperature. Pyrogenic PAHs emission through automobiles depends on various sources like engine type, age of the vehicle, fuel type, etc. Generally un-combusted fuels in diesel engines mostly contributed to the petrogenic PAHs and mixture of pyrogenic and petrogenic formation (Williams et al. 1986, 1989; Miguel et al. 1998).

### 8.3.4 Biogenic Polycyclic Aromatic Hydrocarbons

Most of the PAHs precursor biosynthesized in nature remains uncertain. Biogenic PAHs can be synthesized by natural organisms such as bacteria and plants, and formed during degradation of vegetative organic matter. Several reports showed that the contribution of microorganisms in the production of PAHs. Precursor of PAHs such as abietic acid substantially present in the tree resins e.g. conifers resins. These are formed by diagenesis and combustions of the resins. For instance, Retene is a $C_4$-phenanthrene isomer (l-methyl-7-(l-methylethyl)-phenanthrene) compound, is also of an algal and bacterial origin and also abundant residues of conifer plants which is present in sediments of pristine northern environment (Wen et al. 2000). Similarly, Simoneltite – a substituted PAH compound is abundantly present where conifer residues exist. Biogenic PAHs are mainly found in coastal sediments and conifers residual area of the world. Natural sources of PAHs formation involve volcanoes, bacterial and algal synthesis, forest and bush fires, erosion of petroleum hydrocarbons containing sedimentary rocks and decomposition of vegetative litterfall.

## 8.4 Ecological Impacts of Polycyclic Aromatic Hydrocarbons

Polycyclic aromatic hydrocarbons are omnipresent and can be easily transportable up to a large distance via atmospheric migration (Gong et al. 2018). The fate of PAHs in environment is primarily evaluated by their physico-chemical properties, however natural processes such as temperature, light intensity, biological degradation, type of sorbent, and the concentration of oxidizing pollutants like $O_3$, $NO_x$, $OH^\bullet$ radicals are also essential factors (Matsuzawa et al. 2001). The size of the particles also has direct impact on the accumulation of PAHs in the environment. The size of urban aerosol particles ranges from a nanometer to micrometers and the particles size less than 2.5 μm generally introduced as fine particles (Seinfeld and Pandis 1998). In highly polluted areas, the particles are prone to have a bimodal distribution viz. Mode I ranging from 0.05 to 0.12 μm and Mode II from 0.5 to

1.0 μm in size (Venkataraman and Friedlander 1994). Automobile soot (from diesel and gasoline) also ranged from a few nanometers up to 0.3 μm with a mean diameter of ~0.1 μm (Kim et al. 2001). Five-ringed compound PAHs e.g. B[a]P, benzo-fluoranthenes, and perylene associated with aerosols were mainly adsorbed to particles in the range of 0.1–2.0 μm in size (Allen et al. 1996). Molecular weight of PAHs is inversely proportional to particle size i.e. larger particles of size 0.5–6.0 μm were found mostly associated with PAHs of lower molecular weight. In general, scavenging ratio (gas + particle) is higher for the less volatile compounds which are associated with particles where dry deposition is the main mechanism for the removal of PAHs from the atmosphere (Golomb et al. 2001; Offenberg and Baker 2002). These are hydrophobic which means that they have greater tendency to associate with particles than to dissolve in water. Relative proportion of polycyclic aromatic hydrocarbons (PAHs) in different components of the environment is, however, given in Table 8.2.

Thus, PAHs accumulated in the aquatic environment will tend to associate with settling particles. The adsorption of PAHs on soot particles can weaken their bioavailability, preserving in sediments and biodegradation rates (McElroy et al. 1989; McGroddy et al. 1996). In aquatic, soil and atmospheric environment, PAHs are accumulated at higher concentration and also have long degradation periods. During the winter season, pollutants migration increased by the atmospheric conditions influences smog clouds from air to soil, air to water reserves, and from air to humans (Manzetti 2013). PAHs enter into aquatic environment via atmospheric deposition, regular disposal of industrial effluents, improper discharge of municipal wastewater and some hazardous PAHs found at elevated levels at waste disposal sites. PAHs are detected absorbed into gases and particulates in the air. Figure 8.3 represents the dispersion and impacts of PAHs through air, terrestrial and aquatic environments. PAHs contamination transferred from one area to another via air and deposited back to earth via rain and smog and run off to aquatic environments and also represents the pathways of distribution of PAHs from source to living organisms. The plant can absorb PAHs from soils via roots and translocate them to different other parts. However, the uptake and translocation process of PAHs by plants depend on several factors such as concentration, physico-chemical characteristics, water solubility and soil types. In literature, PAHs induced phytotoxic effects are rare, therefore, no

**Table 8.2** The relative proportion of polycyclic aromatic hydrocarbons (PAHs) in different components of the environment

| | Percentage of total PAHs | |
|---|---|---|
| Source | ΣPAHs (%) | Benzo [a] pyrene (%) |
| Soil | 94.40 | 92.90 |
| Freshwater sediments | 5.40 | 7.10 |
| Water | Less than 0.01 | Less than 0.01 |
| Air | 0.10 | Less than 0.01 |
| Vegetables | 0.10 | Less than 0.01 |
| Soil biota | Less than 0.01 | Less than 0.01 |

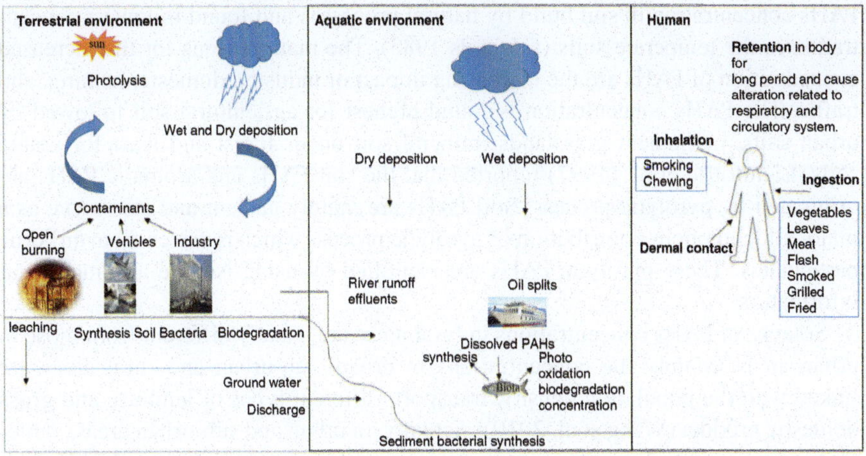


**Fig. 8.3** Dispersion and impacts of polycyclic aromatic hydrocarbons (PAHs) through air, terrestrial and aquatic environments. Migration and accumulation patterns of the most recurring PAHs and also represents the pathways of biodistribution of the PAHs from source to living organisms. (Source: Modified from Suess 1976 and Manzetti 2013)

literature is available on them. A few plants contain substances that can safeguard them against the PAHs, whereas several others can synthesize PAHs which act as their growth hormones (ATSDR 2010).

The process is known as biological degradation which is mainly responsible for depletion of PAHs from soil and microorganisms like fungi and bacteria may breakdown into other inorganic or organic end-products like carbon dioxide and water (Wilson and Jones 1993). Many organic pollutants including PAHs are degraded by the microorganisms. These microorganisms (e.g. fungi) use organic compounds as a source of carbon and energy and metabolize PAHs into water-soluble compounds (Cerniglia 1992). In environment, most of the oxidants like singlet oxygen, ozone, hydrogen peroxides, organic peroxides, and radicals like alkoxy radicals ($RO^•$), hydroxyl radicals ($HO^•$) and peroxy radicals ($RO_2^•$) that frequently initiate the oxidation reactions are directly or indirectly produced by photochemical processes (Neilson 1994).

## 8.4.1 *Effects on Agricultural Soils*

There is an increase in PAHs concentration in soil with natural processes like vegetation fires and volcanic exhalations and also because of anthropogenic activities. Atmospheric runoff water, fallout, oil spillage, etc. could deposit PAHs in the soil, which usually adsorbed onto the soil particles for a long time due to their strong hydrophobicity and difficult degradation (Wang et al. 2015). Its concentration in soil depends on soil texture and properties. 1–10 μg $kg^{-1}$ is the estimated range of

PAHs concentration in soil build by natural processes and found in lowest concentration in the temperate soils (Edwards 1983). The main reasons for the increased concentration of PAHs are the increasing impact of industry, domestic heating, and traffic. The PAHs concentration is found highest for agricultural soil followed by urban soils, permanent grasslands, mineral soil under forest and least for arable soils. Lichtfouse et al. (1997) reported that the soil PAHs and aromatic PAHs are originated by pyrolytic process. Soil PAHs are mainly ancient and also serve as a major allochthonous contribution of pyrolytic process which is driven from airborne particulates. These pyrolytic PAHs are identified by using isotope and molecular information.

Sources of PAHs concentrations in rural areas are usually different from those in urban areas. Mainly the concentrations in urban, sub-urban and rural areas are linked with the population density, transport, different types of land use and gross domestic product (Wang et al. 2010). Actually, in urban and sub-urban areas, vehicular and non-vehicular combustion are considered as the main source of PAHs pollutants, while in rural areas, biomass burning is the major source to the soil (Xie et al. 2012). The level of pollution is also varying among rural and urban areas due to the consumption of different types of fuel (Xiao et al. 2014). In urban areas, PAHs burden to soil increases from soil, coke tar, biomass burning, traffic exhaust to coal combustion (Wang et al. 2013). Soil acts as secondary sources for two to three ringed compound PAHs, whereas a sink for five to six ringed PAHs in winter and summer (Zhong and Zhu 2013). Generally, SOM is an essential parameter that influences the fate of persistent organic compounds, migration, sorption, and sequestration (Agarwal et al. 2009). A similar study has been found showing a good correlation between SOM and soil PAHs concentration (Zhang et al. 2015). Some researchers have reported that PAHs concentrations are substantially higher in concentration found in the temperate soils than tropical soils (Wilcke et al. 2000). The reason behind this may include the tropical climate which enhances the abundant microbial photo-oxidation, degradation and volatilization (Wilcke et al. 1999). Moreover, soil temperature and soil moisture play an important role in soil PAHs decomposition and volatilization. If the level of contaminants increases in the soil, it changes the physical and chemical properties of the soil. All these contaminants are adhered to soil particles and reach out at the rhizosphere, and contaminate groundwater (Wilcke 2000).

Fertilizers and pesticides are the main sources of contamination in the agricultural sector, with an increase in PAHs amount in the soil, which further affects the nitrogen fixation rate, micronutrient absorption and increase in soil toxicity. As soil toxicity is increased, functioning and diversity of microorganisms are also got affected due to the availability of limited food resources and changes in surrounding conditions. In some cases, contamination of soils with PAHs also stimulates the growth of plants e.g. all monocotyledonous plants in soil at PAHs level 100 mg $kg^{-1}$ but in some other plants such as wheat and oats showed considerable inhibition of plant growth to the PAHs level of 100 mg $kg^{-1}$ (Maliszewska-Kordyach and Smreczak 2003). A Dutch study determined the acceptable levels of PAHs. If levels

**Table 8.3** Recommended "limit values" for polycyclic aromatic hydrocarbons (PAHs) in soil

| Criterion | PAHs in soil (mg $kg^{-1)}$ | Recommendation |
|---|---|---|
| Soil quality criterion | Less than 1.5 | The safe level for contact with soil |
| Cut-off criterion | Greater than 1.5 | The level at which all contact with soil should be cut off if the land use of the area sensitive |

Source: Dutch report on the natural degradation of PAHs presents in soil and groundwater (http://www.mst.dk/publica/projects/2001/87-7944-367-2.htm)

go beyond these thresholds, remediation action is vital for the safety of humans. Many means are available to clean up contaminated sites (Table 8.3).

In agricultural soils of the Delhi region, low molecular weight PAHs are present most abundant and it indicates the presence of prominent combustion products due to low temperature pyrolytic processes like petrogenic sources and biomass burning (Ping et al. 2007). In the rural areas, open burning of agricultural residue is more common which lead to the generation of lighter molecular weight PAHs at low temperature due to incomplete combustion of residue. Agricultural soils are prone to incidental inputs of fuel like diesel used in vehicles for several agricultural or farm practices. Higher molecular weight PAHs generated mainly from generator and tractors which are sources of combustion during farm practices (Agarwal et al. 2009).

### 8.4.2 Effects on Soil Biota and Biological Fertility in Agricultural Soils

Soil is a prominent habitat for several microbial populations. It is also the most biodiverseed ecosystems on the earth. The adaptability of soil micro-organisms allows them to play a critical role in regulating plant growth, soil quality and characteristics, the cycle of matter, and balancing the energy flow in the soil ecosystems. Soil biota plays a crucial role in quantification of the ecological impact of chemical contamination of soils and convenient for identifying clean-up priorities and monitoring environmental changes. On wider scale, nematodes are better indicators for polycyclic aromatic hydrocarbons (PAHs) contamination of soil than micro-arthropods due to close contact with the soil particles and contaminants relative knowledge of their taxonomy and trophic groups, and permeable cuticle (Blakely et al. 2002). But on narrow scale, microbial activities are very important in maintaining nutritional food chain, therefore, soil microorganisms are very sensitive towards any change in the ecosystem, their activity and diversity rapidly altered by perturbation (Schloter et al. 2003). Moreover, soil microbial activity plays a major role in biological fertility, biogeochemical cycles, and their stabilization.

Soil bio-chemical procedures are the tools for preserving soil quality, decomposition of harmful substances, formation of SOM, biochemical cycles, and the formation of soil structure. Soil contamination by hazardous metals and PAHs reduce soil microbial activities like enzymatic activities and soil respiration. Soil microbial activities depend on organic matter, soil chemical properties, soil pH, and climatic conditions. Microbial biomass carbon (MBC) was very low in all contaminated soils compared to the control soil and soil microorganism's diversity started decreasing when contamination of persistent organic pollutants increased (Thavamani et al. 2012). Toxic contaminants can reduce soil microbial activity and severely threaten soil ecosystem functioning. Some petroleum hydrocarbons are harmful to animals and mutagenic to soil bacteria.

Some studies revealed that pyrene and phenanthrene are of carcinogenic nature in rodents and mutagenic in bacterial cells (Wilson and Jones 1993). PAHs contaminated soils are remediated by various methods such as physical, chemical and biological that strip out contaminants from waste disposal site. Presently, biological methods are more accepted as compared to physical and chemical methods. Biodegradation is a part of biological remediation that decomposes contaminants via living organisms. Biodegradation of PAHs has been accomplished via co-metabolic degradation, radical oxidation, and complete mineralization (Mahro et al. 1994; Kulik et al. 2006). Soil microorganisms i.e. bacteria and fungi play key role in the degradation of PAHs and use several metabolic pathways for the degradation process. However, involvement of various factors such as pre-exposure time, soil properties, and pollutant concentration affect biodegradation of PAHs. Generally, breakdown of PAHs is done at greater rates by the microbial communities that are already present in contaminated soil as compare to the uncontaminated soil (Kulik et al. 2006). Composition and activities of microorganisms is significantly influenced by contaminants and selected microbial communities which function as active member and survived in contaminated soil.

In unfavourable and stressful environment, highly dynamic microbe-soil-contaminants interaction is a complex process and various metabolic pathways are occurring simultaneously. An alteration in the soil biodiversity may result in the reduction of soil quality. Soil enzyme activity is the active force behind all the biochemical processes which are occurring in the soil, also in the decomposition of organic contaminants. Some studies revealed that the enzymes present in PAHs contaminated soil actively participated in biological transformation process of C, N, and P like sucrase, beta-glycosidase, urease, protease, and phosphatase, and the oxidation of aromatic compounds such as catalase, dehydrogenase, lactase, and phenol oxidase.

As the concentration of PAHs increases in the soil, the number of *k*-strategists decreases due to the elimination or contamination of the food sources. Electrical conductivity is negatively related to the concentration of contaminants in soil bulk density, whereas, positively related to the contaminants (Table 8.4). The increase in bulk density leads to the decrease in the amount of soil oxygen, poor nutrient transportation, elimination of detrital food chain and decomposition rate in the soil which directly affects biological fertility and ecological successional groups decline (Blakely et al. 2002).

**Table 8.4** Factors influencing PAHs concentration in soil and their effects on soil biota and biological fertility

| Concentration of PAHs increase in soil | | |
|---|---|---|
| Factors affected by PAHs in soil | Effect | References |
| Bulk density | Increase | Blakely et al. (2002) |
| Electrical conductivity | Decrease | Canet et al. (2001) |
| Presence of oxygen | Decrease | |
| Nutrient transportation | Decrease | |
| Decomposition rate | Decrease | |
| Ecological successional group | Decrease | |

### 8.4.3 *Effects on Agricultural Crops*

Soil acts as a sink for the storage of contaminants and plants act as translocating agents of these contaminants from soil significantly. Further, the uptake of contaminants is also affected by physico-chemical conditions of soil, types of plant and by physical and chemical properties of the contaminants. If the concentration of toxic elements increases in the soil, the plant uptake of water-soluble toxic elements then directly or indirectly affects plant metabolism and functioning. However, there are two common pathways for the entrancement of PAHs into the plants: (i) soil to plant, root uptake and translocation, and (ii) air to plant, deposition from atmosphere through stomata. Plant uptake of PAHs from soil differs significantly and is affected not only by the types of plants and physico-chemical conditions of soils but also by the physical and chemical properties of the contaminants. Organic chemicals uptake by a plant is of great importance as it affect their environment, transport of minerals and threatens crop growth and safety (Li et al. 2017). The pathways of extraction, transformation, and accumulation of PAHs in the plants from soil include the passive absorption, plant transpiration, and reactions with root lipid content. A vapour phase component of volatile PAHs compounds in the air is exposed to an air-leaf exchange process which moves towards equilibrium with time (Wild et al. 2004).

Plants are extremely sensitive to PAHs which resulted in their quick response to PAHs exposure (Kummerova et al. 2006). Presence of volatile water-soluble low molecular weight hydrocarbons (less than three rings) such as benzene, toluene, styrene, naphthalene, etc. strongly inhibits the plant growth and germination (Henner et al. 1999). The toxicity of PAHs is influenced by plant metabolism. Photo-oxidation affects the toxicity of PAHs and generally they are more toxic in UV light. Phytotoxic effects induced by PAHs are rare and some plants contain substrate that can protect them against the effect, whereas, others can synthesize PAHs which act as growth hormones (ATSDR 2010). Generation of reactive oxygen species in the photosynthetic apparatus may be induced by the deposition of PAHs on the leaf surfaces, which closely resembles the action of the herbicide paraquat (Oguntimehin et al. 2008). Table 8.5 represents the direct and indirect effects of

**Table 8.5** Representation of polycyclic aromatic hydrocarbons (PAHs) toxic effects on agriculture crops

| PAHs toxicity on agricultural crop | | |
|---|---|---|
| Direct effects | Indirect effects | Target organelle |
| Seed germination | Generation of ROS | Mitochondria |
| Root elongation | Lipid oxidation | Plastids |
| Plant biomass | DNA fragmentation | Peroxisomes |
| Mineral uptake | DNA denature | Cell membrane |
| Photosynthesis and respiration | Disturbance in membrane permeability | Nucleus |
| Flowering and fruiting | Enzyme inactivation | |

PAHs toxicity on agricultural crops. PAHs toxicity targets organelles such as mitochondria, plastids, peroxisomes, cell membrane and nucleus which directly affects seed germination, root elongation, plant biomass, mineral uptake, photosynthesis, respiration, flowering and fruiting and indirectly affects the reactive oxygen species (ROS) generation, oxidation of lipids, DNA fragmentation and denature, disturbance in membrane permeability and enzyme inactivation. Some direct toxic effects of potentially toxic elements on the plants are leaf chlorosis, reduced stomatal opening, disturbed water balance, inhibition of cytoplasmic enzymes and damage to the cell structures caused by oxidative stress. At cation exchange sites of plants, essential nutrients are replaced by potentially toxic elements. High PAHs concentration negatively influences plant growth and development and even death by decreasing organic matter decomposition, soil nutrients, and enzymatic activities required for plant metabolism (Bansal 2018).

## 8.5 Health Risk Assessment

A health risk assessment is a health questionnaire which is used to provide individuals with an evaluation of their health risks and quality of life and in order to evaluate the potential of chemicals to cause health problems called as risk assessment (USEPA 2001). It mainly consists of four basic steps:

1. Toxicity identification
2. Exposure assessment
3. Dose-response assessment
4. Risk characterization

In the environment, soil system is a major reservoir of PAHs and behaves like a source of pollutants that can alter the human and ecosystem health due to their potential carcinogenic and mutagenic effects (Wilcke 2007). In comparison to the rural area, urban area soils found to be more contaminated, due to industry, heavy traffic, petrochemical plants, and dense population, and these agricultural soils are profoundly contaminated with different concentration of PAHs but they are still used for the production of rice and vegetables (Tao et al. 2004). Exposure to PAHs through dust has also become a major public concern especially where the population is very dense where incomplete combustion from different activities such as heating, release from heavy traffic, emission from industries, and smoking can also lead to an increase of PAHs in indoor dust (Murillo et al. 2017). However, information of PAHs on indoor dust and the intensity of human exposure are quite limited in comparison to the data available on outdoor sources of PAHs (Ali et al. 2016). Wang et al. (2017) stated that risk of cancer from PAHs contaminated soil is lower than crops. Due to decrease in land resources and high demand of urban agriculture, the agricultural soils quality in suburban regions in the proximity of pollution sources of PAHs becomes major concern (Tong et al. 2018). It has also been observed from the literature that PAHs concentration is comparatively more in adult farmers because they spent more time in the agricultural fields and exposed to PAHs contamination via three pathways: (a) Oral intake (Ingestion), (b) Dermal contact, and (c) Inhalation (Xia et al. 2016; Tong et al. 2018; Cao et al. 2019; Mihankhah et al. 2020).

Quantitative health risks analyses are essential for assessing human health, carcinogenic and non-carcinogenic potential of PAHs, as well as for providing the support for environmental protection policies. Regular approaches for risk assessment involve both deterministic and uncertainty methods (Yang et al. 2015). According to the deterministic method, the health risk is calculated by the highest values of the risk parameters and contaminants that may lead to less informative and stable results (Peng et al. 2016). PAHs concentration for exposure characterization varies temporally, spatially, and dose-response results are inherently variable (Sander et al. 2006).

The PAHs present in soil and environment may enter into the human body via direct or indirect pathways. In an agro-ecosystem, these compounds may pass into humans especially farmers during agricultural activities like sowing, fertilization, and harvesting. PAHs are lipophilic in nature, so they are quickly dissolved and transferred by cell membrane that could lead to the development of gene mutation, malformation, and cancer in human beings and animals (Franco et al. 2008; Skrbic et al. 2019). Moreover, they can also aggregate in vegetation (Kipopoulou et al. 1999) that could indirectly induce human exposure via food consumption (Juhasz and Naidu 2000; Khanal et al. 2018). Dietary exposures are also sources of human exposure to PAHs. These compounds are produced in food as a result of food processing techniques like curing, smoking, drying, roasting, refining, and grilling. All these food processing steps are known to generate and increase the level of PAHs in food.

## 8.6 Recommendations for Future Research

Based on the above literature survey, the following recommendations could be drawn for future research:

1. Urban and rural air pollution should be monitored at a regular time interval and not only by seasonally, particularly in the developing countries.
2. Particle traps on motor vehicles and catalytic convertor should be taken.
3. Treatment of industrial effluents must be carried out strictly and filtration of industrial emission should be taken.
4. To make a healthy surrounding we should adopt eco-friendly cultivation practices like polyculture and crop rotation, use of Integrated Pest Management (IPM), use of bio-pesticides such as parasitoids, predators and insect pathogens and use of the pesticides which are obtained from natural plant products like tobacco extracts, garlic extracts and neem extracts etc.

## 8.7 Conclusions

Based on the literature review, we conclude that the abundance and diversity of soil bacteria, fungi, micro-arthropod, and nematodes tend to decrease with increased concentration of PAHs in the soil. The modified composition of soil-biota compromises several ecological functions like litter decomposition, nutrient cycling, soil organic matter, soil fertility and productivity of the agro-ecosystems. Not only the soil but also the ground water is affected when these compounds reach via leaching into the groundwater through soil surface. These compounds probably enter into the human body or animals when they eat crop plants that are grown on contaminated soils or drink the contaminated water. Since these compounds are lipid soluble, therefore, they can easily pass the cell membrane and enter into the elementary canal. Moreover, PAHs may affect the respiratory system when inhaled from the air and can affect the liver, digestive system, heart and even brain when taken internally through affected food or water. Some higher molecular weight PAHs are also considered as carcinogens which induce cancer in liver, skin, throat, etc.

Though effective management of PAHs has not achieved yet, biodegradation (or bioremediation) of these toxic compounds can prove a better strategy to minimize their impacts. The biodegradation is often achieved by a micro-organism such as algae, fungi, and bacteria. In this process, breakdown of organic compounds occurs through bio-transformation which leads to the formation of less complex metabolites and inorganic minerals such as $H_2O$, $CO_2$ and $CH_4$via mineralization process under aerobic and anaerobic conditions. Although, PAHs compound may undergo several processes like photolysis, volatilization, adsorption and chemical degradation, however, microbial degradation contributes maximum to the overall degradation of PAHs. Overall, it can be concluded that PAHs that are present in the atmosphere has emerged as the potential contaminants of the soils. These contaminants are not only affecting the structure and functioning of the agro-ecosystems but also risking human health with deadliest diseases like cancer, chronic asthma and other respiratory disorders. Therefore, there is a need to devise effective management strategies for these toxic compounds in the agro-ecosystems by stakeholders at individual level and government should pay attention to propose effective policies to check their emission and manage the contaminated soils at country level.

**Acknowledgements** The authors are grateful to the Chairperson, Department of Botany, Panjab University, Chandigarh, for providing all necessary facilities required for work. Abhishek Kumar is thankful to the University Grants Commission, Government of India, New Delhi, for the financial support in the form of Junior Research Fellowship [UGC Ref. No.: 507-(OBC) (CSIR-UGC NET DEC. 2016)]. The corresponding author also acknowledges the Department of Science and Technology, Government of India, for the support in the form of PURSE Grant. We extend our thanks to anonymous reviewers for their critical and constructive comments on earlier version of the manuscript.

## References

Abbas I, Badran G, Verdin A, Ledoux F, Roumié M, Courcot D, Garçon G (2018) Polycyclic aromatic hydrocarbon derivatives in airborne particulate matter: sources, analysis and toxicity. Environ Chem Lett 16:439–475. https://doi.org/10.1007/s10311-017-0697-0

Abdel-Shafy HI, Mansour MS (2016) A review on polycyclic aromatic hydrocarbons: source, environmental impact, effect on human health and remediation. Egypt J Pet 25(1):107–123. https://doi.org/10.1016/j.ejpe.2015.03.011

Agarwal T, Khillare PS, Shridhar V, Ray S (2009) Pattern, sources and toxic potential of PAHs in the agricultural soils of Delhi, India. J Hazard Mater 163(2):1033–1039. https://doi.org/10.1016/j.jhazmat.2008.07.058

Agency for Toxic Substance and Disease Registry (ATSDR) (2010) Public Health Statement

Ali N, Ismail IMI, Khoder M, Shamy M, Alghamdi M, Costa M, Eqani SAMAS (2016) Polycyclic aromatic hydrocarbons (PAHs) in indoor dust samples from cities of Jeddah and Kuwait: levels, sources and non-dietary human exposure. Sci Total Environ 573:1607–1614. https://doi.org/10.1016/j.scitotenv.2016.09.134

Allen JO, Dookeran NM, Smith KA, Sarofim AF, Taghizadeh K, Lafleur AL (1996) Measurement of polycyclic aromatic hydrocarbons associated with size-segregated atmospheric aerosols in Massachusetts. Environ Sci Technol 30(3):1023–1031. https://doi.org/10.1021/es950517o

Alshaarawy O, Elbaz HA, Andrew ME (2016) The association of urinary polycyclic aromatic hydrocarbon biomarkers and cardiovascular disease in the US population. Environ Int 89:174–178. https://doi.org/10.1016/j.envint.2016.02.006

Bansal OP (2018) The influence of potentially toxic elements on soil biological and chemical properties. In: Metals in soil-bioavailability, contamination and remediation. IntechOpen. https://doi.org/10.5771/intechopen.81348

Berkowitz N (1988) Coal aromaticity and average molecular structure. In: Eberg LB (ed) Polynuclear aromatic compounds, Advances in chemistry series 217. American Chemical Society, Washington, DC, pp 217–233

Blakely JK, Neher DA, Spongberg AL (2002) Soil invertebrate and microbial communities, and decomposition as indicators of polycyclic aromatic hydrocarbon contamination. Appl Soil Ecol 21(1):71–88. https://doi.org/10.1016/S0929-1393(02)00023-9

Canet R, Birnstingl JG, Malcolm DG, Lopez-Real JM, Beck AJ (2001) Biodegradation of polycyclic aromatic hydrocarbons (PAHs) by native microflora and combinations of white-rot fungi in a coal-tar contaminated soil. Bioresour Technol 76(2):113–117. https://doi.org/10.1016/S0960-8524(00)00093-6

Cao Y, Yang B, Song Z, Wang H, He F, Han X (2016) Wheat straw biochar amendments on the removal of polycyclic aromatic hydrocarbons (PAHs) in contaminated soil. Ecotoxicol Environ Saf 130:248–255. https://doi.org/10.1016/j.ecoenv.2016.04.033

Cao Z, Wang M, Chen Q, Zhu C, Jie J, Li X, Bu Q (2019) Spatial, seasonal and particle size dependent variations of PAH contamination in indoor dust and the corresponding human health risk. Sci Total Environ 653:423–430. https://doi.org/10.1016/j.scitotenv.2018.10.413

Cerniglia CE (1992) Biodegradation of polycyclic aromatic hydrocarbons. Biodegradation 3:351–368. https://doi.org/10.1016/0958-1669(93)90104-5.

Edwards NT (1983) Polycyclic aromatic hydrocarbons (PAH's) in the terrestrial environment – a review. J Environ Qual 12(4):427–441. https://doi.org/10.2134/jeq1983.00472425001200040001x

Eghbal N, Nasrabadi T, Karbassi AR, Taghavi L (2019) Evaluating the potential of plants (leaves) in removal of toxic metals from urban soils (case study of a district in Tehran city). Pollution 5(2):387–394. https://doi.org/10.22059/poll.2019.272090.555

Emsbo-Mattingly S, Uhler A, Stout SA, McCarthy KS, Douglas GS, Brown JS, Boehm PD (2001) Polycyclic aromatic hydrocarbon (PAH) chemistry of MGP tar and source identification in sediment. In: Sediments guidance compendium, Report No. 1005216. Electric Power Research Institute, Palo Alto, California, pp 1–41

Farooq S, Eqani SAMAS, Malik RN, Katsoyiannis A, Zhang G, Zhang Y, Shinwari ZK (2011) Occurrence, finger printing and ecological risk assessment of polycyclic aromatic hydrocarbons (PAHs) in the Chenab River, Pakistan. J Environ Monit 13(11):3207–3215. https://doi.org/10.1039/C1EM10421G

Fazeli G, Karbassi AR, Nasrabadi T (2018) Anthropogenic share of metal contents in soils of urban areas. Pollution 4(4):697–706. https://doi.org/10.22059/poll.2018.261506.474

Flowers L, Rieth SH, Cogliano VJ, Foureman GL, Hertzberg R, Hofmann EL, Schoeny RS (2002) Health assessment of polycyclic aromatic hydrocarbon mixtures: current practices and future directions. Polycycl Aromat Compd 22(3–4):811–821. https://doi.org/10.1080/10406630290103960

Franco SS, Nardocci AC, Gunther WMR (2008) PAH biomarkers for human health risk assessment: a review of the state-of-the-art. Cad Saude Publica 24:569–580. https://doi.org/10.1590/s0102-311x2008001600009

Franco CFJ, de Resende MF, de Almeida Furtado L, Brasil TF, Eberlin MN, Netto ADP (2017) Polycyclic aromatic hydrocarbons (PAHs) in street dust of Rio de Janeiro and Niterói, Brazil: particle size distribution, sources and cancer risk assessment. Sci Total Environ 599:305–313. https://doi.org/10.1016/j.scitotenv.2017.04.060

Golomb D, Barry E, Fisher G, Varanusupakul P, Koleda M, Rooney T (2001) Atmospheric deposition of polycyclic aromatic hydrocarbons near New England coastal waters. Atmos Environ 35:6245–6258. https://doi.org/10.1016/s1352-2310(01)00456-3

Gong X, Xiao L, Zhao Z, Li Q, Feng F, Zhang L, Deng Z (2018) Spatial variation of polycyclic aromatic hydrocarbons (PAHs) in surface sediments from rivers in hilly regions of southern China in the wet and dry seasons. Ecotoxicol Environ Saf 156:322–329. https://doi.org/10.1016/j.ecoenv.2018.03.004

Haynes BS (1991) Fossil fuel combustion: a source book. Wiley, New York, pp 261–326

Henner P, Schiavon M, Druelle V, Lichtfouse E (1999) Phytotoxicity of ancient gaswork soils. Effect of polycyclic aromatic hydrocarbons (PAHs) on plant germination. Org Geochem 30(8):963–969. https://doi.org/10.1016/S0146-6380(99)00080-7

Hu T, Zhang J, Ye C, Zhang L, Xing X, Zhang Y, Zhang Q (2017) Status, source and health risk assessment of polycyclic aromatic hydrocarbons (PAHs) in soil from the water-level-fluctuation zone of the Three Gorges Reservoir, China. J Geochem Explor 172:20–28. https://doi.org/10.1016/j.gexplo.2016.09.012

Jenkins BM, Jones AD, Turn SQ, Williams RB (1996) Particle concentrations, gas partitioning, and species intercorrelations for polycyclic aromatic hydrocarbons (PAH) emitted during biomass burning. Atmos Environ 30:3825–3835. https://doi.org/10.1016/1352-2310(96)00084-2

Jiang Y, Yves UJ, Sun H, Hu X, Zhan H, Wu Y (2016) Distribution, compositional pattern and sources of polycyclic aromatic hydrocarbons in urban soils of an industrial city, Lanzhou, China. Ecotoxicol Environ Saf 126:154–162. https://doi.org/10.1016/j.ecoenv.2015.12.037

Juhasz AL, Naidu R (2000) Bioremediation of high molecular weight polycyclic aromatic hydrocarbons: a review of the microbial degradation of benzo [a] pyrene. Int Biodeterior Biodegradation 45(1–2):57–88. https://doi.org/10.1016/s0964-8305(00)00052-4

Khanal R, Furumai H, Nakajima F, Yoshimura C (2018) Carcinogenic profile, toxicity and source apportionment of polycyclic aromatic hydrocarbons accumulated from urban road dust in Tokyo, Japan. Ecotoxicol Environ Saf 165:440–449. https://doi.org/10.1016/j.ecoenv.2018.08.095

Kim WS, Kim SH, Lee DW, Lee S, Lim CS, Ryu JH (2001) Size analysis of automobile soot particles using field-flow fractionation. Environ Sci Technol 35(6):1005–1012. https://doi.org/10.1012/es001329n

Kipopoulou AM, Manoli E, Samara C (1999) Bioconcentration of polycyclic aromatic hydrocarbons in vegetables grown in an industrial area. Environ Pollut 106(3):369–380. https://doi.org/10.1016/s0269-7491(99)00107-4

Kulik N, Goi A, Trapido M, Tuhkanen T (2006) Degradation of polycyclic aromatic hydrocarbons by combined chemical pre-oxidation and bioremediation in creosote contaminated soil. J Environ Manag 78(4):382–391. https://doi.org/10.1016/j.jenvman.2005.05.005

Kummerova M, Krulova J, Zezulka S, Triska J (2006) Evaluation of fluoranthene phytotoxicity in pea plants by hill reaction and chlorophyll fluorescence. Chemosphere 65:489–496. https://doi.org/10.1016/j.chemosphere.2006.01.052

Li Q, Li Y, Zhu L, Xing B, Chen B (2017) Dependence of plant uptake and diffusion of polycyclic aromatic hydrocarbons on the leaf surface morphology and micro-structures of cuticular waxes. Sci Rep 7:46235. https://doi.org/10.1038/srep46235

Lichtfouse E, Budzinski H, Garrigues P, Eglinton TI (1997) Ancient polycyclic aromatic hydrocarbons in modern soils: 13C, 14C and biomarker evidence. Org Geochem 26(5–6):353–359. https://doi.org/10.1016/S0146-6380(97)00009-0

Liu G, Niu J, Guo W, An X, Zhao L (2016) Ecological and health risk-based characterization of agricultural soils contaminated with polycyclic aromatic hydrocarbons in the vicinity of a chemical plant in China. Chemosphere 163:461–470. https://doi.org/10.1016/j.chemosphere.2016.08.056

Louda WJ, Baker EW (1984) Perylene occurrence, alkylation and possible sources in deep ocean sediments. Geochim Cosmochim Acta 48:1043–1058. https://doi.org/10.1016/0016-7037(84)90195-9

Mahro B, Schaefer G, Kastner M (1994) Pathways of microbial degradation of PAHs in soil. In: Hinchee RE et al (eds) Bioremediation of chlorinated and PAH compounds. Lewis Publishers, Boca Raton, pp 203–217

Maliszewska-Kordybach B, Smreczak B (2003) Habitat function of agricultural soils as affected by heavy metals and polycyclic aromatic hydrocarbons contamination. Environ Int 28:719–728. https://doi.org/10.1016/S0160-4120(02)00117-4

Maliszewska-Kordybach B, Smreczak B, Klimkowicz-Pawlas A, Terelak H (2008) Monitoring of the total content of polycyclic aromatic hydrocarbons (PAHs) in arable soils in Poland. Chemosphere 73(8):1284–1291. https://doi.org/10.1016/j.chemoshpere.2008.07.009

Manahan SE (1994) Environmental chemistry, 6th edn. Lewis Publisher, New York, p 868

Manzetti S (2013) Polycyclic aromatic hydrocarbons in the environment: environmental fate and transformation. Polycycl Aromat Compd 33(4):311–330. https://doi.org/10.1080/10406638.2013.781042

Matsuzawa S, Nasser-Ali L, Garrigues P (2001) Photolytic behavior of polycyclic aromatic hydrocarbons in diesel particulate matter deposited on the ground. Environ Sci Technol 35(15):3139–3143. https://doi.org/10.1021/es001606q

McElroy AE, Farrington JW, Teal JM (1989) Bioavailability of polycyclic aromatic hydrocarbons in the aquatic environment. In: Varanasi U (ed) Metabolism of Polycyclic Aromatic Hydrocarbons in the Aquatic Environment. CRC Press, Boca Raton, FL, pp 1–39.

McGroddy SE, Farrington JW, Gschwend PM (1996) Comparison of the in situ and desorption sediment water partitioning of polycyclic aromatic hydrocarbons and polychlorinated biphenyls. Environ Sci Technol 30(8):2660. https://doi.org/10.1021/es962003a

Miguel A, Kirchstetter T, Harley R (1998) On-road emissions of particulate polycyclic aromatic hydrocarbons and black carbon from gasoline and diesel vehicles. Environ Sci Technol 32:450–455. https://doi.org/10.1021/es970566w

Mihankhah T, Saeedi M, Karbassi A (2020) Contamination and cancer risk assessment of polycyclic aromatic hydrocarbons (PAHs) in urban dust from different land-uses in the most populated city of Iran. Ecotoxicol Environ Saf 187:109838. https://doi.org/10.1016/j.ecoenv.2019.109838

Murillo JH, Villalobos MC, Marín JFR, Guerrero VHB, Arias DS (2017) Polycyclic aromatic hydrocarbons in PM2. 5 and PM10 atmospheric particles in the metropolitan area of Costa Rica: sources, temporal and spatial variations. Atmos Pollut Res 8(2):320–327. https://doi.org/10.1016/j.apr.2016.10.002

Neilson AH (1994) Organic chemicals in the aquatic environment; distribution, persistency and toxicity. Lewis Publishers, Boca Raton, pp 1–421

Offenberg JH, Baker JE (2002) Precipitation scavenging of polychlorinated biphenyls and polycyclic aromatic hydrocarbons along an urban to over-water transect. Environ Sci Technol 36(17):3763–3771. https://doi.org/10.1021/es025608h

Oguntimehin I, Nakatani N, Sakugawa H (2008) Phytotoxicities of fluoranthene and phenanthrene deposited on needle surfaces of the evergreen conifer, Japanese red pine (Pinus densiflora Sieb. et Zucc.). Environ Pollut 154(2):264–271. https://doi.org/10.1016/j.envpol.2007.10.039

OSWER (2015) FY 2016-2017 National program manager's guidance. Environmental Protection Agency, United States. p.77. https://www.epa.gov/sites/production/files/2015-04/documents/oswer_fy_16_17_npm_guidance_final.pdf

Peng C, Cai Y, Wang T, Xiao R, Chen W (2016) Regional probabilistic risk assessment of heavy metals in different environmental media and land uses: An urbanization-affected drinking water supply area. Sci Rep 6:370–384. https://doi.org/10.1038/srep37084

Ping LF, Luo YM, Zhang HB, Li QB, Wu LH (2007) Distribution of polycyclic aromatic hydrocarbons in thirty typical soil profiles in the Yangtze River Delta region, east China. Environ Pollut 147:358–365. https://doi.org/10.1016/j.envpol.2006.05.027

Ravindra K, Sokhi R, Van Grieken R (2008) Atmospheric polycyclic aromatic hydrocarbons: source attribution, emission factors and regulation. Atmos Environ 42(13):2895–2921. https://doi.org/10.1016/j.atmosenv.2007.12.010

Richter H, Howard J (2000) Formation of polycyclic aromatic hydrocarbons and their growth to soot – a review of chemical reaction pathways. Prog Energy Combust Sci 26:565–608. https://doi.org/10.1016/s0360-1285(00)00009-5

Rong H, Hong-Fu W, Yan-Tun S, Jiang H, Shao-Lin P (2007) Polycyclic aromatic hydrocarbons in agricultural soils of the southern subtropics, China. Pedosphere 17(5):673–680. https://doi.org/10.1016/S1002-0160(07)60081-2

Sander P, Bergback B, Oberg T (2006) Uncertain numbers and uncertainty in the selection of input distributions-consequences for a probabilistic risk assessment of contaminated land. Risk Anal 26(5):1363–1375. https://doi.org/10.1111/j.1539-6924.2006.00808.x

Schloter M, Dilly O, Munch JC (2003) Indicators for evaluating soil quality. Agric Ecosyst Environ 98(1–3):255–262. https://doi.org/10.1016/S0167-8809(03)00085-9

Seinfeld JH, Pandis SN (1998) From air pollution to climate change. Atmos Chem Phys:1–1190

Singh R, Srivastava P, Singh P, Upadhyay S, Raghubanshi AS (2019) Human overpopulation and food security: challenges for the agriculture sustainability. In: Urban agriculture and food systems: breakthroughs in research and practice. IGI Global, pp 439–467. https://doi.org/10.4018/978-1-5225-8063-8.ch022

Skrbic B, Durisic-Mladenovic N, Zivancev J, Tadic D (2019) Seasonal occurrence and cancer risk assessment of polycyclic aromatic hydrocarbons in street dust from the Novi Sad city, Serbia. Sci Total Environ 647:191–203. https://doi.org/10.1016/j.scitotenv.2018.07.442

Soltani N, Keshavarzi B, Moore F, Tavakol T, Lahijanzadeh AR, Jaafarzadeh N, Kermani M (2015) Ecological and human health hazards of heavy metals and polycyclic aromatic hydrocarbons (PAHs) in road dust of Isfahan metropolis, Iran. Sci Total Environ 505:712–723. https://doi.org/10.1016/j.scitotenv.2014.09.097

Srivastava P, Singh R, Tripathi S, Raghubanshi AS (2016) Urgent need for sustainable thinking in agriculture: an Indian scenario. Ecol Indic 67:611–622. https://doi.org/10.1016/j.ecolind.2016.03.015

Suess MJ (1976) The environmental load and cycle of polycyclic aromatic hydrocarbons. Sci Total Environ 6(3):239–250. https://doi.org/10.1016/0048-9697(76)90033-4

Tan YL, Heit M (1981) Biogenic and abiogenic polycyclic aromatic hydrocarbons in sediments from two remote Adirondack lakes. Geochim Cosmochim Acta 45:2267–2279. https://doi.org/10.1016/0016-7037(81)90076-4

Tang L, Tang XY, Zhu YG, Zheng MH, Miao QL (2005) Contamination of polycyclic aromatic hydrocarbons (PAHs) in urban soils in Beijing, China. Environ Int 31(6):822–828. https://doi.org/10.1016/j.envint.2005.05.031

Tao S, Cui YH, Xu FL, Li BG, Cao J, Liu WX, Sun R (2004) Polycyclic aromatic hydrocarbons (PAHs) in agricultural soil and vegetables from Tianjin. Sci Total Environ 320(1):11–24. https://doi.org/10.1016/S0048-9697(03)00453-4

Thavamani P, Malik S, Beer M, Megharaj M, Naidu R (2012) Microbial activity and diversity in long-term mixed contaminated soils with respect to polyaromatic hydrocarbons and heavy metals. J Environ Manag 99:10–17. https://doi.org/10.1016/j.jenvman.2011.12.030

Tong R, Yang X, Su H, Pan Y, Zhang Q, Wang J, Long M (2018) Levels, sources and probabilistic health risks of polycyclic aromatic hydrocarbons in the agricultural soils from sites neighboring suburban industries in Shanghai. Sci Total Environ 616:1365–1373. https://doi.org/10.1016/j.scitotenv.2017.10.179

USEPA (2001) Risk assessment guidance for superfund: volume III. Part a: process for conducting probabilistic risk assessment. EPA 540-R-02-002.

Venkataraman C, Friedlander SK (1994) Size distributions of polycyclic aromatic hydrocarbons and elemental carbon. Ambient measurements and effects of atmospheric processes. Environ Sci Technol 28(4):563–572. https://doi.org/10.1021/es00053a006

Venkatesan MI (1988) Occurrence and possible sources of perylene in marine sediments-a review. Mar Chem 25(1):1–27. https://doi.org/10.1016/0304-4203(88)90011-4

Wang W, Simonich SLM, Xue M, Zhao J, Zhang N, Wang R, Cao J, Tao S (2010) Concentrations, sources and spatial distribution of polycyclic aromatic hydrocarbons in soils from Beijing, Tianjin and surrounding areas, North China. Environ Pollut 158(5):1245–1251. https://doi.org/10.1016/j.envpol.2010.01.021

Wang XT, Miao Y, Zhang Y, Li YC, Wu MH, Yu G (2013) Polycyclic aromatic hydrocarbons (PAHs) in urban soils of the megacity Shanghai: occurrence, source apportionment and potential human health risk. Sci Total Environ 447:80–89. https://doi.org/10.1016/j.scitotenv.2012.12.086

Wang XT, Chen L, Wang XK, Lei BL, Sun YF, Zhou J, Wu MH (2015) Occurrence, sources and health risk assessment of polycyclic aromatic hydrocarbons in urban (Pudong) and suburban soils from Shanghai in China. Chemosphere 119:1224–1232. https://doi.org/10.1016/j.chemosphere.2014.10.019

Wang J, Zhang X, Ling W, Liu R, Liu J, Kang F, Gao Y (2017) Contamination and health risk assessment of PAHs in soils and crops in industrial areas of the Yangtze River Delta region, China. Chemosphere 168:976–987. https://doi.org/10.1016/j.chemosphere.2016.10.113

Wen Z, Ruiyong W, Radke M, Qingyu W, Guoying S, Zhili L (2000) Retene in pyrolysates of algal and bacterial organic matter. Org Geochem 31:757–762. https://doi.org/10.1016/s0146-6380(00)00064-4

White AJ, Bradshaw PT, Herring AH, Teitelbaum SL, Beyea J, Stellman SD, Conway K (2016) Exposure to multiple sources of polycyclic aromatic hydrocarbons and breast cancer incidence. Environ Int 89:185–192. https://doi.org/10.1016/j.envint.2016.02.009

Wilcke W (2000) Synopsis polycyclic aromatic hydrocarbons (PAHs) in soil – a review. J Plant Nutr Soil Sci 163(3):229–248. https://doi.org/10.1002/1522-2624(200006)163:3%3C229::AID-JPLN229%3E3.0.CO;2-6.

Wilcke W (2007) Global patterns of polycyclic aromatic hydrocarbons (PAHs) in soil. Geoderma 141(3–4):157–166. https://doi.org/10.1016/j.geoderma.2007.07.007

Wilcke W, Muller S, Kanchanakool N, Niamskul C, Zech W (1999) Polycyclic aromatic hydrocarbons (PAHs) in hydromorphic soils of the tropical metropolis Bangkok. Geoderma 91:297–309. https://doi.org/10.1016/S0016-7061(99)00012-9

Wilcke W, Amelung W, Martius C, Grarcia MVB, Zech W (2000) Biological sources of polycyclic aromatic hydrocarbons (PAHs) in the Amazonian rain forest. J Plant Nutr Soil Sci 163:27–30. https://doi.org/10.1002/(SICI)1522-2624(200002)163:1%3C27::AID-JPLN27%3E3.0.CO;2-E

Wild E, Dent J, Barber JL, Thomas GO, Jones KC (2004) A novel analytical approach for visualizing and tracking organic chemicals in plants. Environ Sci Technol 38(15):4195–4199. https://doi.org/10.1021/es049915u

Williams PT, Bartle KD, Andrews GE (1986) The relation between polycyclic aromatic hydrocarbons in diesel fuels and exhaust particulates. Fuel 65:1150–1158. https://doi.org/10.1016/0016-2361(86)90184-5

Williams PT, Abbass MK, Andrews GE, Bartle KD (1989) Diesel particulate emissions: the role of uncombusted fuels. Combust Flame 75:1–24. https://doi.org/10.1016/0010-2180(89)90083-7.

Wilson SC, Jones KC (1993) Bioremediation of soil contaminated with polynuclear aromatic hydrocarbons (PAHs): a review. Environ Pollut 81(3):229–249. https://doi.org/10.1016/0269-7491(93)90206-4

Xia Y, Ti C, She D, Yan X (2016) Linking river nutrient concentrations to land use and rainfall in a paddy agriculture–urban area gradient watershed in southeast China. Sci Total Environ 566:1094–1105. https://doi.org/10.1016/j.scitotenv.2016.05.134

Xiao Y, Tong F, Kuang Y, Chen B (2014) Distribution and source apportionment of polycyclic aromatic hydrocarbons (PAHs) in forest soils from urban to rural areas in the Pearl River delta of southern China. Int J Environ Res Public Health 11(3):2642–2656. https://doi.org/10.3390/ijerph110302642

Xie XM, Liao M, Yang J, Chai JJ, Fang S, Wang RH (2012) Influence of root exudates concentration on pyrene degradation and soil microbial characteristics in pyrene contaminated soil. Chemosphere 88:1190–1195. https://doi.org/10.1016/j.chemosphere.2012.03.068

Yang W, Lang YH, Bai J, Li ZY (2015) Quantitative evaluation of carcinogenic and non-carcinogenic potential for PAHs in coastal wetland soils of China. Ecol Eng 74:117–124. https://doi.org/10.1016/j.ecoleng.2014.10.015

Yin CQ, Jiang X, Yang XL, Bian YR, Wang F (2008) Polycyclic aromatic hydrocarbons in the vicinity of Nanjing, China. Chemosphere 73(3):389–394. https://doi.org/10.1016/j.chemosphere.2008.05.041

Zhang J, Liu G, Wang R, Liu J (2015) Distribution and source apportionment of polycyclic aromatic hydrocarbons in Bank soils and river sediments from the middle reaches of the Huaihe River, China. Clean (Weinh) 43(8):1207–1214. https://doi.org/10.1002/clen.201400054

Zhong Y, Zhu L (2013) Distribution, input pathway and soil–air exchange of polycyclic aromatic hydrocarbons in Banshan Industry Park, China. Sci Total Environ 444:177–182. https://doi.org/10.1016/j.scitotenv.2012.11.091

# Chapter 9
# Emerging Nano-agrochemicals for Sustainable Agriculture: Benefits, Challenges and Risk Mitigation

**Disha Mishra and Puja Khare**

**Abstract** With the advent of nanotechnology, the research towards the development of newer nano-agrochemicals has received greater attention from the scientific community. The application of nano-agrochemicals like nanofertilizers, nanopesticides, nanoherbicides or other nanoscale carriers has created a revolution in modern agriculture technology. The benefits of these agrochemicals involve higher yield, crop protection, increased nutrient efficiency, enhanced soil fertility, and water availability. A variety of carbon-based, metal and its oxides, single, and multi-walled carbon nanotubes, nanoparticles of metals (such as Zn, Cu, Fe, Ag, $TiO_2$, $Ce_2O_3$), and nanocomposites are mainly used as a carrier for controlled and site-specific release of active ingredients. But their ecotoxicity, potential residues in the ecosystem, carryover amount in foodstuffs, and phytotoxicity are the major concerns to be addressed carefully. In order to attain sustainability in modern agro-ecosystems, the toxico-kinetics and toxico-dynamics study, exposure assessment, characterization methods, interaction level with other biotic and abiotic components, as well as background levels in food and feed matrices should be considered and managed.

Thus, present chapter has gathered the information about the application of emerging nanoagrochemicals for the sustainable agriculture. The summarized data provide a valuable framework to identify potential benefits, biopersistence and their risk assessment. Furthermore, the chapter has suggested the developmental gaps and guide for future research.

**Keywords** Biopersistence · Crop protection · Controlled release · Nanoagrochemicals · Toxicokinetics

D. Mishra (✉) · P. Khare
Central Institute of Medicinal and Aromatic Plants (CIMAP), Lucknow, India
e-mail: mishra.disha1@gmail.com



V. Kumar Singh et al. (eds.), *Sustainable Agriculture Reviews 50*, Sustainable Agriculture Reviews 50, https://doi.org/10.1007/978-3-030-63249-6_9

## 9.1 Introduction

In the past decades, the outburst of population load in developing countries has forced the increase in crop productivity for balancing the demand-supply chain. For fulfilling the requirements of the population and maximizing crop production, various technological innovations have come in agri-sector like fertilizers, pesticides, genetically manipulated crops, hybrids, and high yielding varieties. Unfortunately, these patterns of modern agriculture have caused a massive threat to the ecological matrices like soil pollution, bioaccumulation of pollutants, and eutrophication. In this regard, new, emerging, fast, cost-effective, and sustainable techniques have replaced the old one and able to tackle the problem associated with human health, hunger, disease susceptibility, and sustainability in agriculture (Singh et al. 2019).

In the current scenario, nanotechnology has been adapted for current research and development to formulate new techniques to combat unfavorable situations in the field of agriculture. Many of the modern techniques have been investigated using nano-enabled agricultural techniques. A nanomaterial, as per the European Commission, is a natural, accompanying, or synthetic material, including particles, in a progressive state or as agglomerate, where more than half of the population of particle size lies in 1–100 nm (Rauscher et al. 2017).

By adopting the nanotechnology for the agricultural purpose, many of the currently existing problems of the developing counties can be eradicated. Extending to the application of technology, the global turnover in the manufacturing of nanoparticles has reached up to $1 trillion by 2015 along with more than 3000 patents have been lodged (Roco 2011). The potential application of nanoparticles for sustainable agriculture can be envisaged in terms of an alteration of the plant production system, crop yield and highly economic status through the smart, controlled and targeted delivery of desired chemicals. The chemical entities used at nano-scale might be a fertilizer, pesticide, nutrient or genetic material which might be turned into nanoarchitects for building sustainable environment (Ghormade et al. 2011). The nanocarriers can modulate the plant gene expression followed by alteration in the overall biological pathway leading to desired growth and development of plant (Nair and Chung 2014). They can be composed of metal-organic frameworks, quantum dots, lipid, polymers, emulsions, silicates, layered hydroxides, ceramics, and dendrimers. The compositional characteristics of nanoparticles are important for delivering the active molecules. Properties of nanomaterials like shape, size, surface area, charge, and aggregation behavior directly control their performance and toxicity in the medium. The major advantage of using nanocarriers is smart, time-controlled, self-regulated, spatially targeted delivery of active ingredient by improving its penetration, solubility, and bioavailability for the specific plant tissue (Kumar 2019). Therefore, rather than applying conventional techniques, the implementation of nano-formulated agrochemicals would help to decrease the harmful effects like decomposition and degradation before achieving the targeted delivery, multiple application, biomagnifications in the food chain and other undesirable effects like phytotoxicity. In this perspectives, present chapter has given a brief insight about the recent research defining formulation, application, and fate of emerging nano-agrochemicals.

## 9.2 Nanomaterials as a Delivery Vehicle for Agrochemicals

The construction of nanomaterials with specific properties like high solubility, stability, degradation, and effectiveness is the major thrust area of exploration. To achieve this goal properly, designed nanomaterials are required. Formulation of nanomaterials used for agricultural practices can be inorganic moiety, organic moiety or combination of both to generate hybrid nanocarriers. For instance, various metal nanoparticles like Ag, $TiO_2$, ZnO, and silica are used for the encapsulation of active ingredients (Peters et al. 2016). The application, distribution and fate of nano-agrochemicals has been summarized in the Fig. 9.1.

The increased application of chemicals and fertilizers due to the green revolution has created huge pressure on the soil biodiversity, induced resistance development in the pathogens, and most importantly groundwater contamination. Nanomaterials have the potential to deliver considerable results for maximizing output with the safe use of chemicals with less environmental implications. The main goal of nanotechnology is to fabricate novel nanocomposite carriers to deliver nutrients,

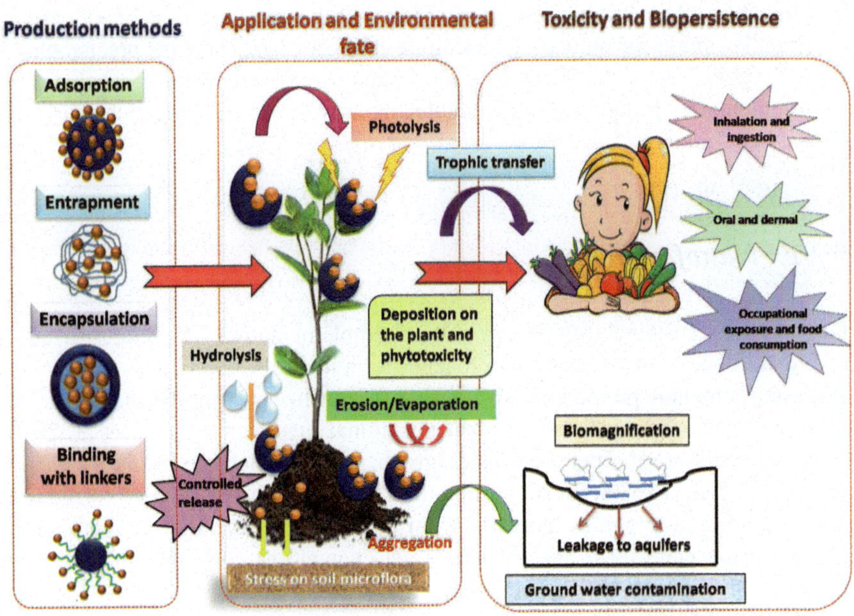


Fig. 9.1 Production methods, application, fate, and toxicity of nanoagrochemicals

fertilizers, and pesticides for construction of modern precision farming. For instance, nanopolymers and nanoclays were applied to soil to retain water holding capacity (Pulimi and Subramanian 2016), and super adsorbent quantum dots, carbon nanotubes, copper, gold, silver nanoparticles, and graphene oxides were utilized as slow release agents for mineral and nutrient supplement as well as crop protection agent (Mazzaglia et al. 2017). Giving more emphasis to environmental sustainability, many of degradable, hydrophilic, organic and polymeric materials like lignin, starch, cellulose derivatives, phospholipids, chitosan, lecithin, derivatives, alginates, albumin, polylactides, poly(propylene glycol), polyacrylamide, and polysorbate have also been exploited for the construction of nano-enabled materials.

Waste processed nanocellulose was employed as slow release urea fertilizers which showed water retention of 7.2% after 30 days and resistant to degradation rate with 30% even after 90 days burial in soil (Zhang et al. 2014). The composite of biogenic nanosilica and nano-fibrillated cellulose were also tried for the slow release of tebuconazole to sustain its biocidal potential for long duration (Mattos and Magalhães 2016). Starch-based hydrogels were explored for controlled release of carbendazim to check its efficacy and water absorption potential and the results showed improved water holding capacity upto 8.2% in soil followed by the prolonged release of active ingredient upto 240 hours in distilled water (Bai et al. 2015). A novel nitrogen based slow release fertilizer was recently developed using leftover rice-g-poly(acrylic acid)/montmorillonite network for minimizing the leaching losses of nitrogen (19.7%) as compared to pure urea (52.3%) (Zhou et al. 2018). Therefore, most of the recent nanocomposites are generally based on the fabrication by utilization of renewable resources, which might be a viable, economical, and ecofriendly option for industrial agriculture purpose also. Thus, for precision agriculture, optimization of nutrients management, yield, and ways to decrease the environmental pollution through nanotechnology-based agrochemicals are discussed here briefly.

### 9.2.1 Nanofertilizers

For attaining sustainability in agriculture, implementation of nanofertilizers is regarded as one of the advanced approaches for nutrient bioavailability to the plants and environmental protection. The term nanofertilizers comprise nanomaterials covering nutrients which acts as plant nutritional mediator (Singhal et al. 2018). Nanofertilizers are the types of nano-agrochemicals which are primarily used to provide nutrients to plant for supporting their growth. They can be synthesized from the bulk material or can be extracted from the different vegetative or reproductive part as in the nano form, the properties get changed as compared to bulk form. Based on the requirement of plant, the categories can be divided into three types i.e. macronutrient fertilizers, micronutrient fertilizers, and nanoparticulate fertilizers. For productive plant growth mainly nitrogen (N), phosphorus (P), potassium (K), magnesium (Mg), sulfur (S) and calcium (Ca) are considered as the macronutrients.

Due to the increased surface area, slow-release properties, and extraordinary features in the nanoform, the fertilizers are more efficient in terms of increasing production, soil health improvement and qualitative output of agricultural practices.

In this regard, nutrients can be encapsulated through nanocoating, nanoporous polymeric materials or supplied in the nanoform of emulsions or particles (DeRosa et al. 2010). But the soft materials developed from suitable polysaccharide biopolymers in the form of gels, capsules, biomimetic films, membranes, and composites could be a better strategy for the formulation of controlled-release fertilizers (Campos et al. 2015). In this way, nanofertilizers with high surface area, higher sorption capacity, and lower aspect ratio of nano-enabled fertilizers would hold the nutrients and release to the target site when required (Duhan et al. 2017). Properties such as controlled and slow-release are beneficial for plants to support the increased nutrient uptake and decrease the problems associated with conventional fertilizers. The rate and pattern of release of fertilizers from nanoshell can be controlled by hydrolysis of polymeric materials and it can be adjusted through choosing appropriate coating materials used for encapsulation. Novel nutrient cargo systems of nanoparticles have potential to infiltrate seeds or move in the root tissue via utilizing the permeable surface of the nano-sized area of the plant for continued release of nutrients (Zafar et al. 2016). The balanced dose of nutrients can be augmented inside nanomaterials through the modification on the synthetic route and later through adsorption of both cationic and anionic nutrients (Manikandan and Subramanian 2016). Consequently, these controlled-release fertilizers maintain crop nutrient management and minimize agricultural input for creating sustainable agriculture.

Manikandan and Subramanian (2016) have developed nano-zeolite based nitrogenous fertilizers and it was tested against conventional urea fertilizers. The results have demonstrated that over the conventional fertilizers the zeolite-based nanofertilizers has served as a slow-releasing system and was able to increase the nitrogen use efficiency and exert a positive effect on plant growth and crop yield. The efficacy of composite Layered Double Hydroxide and phosphate ions (LDH-phosphate) was tested as slow-releasing phosphate as compared to commercial triple superphosphate fertilizer. The kinetic study showed that slow P release has promoted an increase in the soil pH value and availability of P to the plants (Benício et al. 2016). Nitrogen, Phosphorus and Potassium (NPK) loaded chitosan nanofertilizer were applied to coffee seedlings in a greenhouse condition and the application resulted into improved nutrient status as compared to control and also the physiological responses in terms of plant height, leaf area, and the number was also improved (Ha et al. 2019). Abdelsalam et al. (2019) have found that foliar application of NPK based nanoparticles showed an increase in yield and morphological responses in wheat. The genotoxic effect of nanoparticles detected that root tip cells of wheat can readily internalize the NPK based nanoparticles. Even the stress of salt was also reduced through the spray of nanoparticles in various crops due to increase in the surface area of the nanoparticles which are responsible to enhance the enzymatic activity for the reduction in salt stress (Soliman et al. 2015; Kalteh et al. 2018). Impact of Cu nanoparticles on the soil nitrification kinetics has demonstrated a similar pattern of the nitrification to the control with around 9% increase in the average rate of nitrification kinetics (VandeVoort and Arai 2018). Nanofertilizers are designed to release the nutrients on the demand of plant by preventing their interaction with soil, water or microorganisms. A slow-releasing urea hydroxyapatite nanocomposite (6:1 by weight) was employed in the soil to maintain yield and reducing the urea application by lowering its solubility (Kottegoda et al. 2017). In fact, temperature sensitive formulation for releasing fertilizers have shown that increase in temperature causes a decrease in the rate of release as if increase of temperature from 20–40 °C has drifted the nitrogen release to rapid as compared to potassium, while phosphorous release shifted towards the lower side (Du et al. 2006).

Nano-chitosan based on NPK was created to check the effect of carrier system on the plant and the environment. The carrier was found to have an adverse effect on the pea plant seedlings with reduced root elongation rate and accumulation of starch at the root tip with a genotoxic effect. Therefore, the effect of nanocarrier system should be identified before its commercial application (Khalifa and Hasaneen 2018). Similarly, the hybrids of chitosan/cellulose and acrylic acid prepared through graft polymerization showed its potential for controlling the release of NPK into the soil and did not exceed beyond 75% even after one month (Essawy et al. 2016).

Micronutrients are also equally important for plant growth as they often added in the form of composite fertilizers. Plant micronutrients include iron (Fe), manganese (Mn), zinc (Zn), copper (Cu), molybdenum (Mo). The nanoformulation of micronutrients are generally added to the plant for support of growth. The slow releasing polymeric formulation of PVA–starch/carbon nanofibers showed a slow release of the Cu and Zn micronutrient with sustaining the growth of chickpea plant (Kumar et al. 2019). ZnO nanoparticles were tested to improve nutritional quality in a tomato plant and the system has demonstrated increase in Zn biofortification along with an increase in lycopene content by 113% over the control (Raliya et al. 2015). Currently, a new type of mycosynthesized nanoparticles has come up with their potential for application as nanofertilizer and nanopesticide for sustainable agricultural practices. In this regard, *Aspergillus* and *Fusarium* sp. were explored for the synthesis of Au, Ag, Ti, Zn, Ce, Fe, Mg, P, and Pt metal and their metal oxide nanoparticles (Chhipa 2019). However, future research in this sector should focus on the development of newer and more efficient nutrient augmented ecofriendly nanomaterials. In order to identify the beneficial effects as well as the possible toxicological profiles of nanofertilizers, scientific research should explore the relationship of nanocarriers with plant and microbial community for enhancing agricultural productivity.

### 9.2.2 Nanopesticides

The term "nanopesticides" includes the nanoformulation of pesticidal active ingredients or small nanomaterials with pesticidal activities (Chhipa 2017). The construction of nanopesticides through the encapsulation inside the nanomaterials or biocomposite materials is beneficial due to their interesting features like stiffness, permeability, crystallinity, thermal stability, solubility, and biodegradability and they are also helpful in reducing the harmful effect of chemical based pesticides. The higher surface area of the nanomaterials increases the affinity of active ingredients for the target (Zhao et al. 2017). The active ingredients of pesticides has often been delivered in the nanoform of dispersions, emulsions, capsules, containers, and cages (Table 9.1). The durability, solubility, and mobility of nanopesticides minimize the harmful effect of active ingredients, and thus, less resistance was developed in non-target organisms (Kah et al. 2013). Complex formulation of nanopesticides includes solid lipid nanoparticles, liposomes, mesoporous silica, metal oxides, and nanoclay. The fungicidal effect of zineb and mancozeb trapped inside multiwall carbon nanotubes graft-poly(citric acid) hybrid material has shown an effective response against *Alternaria alternata* fungi (Sarlak et al. 2014). The enzyme responsive controlled release formulations based on mesoporous silica prepared via interlocking of α-cyclodextrin led to loading of chlorantraniliprole and was able to show insecticidal activity against *Plutella xylostella* (Kaziem et al. 2017).

The utilization of graphene oxide was also increased for the adsorption of chlorpyrifos, endosulfan, and malathion due to availability of hydrophobic interactions, hydrogen bonding, π-π interactions, and the presence of electron donors (like S, P, and N) (Maliyekkal et al. 2013). Apart from these, now-a-days the trend has shifted towards greener route of synthesis of nanopesticides in which plants extracts and metabolites are used for generation of metallic nanoparticles. However, to overcome the drawbacks of synthetic nanoparticles and for the balancing soil health and ecosystem functioning, currently greener encapsulation is in trend. Thus, certain biodegradable and biocompatible materials like cellulose, chitin, nanoclays, and starch are being employed for the encapsulation of active ingredients. Chitosan-based herbal nanopesticide formulation has shown 88.5% antifeedant activity and 90.2% larvicidal activity against *Helicoverpa armigera* (Paulraj et al. 2017). Apart from these, cyanobacteria are emerging as a new class of nanocarriers for pesticides loading due to abundance, heterogeneous surface groups, and biodegradable nature. They have also been reported for improving soil quality and soil-water interface. Recently a fluorescent photoresponsive nanocarrier perylene-3-ylmethanol was explored for controlled release of 2,4-dichlorophenoxyacetic acid with effective regulation in the release of pesticides (Atta et al. 2015). In the perspective of environmental cleanup, new classes of biopesticides are more beneficial than synthetic pesticides formulation. As a biocontrol agent, nanoformulations of plant-based resources and microbe-based formulation are very effective against insects and pests.

Biopesticides exhibited target specificity, low non-target organism toxicity, and low environmental persistence. However, their complex formulation strategy, production costs, and convenience limit their commercial application. Various biological nanopesticides such as nanoparticles synthesized through plant extracts, nanoemulsion of essential oil have been tried against pest. The application of zein-encapsulated neem oil (Pascoli et al. 2019), polyethylene glycol encapsulated geranium, citrus (González et al. 2015), and Ag nanoparticles fabrication with the *Artemisia herba-alba* extract (Alshehri et al. 2018), and *Pongamia pinnata* leaf

**Table 9.1** Encapsuation of different pesticides, encapsualation strategies adopted and their benefits

| Encapsulation material | Pesticide | Encapsulation method | Stimulus | Action | Reference |
|---|---|---|---|---|---|
| **Nanocapsules** | | | | | |
| Azidobenzaldehyde and chitosan | Methomyl | Photo-cross-linking | pH | Diffusion controlled release with time for ($t_{1/2}$) of 36.3–69.5 h and maintained efficacy against armyworm upto 7 days | Sun et al. (2014) |
| Alginate and chitosan | Acetamiprid | Polyelectrolyte complexation | pH | Controlled release 24 hours at pH 10 and 36 hours at pH 7 and 4 while in soil 93% release in 36 h | Kumar et al. (2015) |
| Silica | Fipronil | Emulsion and biomimetic dual-templating method | Miglyol | Sustained release upto 142 hours upto 40.7% and mortality against subterranean termites | Wibowo et al. (2014) |
| **Nanospheres** | | | | | |
| Silica | Metalaxyl | Sol–gel process | – | Slow release of metalaxyl (76%) of in soil within a period of 30 days | Wanyika (2013) |
| Poly vinyl alcohol | Emamectin-benzoate | Microemulsion polymerization method | – | Sustained release for the period of 200 hours with good antiphotolysis capability and stability of loaded microspheres | Wang et al. (2017) |

(continued)

**Table 9.1** (continued)

| Encapsulation material | Pesticide | Encapsulation method | Stimulus | Action | Reference |
|---|---|---|---|---|---|
| Core-shell PDA@ PNIPA | Imidacloprid | Precipitation polymerization method | NIR-light and temperature | More active than PDA with improved functionality for delivery of pesticides | Xu et al. (2017) |
| **Nanogels** | | | | | |
| CA-CMC-bentonite hydrogel composites | Thiamethoxam | *Ex-situ* encapsulation technique | pH | Improvement of barrier properties of the hydrogel matrix | Sarkar and Singh (2017) |
| **Nanoemulsion** | | | | | |
| PDA | 2,4-D | O/W emulsion | – | Sustained release upto 10 hours in different solvent | Tang et al. (2019) |
| **Polymeric micelles** | | | | | |
| Poly[2-(2-Methoxyethoxy) ethyl methacrylate-co-Octadecyl methacrylate | Pyrethrins | Cooperative assembly | Temperature | Regulate the release pattern at changes in temperature and potent larvicidal activity | Zhang et al. (2019) |
| Poly(butylene 2-methylsuccinate)-poly(ethylene glycol) | Avermectin | Self assembly and dialysis | Temperature | Sustained release stage with 91% pesticides release after a short burst release | Han et al. (2019) |
| **Solid lipid nanoparticles** | | | | | |
| SLN | Carbendazim and tebuconazole | Emulsification and solvent evaporation | – | Modified release upto 51% after 6 days | Campos et al. (2015) |
| **Nanoliposomes** | | | | | |
| Chitosan | Imidacloprid and lambda-Cyhalothrin | Ethanol injection /self assembly | – | Controlled release over a period of 26 days with significant insecticidal property | Moradi et al. (2019) |

*PDA* Polydopamine, *PNIP* Poly(N-isopropylacrylamide), *CA* Citric acid, *CMC* Carboxymethyl cellulose, *SLN* Solid lipid Nanoparticles extract coated zinc oxide nanoparticles (Malaikozhundan and Vinodhini 2018) were successfully explored as herbal nanopesticides with effective retardation in pest growth. However, biopesticides derived toxin, their stability under field conditions, lack of sufficient quantity, and quality and legislative process restrict their commercialization for agricultural application

### *9.2.3 Nanoherbicides*

The holistic concept of sustainable agriculture involves a systematic approach to reduce the loss in agricultural production along with increase in crop yield with minimum negative implications to the environment. For this, early detection and control of weed are required through an ideal mechanism which can prolong the effect of active ingredients, maintain the activity and resistance potential over a long period. In this context, a nanocarrier for encapsulating bioactive agents of herbicidal activity is beneficial to increase their solubility, prevent their degradation and promote sustained release of herbicidal compounds. The herbicidal effect of atrazine loaded poly(lactic-co-glycolic acid) nanoparticles has shown a reduction in stem and root length, fresh and dry weight, and the number of leaves in potato plant (Schnoor et al. 2018). Poly(ε-caprolactone) nanoencapsulated atrazine was developed as highly effective herbicidal formulation which showed herbicidal activity through transportation inside the vascular tissue of the leaves and resulted in chloroplast degradation (Bombo et al. 2019). Alharby et al. (2019) have critically reviewed synthesis and applications of nanoherbicides with special emphasis on their bioavailability, distribution and the possible mechanisms of action in plants.

The mycosynthesis of metal nanoparticles such as Au, Ag, Ti, Zn, Ce, Fe, Mg, P, and Pt as well as their metal oxide nanoparticles, from *Aspergillus* and *Fusarium* species, were found useful for removal of weeds (Chippa 2019). The anionic clay (layered double hydroxide or LDH) and a commercial cationic organoclay (Cloisite 10A) were assayed as host-nanocarriers for imazamox herbicide and they showed minimum soil leaching and thereby maintaining its efficacy (Khatem et al. 2019). The nanocapsules of polycaprolactone were used as a carrier for pretilachlor and it was found significantly effective against *Echinochloa crus-galli* sp. (barnyard grass) (Diyanat et al. 2019). Novel nanoparticles system of pectin, chitosan, and sodium tripolyphosphate was employed to encapsulate paraquat and it was found to decrease the toxicity and soil sorption while considerable increase in overall herbicidal activity (Rashidipour et al. 2019). The encapsulated metazachlor in polylactic acid particles were able to reduce the growth of poppy and barnyard grass likely due to sustained release of herbicides (Salač et al. 2019).

The developed herbicides should be specific to the target weed only through the receptor present at the root of target weed. However, the disproportionate and continuous application of herbicides induces resistance development in the weed, damages the crop and causes soil toxicity. Phosphatidylcholine-clay complex was used to encapsulate diuron and alachlor herbicides; its adverse effect on the soil biological activity and increase in half-life in the soil after the continuous application over 127 days was observed (Tejada et al. 2017). The ametryn and atrazine loaded poly(ε-caprolactone) nanocapsules has demonstrated toxicity against alga *Pseudokirchneriell asubcapitata* and the microcrustacean *Daphnia similis* sp. as compared to the free herbicides (Clemente et al. 2014). Recently plant-derived substances were also explored for the herbicidal activity. In this queue, nanoformulation of essential oil (Khare et al. 2019; Das et al. 2019) and plant extract have proven as potent herbicides (Buhroo et al. 2017). Various researches have been performed in this direction to identify the phytotoxic effect of essential oil-based formulation against several weed crops (Hassannejad and Ghafarbi 2013; Hazrati et al. 2017; Alipour et al. 2019). More endeavoring is still required for the practical and sustained application of nanoherbicides along with the exploration of newer chemicals and carriers.

## 9.3 Benefits of Nano-agrochemicals

Nano-agrochemicals offer tremendous opportunities for sustainable agriculture. It is predicted that nano-agrochemicals will cause a breakthrough in agriculture through new tools for early disease management, smart delivery systems, and most notably crop improvement with nanodevices. For the maximum output of agricultural production, the nanotechnology based techniques would result in maximum nutrient use efficiency, thereby minimizing the cost as well as challenges of environmental pollution (Silva and Bonatto 2019). The benefits of using nano-agrochemicals are discussed here precisely.

### *9.3.1 Balanced Crop Nutrition*

The role of fertilizer is the very pivot for attaining sustainable agriculture. In the developing countries, the use of fertilizers for maximum crop yield is tremendous but the results of the rapid and huge application of chemical fertilizers are very dangerous. Therefore, attempts have been made to synthesize the nanofertilizers for balanced crop nutrition without leaving any major portion of chemicals into soil or water. Many researchers have discussed the nano-nutrition to the plants through the application of nanofertilizers, nanopesticides, nanotubes, nanoclay, and nanocapsules (El-Ramady et al. 2018; Iqbal and Umar 2019; Shang et al. 2019). Nanofertilizer technology is very innovative and huge literature along with patents are available in the scientific journals. By applying the nano-scaled fertilizers, the balanced crop nutrition can be achieved through slow or controlled release technology. However, the fate of nano-carriers and their physiological, biochemical, nutritional, and morphological changes in plants and the fate of nanoproducts in soil and plant systems should be addressed.

### 9.3.2 Plant Disease and Weed Control

Weeds and pest emergence severely damage the crop and reduce the yield upto 30–40%. Although, the application of herbicides and pesticides generally as common trend is applied by farmers to overcome this problem, but the efficiency of these chemicals get lossed due to certain environmental factors. Chitosan based nanocomposites were explored for their fungicidal properties against fusarium wilt (caused by *Fusarium oxysporum* f. sp.) of chickpea plant and they resulted in wilt reduction upto 47% as compared to the marketed formulation (Kaur et al. 2018). Sarlak and Taherifar (2017) has developed a hybrid material of multiwalled carbon nanotubes graft-poly(citric acid) for controlling the emergence of *Alternaria alternata* fungi through the encapsulation of various pesticides. Therefore, encapsulation of these chemicals in nanomaterials could result in maximum output value. However, the efforts have also been made towards the postharvest pest control through the development of advanced nanocarriers (Shukla et al. 2019). The nano-herbicides/pesticides application method would save the time and costs associated with tilling and manual picking (Sousa et al. 2018; Alharby et al. 2019; Balah and Pudake 2019).

### 9.3.3 Sustainable Water Use

For sustainable agricultural production and optimize water use, nano-hydrogel can be applied. A hydrogel is the water absorbing granular polymeric system which can absorb the water through hydrogen bonding. It can absorb, swell, release water and nutrients in cycles, leading to more efficient use of water. Novel superadsorbent hydrogel of neutralized acrylic acid, urea, potassium persulfate, and N,N′-methylenebis(acrylamide) has shown good water retention and slow nitrogen release in soil (Cheng et al. 2018). Similarly *Gum tragacanth* and acrylic acid based hydrogel has shown good water holding capacity in sandy and loamy soil (Saruchi et al. 2014). Furthermore, the biodegradable hydrogel can store between 130 and 190 times its weight of rainwater or irrigation water (Demitri et al. 2013; Abdel-Raouf 2019; Tomadoni et al. 2019).

### 9.3.4 *Seed Emergence*

Seed emergence is necessary for optimizing the productivity of any crop. But the germination and propagation of any seed get limited due to unavailability of sufficient moisture in case of rainfed area. Therefore, treatment of seed with nanomaterials helps to germinate the seed faster and steadier. Despite not only controlling emerging percentage but nanomaterials also help in the production of qualitative seed to support high vegetative growth of the crop. The drought resistance, climatic resilience and longevity of the seeds also increase after treatment with nanomaterials (Singh et al. 2016).

### 9.3.5 *Smart Delivery System*

The instability and non-targeted application of conventional fertilizers generally accounts for 70% loss of the fertilizer during application. The smart delivery system based on nanoscale features is pre-programmed, regulated, targeted, and controlled, which can cross the biological and environmental barriers during applications (Kim et al. 2018; Silva and Bonatto 2019; Camara et al. 2019). Smart delivery systems with broad chemical detection and decision-making capability for self-regulation in order to deliver active chemical molecules or nutrients when needed is highly required. The remotely regulated and self-monitored smart delivery system could assist the farmers to increase agricultural production.

### 9.3.6 *Pollution Reduction*

The utilization of nano-agrochemicals with controlled and targeted delivery devices is able to decrease the run-off, soil pollution, and eutrophication subsequently minimizing the health risk due to contamination. In addition, the nanoparticles applied to the soil are also able to remediate the polluted soil (Pulimi and Subramanian 2016; Sangeetha et al. 2017). Many studies reported that the application of nutrient-based nanocarriers were also able to remediate the contaminants from soil.

## 9.4 Toxico-Kinetics, Biopersistence and Environmental Fate of Nano-agrochemicals

With the rapid growth of nano-agrochemicals, they are continuously being applied to soil for increasing fertility or as plant protection methods. Soil and water are likely to be affected by the end-products of these chemicals. After the application of nano-agrochemicals the end-products directly affect the soil health. The direct entry of nanocarriers in aquatic systems is mainly through industrial discharges, disposal of wastewater treatment effluents, while indirectly through surface run-off from soils. However, the fate of nanocarriers in the soil is directly governed by the surface properties of the nanoparticles, adsorption, aggregation, deposition behavior, stability, and mobility as well as colloidal suspensions. After reaching to the terrestrial or aquatic environment, the nanocarriers may undergo certain transformation through biodegradation of surface coatings of nanocarriers. They can also release potentially toxic chemicals into the environment. The homo or hetro-aggregation of nanoparticles with natural mineral and organic colloids alter their fate and toxicity in different matrices. In case of terrestial ecosystem, the interaction of nanoparticles with the soluble organic matter may lead to change in surface charge and mobility, and thus, change in interactions pattern with biotic factors. Whereas, in the aquatic system their heteroaggregation cause accumulation in bottom sediments. Kah et al. (2016) have demonstrated different sorption and degradation parameters of the nanopesticide (insecticide bifenthrin) in two different soils than that of pure active ingredient.

The mobility and fate of $CeO_2$, ZnO and Cu nanoparticles in agricultural soil by wetting-drying cycles showed that relative amount of $CeO_2$ and ZnO nanoparticles leached from soil decreases due to the immobilization by water-stable soil aggregates as dissolution has controlled their leaching kinetics and mobility in ionic forms (Ermolin et al. 2019). The repeated application of $Cu(OH)_2$ nano-pesticides under two fertilization level upto a year posed detrimental effect on the microbial load due to a significant decrease in phosphatase and beta-glucosidase activity (Simonin et al. 2018a). After 90 days exposure of titanium-dioxide ($TiO_2$) nanoparticles, there were changes in the activity and abundance of nitrifying bacteria, ammonia-oxidizing archaea and bacteria, and nitrite oxidizing bacteria (*Nitrobacter* and *Nitrospira*) in the soil (Simonin et al. 2017). The effect of different nanoparticles in the combination, on the soil micro-ecosystem was investigated through the irrigation with the combination of $TiO_2$ and Ag nanoparticles and results showed that $TiO_2$ nanoparticles alone was not toxic to the soil micro-ecosystem, but Ag nanoparticles significantly reduced the soil microorganisms (Liu et al. 2019). Therefore, the dissolution, agglomeration, and aggregation of nanoparticles were governed by its physicochemical characteristics (e.g. shape, size, and surface charge) and soil properties (e.g. pH, ionic strength, organic matter, and clay content). There is an urgent need of the hour to monitor the leftover nanoparticles in soil, sediments, and water bodies. Apart from the entry of nanocarrier inside the environmental matrices, their direct contact to plant through the soil has to be addressed more clearly. It is necessary to study absorption, uptake, movement, and interaction of nanoparticles inside the plant cell.

The accumulation and translocation of nanoparticles inside the edible part of the crop is a major future concern. The features, type, and characteristics of the nanomaterials, as well as the plant species, significantly influence translocation and accumulation of nanocarriers in plant tissues. The effect of ZnO, CuO, and $CeO_2$ nanoparticles on the accumulation and translocation in Carrot (*Daucus carota*) has reported more accumulation of the elements in the taproot peel than that of the shoots, followed by accumulation in the edible flesh (Ebbs et al. 2016). Tripathi et al. (2017) have critically reviewed physiological, biochemical, and molecular alteration due to interaction with metal nanoparticles. Many studies have reported the issue of phytotoxicity due to engineered nanoparticles applied for agricultural purposes. The induced phytotoxicity in the plant resulted in the production of reactive oxygen species, followed by oxidative stress, lipid peroxidation, as well as proteins and DNA damage in plants. After accumulation inside the plant tissue, the degradation and the quality of crops get lowered due to decreased rate of seed germination followed by reduction in fresh and dry biomass and length of roots and shoots, alteration in the rate of photosynthesis, and fast chromatin condensation, DNA damage, lipid peroxidation, reduced rate of transpiration, uncontrolled regulation of stress-related genes, and eventually apoptosis. Therefore, there is a necessity to pursue a well designed, life cycle assessment including environmental risk assessment of nanocarriers at different level of the ecosystem so that their fate, transportation, transformation, and degradation can be explained more accurately.

## 9.5 Nano-agrochemicals as Emerging Contaminants:Perception and Mitigation

In the perspective of agricultural protection, crop yield improvement, pest protection, and lower environmental impact, mostly engineered nanocarriers are used in the form of nano-agrochemicals. The augumetation of emerging nanocarriers have immense potential to bring sustainable intensification in agricultural sector (Fraceto et al. 2016). They can easily replace conventional farming and generate a new arena of precision farming. The use of nanoformulation for desgining next generation nanocarriers for agrochemicals provides several site-specific benefits like smart and controlled delivery with minimum pollution and maximum eco-balance. Furthermore, the reduction in the use of chemical fertilizers and pesticides has benefited the soil health, balanced geo-biological cycle, and improved nutrient quality and crop yield (Chhipa 2017). Despite these advantages currently emerging trend of nano-agrochemicals have several issues to be addressed. Nano-agrochemicals are eventually a potent group of pollutants for agricultural applications. Contamination of water resource, persistence in soil, magnification in food produce, and nanotoxicity are the major concerns associated with the application of nano-agrochemicals (Ma et al. 2018; Simonin et al. 2018a).

For realistic agricultural applications, the properties of nanocarriers like size, surface charge, concentration, morphology, colloidal state, hydrophobicity, aggregation behavior, and reactivity should be addressed properly before bringing them from laboratory to field level. In this sense, for bringing the sustainable intensification in agri-sector, the effect of these technologies should be addressed in context of biophysical, social, and economic sector. A common nanosystem with multifunctional properties like maintained soil health, controlled release of chemicals, nutrient immobilization, soil remediation, minimized leaching, and euthrophication is still required. For creating agricultural and environmental sustainability, introduction of hybrid, and biodegradable nanocarriers with industrially viable process and acquisition of knowledge of risk assessment would be helpful.

The associated risk-benefit and life cycle assessment of nano-enabled chemicals will be helpful in providing the net benefits and opportunity to improve the agricultural productvity. There should be the integration of scientific, governmental and industrial regulatory affairs to overcome the risk associated with the use of nano-agrochemicals. The fate factor, exposure factor and effect factor should be addressed before its entry into the market. Many of the governmental and regulatory authorities have posed certain legislation which should be adopted by manufactures before its commercial emergence. Although the mandate of regulatory burden and risk of stigmatization has turned the pattern of interest of the agrochemical industry for nano-formulation, but the prohibition of the application of nano-agrochemicals until they prove safe and non-toxic will be helpful to minimize the risk associated with it. Different aspects of nano-formulations like fate and effects of active substances have been recorded in the scientific literature, but the mechanisms behind their effects are poorly understood. Thus, elucidation and derivation of the mechanism of those processes and analysis of their consequences in terms of environmental impact assessment is the need of the hour. A critically evaluated environmental risk assessment plan, standard experimental protocols, analytical techniques, and theories should collaborate for mitigation of the risk associated with application of nano-agrochemicals.

## 9.6 Conclusion and Future Perspectives

The concept of nano-agrochemicals has brought a boom in agricultural production with maximum crop yield and minimum application of synthetic chemicals. The potential use of nano-agrochemicals has precisely reduced the chemicals through controlled release or slow-release smart delivery system. However, the toxicity, accumulation, and retention of these carriers have been noticed in the crop and different environmental matrices. Therefore, for boosting the agrarian economy, the green synthesis and mycosynthesis of nanoparticles are safe and advantageous for precision farming.

For the modernization of agrochemicals uses, potentials, applications, and the input should be made for adaptation of new policies, modern analytical techniques, continuous monitoring, and utilization of biodegradable matrices for their delivery. The databank generation, strict legal policies, guidelines, and roadmaps for environmental risk assessment and international collaboration is still required for the phenomenal growth of nano-enabled chemicals. The systematic study of nano-agrochemicals and associated emerging contaminants with their continuous application should be identified in a precise manner. Along with this, the consequences of their end-products in the environment must be evaluated to mitigate the risk associated with them. In the present scenario scientists, regulatory bodies, and industries should come up together to understand the nano-enabled technology for the development of more engineered nanocarriers for sustainable agriculture.

## References


Abdel-Raouf ME (2019) Guar gum based hydrogels for sustained water release applications in agriculture a review. Cur Res Biopolym 2:111. https://doi.org/10.29011/CRBP-111.000011

Abdelsalam NR, Kandil EE, Al-Msari MA, Al-Jaddadi MA, Ali HM, Salem MZ, Elshikh MS (2019) Effect of foliar application of NPK nanoparticle fertilization on yield and genotoxicity in wheat (*Triticum aestivum L.*). Sci Total Environ 653:1128–1139. https://doi.org/10.1016/j.scitotenv.2018.11.023

Alharby HF, Hakeem KR, Qureshi MI (2019) Weed control through herbicide-loaded nanoparticles. In: Husen A, Iqbal M (eds) Nanomaterials and plant potential. Springer, Cham, pp 507–527. https://doi.org/10.1007/978-3-030-05569-1_20

Alipour M, Saharkhiz MJ, Niakousari M, Damyeh MS (2019) Phytotoxicity of encapsulated essential oil of rosemary on germination and morphophysiological features of amaranth and radish seedlings. Sci Hortic 243:131–139. https://doi.org/10.1016/j.scienta.2018.08.023

Alshehri MA, Panneerselvam C, Murugan K, Trivedi S, Mahyoub JA, Hassan MM, Maggi F, Sut S, Dall'Acqua S, Canale A (2018) The desert wormwood (*Artemisia herba-alba*)–from Arabian folk medicine to a source of green and effective nanoinsecticides against mosquito vectors. J Photochem Photobiol B Biol 180:225–234. https://doi.org/10.1016/j.jphotobiol.2018.02.012

Atta S, Bera M, Chattopadhyay T, Paul A, Ikbal M, Maiti MK, Singh NP (2015) Nano-pesticide formulation based on fluorescent organic photoresponsive nanoparticles: for controlled release of 2, 4-D and real time monitoring of morphological changes induced by 2, 4-D in plant systems. RSC Adv 5:86990–86996. https://doi.org/10.1039/C5RA17121K

Bai C, Zhang S, Huang L, Wang H, Wang W, Ye Q (2015) Starch-based hydrogel loading with carbendazim for controlled-release and water absorption. Carbohydr Polym 125:376–383. https://doi.org/10.1016/j.carbpol.2015.03.004

Balah MA, Pudake RN (2019) Use nanotools for weed control and exploration of weed plants in nanotechnology. In: Pudake R, Chauhan N, Kole C (eds) Nanoscience for Sustainable Agriculture. Springer, Cham, pp 207–231. https://doi.org/10.1007/978-3-319-97852-9_10

Benício LPF, Constantino VRL, Pinto FG, Vergütz L, Tronto J, da Costa LM (2016) Layered double hydroxides: new technology in phosphate fertilizers based on nanostructured materials. ACS Sustain Chem Eng 5:399–409. https://doi.org/10.1021/acssuschemeng.6b01784

Bombo AB, Pereira AES, Lusa MG, de Medeiros Oliveira E, de Oliveira JL, Campos EVR, de Jesus MB, Oliveira HC, Fraceto LF, Mayer JLS (2019) A mechanistic view of interactions of a nanoherbicide with target organism. J Agri Food Chem 67:4453–4462. https://doi.org/10.1021/acs.jafc.9b00806

Buhroo AA, Nisa G, Asrafuzzaman S, Prasad R, Rasheed R, Bhattacharyya A (2017) Biogenic silver nanoparticles from *Trichodesma indicum* aqueous leaf extract against *Mythimna separata* and evaluation of its larvicidal efficacy. J Plant Protect Res 57:194–200. https://doi.org/10.1515/jppr-2017-0026

Camara MC, Campos EVR, Monteiro RA, Santo Pereira ADE, de Freitas Proença PL, Fraceto LF (2019) Development of stimuli-responsive nano-based pesticides:emerging opportunities for agriculture. J Nanobiotechnol 17:100. https://doi.org/10.1186/s12951-019-0533-8

Campos EVR, de Oliveira JL, Fraceto LF, Singh B (2015a) Polysaccharides as safer release systems for agrochemicals. Agron Sustain Dev 35:47–66. https://doi.org/10.1007/s13593-014-0263-0

Campos EVR, De Oliveira JL, Da Silva CMG, Pascoli M, Pasquoto T, Lima R, Abhilash P, Fraceto LF (2015b) Polymeric and solid lipid nanoparticles for sustained release of carbendazim and tebuconazole in agricultural applications. Sci Rep 5:13809. https://doi.org/10.1038/srep13809
Cheng D, Liu Y, Yang G, Zhang A (2018) Water-and fertilizer-integrated hydrogel derived from the polymerization of acrylic acid and urea as a slow-release N fertilizer and water retention in agriculture. J Agri Food Chem 66:5762–5769. https://doi.org/10.1021/acs.jafc.8b00872
Chhipa H (2017) Nanofertilizers and nanopesticides for agriculture. Environ Chem Let 15:15–22. https://doi.org/10.1007/s10311-016-0600-4
Chhipa H (2019) Mycosynthesis of nanoparticles for smart agricultural practice: a green and eco-friendly approach. In: Green synthesis, characterization and applications of nanoparticles. Elsevier, Amsterdam, pp 87–109. https://doi.org/10.1016/B978-0-08-102579-6.00005-8
Clemente Z, Grillo R, Jonsson M, Santos N, Feitosa L, Lima R, Fraceto L (2014) Ecotoxicological evaluation of poly (ε-caprolactone) nanocapsules containing triazine herbicides. J Nanosci Nanotechnol 14:4911–4917. https://doi.org/10.1166/jnn.2014.8681
Demitri C, Scalera F, Madaghiele M, Sannino A, Maffezzoli A (2013) Potential of cellulose-based superabsorbent hydrogels as water reservoir in agriculture. Int J Polymer Sci 2013:1–6
DeRosa MC, Monreal C, Schnitzer M, Walsh R, Sultan Y (2010) Nanotechnology in fertilizers. Nat Nanotechnol 5:91. https://doi.org/10.1155/2013/435073
Diyanat M, Saeidian H, Baziar S, Mirjafary Z (2019) Preparation and characterization of polycaprolactone nanocapsules containing pretilachlor as a herbicide nanocarrier. Environ Sci Pollut Res 26:21579–21588. https://doi.org/10.1007/s11356-019-05257-0
Du C-w, Zhou J-m, Shaviv A (2006) Release characteristics of nutrients from polymer-coated compound controlled release fertilizers. J Polym Environ 14:223–230. https://doi.org/10.1007/s10924-006-0025-4
Duhan JS, Kumar R, Kumar N, Kaur P, Nehra K, Duhan S (2017) Nanotechnology: the new perspective in precision agriculture. Biotechnol Rep 15:11–23. https://doi.org/10.1016/j.btre.2017.03.002
Ebbs SD, Bradfield SJ, Kumar P, White JC, Musante C, Ma X (2016) Accumulation of zinc, copper, or cerium in carrot (*Daucus carota*) exposed to metal oxide nanoparticles and metal ions. Environ Sci Nano 3:114–126. https://doi.org/10.1039/C5EN00161G
El-Ramady H, Abdalla N, Alshaal T, El-Henawy A, Elmahrouk M, Bayoumi Y, Shalaby T, Amer M, Shehata S, Fári M, Domokos-Szabolcsy É (2018) Plant nano-nutrition: perspectives and challenges. In: Gothandam K, Ranjan S, Dasgupta N, Ramalingam C, Lichtfouse E (eds) Nanotechnology, food security and water treatment. Environmental chemistry for a sustainable world. Springer, Cham, pp 129–161. https://doi.org/10.1007/978-3-319-70166-0_4
Ermolin M, Fedyunina N, Katasonova O (2019) Mobility and fate of cerium dioxide, zinc oxide, and copper nanoparticles in agricultural soil at sequential wetting-drying cycles. Materials 12:1270. https://doi.org/10.3390/ma12081270
Essawy HA, Ghazy MB, El-Hai FA, Mohamed MF (2016) Superabsorbent hydrogels via graft polymerization of acrylic acid from chitosan-cellulose hybrid and their potential in controlled release of soil nutrients. Int J Biol Macromol 89:144–151. https://doi.org/10.1016/j.ijbiomac.2016.04.071
Fraceto LF, Grillo R, de Medeiros GA, Scognamiglio V, Rea G, Bartolucci C (2016) Nanotechnology in agriculture: which innovation potential does it have? Front Environ Sci 4:20. https://doi.org/10.3389/fenvs.2016.00020
Ghormade V, Deshpande MV, Paknikar KM (2011) Perspectives for nano-biotechnology enabled protection and nutrition of plants. Biotechnol Adv 29:792–803. https://doi.org/10.1016/j.biotechadv.2011.06.007
González JOW, Stefanazzi N, Murray AP, Ferrero AA, Band BF (2015) Novel nanoinsecticides based on essential oils to control the German cockroach. J Pest Sci 88:393–404. https://doi.org/10.1007/s10340-014-0607-1

Ha NMC, Nguyen TH, Wang S-L, Nguyen AD (2019) Preparation of NPK nanofertilizer based on chitosan nanoparticles and its effect on biophysical characteristics and growth of coffee in green house. Res Chem Intermed 45:51–63. https://doi.org/10.1007/s11164-018-3630-7

Han J, Weng Y, Xu J, Guo B (2019) Thermo-sensitive micelles based on amphiphilic poly (butylene 2-methylsuccinate)-poly (ethylene glycol) multi-block copolyesters as the pesticide carriers. Col Surf A Physicochem Eng Asp 575:84–93. https://doi.org/10.1016/j.colsurfa.2019.04.057

Hassannejad S, Ghafarbi S (2013) Allelopathic effects of some Lamiaceae on seed germination and seedling growth of dodder (*Cuscuta campestris Yunck.*). Int J Biosci 3:9–14. https://doi.org/10.12692/ijb/3.3.9-14

Hazrati H, Saharkhiz MJ, Niakousari M, Moein M (2017) Natural herbicide activity of *Satureja hortensis L.* essential oil nanoemulsion on the seed germination and morphophysiological features of two important weed species. Ecotoxicol Environ Saf 142:423–430. https://doi.org/10.1016/j.ecoenv.2017.04.041

Iqbal M, Umar S (2019) Nano-fertilization to enhance nutrient use efficiency and productivity of crop plants. In: Husen A, Iqbal M (eds) Nanomaterials and plant potential. Springer, Cham, pp 473–505. https://doi.org/10.1016/j.coesh.2018.07.009

Kah M, Beulke S, Tiede K, Hofmann T (2013) Nanopesticides: state of knowledge, environmental fate, and exposure modeling. Critical Rev Environ Sci Technol 43:1823–1867. https://doi.org/10.1080/10643389.2012.671750

Kah M, Weniger A-K, Hofmann T (2016) Impacts of (nano) formulations on the fate of an insecticide in soil and consequences for environmental exposure assessment. Environ Sci Technol 50:10960–10967. https://doi.org/10.1071/EN19057

Kalteh M, Alipour ZT, Ashraf S, Marashi Aliabadi M, Falah Nosratabadi A (2018) Effect of silica nanoparticles on basil (*Ocimum basilicum*) under salinity stress. J Chem Health Risks 4:49–55. https://doi.org/10.22034/jchr.2018.544075

Kaur PJS, Duhan JS, Thakur R (2018) Comparative pot studies of chitosan and chitosan-metal nanocomposites as nano-agrochemicals against fusarium wilt of chickpea (*Cicer arietinum L.*). Biocatal Agri Biotechnol 14:466–471. https://doi.org/10.1016/j.bcab.2018.04.014

Kaziem AE, Gao Y, He S, Li J (2017) Synthesis and insecticidal activity of enzyme-triggered functionalized hollow mesoporous silica for controlled release. J Agri Food Chem 65:7854–7864. https://doi.org/10.1021/acs.jafc.7b02560

Khalifa NS, Hasaneen MN (2018) The effect of chitosan–PMAA–NPK nanofertilizer on *Pisum sativum* plants. Biotech 8:193. https://doi.org/10.1007/s13205-018-1221-3

Khare P, Srivastava S, Nigam N, Singh A, Singh S (2019) Impact of essential oils of *E. citriodora*, *O. basilicum* and *M. arvensis* on three different weeds and soil microbial activities. Environ Technol Innov 14:100343. https://doi.org/10.1016/j.eti.2019.100343

Khatem R, Celis R, Hermosín MC (2019) Cationic and anionic clay nanoformulations of imazamox for minimizing environmental risk. Appl Clay Sci 168:106–115. https://doi.org/10.1016/j.clay.2018.10.014

Kim DY, Kadam A, Shinde S, Saratale RG, Patra J, Ghodake G (2018) Recent developments in nanotechnology transforming the agricultural sector: a transition replete with opportunities. J Sci Food Agri 98:849–864. https://doi.org/10.1002/jsfa.8749

Kottegoda N, Sandaruwan C, Priyadarshana G, Siriwardhana A, Rathnayake UA, Berugoda Arachchige DM, Kumarasinghe AR, Dahanayake D, Karunaratne V, Amaratunga GA (2017) Urea-hydroxyapatite nanohybrids for slow release of nitrogen. ACS Nano 11:1214–1221. https://doi.org/10.1021/acsnano.6b07781

Kumar S, Chauhan N, Gopal M, Kumar R, Dilbaghi N (2015) Development and evaluation of alginate–chitosan nanocapsules for controlled release of acetamiprid. Int J Biol Macromol 81:631–637. https://doi.org/10.1016/j.ijbiomac.2015.08.062

Kumar S, Nehra M, Dilbaghi N, Marrazza G, Hassan AA, Kim KH (2019) Nano-based smart pesticide formulations: emerging opportunities for agriculture. J Control Release 294:131–153. https://doi.org/10.1016/j.jconrel.2018.12.012

Liu J, Williams PC, Goodson BM, Geisler-Lee J, Fakharifar M, Gemeinhardt ME (2019) $TiO_2$ nanoparticles in irrigation water mitigate impacts of aged Ag nanoparticles on soil microorganisms, *Arabidopsis thaliana* plants, and *Eisenia fetida* earthworms. Environ Res 172:202–215. https://doi.org/10.1016/j.envres.2019.02.010

Ma C, White JC, Zhao J, Zhao Q, Xing B (2018) Uptake of engineered nanoparticles by food crops: characterization, mechanisms, and implications. Annu Rev Food Sci Technol 9:129–153. https://doi.org/10.1146/annurev-food-030117

Malaikozhundan B, Vinodhini J (2018) Nanopesticidal effects of *Pongamia pinnata* leaf extract coated zinc oxide nanoparticle against the Pulse beetle, *Callosobruchus maculatus*. Mat Today Com 14:106–115. https://doi.org/10.1016/j.mtcomm.2017.12.015

Maliyekkal SM, Sreeprasad T, Krishnan D, Kouser S, Mishra AK, Waghmare UV, Pradeep T (2013) Graphene: a reusable substrate for unprecedented adsorption of pesticides. Small 9:273–283. https://doi.org/10.1002/smll.201201125

Manikandan A, Subramanian K (2016) Evaluation of zeolite based nitrogen nano-fertilizers on maize growth, yield and quality on inceptisols and alfisols. Int J Plant Soil Sci 9:1–9. https://doi.org/10.9734/IJPSS/2016/22103

Mattos BD, Magalhães WLE (2016) Biogenic nanosilica blended by nanofibrillated cellulose as support for slow-release of tebuconazole. J Nanopart Res 18:274. https://doi.org/10.1007/s11051-016-3586-8

Mazzaglia A, Fortunati E, Kenny JM, Torre L, Balestra GM (2017) Nanomaterials in plant protection. In: Axelos MA, Van de Voorde MH (eds) Nanotechnology in agriculture and food science. Wiley, Weinheim, pp 115–133. https://doi.org/10.1002/9783527697724.ch7

Moradi FG, Hejazi MJ, Hamishehkar H, Enayati AA (2019) Co-encapsulation of imidacloprid and lambda-cyhalothrin using biocompatible nanocarriers: characterization and application. Ecotoxicol Environ Saf 175:155–163. https://doi.org/10.1016/j.ecoenv.2019.02.092

Nair PMG, Chung IM (2014) Impact of copper oxide nanoparticles exposure on Arabidopsis thaliana growth, root system development, root lignificaion, and molecular level changes. Environ Sci Pollut Res 21:12709–12722. https://doi.org/10.1007/s11356-014-3210-3

Pascoli M, Jacques MT, Agarrayua DA, Avila DS, Lima R, Fraceto LF (2019) Neem oil based nanopesticide as an environmentally-friendly formulation for applications in sustainable agriculture: an ecotoxicological perspective. Sci Total Environ 677:57–67. https://doi.org/10.1016/j.scitotenv.2019.04.345

Paulraj MG, Ignacimuthu S, Gandhi MR, Shajahan A, Ganesan P, Packiam SM, Al-Dhabi NA (2017) Comparative studies of tripolyphosphate and glutaraldehyde cross-linked chitosan-botanical pesticide nanoparticles and their agricultural applications. Int J Biol Macromol 104:1813–1819. https://doi.org/10.1016/j.ijbiomac.2017.06.043

Peters RJ, Bouwmeester H, Gottardo S, Amenta V, Arena M, Brandhoff P, Marvin HJP, Mech A, Botelho Moniz F, Quiros Pesudo L, Rauscher H, Schoonjans R, Undas AK, Vittoria Vettori M, Weigel S, Aschberger K (2016) Nanomaterials for products and application in agriculture, feed and food. Trends Food Sci Technol 54:155–164. https://doi.org/10.1016/j.tifs.2016.06.008

Pulimi M, Subramanian S (2016) Nanomaterials for soil fertilisation and contaminant removal. In: Ranjan S, Dasgupta N, Licht FE (eds) Nanoscience in food and agriculture sustainable agriculture reviews. Springer, Cham, pp 229–246. https://doi.org/10.1007/978-3-319-39303-2_8

Raliya R, Nair R, Chavalmane S, Wang W-N, Biswas P (2015) Mechanistic evaluation of translocation and physiological impact of titanium dioxide and zinc oxide nanoparticles on the tomato (*Solanum lycopersicum L.*) plant. Metallomics 7:1584–1594. https://doi.org/10.1039/C5MT00168D

Rashidipour M, Maleki A, Kordi S, Birjandi M, Pajohi N, Mohamadi E, Heydari R, Rezaee R, Rasoulian B, Davari B (2019) Pectin/chitosan/Tripolyphosphate nanoparticles: efficient carriers for reducing soil sorption, cytotoxicity and mutagenicity of Paraquat and enhancement of its herbicide activity. J Agr Food Chem 67:5736–5745. https://doi.org/10.1021/acs.jafc.9b01106

Rauscher H, Rasmussen K, Sokull-Klüttgen B (2017) Regulatory aspects of nanomaterials in the EU. Chemie Ingenieur Technik 89:224–231. https://doi.org/10.1002/cite.201600076

Roco MC (2011) The long view of nanotechnology development: the National Nanotechnology Initiative at 10 years. J Nanopart Res 13:427–445. https://doi.org/10.1007/s11051-010-0192-z

Salač J, Šopík T, Stloukal P, Janásová N, Jursík M, Koutný M (2019) Slow release formulation of herbicide metazachlor based on high molecular weight poly (lactic acid) submicro and microparticles. Int J Environ Sci Technol 16(15). https://doi.org/10.1007/s13762-019-02222-9

Sangeetha J, Thangadurai D, Hospet R, Harish ER, Purushotham P, Mujeeb MA, Shrinivas J, David M, Mundaragi AC, Thimmappa SC, Arakera SB (2017) Nanoagrotechnology for soil quality, crop performance and environmental management. In: Prasad R, Kumar M, Kumar V (eds) Nanotechnology. Springer, Singapore, pp 73–97. https://doi.org/10.1007/978-981-10-4573-8_5

Sarkar DJ, Singh A (2017) Base triggered release of insecticide from bentonite reinforced citric acid crosslinked carboxymethyl cellulose hydrogel composites. Carbohyd Polym 156:303–311. https://doi.org/10.1016/j.carbpol.2016.09.045

Sarlak N, Taherifar A (2017) Encapsulation of nanomaterials and production of Nanofertilizers and Nanopesticides: insecticides for Agri-food production and plant disease treatment. In: Ghorbanpour M, Manika K, Varma A (eds) Nanoscience and Plant–Soil Systems. Soil Biology. Springer, Cham, pp 481–498. https://doi.org/10.1007/978-3-319-46835-8_18

Sarlak N, Taherifar A, Salehi F (2014) Synthesis of nanopesticides by encapsulating pesticide nanoparticles using functionalized carbon nanotubes and application of new nanocomposite for plant disease treatment. J Agri Food Chem 62:4833–4838. https://doi.org/10.1021/jf404720d

Saruchi KBS, Jindal R, Kapur GS, Kumar V (2014) Synthesis, characterization and evaluation of *Gum tragacanth* and acrylic acid based hydrogel for sustained calcium chloride release–enhancement of water-holding capacity of soil. J Chinese Adv Mat Soc 2:40–52. https://doi.org/10.1080/22243682.2014.893412

Schnoor B, Elhendawy A, Joseph S, Putman M, Chacón-Cerdas R, Flores-Mora D, Bravo-Moraga F, Gonzalez-Nilo F, Salvador-Morales C (2018) Engineering atrazine loaded poly (lactic-co-glycolic acid) nanoparticles to ameliorate environmental challenges. J Agri Food Chem 66:7889–7898. https://doi.org/10.1021/acs.jafc.8b01911

Shang Y, Hasan M, Ahammed GJ, Li M, Yin H, Zhou J (2019) Applications of nanotechnology in plant growth and crop protection: a review. Molecules 24:2558. https://doi.org/10.3390/molecules24142558

Shukla P, Chaurasia P, Younis K, Qadri OS, Faridi SA, Srivastava G (2019) Nanotechnology in sustainable agriculture: studies from seed priming to post-harvest management. Nanotechnol Environ Eng 4:11. https://doi.org/10.1007/s41204-019-0058-2

Silva LP, Bonatto CC (2019) Green nanotechnology for sustained release of eco-friendly agrochemicals. In: Vaz S Jr (ed) Sustainable Agrochemistry. Springer, Cham, pp 113–129. https://doi.org/10.1016/j.jclepro.2016.11.117

Simonin M, Martins JM, Le Roux X, Uzu G, Calas A, Richaume A (2017) Toxicity of $TiO_2$ nanoparticles on soil nitrification at environmentally relevant concentrations: lack of classical dose–response relationships. Nanotoxicology 11:247–255. https://doi.org/10.1080/17435390.2017.1290845

Simonin M, Colman BP, Tang W, Judy JD, Anderson SM, Bergemann CM, Rocca JD, Unrine JM, Cassar N, Bernhardt ES (2018a) Plant and microbial responses to repeated $Cu(OH)_2$ nanopesticide exposures under different fertilization levels in an agro-ecosystem. Front Microbiol 9:1769. https://doi.org/10.3389/fmicb.2018.01769

Simonin M, Colman BP, Anderson SM, King RS, Ruis MT, Avellan A, Bergemann CM, Perrotta BG, Geitner NK, Ho M, de la Barrera B (2018b) Engineered nanoparticles interact with nutrients to intensify eutrophication in a wetland ecosystem experiment. Eco Appl 28:1435–1449. https://doi.org/10.1002/eap.1742

Singh P, Singh R, Borthakur A, Srivastava P, Srivastava N, Tiwary D, Mishra PK (2016) Effect of nanoscale $TiO_2$-activated carbon composite on *Solanum lycopersicum (L.)* and *Vigna radiata (L.)* seeds germination. Energy Ecol Environ 1:131–140. https://doi.org/10.1007/s40974-016-0009-8

Singh R, Srivastava P, Singh P, Upadhyay S, Raghubanshi AS (2019) Human overpopulation and food security: challenges for the agriculture sustainability. In: Urban agriculture and food systems: breakthroughs in research and practice. IGI Global, pp 439–467. https://doi.org/10.4018/978-1-5225-1683-5.ch002

Singhal NK, Singh V, Kaushal S (2018) Applications of nanoagrochemicals. NanoAgroceuticals & Nano PhytoChemicals, In: Singh B, Katare OP, Souto EB (eds) Emerging trends in nanobiomedicine, Taylor and Francis, pp 1–15. https://doi.org/10.1201/9781351139281-1

Soliman AS, El-feky SA, Darwish E (2015) Alleviation of salt stress on *Moringa peregrina* using foliar application of nanofertilizers. J Hortic For 7:36–47. https://doi.org/10.5897/JHF2014.0379

Sousa GF, Gomes DG, Campos EV, Oliveira JL, Fraceto LF, Stolf-Moreira R, Oliveira HC (2018) Post-emergence herbicidal activity of nanoatrazine against susceptible weeds. Front Environ Sci 6:1–6. https://doi.org/10.3389/fenvs.2018.00012

Subramanian KS, Manikandan A, Thirunavukkarasu M, Rahale CS (2015) Nano-fertilizers for balanced crop nutrition. In: Rai M, Ribeiro C, Mattoso L, Duran N (eds) Nanotechnologies in food and agriculture. Springer, pp 69–80. https://doi.org/10.1007/978-3-319-14024-7_3

Sun C, Shu K, Wang W, Ye Z, Liu T, Gao Y, Zheng H, He G, Yin Y (2014) Encapsulation and controlled release of hydrophilic pesticide in shell cross-linked nanocapsules containing aqueous core. Int J Pharma 463:108–114. https://doi.org/10.1016/j.ijpharm.2013.12.050

Tang C, Li Y, Pun J, Osman ASM, Tam KC (2019) Polydopamine microcapsules from cellulose nanocrystal stabilized Pickering emulsions for essential oil and pesticide encapsulation. Col Surf A Physicochem Eng Asp 570:403–413. https://doi.org/10.1016/j.colsurfa.2019.03.049

Tejada M, Morillo E, Gómez I, Madrid F, Undabeytia T (2017) Effect of controlled release formulations of diuron and alachlor herbicides on the biochemical activity of agricultural soils. J Hazard Mat 322:334–347. https://doi.org/10.1016/j.jhazmat.2016.10.002

Tomadoni B, Casalongué C, Alvarez VA (2019) Biopolymer-based hydrogels for agriculture applications: swelling behavior and slow release of agrochemicals. In: Gutiérrez T (ed) Polymers for Agri-food applications. Springer, Cham pp, pp 99–125. https://doi.org/10.1163/156855502760151580

Tripathi DK, Singh S, Singh S, Pandey R, Singh VP, Sharma NC, Prasad SM, Dubey NK, Chauhan DK (2017) An overview on manufactured nanoparticles in plants: uptake, translocation, accumulation and phytotoxicity. Plant Physiol Biochem 110:2–12. https://doi.org/10.1016/j.plaphy.2016.07.030

VandeVoort A, Arai Y (2018) Macroscopic observation of soil nitrification kinetics impacted by copper nanoparticles: implications for micronutrient Nanofertilizer. Nano 8:927. https://doi.org/10.3390/nano8110927

Wang Y, Wang A, Wang C, Cui B, Sun C, Zhao X, Zeng Z, Shen Y, Gao F, Liu G (2017) Synthesis and characterization of emamectin-benzoate slow-release microspheres with different surfactants. Sci Rep 7:12761. https://doi.org/10.1038/s41598-017-12724-6

Wanyika H (2013) Sustained release of fungicide metalaxyl by mesoporous silica nanospheres. In: Diallo MS, Fromer NA, Jhon MS (eds) Nanotechnology for sustainable development. Springer, pp 321–329. https://doi.org/10.1007/978-3-319-05041-6_25

Wibowo D, Zhao C-X, Peters BC, Middelberg AP (2014) Sustained release of fipronil insecticide *in-vitro* and *in-vivo* from biocompatible silica nanocapsules. J Agri Food Chem 62:12504–12511. https://doi.org/10.1021/jf504455x

Xu X, Bai B, Wang H, Suo Y (2017) A near-infrared and temperature-responsive pesticide release platform through core–shell Polydopamine@ PNIPAm Nanocomposites. ACS Appl Mat Inter 9:6424–6432. https://doi.org/10.1021/acsami.6b11043

Zafar H, Ali A, Ali JS, Haq IU, Zia M (2016) Effect of ZnO nanoparticles on *Brassica nigra* seedlings and stem explants: growth dynamics and antioxidative response. Front Plant Sci 7:535. https://doi.org/10.3389/fpls.2016.00535

Zhang Y, Liang X, Yang X, Liu H, Yao J (2014) An eco-friendly slow-release urea fertilizer based on waste mulberry branches for potential agriculture and horticulture applications. ACS Sustain Chem Eng 2:1871–1878. https://doi.org/10.1021/sc500204z

Zhang Y, Chen W, Jing M, Liu S, Feng J, Wu H, Zhou Y, Zhang X, Ma Z (2019) Self-assembled mixed micelle loaded with natural pyrethrins as an intelligent nano-insecticide with a novel temperature-responsive release mode. Chem Eng J 361:1381–1391. https://doi.org/10.1016/j.cej.2018.10.132

Zhao X, Cui H, Wang Y, Sun C, Cui B, Zeng Z (2017) Development strategies and prospects of nano-based smart pesticide formulation. J Agri Food Chem 66:6504–6512. https://doi.org/10.1021/acs.jafc.7b02004

Zhou T, Wang Y, Huang S, Zhao Y (2018) Synthesis composite hydrogels from inorganic-organic hybrids based on leftover rice for environment-friendly controlled-release urea fertilizers. Sci Total Environ 615:422–430. https://doi.org/10.1016/j.scitotenv.2017.09.084

# Part III
# Transport of Emerging Contaminants in Agroecosystems

# Chapter 10
# Transport of Emerging Contaminants from Agricultural Soil to Groundwater

**Khalid Muzamil Gani, Muntjeer Ali, Monika Dubey, Absar Ahmad Kazmi, Sheena Kumari, and Faizal Bux**

**Abstract** A wide range of emerging contaminants such as pharmaceuticals, personal care products, flame retardants, synthetic hormones and food additives etc., which were used by society in recent past for various purposes but recently were quantified are becoming a concern for human and animal health care. These compounds are transported in various levels of food chain. In last few decades, emerging contaminants are getting more attention due to their potential impact on human, aquatic and animal health. An essential strategy of wastewater treatment is sewage sludge treatment and application of sludge as manure to enhance soil nutrient levels. However, the incomplete removal of emerging contaminants in wastewater and sorbed portion of contaminants got into the sludge that is applied to soil. There is high probability that these compounds will leach into the groundwater resulting in its pollution.

So far, there has been significant amount of work to understand the transport of emerging contaminants from agriculture soil to groundwater. In this chapter, we would discuss how these compounds are transferred to the groundwater and what factors influence their fate and transport in the agricultural soil. Sources of emerging contaminants to the agriculture system and their uptake into biota are also discussed. The discussion would lead us to have closer look at the need to have regulations for the release of contaminants into the environment, their effect on human, animal and plant health and their fate and transport in the agroecosystem.

**Keywords** Agriculture · Biodegradation · Emerging contaminants · Fate and transport · Groundwater · Sorption · Sources and pathways

---

K. M. Gani (✉) · S. Kumari · F. Bux
Institute for Water and Wastewater Technology, Durban University of Technology, Durban, South Africa
e-mail: khalidmzml@gmail.com

M. Ali · M. Dubey · A. A. Kazmi
Environmental Engineering Group, Department of Civil Engineering, Indian Institute of Technology Roorkee, Roorkee, Uttrakhand, India



V. Kumar Singh et al. (eds.), *Sustainable Agriculture Reviews 50*, Sustainable Agriculture Reviews 50, https://doi.org/10.1007/978-3-030-63249-6_10

## 10.1 Introduction

Requirement of freshwater has put stress on the water resources around the world. Agriculture sector consumes 70% of the total fresh water consumption around the globe. Thus, use of reclaimed wastewater for irrigation purpose is a cost effective solution (European Commission 2018). In water stressed areas such as California and Israel, reuse of treated wastewater for agricultural purpose is estimated to increase 2–3 folds (Wu et al. 2015). Sludge produced during the wastewater treatment have also been used as a manure from very long period. About 40% of the sludge produced in the wastewater treatment plants in Europe is used as manure for crops (Peterson and Staples 2003). All 50 states of USA uses around 50% of the biosolids produced in wastewater treatment plants. However, frequent detection of contaminants have invited concerns about using effluent water and sludge from the treatment plants in the agricultural sector.

Compounds such as pharmaceuticals, steroids, personal care products, flame-retardants, and pesticides are termed as emerging contaminants. As per United States Geological Survey, "Emerging contaminants are any synthetic or naturally occurring chemical or any microorganism that is not commonly monitored in the environment but has the potential to enter the environment and cause known or suspected adverse ecological and or Human Health effects" (USGS 2016). The fate of the organic contaminant in the wastewater treatment plants depends on the nature of the contaminant and the type of treatment process. They can be volatilized, degraded aerobically or anaerobically, adsorbed onto the sludge or discharged along with the effluent (Zuloaga et al. 2012). The contaminants associated with treated wastewater and digested sludge is from the wastewater treatment plants where they either are partially or not at all removed or are transformed into their metabolites. Emerging contaminants have frequently been detected in agriculture soil irrigated with treated wastewater, run-off from such fields, soil amended with bio-solids, and the groundwater of the area. This is the reason that emerging contaminants in the environment has garnered the interest of government and researchers across the globe. Majority of Emerging contaminants in the environment are found in parts per trillion (ppt) or parts per billion (ppb). Consequently, key challenge libels in the quantification of emerging contaminants at environmentally at this concentrations. Furthermore, for many emerging contaminants, reference standards are not available, which needs for the validation of analytical methods in the sophisticated instruments. Notwithstanding, these practical challenges, detection and quantification of emerging contaminants at trace environmental concentrations, had improved in the last two decades.

In recent years, studies on transport of emerging contaminants from treated wastewater and sludge to soil and finally to the crop and their bioaccumulation potential in the animals and human beings have been done. Noguera-oviedo and Aga (2016) studied the activity of free estrogens and their conjugates and reported no change in the concentration after the anaerobic digestion. The conjugates were detected at very less concentration and almost all of it converted into deconjugated form after digestion. Biodegradation is a possible removal mechanism for the degradation of emerging contaminants in the agricultural soil under aerobic and anaerobic conditions due to the presence of microorganisms and oxygen. A study on occurrence of 373 compounds such as pharmaceuticals, pesticides, poly-aromatic hydrocarbons, etc. from 2009 to 2011 indicated the presence of emerging contaminants in the Guadalquiver river, south cast of Spain. The river water quality was found to be heavily impacted by agricultural run-off into the river (Robles-molina et al. 2014). González et al. (2019) reported that, 82% carbamazepine was removed when phloem transport in the treated. For ibuprofen, ketoprofen, naproxen being weak acids, the concentration was predicted to be higher in the roots than in the leaves due to phloem transport downward, however, it was below the detection limit. The dietary consumption of these leafy vegetables was much less than the recommended therapeutic doses. The sorption of some compounds onto the soil results in the formation of bound residues which are thus much less available for biodegradation. Li et al. (2013) and Li et al. (2014a) compared the behaviour of acetaminophen and carbamazepine and found that carbamazepine was very less available (< 4.2%) as bond residue as compared to acetaminophen (73.4–93.3%) under similar conditions. The freely available or unbound residues are available to plant uptake or migration. The free contaminant can leach through the soil into the aquifers. The sources and fate of emerging contaminants in the agroecosystem, will be discussed further in this chapter with an overview of the factors responsible for the fate to understand the response of emerging contaminants towards various degradation/mitigation pathways like sorption, biodegradation etc. One of the key concerns is the leaching of emerging contaminants into the ground water. Factors responsible for this movement and the key findings reported till date will be discussed.

## 10.2 Sources of Emerging Contaminants in the Agricultural Field

After being used by the humans and through the industrial discharge, the wastewater gets contaminated with emerging contaminants. During its treatment in the conventional wastewater treatment plants, the contaminants are not completely removed and find its way in the effluent. The treated wastewater when used for irrigation purpose poses a risk of contamination of the agro-ecosystem with these contaminants. Some of the compounds depending on their chemical properties can be adsorbed onto the sludge. The sludge is generally anaerobically treated to decrease the volume and to obtain the biogas. Some of the contaminants may still be present adsorbed onto the sludge. Hydrophobic compounds such as Bisphenol A, Surfynol 104 with log $K_{ow}$ value more than 3, have higher water partition coefficient ($K_d$) values and tend to adsorb more onto the hydrophilic compounds (Hurtado and Montano-chávez 2017). Emerging contaminants in soil and biota may have potential harmful effects on the ecosystem. In some cases where the parent compound is unstable, the metabolite of emerging contaminants can be more toxic to the environment. Irrigation with emerging contaminants contaminated water leads to accumulation of emerging contaminants in the soil. Kinney et al. (2006) in his study found the 305–4060 and 274–1260 folds more concentration of erythromycin and carbamazepine in the soil than the treated wastewater used for irrigation carried over by previous irrigation events. In a study by Peterson and Staples (2003), the concentration of Di-2-ethylhexyl phthalate (DEHP), a widely used plasticizer in sludge from a conventional activated sludge based treatment plant was found to be 153 mg/kg when the influent and effluent concentration of DEHP was 1.6 and 0.03 μg/L.

Application of cow dung as manure is also found to be one of the prime sources of emerging contaminants into the agricultural soil. Evidence of antibiotic resistance bacteria have been found in the agricultural soil. Wichmann et al. (2014) investigated the Antibiotic Resistant Genes in dairy cow manure associated with antibiotics like chloramphenicol, kanamycin, tetracycline, and β-lactam. The research showed that the cow manure can contain numerous antibiotic resistant genes and can be transferred to the agricultural soil when it comes in direct contact with the soil. Fahrenfeld et al. (2014) concluded that fate of antibiotic resistant genes in the soil is dependent not only on the use of antibiotics but also on the degradation of DNA and death of bacterial host. Application and handling of manure is one of the prime reasons for the distribution of antibiotics in the agricultural soil (Durso and Cook 2014). Pesticides and pharmaceuticals (veterinary medicines) are the main source of application of emerging contaminants in the agricultural soil. In the soil about 95% veterinary medicines directly released through urine and feces of animals (Chen and Xia 2017).

Recently newly emerging contaminants such as Nano-scale pesticides (i.e. 1–100 nm) are deliberately added to the agricultural land (Scrinis and Lyons 2009).

## 10.3 Fate of Emerging Contaminants in Subsurface (Agricultural System)

The behavior of fate and transport of emerging contaminants when added in the agricultural environment are found to be same as other class of contaminants. Emerging contaminants might be degraded by biologically, physically, chemically or stick up to soil particles, adsorption by plants, percolate to groundwater, conveyance to surface waters and overflow to drainage water. Table 10.1 list out selected emerging contaminants from different compound categories that have been reported in groundwater globally. The degree to which any of these processes happens will depend on the fundamental physical properties of the emerging contaminants (including water solubility; Fraction of emerging contaminants attached to the organic matter and other soil components; and volatility of emerging contaminants) as well as the properties of the soil and the climatic conditions.

**Table 10.1** List of selected emerging contaminants in groundwater having concentration > 500 ng/L

| Compound | Concentration (ng/L) | | |
|---|---|---|---|
| | Min | Mean | Max |
| *Pharmaceuticals* | | | |
| Paracetamol | 15 | 15,142 | 120,000 |
| Oxybenzone | 2580 | 36,490 | 70,400 |
| Carbamazepine | 1.64 | 4858 | 99,194 |
| Primidone | 110 | 3380 | 12,000 |
| Ibuprofen | 0.6 | 1386 | 12,000 |
| Clofibric acid | 4 | 988 | 7300 |
| Ketoprofen | 3 | 611 | 2886 |
| Lopamidol | 40 | 527 | 2400 |
| Propyphenazone | 15 | 553 | 1250 |
| DEET | 454 | 2251 | 6500 |
| Phenazone | 25 | 1503 | 3950 |
| Naproxen | 0.2 | 1867 | 5580 |
| Erythromycin | 49 | 1046 | 2340 |
| Gemfibrozil | 45 | 782 | 1960 |
| AMDOPH* | 330 | 927 | 1250 |
| Oxybenzone | 2580 | 36,490 | 70,400 |
| Fenofibrate | 45 | 3673 | 7300 |
| Crotamiton | 6.8 | 1953 | 3900 |
| Monensin | 390 | 1370 | 2350 |
| 1_2-Dichloropropane | 440 | 620 | 800 |
| *Life-style compounds* | | | |
| Caffeine | 13 | 9032 | 110,000 |
| Nicotine | 50 | 4060 | 8070 |
| Cotinine | 60 | 173 | 400 |
| *Industrial compounds* | | | |
| bis(2-ethylhexyl) phthalate | 1200 | 125,600 | 250,000 |
| Nonylphenol-1-carboxylic acid | 260 | 17,630 | 35,000 |
| Bisphenol A | 61 | 2786 | 9300 |
| Nonylphenol | 1000 | 16,258 | 84,000 |
| Tri(2-chloroethyl) phosphate | 495 | 656 | 740 |
| Nonylphenol-1-carboxylic acid | 260 | 17,630 | 35,000 |
| 4-Nonyphenol-monoethoxylate* | 500 | 3750 | 7000 |
| 5-methyl-1H-benzotriazole | 516 | 1298 | 2080 |
| *Preservatives and food additives* | | | |
| Acesulfame | 70 | 2385 | 4700 |
| *Endocrine disruptors* | | | |
| Cholesterol | 44 | 887 | 1730 |
| Coprostanol | 74 | 682 | 1290 |

Adapted from Lapworth et al. (2012)
*Dioxypyramidon

Emerging contaminants accumulate much more in soil with high organic matter in it where they may persist for years and may travel through the soil to reach the groundwater (Gomes et al. 2017). Percolation of contaminant with water through the soil depends on the soil water partition coefficient $K_d$. The Octanol Water Partition coefficient, $K_{ow}$ is the ratio of concentration of contaminant in the octanol phase to the concentration in aqueous phase in a two phase Octanol-Water system. This parameter represents the ability of a contaminant to partition between soil and the aqueous phase. All compounds with $K_d > 1000$ and $K_{ow} > 3.7$ are immobilized in the soil (Duarte-Davidson and Jones 1996). Borgman and Chefetz (2013) concluded that application of treated wastewater for irrigation purposes leads to an increase in the pH of soil which results in increased mobility of weakly acidic compounds such as non-steroidal anti-inflammatory drugs such as Naproxen and Diclofenac.

The dewatered sludge contaminated with emerging contaminants do not leach much through the soil (Topp et al. 2010) unless it is irrigated or there is an episode of rain (Blackwell et al. 2009). In case of extreme rainfall, the concentration of emerging contaminants may increase many folds. Triclosan was observed in the leachate 33 years after the application of sludge on the field (Xia et al. 2010).

Leaching of emerging contaminants from agriculture have been reported by Bondarenko et al. (2012). Leachate containing Sulfamethoxazole, Carbamazepine, Trimethoprim, etc. has been reported which implies high mobility of such contaminants in soil matrix. However, the mobility may get affected with the presence of dissolved organic matter, which in turn increases the sorption capacity of the soil, thus, immobilizing the contaminants in the soil. Emerging contaminants were detected in 47 sites in a nationwide monitoring program to find the concentration of emerging contaminants in groundwater where 35 analytes out of a total of 65 were detected at least once. The most common contaminants detected was highly water soluble antibiotic pharmaceutical compound Sulfamethoxazole (Barnes et al. 2008).

In general, water soluble compounds have shown high mobility with leachate though the soil. However, less water soluble compounds sorbed on the suspended particles can leach as well. Models to predict the transport of emerging contaminants through the soil under various environmental conditions need to be made and validated.

### *10.3.1 Sorption in the Soil*

Sorption is the process, any chemical or particles associated with soil particles by attaching to the surface of the soil surface or moving inside the soil particles. It is a very important factor for determining whether a chemical or ECs is going to move to surface water or groundwater or not and affects the degradation of chemicals or contaminants. The understanding of sorption process of pesticides and POPs in the soil is well known due to significant research happened from the decades. These contaminants typically interact with the organic carbon in the soil and sorption behavior can be predicted from knowledge of the hydrophobicity (measurement of the attractiveness of the contaminant to organic matter) of the contaminants. Many Equations have been developed for estimating the sorption performance of contaminants in soils based on the hydrophobicity of the chemicals and the organic matter of the soil. These equations are also used for environmental exposure modelling and environmental risk assessment. While, these equations are doubtless valid for many emerging contaminants, many other emerging contaminants, including the human and veterinary medicines and nanomaterials, materialize to behave differently.

The sorption behavior of emerging contaminants can varies in different types of soils and these differences in sorption of emerging contaminants in different soils cannot be explained by variations in soil organic carbon but are explained that many pharmaceuticals are polar that they can exist in unionized and ionized form in the natural environments (Ter Laak et al. 2006a, b). Consequently, the chemicals not only interact with the organic carbon in the soil but also with interacted metal oxides and clay particles (Ter Laak et al. 2006b) and the sorption behavior depends on the soil properties like pH. The intricacies of these interactions mean that modelling approaches, developed for predicting the sorption of other groups of compounds are unsuitable for use in many anthropoid and veterinary medicines. Therefore, abundant precaution to be taken when using existing modelling and risk assessment approaches, for emerging contaminants.

Behavior and transport of emerging contaminants may also alter when emerging contaminants will also enter the agricultural environment accompanying with manure and slurry mediums. Studies have demonstrated that the addition of manure or sludge can affect the sorption behavior of veterinary and human medicines and personal care products and that they may also affect perseverance (e.g. Boxall et al. 2002; Thiele-Bruhn and Aust 2004; Monteiro and Boxall 2009; Al Rajab et al. 2009). In some cases, the addition of the sludge or compost increases sorption and persistence and in other cases, they decrease sorption and persistence. These effects have been credited to changes in pH or changes of the nature of dissolved organic carbon in the soil/compost system.

Briones and Sarmah (2018) studied the sorption behaviour of antidiabetic drug metformin and its metabolite guanylurea in the agricultural soil and 90% sorption within 4 h and 13 h was observed, respectively. The study established that sorption of these contaminants in the soil depends on the soil characteristics (types of minerals and ion exchange capacity).

Presence of soil organic matter decreases the mobility of the emerging contaminants or increases the sorption potential of the soil (Arye et al. 2011). Application of biosolids on the soils increases the soil organic matter content, and thus, increases the sorption potential of soil for various emerging contaminants. In case of contaminated wastewater irrigation, the mobility of acidic emerging contaminants like Diclofenac and Naproxen increases primarily due to increase in pH of water it is being irrigated with (approx. 8.4) and not because of the complexes formed with the organic content in the wastewater (Borgman and Chefetz 2013).

Studies have explored the interaction of engineered nanoparticles with soil particles (e.g. Li et al. 2008). However, based on information for other processes, behavior of these materials will be different from non-particulate contaminants and that new models and standards will need to be developed for engineered nanoparticles in the soil environment.

### 10.3.2 Prolonged Existence to the Soil

The determination of emerging contaminants may increase in the presence of sludge. Mineralization of ECs, diuron, glyphosate, and nonylphenol was found to be 40–80 times lower in the compost amended soil than in the contaminated (Ghanem et al. 2006). Hormone such as 17 β-estradiol was found to be stabilized as non-extractable residue in the soil amended with spiked sewage compost (Dubroca et al. 2009). Aerobic degradation (oxygen rich environment) is the main pathway of emerging contaminants corresponding to other pollutants. Degradation and reduction pathway may be occurring such as photolysis and hydrolysis, which depends on characteristics of pollutants. Data available on persistence of emerging contaminants in the soil only for human and veterinary medicines. Data needs to be generated for emerging contaminants such as pharmaceuticals, personal care products, pesticides and their transformation by-products. Subsequently, it will be useful for environmental risk assessment and regulations.

Biosolids or sludge matrix showed good sorption and affect degradation rates compared to soil matrix. For example, Degradation rate of pharmaceuticals such as caffeine (which comes from coffee) increased in soil when soil amended with aerobically digested sludge. In dissimilarity, Caffeine mineralization was not enhanced in presence of anaerobically treated sludge in the soil (Topp et al. 2006). Furthermore, Anti-inflammatory drug such as Neproxen degraded with faster rate by the addition of activated sludge (Topp et al. 2008a).

In the laboratories, researchers generally look at the complete mineralization of emerging contaminants by using sophisticated technologies and identified or detected the metabolites (transformation products) by using analytical extraction and detection techniques. Kolz et al. (2005) and Topp et al. (2008a) was also reported the degradation products of Emerging contaminants for naproxen and hormones like estrone and 17β-estradiol. Although, some of the compounds are significantly mineralized, but others are converted to their metabolites. Nevertheless, during the treatment parent compounds completely absent when it was checked by analytically. This might be due to that the parent emerging contaminant is binding very tightly to the soil particle and cannot be extracted by the analytical methods. This phenomenon is called non-extractable residue which was also found in case of some pesticides. Therefore, non-extractable residues are essential for environmental risk assessment study and cannot be ignored because non-extractable residue could be moved later when changes come in soil condition.

Newly most challenged engineered nanomaterials also come in agricultural soil. These nanomaterials are likely to dissipate through different pathways than conventional chemicals. For example, the particles may combine or agglomerate which will alter the properties of the particle and some nanomaterials may come in suspension. Many nanomaterials are likely to be coated with organic molecules; it is possible that microbes or abiotic processes will degrade these coating. Behaviors understanding of nanomaterials in agricultural soil, it is likely that we will need to begin to develop an understanding of all of these different pathways of dissipation.

### *10.3.3 Transportation into the Soil System*

The applied pollutants or produced in soil can be transported to natural water systems in surface run-off, subsurface flow and channel flow. There are many factors which can affect the extent of these processes such as solubility of pollutants, behavior of sorption pollutants persistence i.e. Physical structure, pH, organic matter, ion exchange capacity of soil material and climatic conditions such as temperature and extent of rainfall and intensity. Recently, few of studies have been reported on fate and transport of emerging contaminants but most of the work was published on pesticides, nutrients, and bacteria transport from agricultural field. For the transport of pharmaceuticals and personal care products and veterinary medicines from the soil surface to field drains, streams, rivers and ground water, an intact and undistributed soil column (Lysimeter), field-plot and full-scale studies have been investigated (Aga et al. 2003; Kay et al. 2005b, c, d; Kreuzig and Holtge 2005; Blackwell et al. 2007, 2009; Burkhardt et al. 2005; Hamscher et al. 2005; Chefetz et al. 2008: Lapen et al. 2008; Topp et al. 2008b). There are many investigations for the transportation of emerging contaminants from soil to other environment. These investigations are described in more detail below.

#### 10.3.3.1 Leaching to Groundwater

Leaching of emerging contaminants from soil to groundwater was investigated in many studies such as laboratory column studies lysimeter studies and full-scale field studies. Leaching of six pharmaceuticals was investigated to transport from soil to groundwater using a column study (Oppel et al. 2004). It was found out that majority of pharmaceuticals were not leached, but two compounds such as Clofibric and Iopromide were moved under experimental conditions.

Hamscher et al. (2000) and Blackwell et al. (2007) investigated the movement of antibiotics from sulfonamide and tetracycline group (a veterinary medicine, most widely used classes of antibiotics in the world) in the profile of soil at the field scale using suction probes. Sulfonamide antibiotics were found at depth but the tetracyclines were not; most likely due to the high potential for tetracyclines to sorption in the soil. Chlortetracycline applied to agricultural soil was detected at the depth of 25 and 35 cm, when applied manure but monensin, a coccidiostat, remained in the upper soil layers (Carlson and Mabury 2006).

Sulfonamide antibiotics residue were detected at a few of the study sites investigated by Hirsch et al. (1999) in Germany. It was attributed that one site irrigated with domestic sewage while other site has become contaminated due to manure application to the soil surface (Hirsch et al. 1999).

#### 10.3.3.2 Run-off

Kay et al. (2005a), Kreuzig et al. (2005) and Topp et al. (2008a) reported the transport of emerging contaminants through run-off for tetracycline antibiotics, sulfonamide antibiotics, anti-inflammatory drugs, antiepileptics, beta blockers, antidepressants, antibacterial drug i.e. triclosan and caffeine. As mentioned in leaching effect, the run-off influenced sorption by these compounds in the presence of manure or sewage sludge in soil and nature of agricultural land in which manure is added. Tetracyclines, a highly sorptive compound showed lower run-off significantly compared to highly mobile sulfonamide (Kay et al. 2005a). However, under actual field conditions total mass losses to the surface of water is very small in case of water-soluble compound sulfonamides. It is less than the 0.6% of the mass applied to be transported to the surface water (Stoob et al. 2007). Burkhard et al. (2005) reported that sulfonamides transported through run-off by 10–40 times higher when manure and slurry is applied to the soil. This might be due to the stopping of soil surface by slurry and changes in pH of the matrix by adding manure which can alters the speciation and fate of drugs (Burkhard et al. 2005). It has also been shown that run-off transport of ECs from cultivated soils is significantly lower than run-off from grasslands (Kreuzig et al. 2005). Run-off of emerging contaminants can also alter by the application of sewage sludge or other materials, for example application of sewage sludge by injecting effectively reduced the run-off of many pharmaceuticals (Topp et al. 2008a). Therefore, the management of agricultural land in proper way by adding manure and biosolids, it liberated the significant impact on the transport of emerging contaminants to the surface water and it is possible, that changes in land management practices could offer a management solution in the occurrence that an emerging contaminant is found to cause impacts in agricultural systems.

#### 10.3.3.3 Drain Flow

The transport of a range of veterinary antibacterial substance (i.e. tetracyclines, macrolides, sulfonamides and trimethoprim, a veterinary drug that is often used in combination with sulfonamide antibiotics) has been investigated using lysimeter and field-based studies in tile-drained clay soils (Kay et al. 2004; Boxall et al. 2006). Following application of pig slurry spiked with oxytetracycline and sulfachloropyridazine, the test compounds were detected in drain flow water (Kay et al. 2004). Concentrations of the sulfonamide were an order of magnitude higher than the tetracycline even though the amount of each test compound applied to the field was similar. These differences are again likely due to differences in sorption behavior. In a subsequent investigation at the same site (Kay et al. 2004), in which the soil was tilled, much lower concentrations were observed in the drain flow suggesting that tillage may be a useful mitigation strategy. While the pig slurry used in these studies was obtained from a pig farm where tylosin was used as a prophylactic treatment, this substance was not detected in any drain flow samples, possibly because it is not persistent in slurry (Loke et al. 2000).

Similar studies have been done on human pharmaceuticals in tile drain water from sludge amended fields (Lapen et al. 2008, 2009). A range of pharmaceuticals and personal care products were observed in the drainage waters including naproxen, acetaminophen, ibuprofen, cotinine, carbamazepine, triclosan, atenolol, triclocarban, gemfibrozil. The observed concentrations were significantly lower than those observed in wastewater effluents.

### *10.3.4 Transport into the Surface Water*

In the water column, substances may be degraded abiotically via photodegradation and/or hydrolysis or biotically by aerobic or anaerobic organisms. Highly sorptive substance may partition to the bed sediment. For example, ivermectin show that when added to water, the compound dissipates quickly from the water column and that this dissipation is observed increase in the concentration of the compound in the bed sediment (Sanderson et al. 2007). While many compounds degrade very quickly, others persist in the sediment for months to years (Boxall et al. 2004).

### *10.3.5 Uptake into Biota*

Emerging contaminants may also be taken up from soil into biota (Migliore et al. 2003; Kumar et al. 2005; Boxall et al. 2006; Dolliver et al. 2007). The potential uptake of veterinary and human medicines into plants is receiving increasing attention (Fig. 10.1). Studies with a range of veterinary medicines (Boxall et al. 2006) showed that a number of antibiotics are taken up by plants following exposure to soil at environmentally realistic concentrations of the compounds, whereas other compounds were not observed to be accumulated. Less work has been done on human pharmaceuticals. Redshaw et al. (2008) have shown that the antidepressant compound fluoxetine is accumulated by brassicas. The factors affecting uptake of emerging contaminants into plants are poorly understood and this is an area that needs much more research. Recently, the uptake of emerging contaminants into other soil organisms has also been explored. For example, Kinney et al. (2006) reported the occurrence of anthropogenic waste indicators, including the pharmaceutical trimethoprim in earthworm tissues.

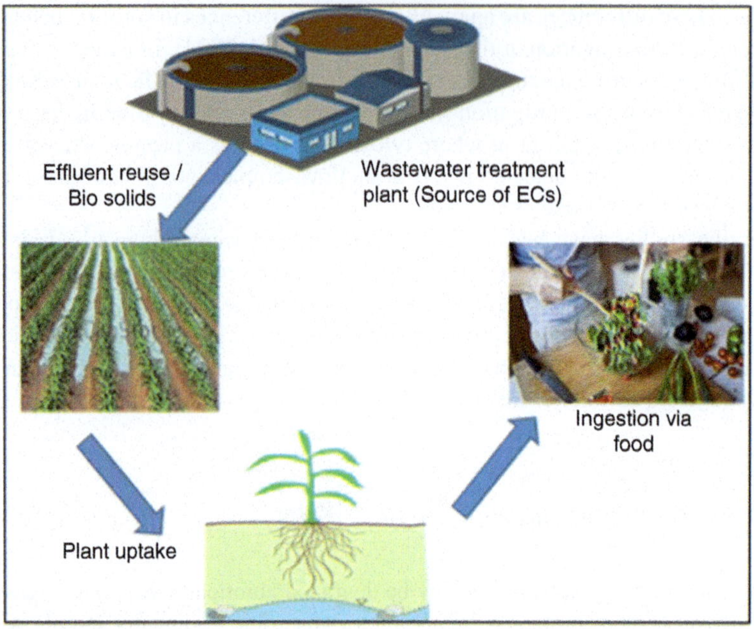


**Fig. 10.1** Transport of emerging contaminants from wastewater treatment plants into agriculture soil and uptake by human beings

## 10.4 Factors Influencing Biodegradation of Emerging Contaminants in Agriculture Soil

It is inevitable that pharmaceuticals will be released to the soil environment. An understanding of those chemical and environmental factors affecting the fate of pharmaceuticals in the soil environment would, therefore be highly beneficial in order to characterize the environmental risks of pharmaceuticals. Studies with single substances in different soil types indicated that degradation rates are variable, but it is not yet possible to correlate persistence with soil properties or soil bioactivity. Pharmaceuticals will enter the environment associated with biosolids and the presence of biosolids significantly reduces degradation rates compared to soil alone. As pharmaceuticals will never be in the environment as single compounds, a consideration of the impacts of mixtures of different pharmaceuticals and other compounds needs to be assessed. Degradation may be significantly slower in mixtures compared to single compound. Overall, the degradation of pharmaceuticals in the environment is a very complex issue and a lot more data on the degradation behavior of pharmaceuticals in a range of well characterized soils with different properties are needed in order to understand what will happen to a pharmaceutical in the real soil environment.

## 10.5 Factors Influencing Sorption of Emerging Contaminants in Agricultural Soil

In semi-arid regions, wastewater reuse in agriculture is an important source of irrigation and this is the most common route of emerging contaminants to agricultural fields (Kinney et al. 2006). Depending upon the mobility of compound, some compounds are leached into the groundwater and some are sorbed to the soil altering microbial community in the soil and up taken by plants (Thiele-Bruhn 2003). Sorption is the main process responsible for attenuation and accumulation of emerging contaminants in the subsurface. This sorption of emerging contaminants onto soil is influenced by dissolved organic matter and soil moisture.

### *10.5.1 Dissolved Organic Matter*

Dissolved organic matter is a main component in agricultural soil and its direct effect on sorption of emerging contaminants is the competition for the sorption sites on soil. The dissolved organic matter has a linear relationship with adsorption of emerging contaminants. Aziz et al. (2016) reported linear relationship between solid-water partition coefficient ($K_d$) of estrone and DOC in soil. Similarly, sorbed amount of 17α-trenbolone, trendione, atrazine and sulfamethazine has shown direct relation to the amount of soil organic carbon (Jenks et al. 1998; Lertpaitoonpan et al. 2009; Robinson et al. 2017).

High sorption of hormones among emerging contaminants is well documented (Lee et al. 2003; Casey et al. 2005) which is affected by particle size of the soil. Higher sorption of testosterone was reported in small particle sized soil than bulk soil particles (Qi and Zhang 2016). However, increase in temperature, water/soil ratio and soil depth lower the sorption to soil (Ma et al. 2015). A competitive advantage over the active sorption sites has been observed on co-occurrence of some emerging contaminants. An antagonistic effect on the sorption of 17 α-ethinylestradiol is reported in the co-occurrence of other compounds (estrone, β-estradiol, and estriol) (Li et al. 2014b). Antibiotics are also sorbed onto soil and tend to retain, persist and accumulate in soil matrix. pH and DOC are two main factors that influence sorption of antibiotics in soil. For instance, higher pH increased rate of sorption of ofloxacin in soil (Wang et al. 2017). However, sorption of sulfonamides (sulfamethoxazole, sulfadimethoxine and sulfamethazine) increases with decreasing pH (Park and Huwe 2016). Sorption of chlortetracycline and tylosin reduced and sorption of sulfamethazine increased, when daily manure derived dissolved organic C was added to the soil (Lee et al. 2014).

### 10.5.2 Soil Moisture

Soil moisture is one of the important parameter, which affects the sorption of the emerging contaminants in the soil. Changes in rainfall intensity and amount directly influence the soil moisture (Brevik 2013). Increase in moisture content lowers the sorption of compounds which is mainly due to increase in bioavailability and dissolution. Breus et al. (2014) investigated sorption of vapor hydrocarbons on soil at different depths under different moisture conditions. In soil layer at depth of 35–45 cm and with moisture content at 0 and 9.8%, concentration of p-xylene was 21.61 and 0.19 μL/g, respectively. Presence of dissolved organic matter lowers the dependency of sorption on moisture content. At higher moisture content, the effect on sorption of emerging contaminants is negligible. For instance, sorption of carbofuran was reported independent of soil moisture content in the range of 12.5–20% (Shelton and Parkin 1991). However, at 10% moisture content in soil, its sorption reduced partially.

### 10.5.3 Soil pH

Adsorption of emerging contaminants onto soil is dependent on the pH of the soil matrix, which makes pH as a determining factor for mobility or retention of emerging contaminants in soil. Different types of soils exhibit different pH values and emerging contaminants can exist as cations, anions or zwitterions depending on the environmental pH (Sassman and Lee 2005). For example, sulfamethazine has lower $K_d$ values when the pH of soil increased showing decrease in sorption (Lertpaitoonpan et al. 2009). Lower sorption at higher pH may be due to dominance of electrostatic repulsion between anionic sulfamethazine and negatively charged clay mineral ions over hydrophobic sorption between unionized sulfamethazine and organic carbon. This lower sorption at higher pH can increase the mobility of emerging contaminants in porous media. For instance, ciprofloxacin is more mobile in sand columns at higher pH of 9.5 than 5.6 due to negative ionic charge (Chen et al. 2011).

## 10.6 Effect and Risks of Emerging Contaminants on Ecosystem and Human

The exposure to emerging contaminants in the agricultural environment is through four pathways: direct exposure, accumulation in the plants, accumulation in the grazing animals and water contamination (Clarke and Porter 2010). Inhalation and other exposure route such as dermal contact has a very little risk involved with respect to emerging contaminants. The short term and long term effects are not yet completely understood. Ahmed (2015) studied the growth of cucumber, tomatoes and lettuce in the soil spiked with tetracyclines and sulfonamides at concentrations of 5, 10, and 20 mg/kg and found that these contaminants were not only inhibiting their growth but also high concentrations of the contaminants were found in the roots and leaves of cucumber and tomato. When grown in soils contaminated with emerging contaminants, root vegetables such as carrot are likely to adsorb emerging contaminants due to their lipophilic nature, where the peel of carrot has been reported to contain 7 to 10 times the concentration of polychlorinated dioxins and polychlorinated furans and poly chlorinated biphenyls (O'Connor et al. 1992).

Ingestion of crops and soil contaminated emerging contaminants by the grazing animals is the source of emerging contaminants in the animals. In fact, 80–95% of the contamination in humans is also thought to be because of the consumption of animal derived products (Fries 2002). Some harmful effects according to Du and Liu (2012) include: reduced crop yield and seed germination capacity, reduced microbial activity, traces of emerging contaminants in the yield and leaching of emerging contaminants to the groundwater through the soil profile.

Some of the harmful effects of emerging contaminants are:

1. Limited crop yield and seed germination capacity
2. Uptake of the contaminants by plant
3. Ingestion by farm animals
4. Alteration in the soil microbial structure
5. Introduction and bioaccumulation of emerging contaminants in the food chain by consumption of contaminated plants and animals

The crops are at the risk of exposure to emerging contaminants and their concentration in plant tissues depends on uptake, translocation and metabolism in plants (Dudley et al. 2019). The chronic exposure to mixture of emerging contaminants can have an impact on seedling germination and growth as well as phytohormone homeostasis. They can serve as carriers of various emerging contaminants into the food chain. Ability of crop plant to uptake emerging contaminants was found to be in the order: leafy vegetables > root vegetables > cereals and fodder crops > fruit vegetables (Christou et al. 2019). The adsorbed compounds can form metabolites with unknown harmful effects. Some antibiotics found in the soil can be a source for building up of antimicrobial resistance genes and shifting the soil microbial structure, and thus, alter the performance of soil microorganisms for nutrients and carbon uptake (Liu et al. 2011). Compounds such as anxiolytic drug Oxazepam is found to alter the behavior of wild fish in their natural habitat (Brodin et al. 2013). Other studies have also revealed harmful effects of emerging contaminants on sensitive species like mussels, fish and tadpoles.

Wu (2013) estimated that consumption of leafy vegetables induces an annual average concentration of 0.04 to 350 μg of pharmaceutical and personal care products in a 70 kg US citizen. This is much less than the prescribed single doses which ranges from 20–200 mg. The average concentration of emerging contaminants in crops varies in the range from 1–7500 ng/kg (Singh et al. 2020). Another study by Carter et al. (2014) concluded that triclosan accumulates in the tissues of the leaves of the crops and consumption of such leaves may account for 83.8% of acceptable daily intake. However, under lack of studies on human health risk assessment, it is still difficult to estimate the risk involved due to the presence of emerging contaminants in agro-ecosystem.

## 10.7 Conclusions

This chapter summarizes the whole process of occurrence of emerging contaminants in agriculture sector, their fate in the subsurface and transport to the groundwater. Introduction of treated wastewater for irrigation and treated sludge for soil nutrition are potential causes of emerging contaminants entry in agriculture soil. Groundwater contamination with these emerging contaminants in agriculture soil is dependent on soil condition as well as compound properties. Proofs of antimicrobial resistance in the microbiological population indicate a shift in the nature of bacterial population responsible for the uptake of nutrients from the soil or fixing of nutrients in the soil. This shift can result into very harmful consequences for the yield. Biodegradation is main pathway for the removal of emerging contaminants from the agro-environment. However, complete mineralization may not be possible. Most of the studies done till date have concluded that the concentration of emerging contaminants to be well below the level to cause acute or chronic risk to human health. However, more studies on the ecological risks involved with the application of biosolids or treated wastewater needs to be done and a clear methodology outlining the concentration of emerging contaminants in the biosolids and treated wastewater that can be used for various agricultural purpose needs to be made. The studies are mostly localized that investigate the groundwater contamination near a point source, for instance wastewater discharge point, but limited knowledge is available on the diffused sources of pollution like sewage leaks, stormwater infiltration etc.

## References

Aga DS, Goldfish R, Kulshrestha P (2003) Application of ELISA in determining the fate of tetracyclines in land-applied livestock wastes. Analyst 128(6):658–662

Ahmed MBM, Rajapaksha AU, Lim JE, Vu NT, Kim IS, Kang HM, Lee SS, Ok YS (2015) Distribution and accumulative pattern of tetracyclines and sulfonamides in edible vegetables of cucumber, tomato, and lettuce. J Agr Food Chem 63(2):398–405. https://doi.org/10.1021/jf5034637

Al Rajab AJ, Sabourin L, Scott A, Lapen DR, Topp E (2009) Impact of biosolids on the persistence and dissipation pathways of triclosan and triclocarban in an agricultural soil. Sci Tot Environ 407:5978–5985

Arye G, Tarchitzky J, Chen Y (2011) Treated wastewater effects on water repellency and soil hydraulic properties of soil aquifer treatment infiltration basins. J Hydrol 397(1–2):136–145. https://doi.org/10.1016/j.jhydrol.2010.11.046

Aziz EA, Omar WW, Zainodin NA, Islam MS (2016) The role of organic carbon in natural and synthetic oestrogens sorption onto soils. Int J Environ Sci Dev 7(7):520–524

Barnes KK, Kolpin DW, Furlong ET, Zaugg SD, Meyer MT, Barber LB (2008) A national reconnaissance of pharmaceuticals and other organic wastewater contaminants in the United States – I groundwater. Sci Tot Environ 402(2–3):192–200. https://doi.org/10.1016/j.scitotenv.2008.04.028

Blackwell PA, Kay P, Boxall ABA (2007) The dissipation and transport of veterinary antibiotics in a sandy loam soil. Chemosphere 67(2):292–299

Blackwell PA, Kay P, Ashauer R, Boxall ABA (2009) Effects of agricultural conditions on the leaching behaviour of veterinary antibiotics in soils. Chemosphere 75(1):13–19. https://doi.org/10.1016/j.chemosphere.2008.11.070

Bondarenko S, Gan J, Ernst F, Green R, Baird J, McCullough M (2012) Leaching of pharmaceuticals and personal care products in Turfgrass soils during recycled water irrigation. J Environ Qual 41(4):1268. https://doi.org/10.2134/jeq2011.0355

Borgman O, Chefetz B (2013) Combined effects of biosolids application and irrigation with reclaimed wastewater on transport of pharmaceutical compounds in arable soils. Water Res 47(10):3431–3443. https://doi.org/10.1016/j.watres.2013.03.045

Boxall ABA, Blackwell PA, Cavallo R, Kay P, Tolls J (2002) The sorption and transport of a sulphonamide antibiotic in soil systems. Toxicol Lett 131:19–28

Boxall AB, Sinclair CJ, Fenner K, Kolpin D, Maund SJ (2004) Peer reviewed: when synthetic chemicals degrade in the environment. Environ Sci Technol 38(19):368A–375A

Boxall ABA, Johnson P, Smith EJ, Sinclair CJ, Stutt E, Levy L (2006) Uptake of veterinary medicines from soils into plants. J Agr Food Chem 54(6):2288–2297

Breus IP, Mishchenko AA, Shinkarev AA, Neklyudov SA, Breus VA (2014) Effect of organic matter on the sorption activity of heavy loamy soils for volatile organic compounds under low moisture conditions. Eurasian Soil Sci 47(12):1216–1226

Brevik EC (2013) The potential impact of climate change on soil properties and processes and corresponding influence on food security. Agriculture 3:398–417

Briones RM, Sarmah AK (2018) Detailed sorption characteristics of the anti-diabetic drug metformin and its transformation product guanylurea in agricultural soils. Sci Tot Environ 630:1258–1268. https://doi.org/10.1016/j.scitotenv.2018.02.306

Brodin T, Fick J, Jonsson M, Klaminder J (2013) Dilute concentrations of a psychiatric drug alter behavior of fish from natural populations. Science 339(6121):814–815. https://doi.org/10.1126/science.1226850

Burkhardt M, Stamm C, Waul C, Singer H, Muller S (2005) Surface runoff and transport of sulfonamide antibiotics and tracers on manured grassland. J Environ Qual 34(4):1363–1371

Carlson JC, Mabury SA (2006) Dissipation kinetics and mobility of chlortetracycline, tylosin, and monensin in an agricultural soil in Northumberland County, Ontario, Canada. Environ Toxicol Chem 25(1):10

Carter LJ, Harris E, Williams M, Ryan JJ, Kookana RS, Boxall ABA (2014) Fate and uptake of pharmaceuticals in soil-plant systems. J Agr Food Chem 62(4):816–825. https://doi.org/10.1021/jf404282y

Casey FX, Šimůnek J, Lee J, Larsen GL, Hakk H (2005) Sorption, mobility, and transformation of estrogenic hormones in natural soil. J Environ Qual 34:1372–1379

Chefetz B, Mualem T, Ben-Ari J (2008) Sorption and mobility of pharmaceutical compounds in soil irrigated with reclaimed wastewater. Chemosphere 73(8):1335–1343

Chen C, Xia K (2017) Fate of land applied emerging organic contaminants in waste materials. Curr Pollut Rep 3:38–54. https://doi.org/10.1007/s40726-017-0048-6

Chen H, Gao B, Li H, Ma LQ (2011) Effects of pH and ionic strength on sulfamethoxazole and ciprofloxacin transport in saturated porous media. J Cont Hydrol 126(1–2):29–36

Christou A, Papadavid G, Dalias P, Fotopoulos V, Michael C, Bayona JM, Piña B, Fatta-Kassinos D (2019) Ranking of crop plants according to their potential to uptake and accumulate contaminants of emerging concern. Environ Res 170:422–432

Clarke BO, Porter NA (2010) Persistent organic pollutants in sewage sludge: Levels, sources, and trends. In: ACS symposium series, vol 1048. American Chemical Society, pp 137–171. https://doi.org/10.1021/bk-2010-1048.ch006

Dolliver H, Kumar K, Gupta S (2007) Sulfamethazine uptake by plants from manure-amended soil. J Environ Qual 36(4):1224–1230

Du L, Liu W (2012) Occurrence, fate, and ecotoxicity of antibiotics in agro-ecosystems. A review. Agron Sustain Dev 32(2):309–327. https://doi.org/10.1007/s13593-011-0062-9

Duarte-Davidson R, Jones KC (1996) Screening the environmental fate of organic contaminants in sewage sludge applied to agricultural soils: II. The potential for transfers to plants and grazing animals. Sci Tot Environ 185:59–70. https://doi.org/10.1016/0048-9697(96)05042-5

Dubroca J, Collignon N, Brault A, Hernandez-Raquet G, Patureau D, Mougin C (2009) Fate of 17β-estradiol in terrestrial model ecosystems amended with contaminated composted biosolids. Environ Chem Lett 7:369–373

Dudley S, Sun C, McGinnis M, Trumble J, Gan J (2019) Impacts and biotransformation of CECs in crops under acute and chronic exposures. In: 257th ACS National Meeting & Exposition, p. ENVR-0566. American Chemical Society Washington, DC, Orlando, FL, United States

Durso LM, Cook KL (2014) Impacts of antibiotic use in agriculture: what are the benefits and risks? Curr Opin Microbiol 19:37–44. https://doi.org/10.1016/j.mib.2014.05.019

Fahrenfeld N, Knowlton K, Krometis LA, Hession WC, Xia K, Lipscomb E, Libuit K, Green BL, Pruden A (2014) Effect of manure application on abundance of antibiotic resistance genes and their attenuation rates in soil: field-scale mass balance approach. Environ Sci Technol 48(5):2643–2650. https://doi.org/10.1021/es404988k

Fries GF (2002) Transport of persistent organic pollutants to animal products: fundamental principles and application to health risk assessment. In Paustenbach DJ (Ed.) Human and ecological risk assessment: theory and practice (Wiley Classics Library) (29th ed., pp. 873–894). New York: Wiley. Retrieved from https://www.wiley.com/en-us/Human+and+Ecological+Risk+Assessment%3A+Theory+and+Practice+%28Wiley+Classics+Library%29-p-9780470253199

Ghanem A, Dubroca J, Chaplin V, Mougin C (2006) Fate of herbicides and nonylphenol in soil – plant – water systems amended with contaminated sewage sludge. Environ Chem Lett 4:63–67. https://doi.org/10.1007/s10311-006-0034-5

Gomes AR, Justino C, Rocha-Santos T, Freitas AC, Duarte AC, Pereira R (2017) Review of the ecotoxicological effects of emerging contaminants to soil biota. J Environ Sci Health A Tox Hazard Subst Environ Eng 52(10):992–1007. https://doi.org/10.1080/10934529.2017.1328946

González M, Fernández-lópez C, Polesel F, Trapp S (2019) Predicting the uptake of emerging organic contaminants in vegetables irrigated with treated wastewater – implications for food safety assessment. Environ Res 172:175–181. https://doi.org/10.1016/j.envres.2019.02.011

Hamscher G, Sczesny S, Abu-Qare A, Hoper H, Nau H (2000) Substances with pharmacological effects including hormonally active substances in the environment: identification of tetracyclines in soil fertilized with animal slurry. Deutsche Tierarztliche Wochenschrift 107(8):332–334

Hamscher G, Pawelzick HT, Hoper H, Nau H (2005) Different behavior of tetracyclines and sulfonamides in sandy soils after repeated fertilization with liquid manure. Environ Toxicol Chem 24(4):861–868

Hirsch R, Ternes T, Haberer K, Kratz KL (1999) Occurrence of antibiotics in the aquatic environment. Sci Tot Environ 225(1–2):109–118

Hurtado C, Montano-chávez YN (2017) Degradation of emerging organic contaminants in an agricultural soil: decoupling biotic and abiotic processes. Water Air Soil Pollut 228(7). https://doi.org/10.1007/s11270-017-3402-9

Jenks BM, Roeth FW, Martin AR, McCallister DL (1998) Influence of surface and subsurface soil properties on atrazine sorption and degradation. Weed Sci 46(1):132–138

Kay P, Blackwell PA, Boxall ABA (2005a) Fate and transport of veterinary antibiotics in drained clay soils. Environ Toxicol Chem 23(5):1136–1144

Kay P, Blackwell PA, Boxall ABA (2005b) A lysimeter experiment to investigate the leaching of veterinary antibiotics through a clay soil and comparison with field data. Environ Pollut 134:333–341

Kay P, Blackwell PA, Boxall ABA (2005c) Transport of veterinary antibiotics in overland flow following the application of slurry to land. Chemosphere 59(7):951–959

Kay P, Blackwell PA, Boxall ABA (2005d) Column studies to investigate the fate of veterinary antibiotics in clay soils following slurry application to agricultural land. Chemosphere 60(4):497–507

Kinney C, Urlong E, Werner S, Cahill J (2006) Presence and distribution of wastewater-derived pharmaceuticals in soil irrigated with reclaimed water. Environ Toxicol Chem 25(2):317–326. https://doi.org/10.1897/05-187R.

Kolz AC, Moorman TB, Ong SK, Scoggin KD, Douglass EA (2005) Degradation and metabolite production of tylosin in anaerobic and aerobic swine-manure lagoons. Water Environ Res 77(1):49–56

Kreuzig R, Holtge S (2005) Investigations on the fate of sulfadiazine in manured soil: laboratory experiments and test plot studies. Environ Toxicol Chem 24(4):771–776

Kreuzig R, Holtge S, Brunotte J, Berenzen N, Wogram J, Schulz R (2005) Test-plot studies on runoff of sulfonamides from manured soils after sprinkler irrigation. Environ Toxicol Chem 24(4):777–781

Kumar K, Gupta SC, Baidoo SK, Chander Y, Rosen CJ (2005) Antibiotic uptake by plants from soil fertilized with animal manure. J Environ Qual 34(6):2082–2085

Lapen DR, Topp E, Metcalfe CD, Li H, Edwards M, Gottschall N, Bolton P, Curnoe W, Payne M, Beck A (2008) Pharmaceuticals and personal care products in tile drainage following land application of sewage biosolids. Sci Tot Environ 399:50–65

Lapen DR, Topp E, Metcalfe CD, Li H, Edwards M, Gottschall N, Bolton P, Curnoe W, Payne M, Beck A (2009) Pharmaceutical and personal care products in tile drainage following land application of municipal biosolids. Sci Tot Environ 399(1–3):50–65

Lapworth DJ, Baran N, Stuart ME, Ward RS (2012) Emerging organic contaminants in groundwater: a review of sources, fate and occurrence. Environ Pollut 163:287–303. https://doi.org/10.1016/j.envpol.2011.12.034

Lee LS, Strock TJ, Sarmah AK, Rao PSC (2003) Sorption and dissipation of testosterone, estrogens, and their primary transformation products in soils and sediment. Environ Sci Technol 37(18):4098–4105

Lee J, Seo Y, Essington ME (2014) Sorption and transport of veterinary pharmaceuticals in soil—a laboratory Study. Soil Sci Soc Am J 78(5):1531. Available at: https://dl.sciencesocieties.org/publications/sssaj/abstracts/78/5/1531

Lertpaitoonpan W, Ong SK, Moorman TB (2009) Effect of organic carbon and pH on soil sorption of sulfamethazine. Chemosphere 76(4):558–564

Li D, Lyon DY, Li Q, Alvarez PJJ (2008) Effect of soil sorption and aquatic natural organic matter on the antibacterial activity of a fullerene water suspension. Environ Toxicol Chem 27(9):1888–1894

Li J, Dodgen L, Ye Q, Gan J (2013) Degradation kinetics and metabolites of carbamazepine in soil. Environ Sci Technol 47(8):3678–3684. https://doi.org/10.1021/es304944c

Li J, Ye Q, Gan J (2014a) Degradation and transformation products of acetaminophen in soil. Water Res 49:44–52. https://doi.org/10.1016/j.watres.2013.11.008

Li Y, Zhang C, Li S, Zhou C, Li X (2014b) Single and competitive adsorption of 17 α-ethinylestradiol and bisphenol a with estrone, β-estradiol, and estriol onto sediment. Mar Drugs 12(3):1349–1360

Liu A, Fong A, Becket E, Yuan J, Tamae C, Medrano L, Miller JH (2011) Selective advantage of resistant strains at trace levels of antibiotics: a simple and ultrasensitive color test for detection of antibiotics and genotoxic agents. Antimicro Agents Chemother 55(3):1204–1210. https://doi.org/10.1128/aac.01182-10

Loke ML, Ingerslev F, Halling-Sorensen B, Tjornelund J (2000) Stability of tylosin A in manure containing test systems determined by high performance liquid chromatography. Chemosphere 40:759–765

Ma R, Zhang TC, Bartelt-Hunt SL, Qi Y, Kranz WL, Snow DD, Mader TL, Shapiro CA, Shelton DP, Van Donk SJ, Tarkalson DD (2015) Influence of soil properties and test conditions on sorption and desorption of testosterone. J Environ Eng (United States), 141(7). Available at: http://

www.scopus.com/inward/record.url?eid=2-s2.084931301635&partnerID=40&md5=218aa5c4863d8a71961bc7ccd36d7455

Migliore L, Cozzolino S, Fiori M (2003) Phytotoxicity to and uptake of enrofloxacin in crop plants. Chemosphere 52(7):1233–1244

Monteiro SC, Boxall ABA (2009) Factors affecting the degradation of pharmaceuticals in agricultural soils. Environ Toxicol Chem 28(12):2546–2554

Noguera-oviedo K, Aga DS (2016) Lessons learned from more than two decades of research on emerging contaminants in the environment. J Hazard Mat 316:242–251. https://doi.org/10.1016/j.jhazmat.2016.04.058

O'Connor G, Chaney R.L, Ryan JA (1992) Bioavailability to plants of sludge-borne toxic organics. Cincinnati, OH (United States). https://doi.org/PB-92-153931/XAB

Oppel J, Broll G, Loffler D, Meller M, Rombke J, Ternes T (2004) Leaching behavior of pharmaceuticals in soil testing systems: a part of an environmental risk assessment for groundwater protection. Sci Total Environ 328:265–273

Park JY, Huwe B (2016) Effect of pH and soil structure on transport of sulfonamide antibiotics in agricultural soils. Environ Pollut 213:561–570

Peterson DR, Staples CA (2003) Degradation of phthalate esters in the environment. In: The Handbook of Environmental Chemistry, vol 3. Springer, Berlin, pp 85–124. https://doi.org/10.1007/b11464

Qi Y, Zhang TC (2016) Sorption of testosterone to soil colloids of different size fractions: development of a centrifugation and mass balance – based methodology. J Environ Eng 142(11) Available at: http://doi.wiley.com/10.1002/etc.3711

Redshaw CH, Wootton VG, Rowland SJ (2008) Uptake of the pharmaceutical fluoxetine hydrochloride from medium by Brassicaceae. Phytochemistry 69(13):2510–2516

Robinson JA, Ma Q, Staveley JP, Smolenski WJ (2017) Sorption and desorption of 17α-trenbolone and trendione on five soils. Environ Toxicol Chem 36(3):613–620

Robles-molina J, Gilbert-lópez B, García-reyes JF, Molina-díaz A (2014) Monitoring of selected priority and emerging contaminants in the Guadalquivir River and other related surface waters in the province of the province of Jaén, South East Spain. Sci Tot Environ 480(1):247–257. https://doi.org/10.1016/j.scitotenv.2014.01.121

Sanderson H, Laird B, Pope L, Brain R, Wilson C, Johnson D, Peregrine A, Bryning G, Boxall A, Solomon K (2007) Assessment of the environmental fate and effects of ivermectin in aquatic mesocosms. Aquatic Toxicol 85:229–240

Sassman SA, Lee LS (2005) Sorption of three tetracyclines by several soils: assessing the role of pH and cation exchange. Environ Sci Technol 39(19):7452–7459

Scrinis G, Lyons K (2009) Nanotechnology and the techno-corporate Agri-food paradigm. Food security, nutrition and sustainability 252

Shelton DR, Parkin TB (1991) Effect of moisture on sorption and biodegradation of carbofuran in soil. J Agr Food Chem 39(11):2063–2068

Singh BN, Hidangmayum A, Singh A, Guru A, Yashu BR, Singh GS (2020) Effect of emerging contaminants on crops and mechanism of toxicity. In: Lichtfouse E (ed) Sustainable agriculture reviews 40. Springer, Cham, pp 217–241. https://doi.org/10.1007/978-3-030-33281-5_6

Stoob K, Singer HP, Mueller SR, Schwarzenbach RP, Stamm CH (2007) Dissipation and transport of veterinary sulphonamide antibiotics after manure application to grassland in a small catchment. Environ Sci Technol 41:7349–7355

Ter Laak TL, Gebbink WA, Tolls J (2006a) Estimation of soil sorption coefficients of veterinary pharmaceuticals from soil properties. Environ Toxicol Chem 25(4):933–941

Ter Laak TL, Gebbink WA, Tolls J (2006b) The effect of pH and ionic strength on the sorption of sulfachloropyridazine, tylosin, and oxytetracycline to soil. Environ Toxicol Chem 25(4):904–911

Thiele-Bruhn S (2003) Pharmaceutical antibiotic compounds in soils–a review. J Plant Nutr Soil Sci 166(2):145–167

Thiele-Bruhn S, Aust MO (2004) Effects of pig slurry on the sorption of sulfonamide antibiotics in soil. Arch Environ Contam Toxicol 47(1):31–39

Topp E, Hendel J, Lu Z, Chapman R (2006) Biodegradation of caffeine in agricultural soils. Can J Soil Sci 86:533–544

Topp E, Hendel JG, Lapen DR, Chapman R (2008a) Fate of the non-steroidal anti-inflammatory drug naproxen in agricultural soil receiving liquid municipal biosolids. Environ Toxicol Chem 27:2005–2010

Topp E, Monteiro SC, Beck A, Ball Coehlo B, Boxall AB, Duenk PW, Kleywegt S, Lapen DR, Payne M, Sabourin L, Li H, Metcalfe CD (2008b) Runoff of pharmaceuticals and personal care products following application of biosolids to an agricultural field. Sci Tot Environ 396:52–59

Topp E, Metcalfe CD, Boxall AB, Lapen DR (2010) Transport of PPCPs and veterinary medicines from agricultural fields following application of biosolids or manure. In: ACS symposium series, vol 1048. American Chemical Society, pp 227–240. https://doi.org/10.1021/bk-2010-1048.ch010

USGS (2016) Drinking water exposure to chemical and pathogenic contaminants: emerging contaminants. Retrieved May 6, 2019, from https://archive.usgs.gov/archive/sites/health.usgs.gov/dw_contaminants/emc.html

Wang P, Zhang D, Zhang H, Li H, Ghosh S, Pan B (2017) Impact of concentration and species of sulfamethoxazole and ofloxacin on their adsorption kinetics on sediments. Chemosphere 175:123–129

Wichmann F, Udikovic-Kolic N, Andrew S, Handelsman J (2014) Diverse antibiotic resistance genes in dairy cow manure. MBio 5(2). https://doi.org/10.1128/mbio.01017-13

Wu X, Ernst F, Conkle JL, Gan J (2013) Comparative uptake and translocation of pharmaceutical and personal care products (PPCPs) by common vegetables. Environ Int 60:15–22. https://doi.org/10.1016/j.envint.2013.07.015

Wu X, Dodgen LK, Conkle JL, Gan J (2015) Plant uptake of pharmaceutical and personal care products from recycled water and biosolids: a review. Sci Total Environ 536:655–666. https://doi.org/10.1016/j.scitotenv.2015.07.129

Xia K, Hundal LS, Kumar K, Armbrust K, Cox AE, Granato TC (2010) Triclocarban, triclosan, polybrominated diphenyl ethers and 4-nonylphenol in biosolids and in soil receiving 33-year biosolids application. Environ Toxicol Chem 29(3):597–605. https://doi.org/10.1002/etc.66

Zuloaga O, Navarro P, Bizkarguenaga E, Iparraguirre A, Vallejo A, Olivares M, Prieto A (2012) Overview of extraction, clean-up and detection techniques for the determination of organic pollutants in sewage sludge: a review. Anal Chim Acta 736:7–29. https://doi.org/10.1016/j.aca.2012.05.016

# Chapter 11
# Fate and Transport of Engineered Nanoparticles as an Emerging Agricultural Contaminant

**Tamanreet Kaur**

**Abstract** The immense pressure on agro-ecosystem due to ever expanding global population coupled with changing environmental conditions has increased the importance of sustainable agricultural practices. The integrating approach of agri-nanotechnology is anticipated to become a major agricultural thrust in near future by increasing productivity without decontamination of soils, water, and offers protection against various pests of plants and microbial diseases. The potential benefits of agri-nanotechnology includes enhancement of food quality, reduction of agricultural inputs, smart delivery systems for controlled release of agrochemicals, nanosensors for monitoring agricultural fields, etc. With the rapid advancement in field of nanotechnology, the unique novel sized engineered nanoparticles have been widely developed in agriculture. Besides rise in diversity and quantities of engineered nanoparticles produced, the probable impact of these emerging agro-ecosystem contaminants also raised global concern about their fate, transport, bioavailability, toxicity that limit the complete acceptance to adopt nanotechnology in agricultural sector.

This chapter highlights recent nanotechnology researches in agriculture with recommendation for adopting engineered nanoparticles along with mitigating the risk assessment factors. Concise description on direct and accidental release of engineered nanoparticles into agro-ecosystem, their bioavailability and phytotoxicity, advantages and disadvantages of nanoparticle's application on crop plants are discussed. The abiotic transformation processes along with environmental factors influencing dynamics of fate and transport of engineered nanoparticles released into agricultural soils are outlined. Furthermore, an emphasis is laid on delivery methods of nanoparticles into crop plants, its penetration and translocation mechanisms. For successful adoption of agri-nanotechnology, future improvements in the analytical tools to understand the fate and transport of engineered nanoparticles in agro-ecosystem and more awareness about their phytoxicity, associated benefits and interrelated issues are required to be explored.

---

T. Kaur (✉)
Department of Zoology, Kanya MahaVidyalaya, Jalandhar, Punjab, India
e-mail: tamanreetkaur@gmail.com


V. Kumar Singh et al. (eds.), *Sustainable Agriculture Reviews 50*, Sustainable Agriculture Reviews 50, https://doi.org/10.1007/978-3-030-63249-6_11

**Keywords** Agri-nanotechnology · Emerging contaminants · Engineered nanoparticles · Smart delivery system · Sustainable agriculture

## Abbreviations

| | |
|---|---|
| DLVO | Derjaguin-Landau-Verwey-Overbeek |
| FITC | Fluorescein isothiocyanate |
| MWCNTs | Multi-walled carbon nanotubes |
| SWNTs | Single-walled carbon nanotubes |
| WWTP | Waste water treatment plant |

## 11.1 Introduction

Nanotechnology plays a leading role in shaping modern agriculture by renovating food production processes. Nanomaterials have potential to reveal an inventive strategy in plant biotechnology and agricultural research (Scrinis and Lyons 2007). Nanotechnology offers the probability of precision agriculture (that means increasing agricultural yield with minimizing input) in the period of developing sustainable agricultural practices (Chen and Yada 2011). Nanoparticles (1–100 nm size) are widely used in the modern agriculture to combat global demand of crop production, food sustainability, enhanced food nutrition, food security, superior food quality and mitigating climate change (Ghormade et al. 2011; Mishra et al. 2014; Handford et al. 2015). Categorization of nanoparticles according to their source of emission includes natural, incidental, and engineered nanoparticles. Natural nanoparticles are formed due to natural processes (volcanic emissions, ocean spray particles, soot released by wildfires) or incidentally released as result of anthropogenic activities (mostly emissions from combustion of fossil fuels, diesel engines) or are chemically synthesized, engineered nanoparticles that are having novel nanoscale size ($\leq$100 nm), have characteristics of distinctive chemical, biological, and physical properties in contrast to larger sized particles of the same chemistry.

In the context of sustainable agriculture, applying engineered nanoparticles could be considered amongst the promising approach to significantly enhance agricultural quality as well as quantity (Sekhon 2014; Liu and Lal 2015). The smart delivery systems must have combination of timely controlled, accurately targeted, automated, and possesses multi-purpose traits required for successful targeting into agro-ecosystems (Nair et al. 2010). Engineered nanoparticles due to minute size have relatively large surface area that results into enhanced, unique properties that can make the agricultural systems smarter. Improper handling of nanoparticles may give rise to significant threats to the agro-ecosystems and human health. For long term agricultural sustainability, consideration of appropriate risk assessment is also required in spite of positive implications of engineered nanoparticles on agriculture. Hence, it is also obligatory to assess the fate and transport of engineered nanoparticles and, their potential toxicity through development of relevant risk evaluation techniques (Jacobs et al. 2015; Jha and Pudake 2016).

The present book chapter focuses on the interaction of engineered nanoparticles with soil-plant system for understanding fate and transport of these particles in agro-ecosystem. The potentiality and advantages for using engineered nanoparticles in agro-ecosystem along with risk factors associated are discussed by emphasizing on: (1) primary interactions between engineered nanoparticles and agricultural crops along with their positive and/or negative effects on crops, (2) exposure pathways of engineered nanoparticles in agro-ecosystem, and (3) environmental implications and physiochemical transformation processes influencing fate and transport of engineered nanoparticles in agro-ecosystem.

## 11.2 Categorization and Implications of Engineered Nanoparticles on Growth, Development and Phytotoxicity of Plants

Engineered nanoparticles are obtained from engineered nanomaterials which are devised and manufactured to have enhanced chemical or biological activity. Engineered nanoparticles are produced through certain chemical processes and/or physical processes, are designed with very specific properties concerning agricultural sustainability. Different engineered nanoparticles show variations in size, shape, physical conformation, charges present on surface, surface area, and coatings or functional groups. Engineered nanoparticles are broadly categorized into four groups:

(i) Engineered inorganic-based nanoparticles: These nanoparticles includes metals (Silver, Titanium dioxide, Zinc oxide), metal oxides, and metal salts i.e. zero-valent metals (Iron, Silver, gold);
(ii) Engineered organic-based nanoparticles: This group of nanaoparticles includes carbonaceous nanomaterials and nanosheets e.g. carbon nanotubes, fullerenes ($C_{60}$ and $C_{540}$);
(iii) Engineered polymer-based nanoparticles: Polymeric nanoparticles are made by reaction of various elements with organic polymers e.g. polyvinyl pyrrolidone-coated (Silver, Titanium oxide, Zinc oxide nanoparticles), polyethylene glycol coated (Silver, Silicon dioxide, Titanium oxide, Gold nanoparticles), Polyvinylpyrrolidine, and polyethylene glycol coated magnetic polymers (Iron, Cobalt); and
(iv) Miscellaneous engineered nanoparticles: These include quantum dots, dendrimers, and graphene nano-foils (Nowack and Bucheli 2007).

The potential toxicity of engineered nanoparticles reported both beneficial and harmful impact on different plants (Siddiqi and Husen 2017; Tripathi et al. 2017). Different positive and negative implications of engineered nanoparticles in different plant species have been overviewed in Table 11.1. Engineered nanoparticles can boost agricultural production by various methods such as pesticides enfolding of nanoparticles for its controlled delivery; nanodevices for the delivery of genetic material into selected plants; nanofertilizer for improvement in crop nutrition, growth, yield and soil toxicity; nanopesticide to combat crop pathogens; nanoherbicide for weed management and nanosensors for rapid detection of plant pathogens and improvement of soil fertility (Campos et al. 2015; Oliveira et al. 2015; Maruyama et al. 2016). As cited in an example of tomato plants (*Lycopersicon esculentum*) grown in soil amended with multi-walled carbon nanotubes (MWCNTs) significantly increased two times the number of flowers and fruits in comparison to plants of un-amended soil, most likely this mechanism is associated with activating the genes or proteins important for plant maturation as well as enhancing plant biomass (Khodakovskaya et al. 2013). Liu and Lal (2015) reported significant increase in growth and seed yield in soybean (*Glycine max*) treated with hydroxyapatite nanoparticles (phosphorous nanofertilizers) as compared to a regular phosphorous fertilizer.

Nanofertilizers besides enhancing plant growth rate are released at slower rates in soil that reduce the surface runoff of these nutrients from fields and decreases risks associated with environmental pollution. Nanopesticides are more stable, soluble, released slowly and effective in pest management as compared to regular pesticides (Duhan et al. 2017). Fluorescent dendrimers, polymeric nanoparticles, liposomes have been recently reported to be used as nanopesticides (Liu et al. 2015). Nanoherbicides like organic and polymeric nanoparticles are distributed in fields via nanocarriers in the form of nanospheres or nanocapsules (Tanaka et al. 2012). Nanomaterials are also developed as biosensors in the crop fields. For instance, single-walled carbon nanotubes (SWNTs) were fluorescent near infrared region for tracking of nitric oxide in mouse-ear cress (*Arabidopsis thaliana*) by Giraldo et al. (2014). Considering the potential benefits of engineered nanoparticles in agricultural sector many countries of the world are giving considerable initiatives for evaluation of suitability and compatibility in integration of nanotechnology with agriculture.

The ever-escalating need for novel engineered nanoparticles leading to their severe release into the environment causes pollution. Moreover, due to minute sizes of engineered nanoparticles and correspondingly enhanced reactivity allow them to interact more efficiently than larger particles thereby posing potential threat to environmental and human health (Rosenfeldt et al. 2014). Also indicated by multiple studies, nanosized particles are more reactive as compared to microsized particles because with decreasing size of nanoparticle, the surface area is increased that displaces a larger population of its atoms or molecules on outer side of nanoparticle instead of its inside (Begum et al. 2011). Therefore, the behavior of surface atoms/molecules dominates the behavior of internal atoms/molecules that affects the

**Table 11.1** Positive/negative impact of engineered nanoparticles on various plant species

| Engineered nanoparticles | Plants | Beneficial impact | Detrimental impact |
|---|---|---|---|
| **Engineered inorganic nanoparticles** | | | |
| Ag ENPs | Brown mustard<br>Common watermelon<br>Pumpkin<br>Radish<br>Corn | Stimulates sapling maturation in common watermelon and pumpkin; stimulates growth and of biomass of brown mustard and improves its photosynthetic quantum yield | Harmful effect on growth of corn and its roots, reduced maturation of radish saplings and induction of its oxidative stress |
| Au ENPs | Climbing lily<br>Barley<br>Rice<br>Tomato | Neutral toxic response in barley and tomato; enhanced seed sprouting of climbing lily and its maturation | Strong accumulation of nanoparticles in the roots of plants |
| CuO ENPs | Mouse ear cress<br>Cucumber<br>Rice<br>Wheat | Provision of vital nutrients required for growing plant due to the occurrence of copper in nanoparticle; enhanced phytotoxicity with increased amount of reactive oxygen species enzymes | Reduction of plant biomass is related to amount of dose of CuO ENPs |
| **Engineered organic nanoparticles** | | | |
| Carbon-based<br>Fullerene ENPs | Soybean<br>Balsam-pear<br>Eastern cottonwood<br>Tomato<br>Corn | Increased fruit yield and biomass in balsam-pear | Reduced the aggregation of synthetic pesticides in tomato, corn, zucchini and soybean; at high concentrations of fullerene ENPs reduced biomass of corn and soybean; enhanced absorption of TCE in eastern cottonwood |
| MWCNTs<br>SWCNTs | Mouse ear cress<br>Tobacco<br>Arugula<br>Cotton<br>Rice<br>Tomato | Increased sprouting, height of plant and flowering of tomato; increased growth of tobacco callus cells and metabolism; enhanced potency in deoxyribonucleic acid delivery into tobacco, cotton and arugula | Chromatin condensation and apoptosis of rice protoplasts and mouse ear cress; stress-inducing ROS accumulation in protoplasts |
| **Engineered polymer type nanoparticles** | | | |
| Ag ENPs | Italian ryegrass<br>Corn<br>Water hyssop<br>Mung bean<br>Common wheat | Enhance plant growth in italian ryegrass and corn | Decreased root elongation and biomass yield of italian ryegrass; damaged epidermal layer and root tip in water hyssop; various concentrations of ENPs influenced the mung bean seed sprouting; drop in weight of shoot, reduction in growth and enhanced yield of biomass in common wheat |

(continued)

Table 11.1 (continued)

| Engineered nanoparticles | Plants | Beneficial impact | Detrimental impact |
|---|---|---|---|
| ZnO ENPs | Onion | Increases size of plant; advantageous for seed sprouting but at lower concentration; modified microbial enzymatic activities of soil; reduced flowering time and improved seed production in onion | Obstruction of root elongation at higher concentration in lettuce, radish, corn, rape, cucumber and ryegrass |
| | Mouse ear cress | | |
| | Rape | | |
| | Cucumber | | |
| | Lettuce | | |
| | Ryegrass | | |
| | Rice | | |
| | Radish | | |
| | Corn | | |
| **Miscellaneous types of nanoparticles** | | | |
| CdSe/ZnS QDs | Onion | Detects plant pathogen efficiently when used as biosensors | Enhanced production of ROS; movement of nanoparticles into trophic chain; reduction in viability of plant cell; decrease in length of root |
| | Mouse ear cress | | |
| | Alfalfa | | |

Sources: Goswami et al. (2019) and Sanzari et al. (2019)

*ENPs* Engineered nanoparticles, *MWCNTs* Multi-walled carbon nanotubes, *SWCNTs* Single-walled carbon nanotubes, *QDs* Quantam dots, *TCE* Trichloroethylene, *ROS* Reactive oxygen species

properties of nanoparticles. Various experimental studies have also confirmed that higher surface to volume ratio of engineered nanoparticles than micro-sized particles changes its structural and physiochemical properties (Currie and Perry 2007; Lee et al. 2012).

It was elucidated in an experiment of application of Fluorescein isothiocyanate (FITC) labeled silica nanoparticles and photo-stable Cadmium-Selenide quantum dots on rice seeds showed that sprouting in seeds was suppressed with quantum dots whereas induction in seed germination was observed influorescein isothiocyanate labeled silica nanoparticles (Torney et al. 2007). Similarly, engineered nanoparticles of 21 nm in size ($TiO_2$) leads to 43 times more inflammation in organisms as compared to particles sized 250 nm having identical mass (Castiglione et al. 2011; Uhram et al. 2013). Greater surface area of smaller particles as compared to larger particles of similar mass of material may result into increased inflammation. The presence of increased concentration of nanoparticles within the farming system severely declines seed germination, photosynthesis activity that reduces the plant growth, transpiration rate that slows down a continued water loss, and ultimately influences total biomass yield (Hossain et al. 2016) by altering its structural, anatomical, biochemical, physiological, and genetic make-up (Tripathi et al. 2016). Engineered nanoparticles could thus be considered as activator for growth of plants along with under a novel category of emerging contaminants.

In an agro-ecosystem, nanoparticles mainly interacts directly with soil and plants, therefore study of engineered nanoparticles in soil–plant system could be reviewed among the primary system for knowing the fate and transport of these particles in an agro-ecosystem. Most of the studies demonstrated that engineered nanoparticles can cause significant phytotoxicity, emphasizing need for further research on effect of these particles on agro-ecosystem.

## 11.3 Pathways of Engineered Nanoparticles into Agricultural Soil

Agro-ecosystem are more vulnerable to engineered nanoparticles than to natural nanoparticles, because these factory-made nanoparticles find their way to different environmental compartments due to unsuitable disposal, or released by accident from landfill sites, sewage sludge, and during generation of wastewater in their process of manufacturing (Cornelis et al. 2014). Engineered nanoparticles can thus enter the agro-ecosystem directly via handling (fertilizers, pesticides for remediation of contaminated soils) or via accidental release (Gottschalk et al. 2009).

One of the important challenges faced by nanotechnology researches is the use of various nanoparticles for soil decontamination resulting from pollution by industrialization and urbanization. The engineered nanoparticles extensively used for soil remediation includes application of nano-iron (e.g. zero-valent iron nanopaticles, bimetallic iron nanoparticles and magnetite nanoparticles), titanium oxide nanoparticles, titanantes, nano-carbon (e.g. carbon nanotubes and graphene), and group of nanoparticles such as gold, silver, palladium, and amphiphilic polyurethane (Li et al. 2016). Besides soil remediation, it also significantly affects soil biology and phytotoxicity of nanoparticles (Fernandes et al. 2017; Jain et al. 2017) because engineered nanoparticles once released into polluted soils also interacts and affects agro-ecosystem components (soil, crops). The co-existence of nanoparticles with contaminants in soil may ultimately change their behavior and modify their phytotoxicity for soil biota. In an another example of crucial unexpected exposure pathway for engineered nanoparticles into agricultural soils is the waste water treatment plant (WWTP) sludge (Gottschalk et al. 2009; Johnson et al. 2011) being practiced in some countries as land reclamation due to presence of high organic matter and higher amount of phosphorus and nitrogen nutrients in sludge (Weggler-Beaton et al. 2003) resulting in unintentional exposure of agricultural soils to nanoparticles. On an average, more than 90% of mass of silver (Ag-nanoparticles), zinc oxide (ZnO-nanoparticles), titania ($TiO_2$-nanoparticles), ceria ($CeO_2$-nanoparticles), fullerenes ($nC_{60}$), carbon nanotubes perforated into waste water treatment plant are found in the sludge (Johnson et al. 2011; Doolette et al. 2013). Once these nanoparticles are discharged into the agricultural soils ultimately their fate is decided by abiotic physiochemical transformation processes that affect their bioavailability to soil biotic components.

As confirmed by many studies that among the main agro-ecosystem compartments (soil, air, water), the amount of engineered nanoparticles in soil is usually high as compared to its quantity in air or water suggesting that the soil is the prime environmental sink for discharge of these nanoparticles into other compartments (Kwak and An 2016; Xie et al. 2017). Nanoparticles interactivity with soil could intensely affect its fate and transport in agro-ecosystem. Soil characteristics of pH value, soil mineralogy, soil organic matter content, and ionic strength have the potential to affect the soluble and toxic properties of engineered nanoparticles to soil biotic factors (Bradfield et al. 2017).

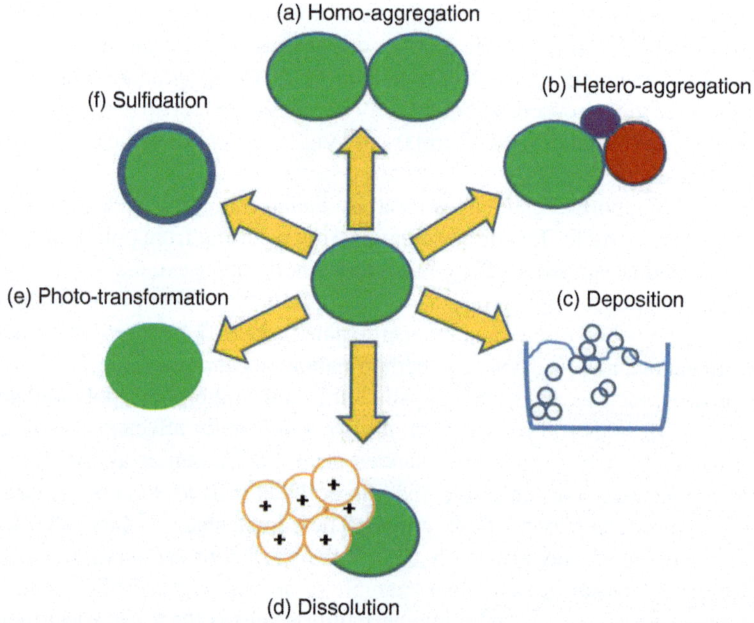


**Fig. 11.1** The fate of engineered nanoparticles released into agricultural soils is determined by various abiotic transformation processes that includes (**a**) homo-aggregation (**b**) hetero-aggregation (**c**) deposition (**d**) dissolution (**e**) photo-transformation (**f**) sulfidation

## 11.4 Fate of Engineered Nanoparticle in Agricultural Soils

Engineered nanoparticles released into the agricultural system, immediately undergo various transformation processes such as aggregation, deposition, dissolution and transformation, photo-transformation and sulfidation that consequently alter the movement and bioavailability of engineered nanoparticles (Lowry et al. 2012; Amde et al. 2017) (Fig. 11.1a–f). Various external factors such as stability, ionic strength, pore size, shape, different capping agents, synthesis method, pH value, valency and surface charge of engineered nanoparticles can play vital roles in its physico-chemical transformation under various conditions of agro-ecosystem. The agricultural key factors which significantly influence the fate as well as bioavailability of engineered nanoparticles for understanding their dynamic transformations are summarized in Table 11.2. Ultimately, the fate of engineered nanoparticles in an agro-ecosystem depends upon abiotic transformation processes and impact of external factors that can accelerate or inhibit the transformation processes are discussed in subsequent sub-sections.

### *11.4.1 Aggregation*

As engineered nanoparticles are discharged into the agro-ecosystem, they associate with array of dissolved or undissolved, inorganic or organic compounds that ultimately influences its dynamics of aggregation and stability of colloids (Sani-kast et al. 2017). Aggregation process proceeds through two distinct pathways of homo-aggregation (when aggregation occurs between identical particles) (Fig. 11.1a) and hetero-aggregation (when aggregation occurs between different particles) (Fig. 11.1b). Engineered nanoparticles being relatively small sized have a comparatively smaller electrostatic barrier in the beginning and hence aggregate rapidly as compared to larger particles, the effect is more pronounced at higher concentration of nanoparticles. Homo-aggregation usually results into a decreased movement of engineered nanoparticles in agricultural soils (Phenrat et al. 2009), larger aggregates have a larger drag coefficients that results into their enhanced deposition (Veerapaneni and Wiesner 1996). Derjaguin-Landau-Verwey-Overbeek (DLVO) theory also explains the aggregation dynamics in natural ecosystems (Liang et al. 2013). It states that homo-aggregation rates decrease if electrostatic (or steric) barriers inhibit collisions between particles so that engineered nanoparticles suspension could be considered as kinetically stable. As compared to homo-aggregation, hetero-aggregation of particles is quantitatively more important process in soil pores (Kim et al. 2012; Zhou et al. 2012). Favorable hetero-aggregation occurs when the surface charge of both the particle types are opposite because of the lack of an electrostatic barrier blocking hetero-aggregation (Kim et al. 2012). Engineered nanoparticles often have a negative surface charge, because of commercial coatings or presence of adsorbed negatively charged dissolved organic matter. The major

**Table 11.2** Effect on different types of engineered nanoparticles by abiotic transformation processes, determining factors, and environmental implications

| Transformation processes | Factors | Environmental implications |
|---|---|---|
| **Aggregation** | | |
| Fullerene ENPs | Hydrophobicity | The hydrophobic nature of carbon species of fullerene allows it to form complexes with other molecules |
| $TiO_2$ ENPs | Humic acid | Mostly –COOH group of $TiO_2$ENPs ligates with humic acids |
| **Homo-aggregation** | | |
| Ag ENPs | Artificial coating | Coating of Ag ENPs with citrate leads to steric stabilization |
| Au ENPs | pH | In alkaline conditions, Au ENPs were stable at pH 8 |
| Au ENPs | Ionic strength | During conditions of higher ionic strength leads to neutralization of negative surface charge and formation of homo-aggregates of Au ENPs |
| **Hetero-aggregation** | | |
| $CeO_2$ ENPs | Inorganic colloids of soil | $CeO_2$ ENPs forms hetero-aggregates with inorganic colloids present in soil |
| Fe EPNPs | pH value | Hetero-aggregation of Fe EPNPs are produced under acidic conditions |
| **Deposition** | | |
| MWCNTs ENPs | Surface charge, hydrophobicity, $Na^+$ | MWCNTs stabilized by soil mineral smectite in aqueous phase whereas another soil mineral kaolinite, having less surface charge for each sodium concentration induced its removal |
| ZnO ENPs | Ionic strength | With the rise in ionic strength, ZnO ENP deposition rate increases in soil |
| **Dissolution** | | |
| Ag ENPs | Artificial coatings | The oxidative dissolution is reduction in Ag ENPs with metallic core |
| Ag ENPs | Size | Oxidation kinetics is faster for smaller than larger particles |
| **Sulfidation** | | |
| Ag ENPs | Anaerobic environment | Ag ENPs along with $HS^-$ ions forms ligand with $Ag^+$ions |
| **Phototransformation** | | |
| Ag ENPs, Au ENPs | Sunlight | Photo-transformation under sunlight leads to reduction of $Ag^+$ and $Au^{3+}$ ions |

Sources: Goswami et al. (2017) and Bundschuh et al. (2018)
*ENPs* Engineered nanoparticles, *ENPNPs* Engineered polymeric nanoparticles, *MWCNTs* Multi-walled carbon nanotubes

mechanisms of hetero-aggregation besides electrostatic forces include bridging, hydrogen bonding, and chemical bonding.

The influence of various factors like soil pH, ionic strength, and ionic balance largely affects distribution as well as aggregation of engineered nanoparticles in the agricultural soil. Hetero-aggregation process more often leads to reduced particle mobility than homo-aggregation (Zhao et al. 2012; Cornelis et al. 2013). On contrary, both homo-aggregation and hetero-aggregation may also enhance the mobility of nanoparticles (due to size-exclusion effect discussed in Sect. 11.5.6).

### *11.4.2 Deposition*

When a colloid or engineered nanoparticle approaches a solid phase or surface of soil aggregates, it attaches by Brownian diffusion or direct interception, and/or by gravitational sedimentation, collectively called as deposition (Fig. 11.1c). The brownian motion is more profound for smaller particles like engineered nanoparticles that have a high diffusion coefficient, while interception takes place as these particles encounters a larger particle or pore wall. Favorable deposition occurs when charges of both particle and surface are opposite as in case of hetero-aggregation. Favorable deposition also takes place when the repulsive energy barrier that prevents particle deposition is small or absent. Many governing factors such as clay content, extractable major cations (such as Iron, Aluminum and Manganese), dissolved organic carbon, and soil pH control the deposition of engineered nanoparticles in the agricultural soils.

### *11.4.3 Dissolution and Transformation*

The most widely investigated transformation reaction is dissolution (Fig. 11.1d) may be due to increased solubility of engineered nanoparticles with the decrease in size that have also been experimentally verified for Ag and Zinc oxide engineered nanoparticles (Zhang et al. 2011). However, size of nanoparticles affects the dissolution rate that can be altered by processes of deposition, aggregation, coating by organic matter, and/or transformation to sparingly soluble materials. The coating of Ag nanoparticles with organic matter may assist or block its dissolution, as it further depends on the presence of surface ligands and coating (Sharma et al. 2014). Moreover, the aged nanoparticles that are deposited in agricultural soils can lead to dissolution under various conditions of environment. Increasing concentration of anions ($Cl^-$, $SO_4^{2-}$, $NO_3^-$) of engineered nanoparticles is usually adsorbed by soil particles or chelation can occur that can stimulate the dissolution and/or transformation of nanoparticles (Cornelis et al. 2012). However, presence of anions like $Cl^-$ and $PO_4^{3-}$ with organic coating can delay the dissolution of engineered nanoparticles (Ho et al. 2010). Engineered nanoparticles like $CeO_2$ being comparatively stable in agricultural soils can accumulate in soil if not leached out (Cornelis et al. 2011).

### 11.4.4 Photo-Transformation

Photo-transformation (Fig. 11.1e) is another significant way of engineered nanoparticles physiochemical transformation. Sunlight plays a major role during photo-transformation by influencing nanoparticle's oxidation state, formation of reactive oxygen species, and its perseverance (Cheng et al. 2011; Lowry et al. 2012). For instance, coating of Ag nanoparticles with gum arabic and polyvinylpyrrolidone showed aggregation and sedimentation under strong sunlight irradiation as it induced oscillating dipole–dipole interaction (Cheng et al. 2011).

### 11.4.5 Sulfidation

Sulfidation (Fig. 11.1f) is a chemical reaction of nanoparticles with sulfur under various environmental conditions; it commonly includes nanoparticles of Ag, ZnO, and CuO (Nair et al. 2010; Thuesombat et al. 2014). For instance, Ag nanoparticle reaction with sulfide may form core–shell $Ag_0/Ag_2S$ structures or forms hollow structures of $Ag_2S$ (Zhang et al. 2011). Sulfidation is one of the main engineered nanoparticle's transformations that occurs in waste water treatment plants, but also occurs in sulphur-rich agricultural soils. The study by Ma et al. (2013) on sulfidation of ZnO nanoparticles indicated it can reach upto 100% conversion within 5 days of incorporation of adequate amount of sulfide through mechanisms of dissolution and re-precipitation. This process may cause aggregation and decreased surface charge of nanoparticles.

During rainy season and/or irrigation, engineered nanoparticles can also interact with dissolved organic matter rich in humic and fulvic acids of soil pore water (Rosenfeldt et al. 2014) that may cause groundwater contamination as the soil pore water containing dissolved organic matter attaches to dissolved metal ions resulting into formation of humic acid-mediated complexes (Rodrigues et al. 2016). The complexes formed coats the nanoparticle leading to stabilization of particle size (Hall et al. 2009) via repulsive forces such as steric and/or electrostatic (Hyung et al. 2007) forces further precipitating engineered nanoparticles.

Hence, the fate and bioavailability of engineered nanoparticles in an agro-ecosystem depends on their characteristic properties and their respective environment. However, for efficient application of engineered nanoparticles in agro-ecosystem injected nanoparticles should have sufficient mobility for its proper distribution around the injection point. Therefore, it is equally important to understand the transportation behavior of engineered nanoparticles in agro-ecosystem.

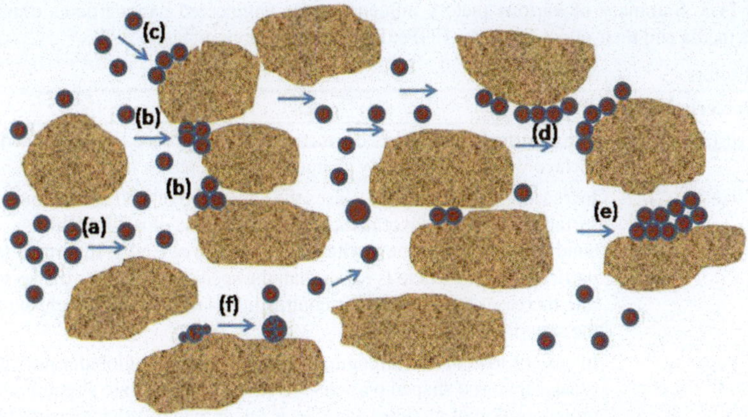


**Fig. 11.2** Engineered nanoparticles released into agricultural soils are transported from their pont of application for uniform distribution undergoes various transformation processes that includes (**a**) advection and dispersion (**b**) straining (**c**) adsorption (**d**) blocking (**e**) ripening (**f**) aggregation

## 11.5 Transport of Engineered Nanoparticles in the Agricultural Soils

Transportation of engineered nanoparticles in agricultural soil plays a major role in determination of the fate of these nanoparticles in agro-ecosystem. The chemical stability of engineered nanoparticle changes with its ageing that can further govern their direction of movement or linkage with additional particles (Scheckel et al. 2009). Majority of the reported data on nanoparticles movement is based on porous media (Solovitch et al. 2010) while only a few studies involved agricultural soil (Cornelis et al. 2013). The main mechanisms of transport of engineered nanoparticle in agro-ecosystem includes advection, dispersion, straining, adsorption, blocking, ripening, aggregation (Fig. 11.2a–f) and size exclusion that are further influenced by properties of engineered nanoparticle, agricultural soil/porous media and, state of water flow through soil/porous media (Chang et al. 2016; Babakhani et al. 2017). Various factors affecting transport of engineered nanoparticles in porous media and their main findings are summarized in Table 11.3.

### *11.5.1 Advection and Dispersion*

The advection is a phenomenon in which the movement of nanoparticles is along the motion of flowing water (Zhong et al. 2017) whereas hydrodynamic dispersion includes the processes of mechanical dispersion and diffusion (Fig. 11.2a). Mechanical dispersion refers to spreading of engineered nanoparticles with the direction of flowing water because of heterogenous porous medium and

**Table 11.3** Summary of various factors influencing the engineered nanoparticles transport in porous media and their major findings (ENPs: Engineered nanoparticles)

| Key Factors | Major findings |
|---|---|
| **Characteristics of ENPs** | |
| ENP surface charge | The positively charged ENPs increases its adherence to porous media surface whereas negatively charged ENPs increases its mobility |
| ENP size | Increasing particle size increases the adherence of ENPs to porous media surface only under favorable conditions. In case of not favorable conditions, there can be a critical particle size of ENPs. If particle size of particular nanoparticle is lesser than the specified value, its transportation rate increases but condition is contrasting as the size of particle exceeds the specified value |
| ENP shape | In case of lower ionic strength, the transportation of globular particles is greater than rod shaped particles. In case of higher ionic strength, the transportation of globular particles is lesser than rod shaped particles. |
| ENP density | With the increase in nanoparticle density, more nanoparticles attaches to porous media |
| ENP concentration | With the increase in nanoparticle concentration, the attachment of nanoparticle with porous media is also enhanced that may also trigger blocking and ripening process |
| **Characteristics of porous media** | |
| Media type | Transportation of nanoparticles is slower in natural porous media as compared to homogenous media. |
| Media size | With the increase in medium size the transportation of ENPs is enhanced whereas with decrease in medium size in few instances can also increase the rate of transportation |
| Media surface roughness | With the increase in medium surface roughness, the adherence of ENPs is also enhanced |
| Moisture content | With the decrease in moisture content the adherence of ENPs is enhanced |
| Organic matter content | Adsorption of organic matter on porous medium surface increases the adherence of ENPs |
| Mineral content | Occurrence of mineral coating on the surface of medium increases adherence of ENPs |
| Clay particles, biofilms, plastic particles, bacteria cells | Occurrence of clay particles, plastic particles, biofilms, and bacteria cells also increases rate of transportation of ENPs |
| **Flow conditions** | |
| Flow velocity | With the increase in flow velocity the rate of transportation of nanoparticles also increases |
| Flow direction | The transportation of nanoparticles is greater in horizontal column as compared to vertical column |
| Flow interruption | The interruption in flow increases adherence of nanoparticles |

Source: Liu et al. (2019)

pore-structure (Mehmani and Balhoff 2015). Factors of particle size and flow velocity have a significant impact on advection and dispersion processes. For nanoparticles having diameter above 1 μm, the mean free path length be one or two particle diameters, such viscous liquids limits the mobility, and therefore, the diffusion is insignificant. Similarly, during slow velocity of fluid, dispersion is of lesser impact but movement may occur from the diffusion process on account of differing concentration gradients.

### *11.5.2 Straining*

Straining (Fig. 11.2b) involves confinement of nanoparticles in narrow pore spaces that do not allow its passage (Porubcan and Xu 2011). Straining can be subdivided into surface straining, wedge, and bridge. In surface straining increased nanoparticle concentration causes arrival of numerous particles simultaneously at surface of pore opening and stuff in it by arching action, it can be ameliorated by increasing flow velocity. Wedging refers to confinement of nanoparticles at two connecting surfaces of porous media excluding any intervention of particles, whereas bridging results from simultaneous arrival of particles that plug a pore throat (Zhang et al. 2012; Dong et al. 2015).

### *11.5.3 Adsorption*

In addition to physical straining, nanoparticles may also be intercepted and retained due to adsorption (Fig. 11.2c) in agricultural soil by action of interfacial forces (dipole and hydrogen force) in-between particles and adsorbent surface. Dipole forces are also known as short-range forces as they are effective only if distance between particles and porous medium surface is in the range of 5 nm. Hydration forces are called as long range forces because they are effective even if particle is separated by 100 nm from a porous medium. The DLVO theory (Liu et al. 2019) is also extensively utilized to understand the interaction forces between particles and surface that control the nanoparticles adsorption. This theory states that particles attach to the porous medium by irreversible and reversible adsorption. In irreversible adsorption, particles associated with porous medium surfaces are sufficiently solid enough that enable them to be kept permanent, this occurs mainly under condition of high ionic strength of particles, leading to disappearance of energy barrier and the particles reaches the DLVO primary minimum (Zhong et al. 2017). On contrary, in reversible adsorption, the nanoparticles associated with porous medium are not solid enough to be kept permanent, these may be removed, the adsorption is kept temporary and the nanoparticles reaches the DLVO secondary minimum (Phenrat et al. 2010) or due to nanoscale surface heterogeneities at shallow primary minimum (Torkzaban and Bradford 2016; Wang et al. 2016). Mostly reports on transportation of engineered nanoparticles have stated that adsorption has a major role in their transportation in either form of reversible or irreversible mechanism (Braun et al. 2015; Babakhani et al. 2018). Engineered nanoparticles having abundant atoms or functional groups on surface that interact with porous media by various processes and it usually results into irreversible adsorption (Babakhani et al. 2017), these nanoparticles are normally not removed until there is remarkable physical or chemical disturbance in agro-ecosystem.

### 11.5.4 Blocking and Ripening

Blocking effect (Fig. 11.2d) is the limited capacity of porous media surface for particle adsorption and as this capacity is filled, the plateau of breakthrough curve starts rising gradually due to reduced deposition rate with the increased amount of adhered nanoparticles (Saiers et al. 1994). Blocking phenomenon has been documented in various engineered nanoparticles like carbon nanoparticles, silver nanoparticles, quantum dots, graphene oxide nanoparticle, etc. (Braun et al. 2015; Sun et al. 2015a; Hu et al. 2017). Whereas in case of ripening (Fig. 11.2e), the interaction among the particles on soil surface are stronger than the particle-surface interactions. During ripening process, the nanoparticles that are attached to soil particles will act as supplementary accumulators for adding-on of mobile particles that ultimately leads to multilayer assemblage of particles on the surface of agricultural soil. Ripening phenomenon is in contrast to the blocking where the rate of deposition is enhanced with time and it gives gradual descending plateau of breakthrough curve (Bradford and Torkzaban 2008). With increased inflow concentration of engineered nanoparticles to the porous media may finally result into clogging of pores (Tosco and Sethi 2010). Ripening phenomenon has been observed for various engineered nanoparticles such as hydroxyapatite nanoparticles (nHAP), graphine oxide nanoparticles (GONPs), nano-scale zero valent iron (nZVI), etc. (Lanphere et al. 2014; Wang et al. 2014; Basnet et al. 2015).

### 11.5.5 Aggregation

Aggregation (Fig. 11.2f) is a pervasive process that occurs between the particles of liquid and solid phase. Aggregation is principally significant for transportation of engineered nanoparticles in porous media as it causes increased particle size, the increased size can further reduce the collision frequency arising between the particles and the surface of porous media (Lin and Wiesner 2012). The main mechanisms involved in particle aggregation are perikinetic aggregation, orthokinetic aggregation, and differential sedimentation (Babakhani et al. 2018). In perikinetic aggregation collisions occurs among the nanoparticles involving Brownian movement while in orthokinetic aggregation collisions among nanoparticles areas a result of any motion or flow in a fluid. The process of differential sedimentation involves settlement of larger particles faster as compared to smaller ones, collisions happen with particles that are present in their track (Babakhani et al. 2017). Aggregation of nanoparticles is influenced by physicochemical properties of engineered nanoparticles (particle size, chemical composition surface, and characteristics) and water chemistry parameters (ionic strength, pH, and organic matter content) (Raychoudhury et al. 2012; Sygouni and Chrysikopoulos 2015). The development of nanoparticle clusters has been demonstrated practically for engineered nanoparticles of Ag, $nTiO_2$, nZnO, etc. (Neukum et al. 2014; Sygouni and Chrysikopoulos 2015; Sun et al. 2015b).

### *11.5.6 Size Exclusion*

The size exclusion is defined as a process for rapid rate of transportation of bigger particles than smaller ones in porous media as the particle having bigger size are confined by size exclusion effect, these particles scatter less and hence moves quickly (Keller et al. 2004). Intuitively, in case of small sized engineered nanoparticles, it is easy to dismiss size exclusion but these particles during transportation process have pronounced capacity for aggregation into large sized particles that further undergoes effect of size exclusion. For instance, an engineered nanoparticle can link with another kind of engineered nanoparticles or can link with clay particles in agricultural soil that is called as hetero-aggregation (Abdel-Fattah et al. 2013). Thus, size exclusion of engineered nanoparticles is particularly noticeable in co-transport/hetero-aggregation of engineered nanoparticles and biofilm coated porous media (Afrooz et al. 2016). Investigations on co-transport of single walled carbon nanotubes with Au nanospheres (Afrooz et al. 2016) or multi-walled carbon nanotubes with $nTiO_2$ (Wang et al. 2014) showed that by forming aggregates these engineered nanoparticles were able to move in porous media.

Ultimately, the transportation of engineered nanoparticles in agricultural soils alters their bioavailability to crop plants. The plant uptake of nanaoparticles depends on various characteristics of naoparticles (size, chemical composition, net charge, and surface functionalization), route of application, interactions with agricultural components (soil, water, microbiota) and the constraints due to physiology and the multi-facet anatomy of plants.

## 11.6 Delivery Methods and Primary Interactions Between Engineered Nanoparticles and Agricultural Plants

Engineered nanoparticles could be injected to vegetative plant part like roots or to the leaves. Atmospheric engineered nanoparticles have the ability to pierce through leaf surfaces by pores of stomata (Eichert et al. 2008; Larue et al. 2014) and this stomatal pathway is of high capacity as pores have size of above 10 nm and rate of translocation is also higher in comparison to other aerial parts of plant (surfaces of shoot, bark) but this pathway also shows variability in permeability that makes it mostly uncertain (Nair 2016). In an experimental analysis of leaf-root translocation pathway, engineered nanoparticle of ceria (Ce) was found in cucumber (*Cucumis sativus*) roots as its leaves were exposed to nanoparticle $CeO_2$ indicated translocation of either nanoparticles or cerium ions (Hong et al. 2014). Translocation in plants coincides with solubility of nanoparticles. Also cited in an experimental studies on the formation of accumulates of engineered nanoparticles (e.g. application of insoluble nanoparticle $TiO_2$) on photosynthetic surface that leads to heating of foliar surface. The result is alteration of gas exchanging capacity due to obstruction in stomata which further alters physiological and cellular functions of plants. The shoot surfaces mostly function as lipophilic barrier as they are normally enveloped by a cuticle made up of biopolymers (cutin, cutan) and associated waxes to protect above-ground plant primary organs (Eichert et al. 2008; Kurepa et al. 2010). However, the surfactants can also improve the penetrability of cuticle for nanoparticles (Nadiminti et al. 2013). For instance, adding a surfactant in maize enabled uptake of quantum dots (4–7 nm) (Hu et al. 2010).

The dynamics of nanoparticle uptake is more complex in roots as compared to the aerial part of plant due to presence of several factors such as mucilage and exudates, symbiotic organisms and, physiochemical-transformation processes in agricultural soil. For example, root mucilage and exudates generally passed into the rhizosphere and can promote nanoparticle adhesion to the root surface that further can increase the internalization of nanoparticle or it can alternatively trigger nanoparticle confinement and its accumulation (Avellan et al. 2017; Milewska-Hendel et al. 2017). Organic matter like presence of humic acids in the agricultural soil may lead to an enhanced stability and better bioavailability of nanoparticles, while salt ions may trigger contrary effects and can induce precipitation (Navarro et al. 2008). Once the engineered nanoparticles penetrate into roots are translocated into leaves with flow of water during transpiration. For instance, exposure of $CeO_2$ engineered nanoparticles to wheat (*Triticum aestivum*) roots was detected in leaf veins (Lin et al. 2009). Damaged plant parts and injuries may also act as feasible pathways for internalizing nanoparticles in aerial and sunken plant parts (Al-Salim et al. 2011). Trichomes (fine outgrowths) on plant organs can influence the dynamic interaction at the plant's surface by trapping nanoparticles on its surface, therefore enhancing its duration of persistence into plant tissues. Nanoparticles may penetrate into plant epidermis by either of methods discussed above and before entering plant's vascular system, there are two major pathways for particle's movement throughout tissues: the apoplastic pathway and the symplastic pathway (Pacheco and Buzea 2017). Thereafter, subsequent cell internalization occurs especially by process of endocytosis (Valletta et al. 2014; Palocci et al. 2017). The ion channels and transport proteins that are embedded could even mediate the passage of engineered nanoparticles across the cell membrane. Inside cytoplasm, these nanoparticles can also adhere to various organelles present inside cytoplasm and interferes with metabolic activity at that particular region of the plant (Zhang et al. 2008).

Among the important factors that ultimately affect internalization of engineered nanoparticles into plants includes stomatal aperture, cell wall composition, presence/absence of cuticle, mucilage and symbiotic microbes. Therefore, for assessment of the risks and/or potential benefits associated with engineered nanoparticles, it is also necessary to explore interactions between nanoparticles and plants. Indeed, the "nano-bio" interface governs the nanotoxicity between nanoparticles and the surface of plant comprising of various biological components like membranes, phospholipids, endocytic vesicles. Finally, it is essential to find the balance point between safety and application of nanoparticles.

## 11.7 Conclusion

The escalating demand of engineered nanoparticles for sustainable agricultural production has also lead to agro-ecosystem contamination. Increased application of novel engineered nanoparticles in agriculture requires a regular survey because vulnerability to these nanoparticles may create a broad environmental risk. Engineered nanoparticles are mobilized and accumulated in agricultural soils via direct release from nanofertilizers, nanoherbicides or released accidentally from wastewater treatment plants potentially resulting into the impact on growth and productivity of crops. Recent approaches suggest that more systemic approach is urgently required to optimize the bio-functionalization of nanoparticles for application in plants but also to completely uncover the role of soil properties affecting fate, transport, bioavailability and phytotoxicity of nanoparticles that limits the absolute approval and readiness to accept the agri-nanotechnology. Moreover, nanoparticle surface properties and its concentration are also needed to be characterized carefully throughout the experiments. At present, measurement methods also require special attention for assessment of fate and transport of engineered nanoparticles in agricultural soils. Furthermore, strong emphasis is required on restricted data and vast knowledge gaps of engineered nanoparticles trans-generational and trophic chain transmission and risk factors associated with the chain transfer.

## References


Abdel-Fattah AI, Dongxu Z, Hakim B, Sowmitri T, S Doug W, Keller AA (2013) Dispersion stability and electrokinetic properties of intrinsic plutonium colloids: implications for subsurface transport. Environ Sci Technol 47:5626–5634. https://doi.org/10.1021/es304729d

Afrooz AR, Das D, Murphy CJ, Vikesland P, Saleh NB (2016) Co-transport of gold nanospheres with single-walled carbon nanotubes in saturated porous media. Water Res 99:7–15. https://doi.org/10.1016/j.watres.2016.04.006

Al-Salim N, Barraclough E, Burgess E, Clothier B, Deurer M, Green S, Malone L, Weir G (2011) Quantum dot transport in soil, plants, and insects. Sci Total Environ 409:3237–3248. https://doi.org/10.1016/j.scitotenv.2011.05.017

Amde M, Liu JF, Tan ZQ, Bekana D (2017) Transformation and bioavailability of metal oxide nanoparticles in aquatic and terrestrial environments. A review. Environ Pollut 230:250–267. https://doi.org/10.1016/j.envpol.2017.06.064

Avellan A, Schwab F, Masion A, Chaurand P, Borschneck D, Vidal V, Rose J, Santaella C, Levard C (2017) Nanoparticle uptake in plants: gold nanomaterial localized in roots of *Arabidopsis thaliana* by X-ray computed nanotomography and hyperspectral imaging. Environ Sci Technol 51:8682–8691. https://doi.org/10.1021/acs.est.7b01133

Babakhani P, Bridge J, Doong RA, Phenrat T (2017) Continuum-based models and concepts for the transport of nanoparticles in saturated porous media: a state-of the-science review. Adv Colloid Interf Sci 246:75–104. https://doi.org/10.1016/j.cis.2017.06.002

Babakhani P, Fagerlund F, Shamsai A, Lowry GV, Phenrat T (2018) Modified MODFLOW-based model for simulating the agglomeration and transport of polymer-modified FeO nanoparticles in saturated porous media. Environ Sci Pol 25:7180–7199. https://doi.org/10.1007/s11356-015-5193-0

Basnet M, Tommaso CD, Ghoshal S, Tufenkji N (2015) Reduced transport potential of a palladium-doped zero valent iron nanoparticle in a water saturated loamy sand. Water Res 68:354–363. https://doi.org/10.1016/j.watres.2014.09.039

Begum P, Ikhtiari R, Fugetsu B (2011) Graphene phytotoxicity in the seedling stage of cabbage, tomato, red spinach, and lettuce. Carbon 49:3907–3919. https://doi.org/10.1016/j.carbon.2011.05.029

Bradfield SJ, Kumar P, White JC, Ebbs SD (2017) Zinc, copper, or cerium accumulation from metal oxide nanoparticles or ions in sweet potato: yield effects and projected dietary intake from consumption. Plant Physiol Biochem 110:128–137. https://doi.org/10.1016/j.plaphy.2016.04.008

Bradford SA, Torkzaban S (2008) Colloid transport and retention in unsaturated porous media: a review of interface-, collector-, and pore-scale processes and models. Vadose Zone J 7:667–681. https://doi.org/10.2136/vzj2007.0092

Braun A, Klumpp E, Azzam R, Neukum C (2015) Transport and deposition of stabilized engineered silver nanoparticles in water saturated loamy sand and silty loam. Sci Total Environ 535:102–112. https://doi.org/10.1016/j.scitotenv.2014.12.023

Bundschuh M, Filser J, Luderwald S, McKee MS, Metreveli G, Schaumann GE, Schulz R, Wagner S (2018) Nanoparticles in the environment: where do we come from, where do we go to? Environ Sci Eur 30:6. https://doi.org/10.1186/s12302-018-0132-6

Campos EV, Oliveira JL, Fraceto LF, Singh B (2015) Polysaccharides as safer release systems for agrochemicals. Agron Sustain Dev 35:47–66. https://doi.org/10.1016/j.carbpol.2013.10.025

Castiglione MR, Giorgetti L, Geri C, Cremonini R (2011) The effects of nano-$TiO_2$ on seed germination, development and mitosis of root tip cells of *Vicia narbonensis* L. and *Zea mays* L. J Nanopart Res 13:2443–2449. https://doi.org/10.1007/s11051-010-0135-8

Chang MP, Chu KH, Heo J, Her N, Min J, Son A, Yoon Y (2016) Environmental behavior of engineered nanomaterials in porous media: a review. J Hazard Mater 309:133–150. https://doi.org/10.1016/j.jhazmat.2016.02.006

Chen H, Yada R (2011) Nanotechnologies in agriculture: new tools for sustainable development. Trends Food Sci Technol 22:585–594. https://doi.org/10.1002/ps.1732

Cheng Y, Yin L, Lin S, Weisner M, Bernhardt E, Liu J (2011) Toxicity reduction of polymer-stabilized silver nanoparticles by sunlight. J Phys Chem C 115:4425–4432. https://doi.org/10.1021/jp109789j

Cornelis G, Ryan B, McLaughlin MJ, Kirby JK, Beak D, Chittleborough D (2011) Solubility and batch retention of $CeO_2$ nanoparticles in soils. Environ Sci Technol 45:2777–2782. https://doi.org/10.1366/14-07495

Cornelis G, Doolette C, Thomas M, McLaughlin MJ, Kirby JK, Beak DG, Chittleborough D (2012) Retention and dissolution of engineered silver nanoparticles in natural soils. Soil Sci Soc Am J 76:891–902. https://doi.org/10.2136/sssaj2011.0360

Cornelis G, Pang L, Doolette C, Kirby JK, McLaughlin MJ (2013) Transport of silver nanoparticles in saturated columns of natural soils. Sci Total Environ 463–464:120–130. https://doi.org/10.1016/j.scitotenv.2013.05.089

Cornelis G, Hund-Rinke K, Kuhlbusch T, Brink N, Nickel C (2014) Fate and bioavailability of engineered nanoparticles in soils: a review. Crit Rev Environ Sci Technol 44:2720e2764. https://doi.org/10.1080/10643389.2013.829767

Currie HA, Perry CC (2007) Silica in plants: biological, biochemical and chemical studies. Ann Bot 100:1383–1389. https://doi.org/10.1093/aob/mcm247

Dong H, Zeng G, Zhang C, Liang J, Ahmad K, Xu P, He M, Lai M (2015) Interaction between $Cu^{2+}$ and different types of surface-modified nanoscale zero-valent iron duringtheir transport in porous media. J Environ Sci 32:180–188. https://doi.org/10.1016/j.jes.2014.09.043

Doolette C, McLaughlin MJ, Kirby JK, Batstone DJ, Harris HH, Ge H, Cornelis G (2013) Transformation of PVP coated silver nanoparticles in a simulated wastewater treatment process and the effect on microbial communities. Chem Cent J 7:46. https://doi.org/10.1186/1752-153X-7-46

Duhan JS, Kumar R, Kumar N, Kaur P, Nehra K, Duhan S (2017) Nanotechnology: the new perspective in precision agriculture. Biotechnol Rep 15:11–23. https://doi.org/10.1016/j.btre.2017.03.002

Eichert T, Kurtz A, Steiner U, Goldbach HE (2008) Size exclusion limits and lateral heterogeneity of the stomatal foliar uptake pathway for aqueous solutes and water-suspended nanoparticles. Physiol Plant 134:151–160. https://doi.org/10.1111/j.1399-3054.2008.01135.x

Fernandes JP, Almeida CMR, Andreotti F, Almeida T, Mucha AP (2017) Response of microbial communities colonizing salt marsh plants rhizosphere to copper oxide nanoparticles contamination and its implications for phytoremediation processes. Sci Total Environ 581–582:801–810. https://doi.org/10.1016/j.scitotenv.2017.01.015

Ghormade V, Deshpande MV, Paknikar KM (2011) Perspectives for nanobiotechnology enabled protection and nutrition of plants. Biotechnol Adv 29:792–803. https://doi.org/10.1016/j.biotechadv.2011.06.007

Giraldo JP, Landry MP, Faltermeier SM, McNicholas TP, Iverson NM, Boghossian AA, Reuel NF, Hilmer AJ, Sen F, Brew JA, Strano MS (2014) Plant nanobionics approach to augment photosynthesis and biochemical sensing. Nat Mater 13:400–408. https://doi.org/10.1038/nmat3890

Goswami L, Kim KH, Deep A, Das P, Bhattacharya SS, Kumar S, Adelodun AA (2017) Engineered nano particles: nature, behavior, and effect on the environment. J Environ Manag 196:297–315. https://doi.org/10.1016/j.jenvman.2017.01.011

Goswami P, Yadav S, Mathur J (2019) Positive and negative effects of nanoparticles on plants and their applications in agriculture. Plant Sci Today 6:232–242. https://doi.org/10.14719/pst.2019.6.2.502

Gottschalk F, Sonderer T, Scholz RW, Nowack B (2009) Modeled environmental concentrations of engineered nanomaterials (TiO2, ZnO, Ag, CNT, fullerenes) for different regions. Environ Sci Technol 43:9216–9222. https://doi.org/10.1021/es9015553

Hall S, Bradley T, Moore JT, Kuykindall T, Minella L (2009) Acute and chronic toxicity of nano-scale $TiO_2$, particles to freshwater fish, cladocerans, and green algae, and effects of organic and inorganic substrate on $TiO_2$ toxicity. Nanotoxicology 3:91–97. https://doi.org/10.1080/17435390902788078

Handford CE, Dean M, Spence M, Henchion M, Elliott CT, Campbell K (2015) Awareness and attitudes towards the emerging use of nanotechnology in the agri-food sector. Food Control 57:24–34. https://doi.org/10.1016/j.foodcont.2015.03.033

Ho CM, Yau SKW, Lok CN, So MH, Che CM (2010) Oxidative dissolution of silver nanoparticles by biologically relevant oxidants: a kinetic and mechanistic study. Chem Asian J 5:285–293. https://doi.org/10.1002/asia.200900387

Hong J, Peralta-Videa JR, Rico C, Sahi S, Viveros MN, Bartonjo J, Zhao L, Gardea-Torresdey JL (2014) Evidence of translocation and physiological impacts of foliar applied $CeO_2$ nanoparticles on cucumber *(Cucumis sativus)* plants. Environ Sci Technol 48:4376–4385. https://doi.org/10.1021/es404931g

Hossain Z, Mustafa G, Sakata K, Komatsu S (2016) Insights into the proteomic response of soybean towards $Al_2O_3$, ZnO, and Ag nanoparticles stress. J Hazard Mater 304:291–305. https://doi.org/10.1016/j.jhazmat.2015.10.071

Hu Y, Li J, Ma L, Peng Q, Feng W, Zhang L, He S, Yang F, Huang J, Li L (2010) High efficiency transport of quantum dots into plant roots with the aid of silwet L-77. Plant Physiol Biochem 48:703–709. https://doi.org/10.1016/j.plaphy.2010.04.001

Hu Z, Zhao J, Gao H, Nourafkan E, Wen D (2017) Transport and deposition of carbon nanoparticles in saturated porous media. Energies 10:1151–1168. https://doi.org/10.3390/en10081151

Hyung H, Fortner JD, Hughes JB, Kim JH (2007) Natural organic matter stabilizes carbon nanotubes in the aqueous phase. Environ Sci Technol 41:179–184. https://doi.org/10.1021/es061817g

Jacobs R, van der Voet H, ter Braak CJF (2015) Integrated probabilistic risk assessment for nanoparticles: the case of nanosilica in food. J Nanopart Res 17:251. https://doi.org/10.1007/s11051-015-2911-y

Jain N, Bhargava A, Pareek V, Akhtar MS, Panwar J (2017) Does seed size and surface anatomy play role in combating phytotoxicity of nanoparticles? Ecotel 26:238–249. https://doi.org/10.1007/s10646-017-1758-7

Jha S, Pudake RN (2016) Molecular mechanism of plant–nanoparticle interactions. In: Kole C, Kumar DS, Khodakovskaya MV (eds) Plant nanotechnology. Springer, Cham, pp 155–181. https://doi.org/10.1007/978-3-319-42154-4_7

Johnson AC, Bowes MJ, Crossley A, Jarvie HP, Jurkschat K, Juergens MD, Lawlor AJ, Park B, Rowland P, Spurgeon D, Svendsen C, Thompson IP, Barnes RJ, Williams RJ, Xu N (2011) An assessment of the fate, behavior and environmental risk associated with sunscreen $TiO_2$ nanoparticles in UK field scenarios. Sci Total Environ 409:2503–2510. https://doi.org/10.1016/j.scitotenv.2011.03.040

Keller AA, Sirivithayapakorn S, Chrysikopoulos CV (2004) Early breakthrough of colloids and bacteriophage MS2 in a water-saturated sand column. Water Resour Res 40:W08304. https://doi.org/10.1029/2003WR002676

Khodakovskaya MV, Kim B, Kim JN, Alimohammadi M, Dervishi E, Mustafa T, Cernigla CE (2013) Carbon nanotubes as plant growth regulators: effects on tomato growth, reproductive system, and soil microbial community. Small 9:115–123. https://doi.org/10.1002/smll.201201225

Kim HJ, Phenrat T, Tilton RD, Lowry GV (2012) Effect of kaolinite, silica fines and pH on transport of polymer-modified zero valent iron nanoparticles in heterogeneous porous media. J Colloid Interface Sci 370:1–10. https://doi.org/10.1016/j.jcis.2011.12.059

Kurepa J, Paunesku T, Vogt S, Arora H, Rabatic BM, Lu J, Wanzer MB, Woloschak GE, Smalle JA (2010) Uptake and distribution of ultra small anatase $TiO_2$ alizarin red S nanoconjugates in *Arabidopsis thaliana*. Nano Lett 10:2296–2302. https://doi.org/10.1021/nl903518f

Kwak J, An Y-J (2016) The current state of the art in research on engineered nanomaterials and terrestrial environments: different-scale approaches. Environ Res 151:368–382. https://doi.org/10.1016/j.envres.2016.08.005

Lanphere JD, Rogers B, Luth C, Bolster CH, Walker SL (2014) Stability and transport of graphene oxide nanoparticles in groundwater and surface water. Environ Eng Sci 31:350–359. https://doi.org/10.1089/ees.2013.0392

Larue C, Castillo-Michel H, Sobanska S, Ceillon L, Bureau S, Barthes V, Ouerdane L, Carriere M, Sarret G (2014) Foliar exposure of the crop *Lactuca sativa* to silver nanoparticles: evidence for internalization and changes in Ag speciation. J Hazard Mater 264:98–106. https://doi.org/10.1016/j.jhazmat.2013.10.053

Lee W, Kwak JI, An Y (2012) Effect of silver nanoparticles in crop plants *Phaseolus radiatus* and *Sorghum bicolor*: media effect on phytotoxicity. Chemosphere 86:491–499. https://doi.org/10.1016/j.chemosphere.2011.10.013

Li Q, Chen X, Zhuang J, Chen X (2016) Decontaminating soil organic pollutants with manufactured nanoparticles. Environ Sci Pollut Res 23:11533–11548. https://doi.org/10.1007/s11356-016-6255-7

Liang Y, Bradford SA, Simunek J, Heggen M, Vereecken H, Klumpp E (2013) Retention and remobilization of stabilized silver nanoparticles in an undisturbed loamy sand soil. Environ Sci Technol 47:12229–122237. https://doi.org/10.1021/es402046u

Lin S, Wiesner MR (2012) Deposition of aggregated nanoparticles — a theoretical and experimental study on the effect of aggregation state on the affinity between nanoparticles and a collector surface. Environ Sci Technol 46:13270–13277. https://doi.org/10.1021/es3041225

Lin SJ, Reppert J, Hu Q, Hudson JS, Leid ML, Ratnikova TA, Rao AM, Luo H, Ke PC (2009) Uptake, translocation and transmission of carbon nanomaterials in rice plants. Small 5:1128––1132. https://doi.org/10.1002/smll.200801556

Liu R, Lal R (2015) Potentials of engineered nanoparticles as fertilizers for increasing agronomic productions. Sci Total Environ 514:131–139. https://doi.org/10.1016/j.scitotenv.2015.01.104

Liu X, He B, Xu Z, JSM YW, Zhang H, Cao J (2015) A functionalized fluorescent dendrimer as a pesticide nanocarrier: application in pest control. Nanoscale 7:445–449. https://doi.org/10.1039/C4NR05733C

Liu G, Zhonga H, Ahmada Z, Yangb X, Huoa L (2019) Transport of engineered nanoparticles in porous media and its enhancement for remediation of contaminated groundwater. Crit Rev Environ Sci Technol. https://doi.org/10.1080/10643389.2019.1694823

Lowry GV, Gregory KB, Apte SC, Lead JR (2012) Transformations of nanomaterials in the environment. Environ Sci Technol 46:6893–6899. https://doi.org/10.1021/es300839e

Ma R, Levard CM, Michel FM, Brown GE, Lowry GV (2013) Sulfidation mechanism for zinc oxide nanoparticles and the effect of sulfidation on their solubility. Environ Sci Technol 47:2527–2534. https://doi.org/10.1021/es3035347

Maruyama CR, Guilger M, Pascoli M, Bileshy-Jose N, Abhilash PC, Fraceto LF, de Lima R (2016) Nanoparticles based on chitosan as carriers for the combined herbicides imazapic and imazapyr. Sci Rep 6:19768. https://doi.org/10.1038/srep19768

Mehmani Y, Balhoff MT (2015) Mesoscale and hybrid models of fluid flow and solute transport. Rev Mineral Geochem 80:433–459. https://doi.org/10.2138/rmg.2015.80.13

Milewska-Hendel A, Zubko M, Karcz J, Stroz D, Kurczynska E (2017) Fate of neutral-charged gold nanoparticles in the roots of the *Hordeum vulgare* L. cultivar. Karat. Sci Rep 7:3014. https://doi.org/10.1038/s41598-017-02965-w

Mishra S, Singh A, Keswani C, Singh HB (2014) Nanotechnology: exploring potential application in agriculture and its opportunities and constraints. Biotech Today 4:9–14. https://doi.org/10.5958/2322-0996.2014.00011.8

Nadiminti PP, Dong YD, Sayer C, Hay P, Rookes JE, Boyd BJ, Cahill DM (2013) Nanostructured liquid crystalline particles as an alternative delivery vehicle for plant agrochemicals. ACS Appl Mater Interfaces 5:1818–1826. https://doi.org/10.1021/am303208t

Nair R (2016) Effects of nanoparticles on plant growth and development. In: Kole C, Kumar DS, Khodakovskaya MV (eds) Plant nanotechnology. Springer, Cham, pp 95–118. https://doi.org/10.1007/978-3-319-42154-4_5

Nair R, Varghese SH, Nair BG, Maekawa T, Yoshida Y, Kumar DS (2010) Nanoparticulate material delivery to plants. Plant Sci 179:154–163. https://doi.org/10.1016/j.plantsci.2010.04.012

Navarro E, Piccapietra F, Wagner B, Marconi F, Kaegi R, Odzak N, Sigg L, Behra R (2008) Toxicity of silver nanoparticles to *Chlamydomonas reinhardtii*. Environ Sci Technol 42:8959–8964. https://doi.org/10.1021/es801785m

Neukum C, Braun A, Azzam R (2014) Transport of stabilized engineered silver (Ag) nanoparticles through porous sandstones. J Contam Hydrol 158:1–13. https://doi.org/10.1016/j.jconhyd.2013.12.002

Nowack B, Bucheli TD (2007) Occurrence, behavior and effects of nanoparticles in the environment. Environ Pollut 150:5–22. https://doi.org/10.1016/j.envpol.2007.06.006

Oliveira HC, Stolf-Moreira R, Martinez CBR, Grillo R, de Jesus MB, Fraceto LF (2015) Nanoencapsulation enhances the post-emergence herbicidal activity of atrazine against mustard plants. PLoS One 10:e132971. https://doi.org/10.1371/journal.pone.0132971

Pacheco I, Buzea C (2017) Nanoparticle interaction with plants. In: Ghorbanpour M, Manika K, Varma A (eds) Nanoscience and plant-soil systems. Springer, Cham, pp 323–355. https://doi.org/10.1007/978-3-319-46835-8_12

Palocci C, Valletta A, Chronopoulou L, Donati L, Bramosanti M, Brasili E, Baldan B, Pasqua G (2017) Endocytic pathways involved in PLGA nanoparticle uptake by grapevine cells and role of cell wall and membrane in size selection. Plant Cell Rep 36:1917–1928. https://doi.org/10.1007/s00299-017-2206-0

Phenrat T, Kim HJ, Fagerlund F, Illangasekare T, Tilton RD, Lowry GV (2009) Particle size distribution, concentration, and magnetic attraction affect transport of polymer-modified Fe(o) nanoparticles in sand columns. Environ Sci Technol 43:5079–5085. https://doi.org/10.1021/es900171v

Phenrat T, Kim HJ, Fagerlund F, Illangasekare T, Lowry GV, Kibbey TCG, O'Carroll DM (2010) Empirical correlations to estimate agglomerate size and deposition during injection of a polyelectrolyte-modified FeO nanoparticle at high particle concentration in saturated sand. J Contam Hydrol 118:152–164. https://doi.org/10.1016/j.jconhyd.2010.09.002

Porubcan AA, Xu S (2011) Colloid straining within saturated heterogeneous porous media. Water Res 45:1796–1806. https://doi.org/10.1016/j.watres.2010.11.037

Raychoudhury T, Tufenkji N, Ghoshal S (2012) Aggregation and deposition kinetics of carboxymethyl cellulose-modified zero-valent iron nanoparticles in porous media. Water Res 46:1735–1744. https://doi.org/10.1016/j.watres.2011.12.045

Rodrigues SM, Trindade T, Duarte AC, Pereira E, Koopmans GF, Romkens PFAM (2016) A framework to measure the availability of engineered nanoparticles in soils: trends in soil tests and analytical tools. TrAC-Trends Anal Chem 75:129–140. https://doi.org/10.1016/j.trac.2015.07.003

Rosenfeldt RR, Seitz F, Schulz R, Bundschuh M (2014) Heavy metal uptake and toxicity in the presence of titanium dioxide nanoparticles: a factorial approach using *Daphnia magna*. Environ Sci Technol 48:6965–6972. https://doi.org/10.1021/es405396a

Saiers JE, Hornberger GM, Liang L (1994) First- and second-order kinetics approaches for modeling the transport of colloidal particles in porous media. Water Resour Res 30:2499–2506. https://doi.org/10.1029/94WR01046

Sani-Kast N, Labille J, Ollivier P, Slomberg D, Hungerbuhler K, Scheringer M (2017) A network perspective reveals decreasing material diversity in studies on nanoparticle interactions with dissolved organic matter. Proc Natl Acad Sci U S A 114:E1756–E1765. https://doi.org/10.1073/pnas.1608106114

Sanzari I, Leone A, Ambrosone A (2019) Nanotechnology in plant science: to make a long story short. Front Bioeng Biotechnol. https://doi.org/10.3389/fbioe.2019.00120

Scheckel KG, Chaney RL, Basta NT, Ryan JA (2009) Advances in assessing bioavailability of metal(loid)s in contaminated soils. Adv Agron 104:1–52. https://doi.org/10.1016/s0065-2113(09)04001-2

Scrinis G, Lyons K (2007) The emerging nano-corporate paradigm: nanotechnology and the transformation of nature, food and agri-food systems. Int J Sociol Food Agri 15:22–44

Sekhon BS (2014) Nanotechnology in agri-food production: an overview. Nanotechnol Sci Appl 7:31–53. https://doi.org/10.2147/NSA.S39406

Sharma VK, Siskova KM, Zboril R, Gardea-Torresdey JL (2014) Organic-coated silver nanoparticles in biological and environmental conditions: fate, stability and toxicity. Adv Colloid Interf Sci 204:15–34. https://doi.org/10.1016/j.cis.2013.12.002

Siddiqi KS, Husen A (2017) Plant response to engineered metal oxide nanoparticles. Nanoscale Res Lett 12:92. https://doi.org/10.1186/s11671-017-1861-y

Solovitch N, Labille J, Rose J, Chaurand P, Borschneck D, Wiesner MR, Bottero JY (2010) Concurrent aggregation and deposition of $TiO_2$ nanoparticles in a sandy porous media. Environ Sci Technol 44:4897–4902. https://doi.org/10.1021/es1000819

Sun Y, Gao B, Bradford SA, Lei W, Hao C, Shi X, Wu J (2015a) Transport, retention, and size perturbation of graphene oxide in saturated porous media: effects of input concentration and grain size. Water Res 68:24–33. https://doi.org/10.1016/j.watres.2014.09.025

Sun P, Shijirbaatar A, Fang J, Owens G, Lin D, Zhang K (2015b) Distinguishable transport behavior of zinc oxide nanoparticles in silica sand and soil columns. Sci Total Environ 505:189–198. https://doi.org/10.1016/j.scitotenv.2014.09.095

Sygouni V, Chrysikopoulos CV (2015) Characterization of $TiO_2$ nanoparticle suspensions in aqueous solutions and $TiO_2$ nanoparticle retention in water-saturated columns packed with glass beads. Chem Eng 262:823–830. https://doi.org/10.1016/j.cej.2014.10.044

Tanaka Y, Kimura T, Hikino K, Goto S, Nishimura M, Mano S, Nakagawa T (2012) Gateway vectors for plant genetic engineering: overview of plant vectors, application for bimolecular fluorescence complementation (BiFC) and multigene construction. In: Barrera-Saldana H (ed) Genetic engineering - basics, new applications and responsibilities. Intechopen, London, p 64. https://doi.org/10.5772/32009

Thuesombat P, Hannongbua S, Akasit S, Chadchawan S (2014) Effect of silver nanoparticles on rice (*Oryza sativa* L. cv. KDML 105) seed germination and seedling growth. Ecotoxicol Environ Saf 104:302–309. https://doi.org/10.1016/j.ecoenv.2014.03.022

Torkzaban S, Bradford SA (2016) Critical role of surface roughness on colloid retention and release in porous media. Water Res 88:274–284. https://doi.org/10.1016/j.watres.2015.10.022

Torney F, Trewyn BG, Lin VS-Y, Wang K (2007) Mesoporous silica nanoparticles deliver DNA and chemicals into plants. Nat Nanotechnol 2:295–300. https://doi.org/10.1038/nnano.2007.108

Tosco T, Sethi R (2010) Transport of non-Newtonian suspensions of highly concentrated micro- and nanoscale iron particles in porous media: a modeling approach. Environ Sci Technol 44:9062–9068. https://doi.org/10.1021/es100868n

Tripathi DK, Singh S, Singh VP, Prasad SM, Dubey NK, Chauhan DK (2016) Silicon nanoparticles more effectively alleviated UV-B stress than silicon in wheat (*Triticum aestivum*) seedlings. Plant Physiol Biochem 110:70–81. https://doi.org/10.1016/j.plaphy.2016.06.026

Tripathi DK, Shweta SS, Singh S, Pandey R, Singh VP, Sharma NC, Prasad SM, Dubey NK, Chauhan DK (2017) An overview on manufactured nanoparticles in plants: uptake, translocation, accumulation and phytotoxicity. Plant Physiol Biochem 110:2–12. https://doi.org/10.1016/j.plaphy.2016.07.030

Uhram S, Minjoo S, Gisuk L, Jinkyu R, Younghun K, Eun JL (2013) Functional analysis of $TiO_2$ nanoparticle toxicity in three plant species. Biol Trace Elem Res 155:93–103. https://doi.org/10.1007/s12011-013-9765-x

Valletta A, Chronopoulou L, Palocci C, Baldan B, Donati L, Pasqua G (2014) Poly(lactic-co-glycolic) acid nanoparticles uptake by *Vitis vinifera* and grapevine-pathogenic fungi. J Nanopart Res 16:1917–1928. https://doi.org/10.1007/s11051-014-2744-0

Veerapaneni S, Wiesner MR (1996) Hydrodynamics of fractal aggregates with radially varying permeability. J Colloid Interface Sci 177:45–57. https://doi.org/10.1006/jcis.1996.0005

Wang D, Su C, Liu C, Zhou D (2014) Transport of fluorescently labeled hydroxyapatite nanoparticles in saturated granular media at environmentally relevant concentrations of surfactants. Colloids Surf 457:58–66. https://doi.org/10.1016/j.colsurfa.2014.05.041

Wang M, Gao B, Tang D (2016) Review of key factors controlling engineered nanoparticle transport in porous media. J Hazard Mater 318:233–246. https://doi.org/10.1016/j.jhazmat.2016.06.065

Weggler-Beaton K, Graham RD, McLaughlin MJ (2003) The influence of low rates of air-dried biosolids on yield and phosphorus and zinc nutrition of wheat (*Triticum durum*) and barley (*Hordeum vulgare*). Aust J Soil Res 41:293–308. https://doi.org/10.1071/SR02074

Xie Y, Dong H, Zeng G, Tang L, Jiang Z, Zhang C, Deng J, Zhang L, Zhang Y (2017) The interactions between nanoscale zero-valent iron and microbes in the subsurface environment: a review. J Hazard Mater 321:390–407. https://doi.org/10.1016/j.jhazmat.2016.09.028

Zhang L, Feng C, Chen Z, Liu L, JiangK LQ, Fan S (2008) Superaligned carbon nanotube grid for high resolution transmission electron microscopy of nanomaterials. Nano Lett 8:2564–2569. https://doi.org/10.1021/nl8012727

Zhang W, Yao Y, Sullivan N, Chen Y (2011) Modeling the primary size effects of citrate-coated silver nanoparticles on their ion release kinetics. Environ Sci Technol 45:4422–4428. https://doi.org/10.1021/es104205a

Zhang W, Isaacson CW, Rattanaudompol US, Powell TB, Bouchard D (2012) Fullerene nanoparticles exhibit greater retention in freshwater sediment than in model porous media. Water Res 46:2992–3004. https://doi.org/10.1016/j.watres.2012.02.049

Zhao LJ, Peralta-Videa JR, Ren MH, Varela-Ramirez A, Li CQ, Hernandez-Viezcas JA, Aguilera RJ, Gardea-Torresdey JL (2012) Transport of Zn in a sandy loam soil treated with ZnO NPs and uptake by corn plants: Electron microprobe and confocal microscopy studies. Chem Eng J 184:1–8. https://doi.org/10.1016/j.cej.2012.01.041

Zhong H, Liu G, Jiang Y, Yang J, Liu Y, Yang X, Liu Z, Zeng G (2017) Transport of bacteria in porous media and its enhancement by surfactants for bioaugmentation: a review. Biotechnol Adv 35:490–504. https://doi.org/10.1016/j.biotechadv.2017.03.009

Zhou D, Abdel-Fattah AI, Keller AA (2012) Clay particles destabilize engineered nanoparticles in aqueous environments. Environ Sci Technol 46:7520–7526. https://doi.org/10.1021/es3004427

# Part IV
# Threats and Mechanism of Toxicity of Emerging Contaminants

# Chapter 12
# Threats of Nano-material Contamination in Agroecosystem: What We Know and What We Need to Know

**Amita Shakya and Faraz Ahmad**

**Abstract** The use of nanomaterials has integrated in day to day human life. The application of nanomaterials is widespread in the range of sectors. Nanomaterials possess peculiar physical, chemical and biological properties in contrast to macro- and micro-particles of the same precursor material primarily due to their unique nanoscale dimensions. Nanomaterials are naturally present in the environment but can also be incidental and engineered for specific purposes. With their sub-micron size, they can free float in the air and may easily penetrate the animal or plant cells with potential to cause range of unidentified effects. The lesser known and many unknown potential hazards raised the concerns regarding nanomaterials and flagged them as "emerging contaminants". In agriculture, nanomaterials can be used as nano-pesticides, nano-fertilizers, nano-fungicides or as catalytic substances for plant germination and plant growth. They can also be used as biomarkers/sensors for plant pathogen detection. Toxic effects to the components of agro-ecosystem, including plants and soil microbes, have been documented with nanomaterials used in agriculture. However, the studies of the systemic effects of altered reactivity of nanomaterials on plants, humans and microbes are still in infancy. Cellular DNA damage, biochemical dysfunction, heavy metal release from metallic nanomaterials, generation of reactive oxygen species etc. are some of the reported toxicities associated with the use of nanomaterials.

Amita Shakya and Faraz Ahmad contributed equally.

A. Shakya (✉)
Department of Agriculture and Environmental Sciences, National Institute of Food Technology Entrepreneurship and Management, Kundli, Sonipat, Haryana, India
e-mail: amita.niftem@gmail.com

F. Ahmad
ICMR-National Institute of Pathology, New Delhi, India
e-mail: farazamu2014@gmail.com



V. Kumar Singh et al. (eds.), *Sustainable Agriculture Reviews 50*, Sustainable Agriculture Reviews 50, https://doi.org/10.1007/978-3-030-63249-6_12

The present chapter deals with the various sources and routes of introduction of exposure to nanomaterials into agro-ecosystem, their fate and potential bearings to environment and human health in near future. Numerous factors, such as plant species, growth conditions, physico-chemical properties of nanomaterials, abiotic and biotic conditions are the elements which decide the interactions and fate of nanomaterials in agro-ecosystem. Furthermore, use of non-biodegradable inorganic chemicals like metals, ceramic, silica, metal oxides etc., is also a matter of grave concern. The nanomaterials induced adverse impacts on plants, animal and human lives are being unraveled and yet to be explored in full. Distinction between natural and engineered nanoparticles is the foremost challenge for the detection of nanomaterials in the environment. The end recipient, duration of activity in the environment, reactions with target and non-target entities, are some of the pressing concerns associated with the use of nanomaterials. Intense research efforts are required to address crucial issues associated with nanomaterials, for example; toxicity levels compared to their microsized counterparts or their transformation in the environment, to unveil their ultimate fate in ecosystem which may help in devise interventional management plans.

**Keywords** Agro-ecosystem · Environmental security · Nanomaterials · Nanoparticles · Nanotoxicity

## Abbreviations

| | |
|---|---|
| AgNPs | Silver nanoparticles |
| DOM | Dissolved organic matter |
| ENMs | Engineered nanomaterials |
| ENPs | Engineered nanoparticles |
| $TiO_2$ | Titanium dioxide |
| ZnO | Zinc oxide |

## 12.1 Introduction

The lecture "There's Plenty of Room at the Bottom: An Invitation to Enter a New Field of Physics" given by physicist Richard Feynman on December 29, 1959, at the annual American Physical Society meeting at Caltech, opened the room for nanotechnology. Norio Taniguchi from Tokyo Science University, very first used the term "Nanotechnology" in 1974. Nanotechnology is an interdisciplinary science emerged as an anchorage for various technological advancements in different fields that supposed to contribute for well-being of the human life. In concomitance with biotechnology, nanotechnology has extended its applications to various domains with a significant impact on daily life, alongside economical and industrial aspects (Khot et al. 2012).

The principal facet of nanotechnology is to either fabricate or utilize the materials with a particle size less than 100 nm in at least one dimension. Such materials are generally classified as "nanomaterials". Application of nanotechnology in agriculture and associated activities has gained a momentum with some impressive advantages. These are ranging from germplasm manipulation and seed germination to field preparation and food packaging. In agrarian sector, the use of nanomaterials primarily aims to reduce the loss of nutrients during fertilizer application. It minimizes the application of plant protection product, which in turn help increase the revenue *via* nutrient management (Ruttkay-Nedecky et al. 2017). Thus, post-harvest management and development of new pesticide formulations are the most focused research endeavors within the domain. With continuing exploration of unique properties and consequent rapid technological advancements, the nanomaterials have found utility in several fields. They have found their passage in areas such as waste water treatment, organic and inorganic pollutant adsorption (from water and soil), construction material coating and fabrication of energy/information storage devices among others. Variety of nanomaterials is being used in microelectronics, biomedical devices, cosmetics, catalysts, semiconductors etc. The C-containing nanomaterials (fullerenes and nanotubes), oxides of silver, titanium, zinc, cerium nano-scale zerovalent iron are reportedly the most common materials with various commercial applications (e.g. personal care products, fertilizers, pesticides, textiles etc.) (Vance et al. 2015). The advancement of research in nanotechnology and use of nanomaterials, not only come up with opportunities but also brings in some challenges for the environment, ecosystem, health, food safety and security, in addition to challenges related to economy.

The classical feature of nanomaterials is the drastic change in their properties with reduction in size. When released in the environment, nanomaterials can undergo several potential transformations and the extent of transformation highly depends upon the properties of nanomaterials and the receiving medium (Batley et al. 2013). Most, if not all, NMs cause toxicity (above a certain concentration) to living organism by altering their physiochemical, biochemical, morphological-anatomical and genetic constituents (Fan and Alexeeff 2010; Pandey et al. 2016; Lewis et al. 2019). However, little is known about the ecotoxicology of NMs towards the non-target organisms. In this chapter we have attempted to critically analyze the fate, behavior, and potential scope of the use of nanomaterials with discussion on influencing factors and limitations associated with their use, particularly in agro-ecosystem. For the sake of easy reading, the term nanomaterials has been used particularly for engineered nanomaterials/nanoparticles in this chapter.

### *12.1.1 Nanomaterials: A Brief Description*

Human eye cannot see nanomaterials without the aid of specialized equipment, but their presence is making a revolution by leaps and bounds in fields like health, agriculture, energy, engineering, and environment. Nanomaterials can be defined as the engineered particles with tiny dimensions (at least one dimension less than 100 nm). In this chapter the term "Nanomaterials" is used for the two categories of nanoscale materials: first, nano-objects, having any external dimension in nanoscale and the second for nanostructure materials, having internal structure or surface structure in nanoscale as described by Hatto (2011). The minuscule size of the nanomaterials makes them unique as compared to their large scale particles, which endows to advantage of their novel and unique attributes at the nanoscale. The basic building block of a nanomaterial is nanoparticle which is ten thousand times smaller than the diameter of a human hair (Schultz 2007). The change in size of a material to atomic level to produce nanoparticles enormously changes the crystalline order or structure, physicochemical composition, surface chemistry, catalytic activity, chemical reactivity, stability, optical, magnetic, electrical, mechanical and thermal properties of the resulting nanomaterials, from their larger sized analogue (Stuart and Compton 2015).

Nanoparticles can overlap with colloids if distinguished according to size (1 nm to 1 mm in diameter) and have distinguished mechanism of action from molecules and ions. It is recommended by various scientific bodies that nanomaterials should be treated and regulated as new materials. For detection and analysis of size distribution of various nanomaterials, crucial and very accurate techniques such as scanning electron microscopy (SEM), dynamic light scattering (DLS), atomic force microscopy (AFM) and epifluorescence microcopy (EFM) etc. are generally employed. While techniques such as atomic absorption spectroscopy (AAS), inductively coupled plasma emission spectrophotometry (ICP), extended X-ray absorption fine structure (EXAFS) and X-ray absorption near edge structure (XANES) spectroscopies, flow field-flow fractionation coupled to an inductively coupled plasma mass spectrometer (FlFFF-ICP-MS) and X-ray diffraction (XRD) are generally used for the compositional understanding of nanomaterials.

### *12.1.2 Types of Nanomaterials and Their Environmental Occurrence*

Depending upon the source of origin, nanomaterials can be divided as natural, incidental and engineered. Depending upon the properties and materials, nanomaterials can be composed of nanoparticle made up of carbon, ceramic, metal and polymeric substances. They can be remaining in the environment as single entity (unbound state) or as aggregates. Nanomaterials can be found in natural environment in the form of volcanic dust and forest fire emissions, humic materials and fulvic acids in soil (dissolved organic fractions), fine particles of clay colloids, mineral precipitates of hydroxides and oxides of aluminum (Al), iron (Fe), manganese (Mn), particles in sea salts, biogenic uraninite ($UO_2$), biological entities such as virus particles (Batley et al. 2013; Stuart and Compton 2015). Various nanominerals such as ferrihydrite and mineral nanoparticles of Fe and Mn (hydr)oxides, uraninite ($UO_2$), and others, bioprecipitated by microorganisms are often associated with microbial redox processes, and sometimes with nanowire (carbon or metal) (Maurice and Hochella 2008). Mineral formation due to terrestrial weathering is the largest source of addition of nanominerals and nanoparticles in soil estimated as $10^{7-8}$ Tg (1 Tg = $10^{12}$ g) (Hochella Jr. et al. 2012).

Perhaps unintentionally, nanomaterials can be produced (incidental) in the particles of the exhaust from a diesel engine and other automobile exhausts (Schultz 2007). Apart from this, many man-made surfaces spontaneously generate incidental nanoparticle. The major sources of nanomaterials are their intentional production also called as "engineered nanomaterials" includes nano-fiber (nanowire, nanotube, nanoroad), nanoparticle and nanoplates which further divided in several classes consist of carbon, polymer, clay and metal (metal oxides, zero-valent) based nanomaterials, semiconductors (e.g., quantum dots) and emulsions (liposome and dendrimers as controlled delivery systems for various nano-formulations). The key difference between natural nanoparticles and engineered nanoparticles lies in that engineered nanoparticles having a large, persisting organic coatings on its surface (Wagner et al. 2014).

The International Organization for Standardization (ISO) classified nanomaterials into three main groups: nanoparticles (all three dimensions between 1 and 100 nm); nanoplates (one dimension between 1 and 100 nm); and nanofibers (two dimensions between 1 and 100 nm) (ISO 2008). On a larger scale, nanomaterials are produced commercially as well as for research purpose worldwide. The use of ENMs have become an integral part of human life as a constituent of food packaging material, drug delivery systems, cosmetics, electronic and biosensors, detergents etc. In pharma and therapeutics, ENMs are used in preparation of wound dressing, and antimicrobial coatings (Ray et al. 2009).

## 12.2 Sources of Nanomaterials in Agroecosystem

In this modern era, nanomaterials are the key players behind numerous technological innovations, which escalate the production and use of nanomaterials for several domains. Sufficient data is available on the toxicity aspects of nanomaterials, but little is known about the present concentration, possible future accumulation and the heterogeneous distribution of nanomaterials in the environment. In the life cycle of nanomaterials, various circumstantial routes and stages, which actually are difficult to predict, contribute significantly to their accumulation in the environment. Agriculture is one of the sectors which is cashing in on the use of nanomaterials. However, the increasing use and production of nanomaterials is simultaneously raising the concerns related to their release in environment, which could affect the health of the ecosystem. In agro-ecosystem, nanomaterials can be entered via aerial route either in a direct/deliberate or indirect and accidental manner via point or non-point sources (Rienzie and Adassooriya 2018).

The contribution of atmospheric or natural deposition of nanomaterials to agro-environment is small. Use of pesticides, herbicides, fungicides, fertilizers having nanomaterials as their core or adjunct ingredient is the 'direct' and major source for nanomaterials deposition to agriculture land (Usman et al. 2020). Nanomaterials not only used as nano-formulations for above said applications in agro-lands but also applied for their fast degradation purposes. Intentional or unintentional release of engineered nanomaterials from photography, paint, textile, cosmetics, personal and health care products either on commercial or industrial level enters the environment as per their potential usage (Ray et al. 2009). More than 15% of total consumer products are assumed to be "nano-enabled" (Stuart and Compton 2015). Various nanoparticles such as ZnO and $TiO_2$ used in paint and sunscreens, Ag nanoparticle used in socks to reduce odor, tend to remain in and accumulate in sewage sludge. At present, the disposal of wastewater treatment plant (WWTP) sewage sludge is the major source for deposition of nanomaterials into land. It was found that 90% of silver nanoparticles remain bound to sewage sludge (Schlich et al. 2013). Various nanomaterials released from different consumer products into wastewaters get deposited into sewage sludge during the treatment process. The leached or applied nanomaterials on soil surface will migrate to various soil zones, although the pathway is unclear due to limited research in the area. A thousand of tons of alumina, silica and ceria has been used for nanomaterial preparations, with different kind of conjugated metal ions as active ingredients like zinc (Zn), copper (Cu), titanium (Ti), silver (Ag), gold (Au) to increase the solubility of these nutrients in agro-ecosystem (Mukhopadhyay 2014; Tourinho et al. 2012; Tripathi et al. 2017a). The particles of nanomaterials present in solid waste and effluents, in conjugation or in free form can be transported to aquatic environment by wind or rainwater runoff.

Despite the increasing use of these nanomaterials, information about their direct and indirect sources with subsequent effects is little known. These nanomaterials ultimately contaminate the air, land and water resources. However, it is not always necessary that the nanomaterials enter in the system should always possess an ability to translocate as it is assumed that many unknown processes can take place between entry and translocation of the nanomaterials. In general, engineered nanomaterials are prepared in controlled preparation facilities from where the uncontrolled release rarely happens. However, during the transportation process from manufacturing sites to intended use sites, the accidental/unintentional/circumstantial release of nanomaterials can be another major source of their release into environment. In many developed countries, any effluents containing nanomaterials will require discharge licenses and will be more likely to go to a wastewater treatment plant (Batley et al. 2013). Such control measures are the necessity of the time with more attention as the possible toxicities due to exposure of nanomaterials are still unknown.

## 12.3 Emerging Concerns Posed by Nanomaterials in Agroecosystems

The little knowledge about the reactivity, consequences or serendipity of nanomaterials when they enter into the environment creates a big challenge to understand the behavioral changes of nanomaterials toward the environment. The physicochemical and morphological properties which make application of nanomaterials superior in various fields compared to their larger particles may transform them into potential toxicants when they get into the food chain. Transformation and translocation of nanomaterials in agro-ecosystem is also a key concern. The fate, effects and deportment of nanomaterials during long term accumulation within agro-ecosystem is unclear. In the following sections the concerns on agro-ecosystem due to nanomaterials are discussed.

### *12.3.1 Agricultural Abiotic Environment*

The most sought environmental application of nanomaterials is contaminant remediation from groundwater and subsurface source areas of contamination at hazardous waste sites (EPA 2008). Use of this water in secondary applications like farmland irrigation may cause accumulation of nanomaterials in land and unintentional/occupational exposure to humans. The fate of nanomaterials in wastewater treatment plants is not sufficiently clear. The impact of nanomaterials in different components of agricultural systems is outlined as:

#### 12.3.1.1 Soil Matrix

Soil is the largest component of an agro-ecosystem. The behavior of nanomaterials in soil system is not only dictated by the properties of soil and pore water, but also impacted due to the characteristics of dissolved nanomaterials. Soil matrix itself is a complex system of solid, solution and gas phases which makes it complicated and challenging to understand the behavior of nanomaterials. The nanomaterials are small enough to fit in the macro- as well as micro-porous environment of the soil. Through Brownian movement, they can travel within the matrix easily. The toxicity and behavior of nanomaterials in soil system is largely influenced by their transport, bioavailability, stability and interactions with the components of soil matrix. Presence of natural colloids and particles in soil system may react with the engineered nanomaterials, which again is a challenge to understand the fate and extent of the toxicity of nanomaterials. With reference to ecotoxicity, another important issue is the limited understanding about the exposure and impact of the nanomaterials to specific organism present in different phases of soil system (soil particles, pore water) (Tourinho et al. 2012). The attachment (immobilization) and detachment of nanomaterials to soil matrix are the key processes which affect the mobility and bioavailability of nanomaterials. Though, the exact concentration of nanomaterials in environment is unknown, but studies suggested a higher accumulation of nanomaterials in soil system compared to water and air (Gottschalk et al. 2009). To assess the impact of nanomaterials on soil microbiota, soil respiration and enzymatic activities are usually measured. Studies showed greater detrimental effects of metal based nanomaterials, as compared to carbon, clay and polymer based nanomaterials (Johansen et al. 2008; Maurice and Hochella 2008).

#### 12.3.1.2 Soil Properties

The soil components like dissolved organic matter (DOM) is composed of humic acid, hydrophilic acid, fatty acid, carbohydrate acid, amino acid, hydrocarbon and clay particles. The origination source of DOM are plant based biomass, root exudates, organic amendments, humus, manure and litter which are most reactive, mobile and charged particles that influence the association of nanomaterials to soil system (Tourinho et al. 2012). These particles have ability to make colloids with components of nanomaterials such as metal, metal oxides, clay or polymer via different pathways (such as complexation, electrostatic attraction, Van der Waals interactions, hydrophobic attraction). The pH and ionic strength of the soil matrix are also the factors which change the reactivity and properties of the nanomaterials leading to the uncertainty of the behavior of nanomaterials in the soil matrix. Increased ionic strength increases the unfavorable deposition and hetero/homo aggregation of nanomaterials. These aggregated nanomaterials physically filtered out and cannot pass through the soil pores and unlikely to the biological uptake which decrease the bioavailability of nanomaterials (Cornelis et al. 2014; Loureiro et al. 2018). However, in case of qualitative description of nanomaterials in soil with the help of mathematical modeling with reference to most accepted colloid filtration theory (Yao et al. 1971), the attachment efficiency ($\alpha$) of nanoparticals found high when ionic strength of the soil matrix is high due to a decrease of the energy barrier between particles and surfaces (Loureiro et al. 2018). Though, these assumptions may not be the same as postulated for transport of nanoparticles in field conditions, as many factors such as pore water composition, weather conditions and soil physicochemical composition (Loureiro et al. 2018). These differences in qualitative research studies may create uncertainties and knowledge gap with reference to assumptions of fate, transport and translocation of nanomaterials in real field conditions.

In contrast to the ionic strength, soil pH can either increases or decrease the mobility of engineered nanomaterials in the soil system. The engineered nanoparticles are generally anionic in soil system either due to coating with dissolved organic substances or some other anionic compounds or due to adsorption of negatively charged organic matter of soil or difference in zero point charge when comes in contact with soil bound water (Loureiro et al. 2018). The soil texture is also an important criterion which influences the retention and bioavailability of nanomaterials. The clay particles in soil are composed of oxides of iron and alumina with pH-dependent surface charge (Cornelis et al. 2014). Change in pH induces nanomaterials to make hetero-aggregate with soil organic matter which increases the size of nanomaterials leading to a decrease in mobility. Increased concentration of nanoparticles in soil adversely affects the soil self-cleaning process and nutrient balance which is the basis for the regulation of the processes of plant nutrition and soil fertility improvement (Javed et al. 2019).

In conclusion, not only the environmental behavior of nanomaterials, but also the properties of physical and chemical constituents of soils, also play important role in mobility, translocation and transformation of nanomaterials. The physical and chemical interactions of nanomaterials with components of soil and the complexity of the soil system are the challenges to understand the bioavailability and toxicity of nanomaterials.

#### 12.3.1.3 Influence of Enzyme Activity on Soil

Enzymes that play a major role in maintaining soil health are amylase, arylsulphatase, β-glucosidase, cellulase, chitinase, dehydrogenase, phosphatase, protease, and urease. Soil, enzymes can be found as endo-enzyme (intracellular/inside microorganisms) and exoenzymes (released by microorganisms), either in free form in soil solution or in the bound form with soil particles (Karimi and Fard 2017). The enzymes bound to clay particles and dissolve organic matter as an immobilized/accumulated fraction have a residual activity which does not found in enzymes free in the soil aqueous phase (Burns 1982).

Soil contamination with nanomaterials poses adverse impact on enzyme activity. Peyrot et al. (2014) reported that increased concentration of Ag nanoparticles inhibit the activity of soil enzymes including phosphomonoesterase, arylsulfatase, β-D-glucosidase, and leucine-aminopeptidase. Impairment of different soil enzymes activities (cellobiohydrolase, β-1,4-glucosidase, β-1,4-xylosidase, β-1,4-N-acetylglucosaminidase, L-leucineaminopeptidase, and acid phosphatase) due to single-walled carbon nanotubes (SWCNTs) was also reported (Jin et al. 2013). Application of zinc oxide (ZnO), copper oxide (CuO), chromium oxide ($Cr_2O_3$), nickel (Ni) nanoparticles in two different kinds of soils showed either inhibitory or a stimulating effect on enzymes such as dehydrogenase, urease, acidic and alkaline phosphatase; and among all, CuO-nanoparticles had most negative impacts (Rahmatpour et al. 2017). Soil acid phosphatase, $\beta$-glucosaminidase, $\beta$-glucosidase, and arylsulfatase activities were reported to be reduced significantly when Ag-nanoparticles were applied to silt loam and sandy soils (Eivazi et al. 2018). However, the level of impact of the nanomaterials is likely to be dependent upon the variation in the soil type, duration of contact between nanomaterials and the soil, and particle size (nano, bulk) (Rahmatpour et al. 2017).

## *12.3.2 Agriculture Biotic Environment*

### 12.3.2.1 Soil Microbiota

In agriculture, soil microorganisms are crucial for the maintenance of various soil functions such as soil structure formation, decomposition of organic matter, toxin removal, cycling of carbon, nitrogen, phosphorus, promoting plant growth, and changes in vegetation (Karimi and Fard 2017). The structural and compositional changes in micro-environment of agricultural lands can be critical for the functional integrity of the soil. The 'dose-response' is also a critical factor deciding the impact on the enzymatic and microbial community of soil. Different nanomaterials react differently to each microbial entity and the response of microbe is fairly different to various nanomaterials. Figure 12.1 briefly explains the proposed mechanism of action (that can be a synergic process) of nanomaterials to soil microorganisms, however the proper toxicity mechanisms of many nanomaterials are still undetermined. In major studies, effect of metal based nanoparticles on soil microbiota has been explored in comparison to organic active ingredient based nanomaterials. Though, metal-based nanomaterials are believed to be more hazardous due to the non-biodegradable and bioaccumulative nature of metal ions. Negative impact of added nanomaterials to the soil has been reported to almost all types of microbes including arbuscular mycorrhizal fungi, rhizobacteria, nitrogen-fixing and nitrifying bacteria of soil system (Holden 2016; Niazi and Gu 2009). The toxicity of nanomaterials can affect the activity, stability, and specificity of microbial enzymes in soil. In long term, reduction in soil microbial diversity and microbial mass has been reported due to the application of nanomaterials in high doses (Zheng et al. 2011).

The polymers of carboxylmethyl-cellulose (CMC), of hydrophobically modified CMC (HM-CMC), and hydrophobically modified polyethylglycol (HM-PEG); the vesicles of sodium dodecyl sulphate/didodecyldimethylammonium bromide (SDS/DDAB) and of monoolein/sodium oleate (Mo/NaO); titanium oxide ($TiO_2$), titanium silicon oxide ($TiSiO_4$), CdSe/ZnS quantum dots, gold nanorods, and Fe/Co magnetic fluid nanomaterials have been checked for their possible impact on soil environment for a period of 30 days and showed significant impacts on the structural diversity of the soil bacterial community (Nogueira et al. 2012). $C_{60}$ fullerene nanoparticles decreased the number of fast-growing bacteria by three to four-folds (Johansen et al. 2008). Differing in their properties, nanoparticles have different effects on soil microorganism.

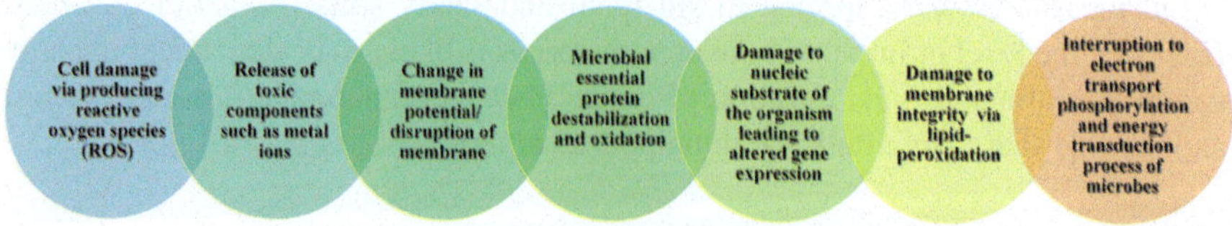


**Fig. 12.1** Potential mechanisms of toxicity of different NMs on soil microbes

You et al. (2018) reported, strong effect of nZnO nanoparticles on soil enzymatic activities than $nTiO_2$, $nCeO_2$, and $nFe_3O_4$ nanoparticles. They found that saline-alkali soil was more susceptible to metal oxide nanoparticles than black soil and found a significant decrease in total bacterial population in saline-alkali soil with nZnO treatment (0.5, 1.0, and 2.0 mg/g dose). Treatment of nZnO to saline-alkali soil showed high variance in their bacterial community (Bacilli, Alphaproteobacteria, and Gammaproteobacteria) composition. In another study, $nTiO_2$ and nZnO showed reduced microbial biomass, diversity and composition of the soil (grassland soil) bacterial community with 60 days incubation period in small amount as well (0, 0.5–2.0 mg/g soil). Similar to the previous results, the effect of nZnO was stronger than that of $nTiO_2$ and stronger shifts in bacterial community composition for nZnO at the same exposure concentration (Ge et al. 2011). Soil incubation with $TiO_2$ nanoparticles decreased the nitrifying bacteria (ammonia-oxidizing) and inhibited the activities of ammonia monooxygenase and nitrite with reduction in the diversity of microbial community (Zheng et al. 2011).

Silver nanoparticles and $AgNO_3$ also found to be responsible for reduced soil respiration and enzyme activities and key parameter for the toxicity majorly depended on silver dose and soil type. High doses (>20 mg Ag/kg) inhibit the soil respiration whereas a small concentration of nanoparticles inhibits the soil urease and phosphatase activities. Soil with lower clay content and ionic strength had more reduced microbial and enzyme activities when Ag-nanoparticles and Ag ions were applied to soil (Rahmatpour et al. 2017). Ag-nanoparticles also affect the C and nutrient cycle by altering the microbial activities (Eivazi et al. 2018). Significant increase in the metabolic quotient ($qCO_2$), increased microbial stress, changes in the bacterial/fungal biomass ratio of soil was noticed when $CeO_2$, $Fe_3O_4$, $SnO_2$-nanoparticles were incubated in soil (Vittori Antisari et al. 2013). A big research gap is still lying there as many a type of nanomaterials have not been tested with difference in type of soil and difference in environmental conditions, which need to be acknowledged.

#### 12.3.2.2 Soil Macro-organisms

Besides microorganism, soil is the niche of different arthropods, nematodes, annelids and vertebrates. However, a little is known about the eco-toxicological impacts of nanomaterials (nano-pesticides, nano-fertilizers) to these organisms. Among all, earthworms are most important and considered as the good indicator of soil health. The earliest study of the impact of nanomaterials on earthworms confirmed the toxicity of Ag, Cu and $TiO_2$-nanoparticles (Heckmann et al. 2011). Toxicity impact of exposure of earthworm (*Eudrilus eugeniae*) to magnetite nanoparticles (17 and 28 nm size) found to be concentration dependent (100–400 ng). Change in animal skin color from brown to black with an increase in the nanoparticle concentration was reported with accumulation of nanoparticles in animal body (Samrot et al. 2017). In other study, similar dose-dependent increase in mortality was observed in earthworms with almost 100% mortality after 96 h exposure to the highest concentration (1000 mg/kg) of ZnO-nanoparticles (Li et al. 2011). Significant reduction in reproduction of earthworms was reported when they exposed to Ag-nanoparticles (94.21 mg/kg) (Shoults-Wilson et al. 2011). However, earthworm can sense the toxic nanoparticles in soil suggesting the behavioral reflex of organism towards unpredictable effects (Usman et al. 2020). An urgent need of comprehensive risk assessment approaches for nanomaterials are necessary to identify the possible hazards of nanomaterials to non-target animals at cellular and biochemical level.

#### 12.3.2.3 Plant System

In agro-ecosystem, after soil, plants (especially crops) are the major contact entities of nanomaterials applied to the agriculture-land. Extensive research data is available which indicates toxic and detrimental effects of nanomaterials on various crop plants belong to the plant families' including Poaceae (Gramineae), Leguminosae (Fabaceae), Solanaceae, Amaryllidaceae, Cucurbitaceae and Brassicaceae (Rajput et al. 2018; Usman et al. 2020). The studies on crops like onion, spinach, coriander, wheat, rice, soybean, mung bean, radish, lettuce, barley, cucumber and tobacco showed negative impacts of nanomaterials. Nanomaterials produce reactive oxygen species (ROS), which subsequently results into oxidative stress, lipid peroxidation, proteins and DNA damage in plants (Ma et al. 2010). Defying the inherent plant defense system (e.g. enzymatic actions), nanomaterials cause stress and phytotoxicity when their concentration increased beyond a certain level (Table 12.1). There can be two routes of entry of nanomaterials in a plant; from roots (bottom up translocation) or through aerial parts (top down movement).

In bottom up translocation nanomaterials enters via plant roots, penetrate into the cell wall of the epithelial cells and enter into the cytoplasm. Subsequently, these nanomaterials enter to the vascular tissue system of the plant through xylem and spread into the other parts of the plant body. In this process, chances of accumulation of nanomaterials in different plant organs in varying amount are high with the possibility of cycling these nanomaterials in agro-ecosystem. The aerial routes of nanoparticles entry into plant are cuticles, trichomes, stomata, stigma, hydathodes, nectarithodes, and lenticels, etc. (Rienzie and Adassooriya 2018). After accumulation, nanomaterials have shown phytotoxicity as decrease in fresh and dry biomass, shoot reduction, retarded root growth, inhibition of germination, photosynthesis process alteration, enhanced lipid peroxidation, chromatin condensation, DNA damage, reducing rate of transpiration decrease in photosynthetic rate and chlorophyll concentrations (Cox et al. 2017; Hossain et al. 2015; Ma et al. 2010). Beyond the limit, all these factors resulted in apoptosis activity in plant cell by shattering the plant internal detoxification mechanism. In literature, data for studies focused on the toxicity of the metal/metal oxide nanomaterials on plant system is available, however; the toxic impact studies with organic, polymeric nanomaterials are very few. It is evident that the effect of same nanomaterial on different plant species is different (Table 12.1). Time of exposure and concentration of nanomaterials are

**Table 12.1** Studies showing toxic effects of nanomaterials on vegetable/crops

| Type of nanomaterials | Concentration of nanomaterials | Crop/vegetable/plant | Time of exposure | Toxicity/harmful effects | References |
|---|---|---|---|---|---|
| Ag-nanoparticles | 5, 10, 20 and 25 μg/ml | *Peltophorumpterocarpum* | 15–90 min | Significantly reduced the pollen germination and retarded the tube growth more potent toxic effect of Ag-nanoparticles than $AgNO_3$ | Dutta Gupta et al. (2020) |
| Ag-nanoparticles | 0.5–100 ng/mL | *Allium cepa* seeds | 4–5 days | Genotoxicity and biochemical alterations in oxidative stress parameters (lipid peroxidation) and activities of antioxidant enzymes (superoxide dismutase and catalase) in light and dark conditions | Souza et al. (2020) |
| Ag-nanoparticles (*Eichhorniacrassipes*) | 5–80 mg/L | *Allium cepa* | 2, 4 and 6 h | Maximum cell death of root tips of *A. cepa* after 20 mg/L treatment<br>Alteration of DNA repair kinetics | Heikal et al. (2020) |
| Ag-nnaoparticles | 100 mg/L | *Allium cepa* roots | | Significant effect on germination, root elongation, mitotic index, nuclear abnormality and micronucleus index (MNI) inmeristematic cells. Cytotoxic and genotoxic effects found to be increased with the reduction of the particle diameter. | Scherer et al. (2019) |
| $TiO_2$ nanoparticles | 0, 20, 40, 60, 80, and 100 mg/kg | Wheat (*Triticumaestivum* L.) | 60 days | At low concentration levels can boost plant chlorophyll content and growth, but cause toxicity at higher concentrations | Rafique et al. (2018) |
| Ag-nanoparticles | 1000 μM and 3000 μM | *Pisiumsativum* seedlings | | Significant decline in photosynthetic pigments and chlorophyll content decline in the total ascorbate and glutathione contents and severely damaged leaf and root anatomical structures due to accumulation of nanoparticles | Tripathi et al. (2017b) |

(continued)

**Table 12.1** (continued)

| Type of nanomaterials | Concentration of nanomaterials | Crop/vegetable/plant | Time of exposure | Toxicity/harmful effects | References |
|---|---|---|---|---|---|
| Ag-nanoparticles (*Swertiachirata*) | 5, 10 and 20 μg/ml | *Allium cepa* | 24 h (sand medium) | Decrease in mitotic and meiotic indexes with increasing concentrations of nanoparticles, different chromosomal aberrations and abnormalities in post meiotic products were observed.<br>Biogenic Ag-nanoparticles were found to be less toxic than chemically synthesized Ag-nanoparticles | Saha and Dutta Gupta (2017) |
| CuO nanoparticles (*Morus alba* leaf extract) | 10, 50, 100, and 500 mg | *Solanumlycopersicum*and *Brassica oleracea var. botrytis* | 24 h (sand medium) | Significant reduction of total chlorophyll and sugar content, increase in lipid peroxidation, electrolyte leakage, and antioxidant enzyme activity, increase in superoxide and hydrogen peroxide formation in leaves (at high concentration of nanoparticles) | Singh et al. (2017) |
| $Al_2O_3$, ZnO, and Ag nanoparticles | 50–500 mg/L | Soybean seedling | 3 days | ZnO- and Ag-nanoparticles affects the plant growth, rigidity of roots, and root cell viability were $Al_2O_3$-nanoparticles challenged normal seedling growth. Severe oxidative burst due to ZnO-nanoparticles and Ag-nanoparticles treatments.<br>Changed proteins primarily associated with secondary metabolism, cell organization, and hormone metabolism. Up-regulation in the oxidation-reduction cascade related genes. Photosystem and protein degradation. | Hossain et al. (2016) |
| $Fe/Fe_3O$ | 10, 20 mg L | *L. sativa* | 15 days | Increased antioxidant activities, reduced root size, affected chlorophyll, particle aggregation on the root, affected water entrance | Trujillo-Reyes et al. (2014) |

| | | | | | |
|---|---|---|---|---|---|
| Multi-walled carbon nanotubes (MWCNTs) | 10 mg/L, 60 mg/L, 100 mg/L, 600 mg/L | *Arabidopsis* T87 suspension cells | 5–7 days | Toxicities increased sharply as the diameters of the agglomerates of the MWCNTs became smaller | Lin et al. (2009) |
| Multi-walled carbon nanotubes (MWCNTs) | 20 mg/L | *Oryza sativa* L. | 6 days | Reactive oxygen species (ROS) increased and cell viability decreased | Tan et al. (2009) |
| Single-walled carbon nanotubes functionalized with poly-3-amino benzenesulfonic acid | 28, 160, 900, and 5000 mg/L | Cabbage, carrot, cucumber, lettuce, onion, and tomato | 24 h and 48 h | Affected root elongation cucumber, onion, lettuce, and tomato | Cañas et al. (2008) |

critical parameters governing the potential toxicity, bioavailability and hazards to the plant species.

### 12.3.3 *Behavior of Nanomaterials in Agro-ecosystem*

In the size reduction process of material to produce nano-sized materials, factors such as surface relaxation and reconstruction, changes the overall different coordination environments of nanomaterials leading to the generation of nanoparticles with different type, surface composition, densities of sites, and different reactivates with respect to processes (Waychunas et al. 2005). In the present section, some properties of nanomaterials have been discussed which are important in view of their behavior in the environment.

Nanomaterials are too small in size because of which they have slow rate of gravitational settling. Due to this, many nanomaterials residues may remain suspended in the environment for a longer period of time. Their small size and light weight also contribute to their transport to various places by air as compared to the larger particles of the same material (Morris and Willis 2007). With increasing production for application in various fields, the exposure of population to nanomaterials is also increasing proportionately. Their different nature and gradual exposure, raises the concern about the impact of nanomaterials over environment and humans, which needs to be addressed. The size reduction leads to the high surface area to volume ratio. High surface area and transformed physicochemical properties of nanomaterials help reorient the electrical conductivity, heat conduction, catalytic activity, surface chemistry, mechanical strength and solubility of nanomaterials.

The nanomaterials may undergo aggregation then precipitated onto soil matrix, or their dispersion in soil might leads to their stabilization in soil solution, or various physical or chemical reactions with other environmental components might change their properties (Chen 2018). The synthesis process, chemical composition, size, morphology, surface coating and functionality with additional modification in nanoparticles to enhance their effectiveness also have an impact on the behavior of nanomaterials in soil matrix.

#### 12.3.3.1 Aggregation and Agglomeration

The nanomaterials showed potential of aggregation (strong bonding between particles) or agglomeration (weak association due to Van der Waals forces) influenced by the factors such as physical forces/mechanical bindings (e.g. Brownian motion, gravity, and fluid motion), chemical forces (electrostatic and covalent interactions), nanomaterials characteristics (e.g., surface properties, particle size, hydrophobicity, size of nanomaterials, zeta potential etc.) and properties of the matrix (viscosity, polarization etc.) (Tourinho et al. 2012; Brar et al. 2015). Nanomaterials can undergo homo-aggregation (within the nanomaterials) or hetero-aggregation (with soil organic matter, heavy metals, microbes or other compounds present in the environment). Agglomeration can be reversible or irreversible as it is a dynamic process affected by surface charge, size, pH of soil, zero point charge, zeta potential of nanomaterials, soil organic matter, ionic strength, and concentration of nanomaterials along with various physicochemical reactions with the environment. However, not all the particles make aggregates and nanomaterials as individual entity are always present in the environment. The homo- or hetero-aggregation or agglomeration of engineered nanoparticles increases their size, and therefore, governs the mobility of nanomaterials in soil matrix leading to reduced bioavailability. Formation of nano-aggregates might cause the reduction in reactivity of nanomaterials (Batley et al. 2013). However, it increases the persistence of nanomaterials in the environment via decreasing the rate of decomposition or dissolution. Larger size of homo- or hetero-aggregates of nanomaterials prevents their translocation not only in soil matrix but also in the cell/cell organelles.

#### 12.3.3.2 Dissolution

Dissolution is a dynamic process generally related to the environmental toxicity, persistence and reactivity of nanomaterials having metal based active ingredient and largely depends upon the intrinsic properties of nanomaterials. The soil properties also influence the dissolution rate; vice-versa, dissolution rate dictates soil properties as well (Chen 2018). The nanomaterials synthesized via using class B soft metal cations (e.g. Au, Ag, Cu, Pt, Zn) having high affinity towards inorganic and organic sulfide ligands and form partially soluble metaloxides. When enter into the environment, these nanomaterials release toxic cations via complete or incomplete dissolution (Lowry et al. 2012).

#### 12.3.3.3 Surface Modifications and Coating

The surface of nanomaterials can be changed in natural conditions due to various biotic and abiotic process such as adsorption and desorption of organic and inorganic compounds, chemical reactions (e.g. reduction and oxidation), re-crystallization and oriented aggregation (Wagner et al. 2014). All these processes are strongly influenced by the presence of reducing or oxidizing agents, soil organic matter, and interaction with soil pore water.

Surface coating significantly changes the properties of nanomaterials. It helps for long term maintenance of properties of nanomaterials in the environment and prevents homo-aggregation of nanoparticles by increasing hydrophobicity, providing electrostatic, steric, or electrostatic repulsive forces between particles. However environmental processes such as dissolution and displacement (with natural compounds in soil matrix) may induce the reversibility of coating process and biodegradation of the particle leading to hetero-aggregation (Tourinho et al. 2012). All this affects the stability of nanomaterials in the environment leading to uncertainty in the fate and behavior of nanomaterials.

#### 12.3.3.4 Transformation of NMs in the Environment

All kind of nanomaterials have very diverse chemical nature which majorly depends upon the synthesis process and surface modifications. This leads to the transformation (physical, biological, chemical) of nanomaterials in the environment via various pathways (reactions with biomacromolecules, redox reactions, aggregation, photochemical reactions, hydrolysis, sulfidation, adsorption of macromolecules and molecules/ions and dissolution) (Lowry et al. 2012). Once released, nanomaterials physically and chemically interacts with the various components of the environment which could result in their transformation, leading to uncertainty about the fate and possible impacts of nanomaterials to the environment. The transformation of nanomaterials in the environment is a non-reluctant process which could be slow or fast making the possibility of exposure and toxicity more fatal.

#### 12.3.3.5 Stability and Mobility

The surface charge properties of nanomaterials largely decide their stability and mobility. The change in surface charge due to contact with external environment leads to agglomeration or aggregation of nanomaterials which affects the transport and translocation behavior of nanomaterials in soil matrix. The stability of nanomaterials in soil significantly modifies with the characteristic properties such as concentration, ionic strength, molecular mass, hydrophobicity, polarity of dissolved organic matter (Chen 2018). The stability and mobility of nanomaterials decide their fate, possible interactions (aggregation or dissolution), mobility or immobilization in environment. The bonding (bridging effect) between nanomaterials and soil dissolved organic matter (DOM) can lead to immobilization of nanomaterials resulting in possible accumulation at surface of soil matrix, or preventing it to translocation deep into the soil matrix preventing soil core contamination.

The intrinsic properties of soil matrix (pH, ionic strength), presence of DOM, natural colloids in association with the characteristic properties of nanomaterials, affects the aggregation, dissolution translocation and mobility of nanomaterials in soil matrix. Also these processes itself create complex interrelated chain of reactions which affect the potential toxic and hazardous effect to the environment due to nanomaterials (Fig. 12.2).

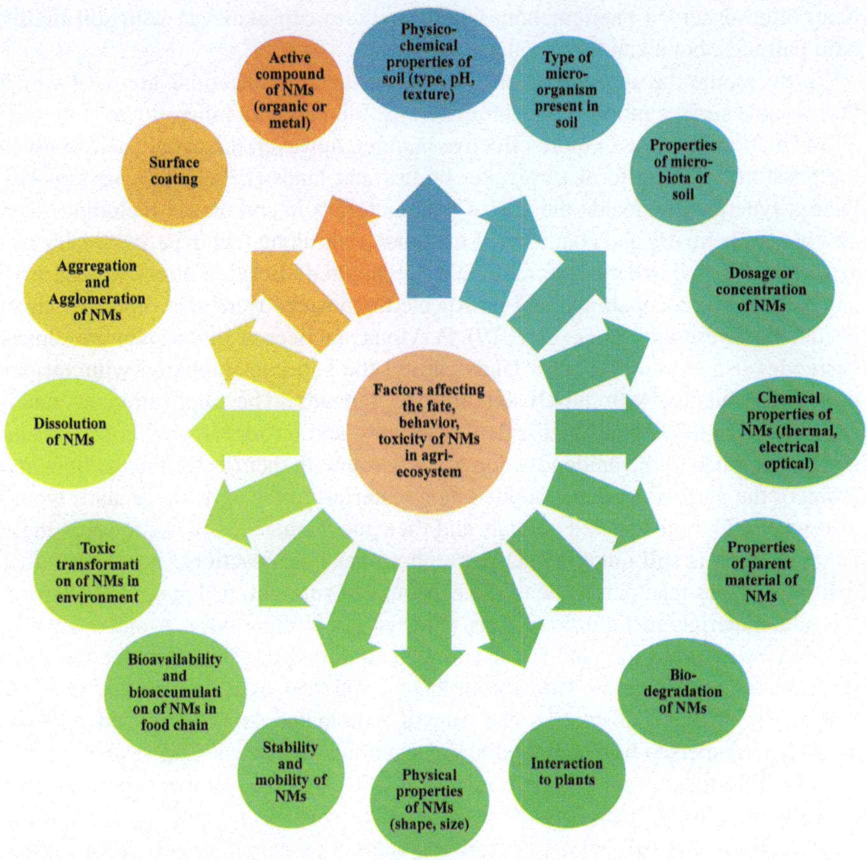


**Fig. 12.2** Factors influencing the impacts of nanomaterials behavior in agro-ecosystem

## 12.4 Fate of Nanomaterials in Agroecosystem

Extensive benefits and various advantages have already brought nanomaterials to our daily life but knowledge is limited on the fate and transport of these nanomaterials to the environment. The research on the fundamental properties that actually affect the fate of nanomaterials in environment is still in its infancy. A research gap about the potential toxicity and bioaccumulation of the nanomaterials is another limitation. Due to their tiny size and high surface area there could be a strong possibility of sorption of nanomaterials in soil system, which could restrict the further mobilization of nanomaterials. However, the small size of nanomaterials also favor them to fit into small spaces between soil particles, leading to their migration to far enough before getting trapped in soil matrix (Morris and Willis 2007). It is recommended that the nanomaterials or applied nano-formulations should decompose fast in soil and slowly in plant with remnant level meets regulatory criteria in produced food commodity (Khot et al. 2012). Agglomeration of nanomaterials on the surfaces is an often observed phenomenon, where particles can associate with soil matrix (soil particles, bound and free water).

To overcome the agglomeration, various surface modifications are used which increase the surface mobility of nanomaterials. Increased mobility of nanomaterials allow them to disperse in more effective manner, however, this would also leads to the possible leaching to nearby water bodies and lands (EPA 2008). Various key factors synergically decide the fate of nanomaterials in soil matrix including, size, treatment, chemistry and condition of nanomaterials along with type, chemistry and properties of soil where nanomaterials are applied. Besides direct exposure to nanopesticides, occupational and unexpected exposures were also reported which need to be investigated (Li et al. 2019). Pervious practices of sludge, manure, slurry, pesticides usage in agricultural lands caused the soil contamination with various organic pollutants with highly lethality to humans. The application of nano-compounds in the form of herbicides, pesticides, and fertilizers would raise crucial concerns related to the unidentified interactions among them as well to the flora and fauna of the agro-ecosystem. Applied nanomaterials to the agriculture lands would also react with non-targeted animals and their biochemical response towards these nanomaterials is still unknown. Diverse photo-catalytic reactions in soil and other natural surfaces take place due to the exposure of sunlight, which could transform the nanomaterials in field conditions (Morris and Willis 2007). Along with this, presence of natural nanomaterials in soil with similar size scale and molecular composition is another factor which complicates the fate of nanomaterials in agro-ecosystem. In aqueous matrix, the fate of nanomaterials is controlled by their solubility, dispersability, and interactions among biological and anthropogenic chemicals in the system (Morris and Willis 2007). Photochemical reactions may alter the physicochemical characteristics of nanomaterials in environment, leading to changes in their behavior in agro-ecosystem. The nanomaterials can be transformed in natural environment which could imitate the naturally occurring nanomaterials of same size, shape and molecular structure. These behaviors of nanomaterials in field conditions make it complicated to study about their fate in the environment. Limited site-specific applications of nanomaterials could also be an approach to reduce the possible risk of their transformation in the environment. Another limitation about the fate of nanomaterials is that the data and knowledge is scarce about the fate of nanomaterials in soil as most of research has been carried out in water system (Usman et al. 2020). Thus, further exploration into the fate of nanomaterials in soil is highly warranted in near future.

## 12.5 Risk Assessment Approaches Towards Engineered Nanomaterials

Kahru and Dubourguier (2010) reviewed different kind of nanomaterials using L(E) $C_{50}$ as indicator and found none of them as "not harmful". In studied nanomaterials some were classified as "extremely toxic" (nAg and nZnO); "very toxic" ($C_{60}$fullerenes and nCuO); "toxic" (SWCNTs and MWCNTs), and "harmful", ($nTiO_2$). This indicates the urgency of proper risk assessment strategies of NMs.

"A risk assessment is the evaluation of scientific information on the hazardous properties of environmental agents, the dose-response relationship, and the extent of exposure of humans or environmental receptors to those agents. The product of the risk assessment is a statement regarding the probability that humans (populations or individuals) or other environmental receptors so exposed will be harmed and to what degree (risk characterization)" (Morris and Willis 2007). The environmental risk assessment of nanomaterials should be a multidisciplinary approach to understand the potential risks and strategic development of remedial outcomes.

Since the research of possible toxic effects on environment due to nanomaterials is in its infancy and data is insufficient to reach on conclusion about the possible threats. The most appropriate assessment of risk is a simple comparison of predicted environmental concentrations (PECs) with predicted no effects concentrations (PNECs) derived from eco-toxicological studies (Batley et al. 2013). The probabilistic risk assessment of engineered nanomaterials can also be estimated with Species Sensitivity Distributions (SSDs) for NMs with all the available toxicological data for specific material (Gottschalk et al. 2009).

The Predicted Environmental Concentration is "an indication of the expected concentration of a material in the environment, taking into account the amount initially present (or added to) the environment, its distribution, and the probable methods and rates of environmental degradation and removal, either forced or natural" (https://www.nanopartikel.info/en/glossary/205-pec).

The Predicted No Effect Concentration (PNEC) is "the concentration of a material which marks the limit below which no adverse effects of exposure in an ecosystem are measured" (https://www.nanopartikel.info/en/glossary/205-pec). With the help of PEC and PNEC the risk quotient (RQ) can be calculated as following:

$$RQ = \frac{PEC}{PNEC}$$

The value of RQ > 1 suggests substance can pose risk to the environment (Mohamed and Paleologos 2018).

However, the required parameters for the estimation of PECs of nanomaterials are again a challenge. To measure the PECs, the life cycle, end points, possible routes of release of nanomaterials to environment (aquatic, terrestrial, aerial), deducing mass flows, estimates of total product usage and release rates to correlate with time of exposure, coatings, concentration and size of nanomaterials with the knowledge of the behavior (used in some models) of nanomaterials in the environment (colloidal forms, attachment to particles, etc.) are needed as the presently available models require these information (Batley et al. 2013). Limited availability of data about characterization, exposure type and duration, toxicity, degradation and dissolution for different kinds of nanomaterials makes it a difficult task to estimate the PECs and PNECs of nanomaterials. Need of life cycle, characterization and fate studies on different nanomaterials are required to estimate the PECs and PNECs.

Eisenberg et al. (2015) reviewed eight engineered nanomaterial risk assessment frameworks (IRGC's Risk Governance Framework, CEA, Nano Risk Framework, Nano Screening Level Life Cycle RA framework, MCDA, CENARIOS, Precautionary Matrix, and XL Insurance Database Protocol) proposed by various authors and organizations under ten critical points and found non as perfect as most of frameworks were applicable on occupational settings with minimal ecological risk considerations. Also, these frameworks were not examined with the different kind of engineered nanomaterials and their products.

To keep an eye on the assessment of possible hazards due to nanomaterials and to make policies for the preventive measures, the Environmental Protection Agency of USA (US EPA) assessed nanomaterials from a life cycle perspective and follows the risk assessment paradigm (Fig. 12.3) described by the National Academy of Sciences (NRC, 1983 and 1994) (Morris and Willis 2007).

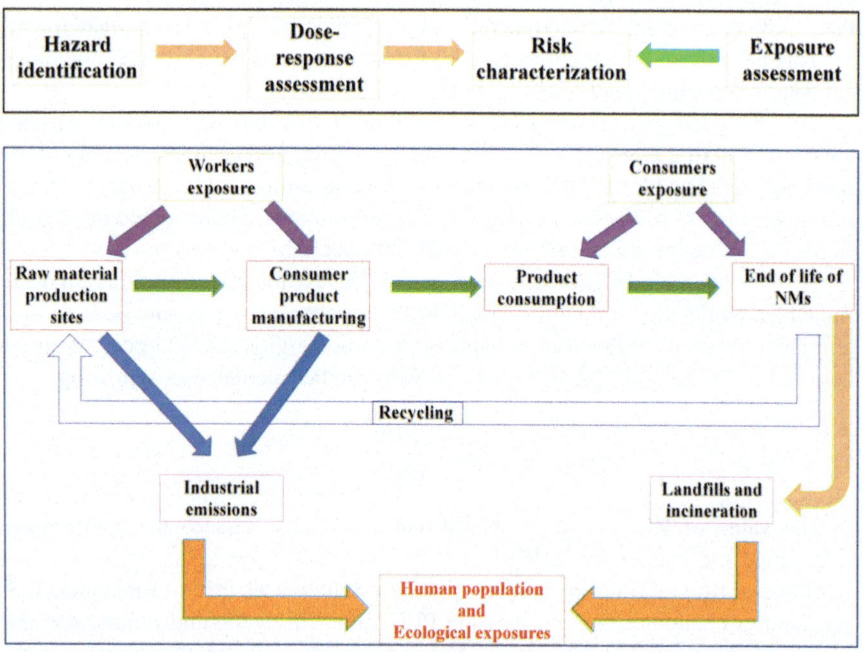


**Fig. 12.3** Risk assessment approach (upper) and life cycle perspective to risk assessment (down) adopted by Environmental Protection Agency (EPA). (Adopted and modified from EPA's white paper 2007)

## 12.6 Limitations and Future Research Scope

The knowledge about behavioral changes of nanomaterials is very limited when they comes in contact with external environment. Extensive toxicology research is needed to apprehend the influence of nanomaterials to the environment to avoid their undesirable effects before they are allowed for commercial use. Attention is required towards crucial issues like toxicity level of nanomaterials compared to their large sized particles, their transformation in the environment (toxic or non-toxic) to unveil their fate in ecosystem. From the analytical point of view, distinction between natural and engineered nanoparticles is the biggest challenge for detection of nanomaterials in the environment (Coll et al. 2016).

In general, most of research about the impact of nanomaterial application is conducted in a single type of soil with relatively high concentration of nanomaterials which is generally not found in soil. It is recommended that future studies should focus on use of concentration of nanomaterials found in natural environment and study about the effects of same nanomaterial towards the different type of soils. The information about the impacts of environmental factors and different organisms onto uptake and accumulation of engineered nanomaterials is scarce.

Potential toxicity of nanomaterials to ecosystem, phytotoxicity and their migration (as such or alerted) in produced agro-commodity is the colossal concern of present era. Knowledge gap about the recalcitrant nature of nanomaterials in the environment is another limitation of the area. An urgent need of in-depth study about 'toxicokinetics' and 'toxicodynamics' of nanomaterials in agricultural usage is required to understand and minimize the hazards that may likely to emerge in near future. Same nanomaterial could behave differently to different plant species, animals and microbiota, which could significantly alter the cropping pattern in agriculture (Cañas et al. 2008; Khot et al. 2012). To address the issues of *in-situ* quantification of nanomaterials in soil matrix, development of analytical methods and tools are required. These aspects need exploration to avoid possible harm of nanomaterials implementation in agriculture. The studies about the biochemical responses and physiological activities of organism towards nanomaterials compared to their large sized particles are critical area of research to understand the possible metabolic risk of nanomaterials.

Despite the continuous efforts in the field of phytotoxicity, innumerable questions of scientific and practical importance are still need to be investigated. The information about the uptake and accumulation of engineered nanomaterials, interrelation of nanomaterials intrinsic properties with respective phytotoxicity, role of difference in plant species towards the resistivity for nanomaterials, correlation of plant root-shoot system towards the combat strategies of phytotoxicity due to nanomaterials is insufficient. The concentration of nanomaterials in trace level against a high background of natural nanomaterials and colloids in the environment is the biggest limitation for tractability of engineered nanomaterials (Klaine et al. 2008).

In view of ever increasing production and consumption, life cycle assessment of nano-formulations or nanomaterials in real conditions, bio-degradation pattern, in-depth studies of toxicity of nanomaterials in biological system, *in-vivo* and *in-vitro* studies of nanomaterials, the dose-response mechanism of nanomaterials, exposure assessment to various subjects, the level of nanomaterials in food and fodders due to application in agricultural practices must need to be explored exponentially with rigorous research (WHO 2010). Studies focus on bioaccumulation, biomagnification of nanomaterials in food chain and food web is required to access the potential human exposure from nanomaterials usage in agro-ecosystem.

The material flow modeling to estimate the life cycle and likely emission of nanomaterials into the environment and landfills could also help to understand the environmental implications of nanomaterials (Keller et al. 2013). Comprehensive environmental assessment (CEA) approaches also help to understand the fate of nanomaterials.

The response of cells/organisms towards nanomaterials if they biodegrade into cells, is also of pivotal concern. The interaction of nanomaterials with other soil amendments such as compost, organic fertilizers, biochar is lesser known. Thus, it is necessary to explore the cross effect and prospects of possible risk assessment about the fate of nanomaterials when they interact with other organic pollutants, heavy metals, pharmaceuticals and personal care products. Also very little knowledge is available regarding occupational hazards associated with exposure to nanomaterials which throw challenge to the future scope of nanomaterials usage. The possible interactions of nanomaterials to various organic molecules, their intercellular behavior (such as: accumulation, release, binding etc.), their biocompatibility towards target and non-target organisms are the subjects of critical concerns with immediate attention. Some nanomaterials have shown photoactive nature, but their susceptibility to photodegradation in the atmosphere has not been studied yet (EPA 2008).

Longitudinal follow-up field application studies of nanomaterials requires more attention to amputate the gap between laboratory scale and real-time application in order to assess the reliable data about behavior and risk assessment of nanomaterials in natural environment. Formulation of appropriate guidelines about the usage and discharge of nanomaterials to environment is a pressing need with strict legislation. It would be more desirable to keep a country wise directory about the nanoparticle-producing companies, products, and end-users. The effect of aged nanomaterials, mixed nanomaterials in agro-ecosystem, and their interactions with freshly applied nanomaterials are the hotspots for parallel research in view of increasing application of nanomaterials in agriculture. Possibility of recovery and recycling of used nanomaterials is needed to be investigated to reduce the burden and hazards posed by nanomaterials. It is also necessary for governing and research bodies to confront challenges of safe and sustainable development and usage of nanomaterials. The commercial, industrial and research laboratories are producing a new waste with continuous research and development, this could provoke the challenges for current waste management efforts. In this context special attention to the potential release of nanomaterials from various nano-enable devices such as sensors to the possible disposal pathways for the produced nano-waste (e.g. waste-water, landfill, incineration or recycling), the bioavailability and persistence of nanomaterials, and subsequent effect in and across the disposal media (air, soil and water) is required (Iavicoli et al. 2014). Apart from it, the social and economical outlook about the feasibility of nanomaterial production and consumption within sustainable technological manner is also momentous. In commercial use, correct labeling on the products about the type, properties and amount of nanomaterials should be applied for the prediction of the fate of nanomaterials used in the product.

Apart from the above discussed issues and identified potential niche for future research, the following questions should kept into consideration to motivate the basic research about the nanomaterials to understand their fate, behavior and bioavailability in the environment:

1. The retention of original shape, structure, size and reactivity of nanomaterials after the application to the system.
2. The transformation of nanomaterials in environmental conditions.
3. To assess whether the desired effect of nanomaterials is really different from the larger particle of the analogous material in the environment.
4. The possible controls on the toxicity of nanomaterials to the biotic components of the environment.

In conclusion, it is necessary to focus on the development of easy to synthesize (green synthesis) stable (but biodegradable) nanomaterials with good dispersibility and wettability. At the same time the nanomaterials synthesized should be less toxic and more photo-generative, with well understood toxico-kinetics and toxicodynamics for their effective use in agricultural production. To organize the process, standard characterized materials with reference to nanomaterials as critical reference materials (CRM) must be available for efficient execution of future research in the area.

## References

Batley GE, Kirby JK, McLaughlin MJ (2013) Fate and risks of nanomaterials in aquatic and terrestrial environments. Acc Chem Res 46:854–862. https://doi.org/10.1021/ar2003368

Brar SK, Zhang TC, Verma M, Surampalli RY, Tyagi RD (2015) Nanomaterials in the environment. American Society of Civil Engineers, Reston

Burns RG (1982) Enzyme activity in soil: location and a possible role in microbial ecology. Soil Biol Biochem 14:423–427. https://doi.org/10.1016/0038-0717(82)90099-2

Cañas JE et al (2008) Effects of functionalized and nonfunctionalized single-walled carbon nanotubes on root elongation of select crop species. Environ Toxicol Chem 27:1922–1931. https://doi.org/10.1897/08-117.1

Chen H (2018) Metal based nanoparticles in agricultural system: behavior, transport, and interaction with plants. Chem Spec Bioavailab 30:123–134. https://doi.org/10.1080/09542299.2018.1520050

Coll C, Notter D, Gottschalk F, Sun T, Som C, Nowack B (2016) Probabilistic environmental risk assessment of five nanomaterials (nano-TiO2, nano-Ag, nano-ZnO, CNT, and fullerenes). Nanotoxicology 10:436–444. https://doi.org/10.3109/17435390.2015.1073812

Cornelis G, Hund-Rinke K, Kuhlbusch T, Van den Brink N, Nickel C (2014) Fate and bioavailability of engineered nanoparticles in soils: a review. Crit Rev Environ Sci Technol 44:2720–2764. https://doi.org/10.1080/10643389.2013.829767

Cox A, Venkatachalam P, Sahi S, Sharma N (2017) Reprint of: silver and titanium dioxide nanoparticle toxicity in plants: a review of current research. Plant Physiol Biochem 110:33–49. https://doi.org/10.1016/j.plaphy.2016.08.007

Dutta Gupta S, Saha N, Agarwal A, Venkatesh V (2020) Silver nanoparticles (AgNPs) induced impairment of in vitro pollen performance of Peltophorum pterocarpum (DC.) K. Heyne. Ecotoxicology 29:75–85. https://doi.org/10.1007/s10646-019-02140-z

Eisenberg D, Grieger K, Hristozov D, Bates M, Linkov I (2015) Risk assessment, life cycle assessment, and decision methods for nanomaterials. In: Nanomaterials in the environment. American Society of Civil Engineers, Reston, pp 383–419. https://doi.org/10.1061/9780784414088.ch15

Eivazi F, Afrasiabi Z, Jose E (2018) Effects of silver nanoparticles on the activities of soil enzymes involved in carbon and nutrient cycling. Pedosphere 28:209–214. https://doi.org/10.1016/S1002-0160(18)60019-0

EPA (2008) Nanotechnology for site remediation: fact sheet. EPA 542-F-08-009, US Environmental Protection Agency, Washington, DC

Fan AM, Alexeeff G (2010) Nanotechnology and nanomaterials: toxicology, risk assessment, and regulations. J Nanosci Nanotechnol 10:8646–8657. https://doi.org/10.1166/jnn.2010.2493

Ge Y, Schimel JP, Holden PA (2011) Evidence for negative effects of TiO2 and ZnO nanoparticles on soil bacterial communities. Environ Sci Technol 45:1659–1664. https://doi.org/10.1021/es103040t

Gottschalk F, Sonderer T, Scholz RW, Nowack B (2009) Modeled environmental concentrations of engineered nanomaterials (TiO2, ZnO, Ag, CNT, fullerenes) for different regions. Environ Sci Technol 43:9216–9222. https://doi.org/10.1021/es9015553

Hatto P (2011) ISO concensus definitions relevant to nanomaterials and nanotechnologies 4th annual nano safety for success dialogue ISO TC 229and BSI NTI/1 Nano-technologies Standardization Committees 29th and 30th March

Heckmann L-H, Hovgaard MB, Sutherland DS, Autrup H, Besenbacher F, Scott-Fordsmand JJ (2011) Limit-test toxicity screening of selected inorganic nanoparticles to the earthworm Eisenia fetida. Ecotoxicology 20:226–233. https://doi.org/10.1007/s10646-010-0574-0

Heikal YM, Şuţan NA, Rizwan M, Elsayed A (2020) Green synthesized silver nanoparticles induced cytogenotoxic and genotoxic changes in Allium cepa L. varies with nanoparticles doses and duration of exposure. Chemosphere 243:125430. https://doi.org/10.1016/j.chemosphere.2019.125430

Hochella MF Jr, Aruguete D, Kim B, Madden AS (2012) Naturally occurring inorganic nanoparticles: general assessment and a global budget for one of Earth's last unexplored major geochemical components. In: Nature's nanostructures. Pan Stanford, London, pp 1–31

Holden YGAMHJKJHPZSWPA (2016) Toxicity of manufactured nanomaterials to microorganisms. In: Engineered nanoparticles and the environment: biophysicochemical processes and toxicity. Wiley, Hoboken, pp 320–346. https://doi.org/10.1002/9781119275855.ch17

Hossain Z, Mustafa G, Komatsu S (2015) Plant responses to nanoparticle stress. Int J Mol Sci 16:26644–26653. https://doi.org/10.3390/ijms161125980

Hossain Z, Mustafa G, Sakata K, Komatsu S (2016) Insights into the proteomic response of soybean towards Al2O3, ZnO, and Ag nanoparticles stress. J Hazard Mater 304:291–305. https://doi.org/10.1016/j.jhazmat.2015.10.071

Iavicoli I, Leso V, Ricciardi W, Hodson LL, Hoover MD (2014) Opportunities and challenges of nanotechnology in the green economy. Environ Health 13:78–78. https://doi.org/10.1186/1476-069X-13-78

ISO (2008) Nanotechnologies: terminology and definitions for nano-objects – nanoparticle, nanofibre and nanoplate. International Organization for Standardization

Javed Z, Dashora K, Mishra M, Fasake VD, Srivastva A (2019) Effect of accumulation of nanoparticles in soil health-a concern on future doi:https://doi.org/10.15761/FNN.1000181

Jin L, Son Y, Yoon TK, Kang YJ, Kim W, Chung H (2013) High concentrations of single-walled carbon nanotubes lower soil enzyme activity and microbial biomass. Ecotoxicol Environ Saf 88:9–15. https://doi.org/10.1016/j.ecoenv.2012.10.031

Johansen A, Pedersen AL, Jensen KA, Karlson U, Hansen BM, Scott-Fordsmand JJ, Winding A (2008) Effects of C60 fullerene nanoparticles on soil bacteria and protozoans. Environ Toxicol Chem 27:1895–1903. https://doi.org/10.1897/07-375.1

Kahru A, Dubourguier H-C (2010) From ecotoxicology to nanoecotoxicology. Toxicology 269:105–119. https://doi.org/10.1016/j.tox.2009.08.016

Karimi E, Fard EM (2017) Nanomaterial effects on soil microorganisms. In: Nanoscience and plant–soil systems. Springer, Cham, pp 137–200. https://doi.org/10.1007/978-3-319-46835-8

Keller AA, McFerran S, Lazareva A, Suh S (2013) Global life cycle releases of engineered nanomaterials. J Nanopart Res 15:1692. https://doi.org/10.1007/s11051-013-1692-4

Khot LR, Sankaran S, Maja JM, Ehsani R, Schuster EW (2012) Applications of nanomaterials in agricultural production and crop protection: a review. Crop Prot 35:64–70. https://doi.org/10.1016/j.cropro.2012.01.007

Klaine SJ et al (2008) Nanomaterials in the environment: behavior, fate, bioavailability, and effects. Environ Toxicol Chem 27:1825–1851. https://doi.org/10.1897/08-090.1

Lewis RW, Bertsch PM, McNear DH (2019) Nanotoxicity of engineered nanomaterials (ENMs) to environmentally relevant beneficial soil bacteria – a critical review. Nanotoxicology 13:392–428. https://doi.org/10.1080/17435390.2018.1530391

Li L-Z, Zhou D-M, Peijnenburg WJGM, van Gestel CAM, Jin S-Y, Wang Y-J, Wang P (2011) Toxicity of zinc oxide nanoparticles in the earthworm, Eisenia fetida and subcellular fractionation of Zn. Environ Int 37:1098–1104. https://doi.org/10.1016/j.envint.2011.01.008

Li L, Xu Z, Kah M, Lin D, Filser J (2019) Nanopesticides: a comprehensive assessment of environmental risk is needed before widespread agricultural application. ACS Publications. https://doi.org/10.1021/acs.est.9b03146

Lin C, Fugetsu B, Su Y, Watari F (2009) Studies on toxicity of multi-walled carbon nanotubes on Arabidopsis T87 suspension cells. J Hazard Mater 170:578–583. https://doi.org/10.1016/j.jhazmat.2009.05.025

Loureiro S et al (2018) Nanomaterials as soil pollutants. In: Soil pollution. Elsevier, pp 161–190. https://doi.org/10.1016/B978-0-12-849873-6.00007-8

Lowry GV, Gregory KB, Apte SC, Lead JR (2012) Transformations of nanomaterials in the environment. ACS Publications. https://doi.org/10.1021/es300839e

Ma X, Geiser-Lee J, Deng Y, Kolmakov A (2010) Interactions between engineered nanoparticles (ENPs) and plants: phytotoxicity, uptake and accumulation. Sci Total Environ 408:3053–3061. https://doi.org/10.1016/j.scitotenv.2010.03.031

Maurice PA, Hochella MF (2008) Nanoscale particles and processes: a new dimension in soil science. In: Advances in agronomy, vol 100. Academic, pp 123–153. https://doi.org/10.1016/S0065-2113(08)00605-6

Mohamed A-MO, Paleologos EK (2018) Chapter 10 – Emerging pollutants: fate, pathways, and bioavailability. In: Mohamed A-MO, Paleologos EK (eds) Fundamentals of geoenvironmental engineering. Butterworth-Heinemann, Oxford, pp 327–358. https://doi.org/10.1016/B978-0-12-804830-6.00010-7

Morris J, Willis J (2007) EPA Whitepaper on nanotechnology

Mukhopadhyay SS (2014) Nanotechnology in agriculture: prospects and constraints. Nanotechnol Sci Appl 7:63. https://doi.org/10.2147/NSA.S39409

Niazi JH, Gu MB (2009) Toxicity of metallic nanoparticles in microorganisms-a review. In: Atmospheric and biological environmental monitoring, Springer, pp 193–206. https://doi.org/10.1007/978-1-4020-9674-7_12

Nogueira V et al (2012) Impact of organic and inorganic nanomaterials in the soil microbial community structure. Sci Total Environ 424:344–350. https://doi.org/10.1016/j.scitotenv.2012.02.041

Pandey R, Singh VP, Sharma N, Chauhan DK (2016) An overview on manufactured nanoparticles in plants: uptake, translocation, accumulation and phytotoxiciy. https://doi.org/10.1016/j.plaphy.2016.07.030

Peyrot C, Wilkinson KJ, Desrosiers M, Sauvé S (2014) Effects of silver nanoparticles on soil enzyme activities with and without added organic matter. Environ Toxicol Chem 33:115–125. https://doi.org/10.1002/etc.2398

Rafique R et al (2018) Dose-dependent physiological responses of Triticum aestivum L. to soil applied TiO2 nanoparticles: alterations in chlorophyll content, H2O2 production, and genotoxicity. Agric Ecosyst Environ 255:95–101. https://doi.org/10.1016/j.agee.2017.12.010

Rahmatpour S, Shirvani M, Mosaddeghi MR, Nourbakhsh F, Bazarganipour M (2017) Dose–response effects of silver nanoparticles and silver nitrate on microbial and enzyme activities in calcareous soils. Geoderma 285:313–322. https://doi.org/10.1016/j.geoderma.2016.10.006

Rajput VD et al (2018) Effect of nanoparticles on crops and soil microbial communities. J Soils Sediments 18:2179–2187. https://doi.org/10.1007/s11368-017-1793-2

Ray PC, Yu H, Fu PP (2009) Toxicity and environmental risks of nanomaterials: challenges and future needs. J Environ Sci Health Part C 27:1–35. https://doi.org/10.1080/10590500802708267

Rienzie R, Adassooriya NM (2018) Toxicity of nanomaterials in agriculture and food. In: Nanomaterials: ecotoxicity, safety, and public perception. Springer, pp 207–234. https://doi.org/10.1007/978-3-030-05144-0

Ruttkay-Nedecky B, Krystofova O, Nejdl L, Adam V (2017) Nanoparticles based on essential metals and their phytotoxicity. J Nanobiotechnol 15:33. https://doi.org/10.1186/s12951-017-0268-3

Saha N, Dutta Gupta S (2017) Low-dose toxicity of biogenic silver nanoparticles fabricated by Swertia chirata on root tips and flower buds of Allium cepa. J Hazard Mater 330:18–28. https://doi.org/10.1016/j.jhazmat.2017.01.021

Samrot AV, Justin C, Padmanaban S, Burman U (2017) A study on the effect of chemically synthesized magnetite nanoparticles on earthworm: Eudrilus eugeniae. Appl Nanosci 7:17–23. https://doi.org/10.1007/s13204-016-0542-y

Scherer MD et al (2019) Cytotoxic and genotoxic effects of silver nanoparticles on meristematic cells of Allium cepa roots: a close analysis of particle size dependence. Sci Total Environ 660:459–467. https://doi.org/10.1016/j.scitotenv.2018.12.444

Schlich K, Klawonn T, Terytze K, Hund-Rinke K (2013) Hazard assessment of a silver nanoparticle in soil applied via sewage sludge. Environ Sci Eur 25:17. https://doi.org/10.1186/2190-4715-25-17

Schultz AC (2007) Nanotechnology: industrial revolution or emerging hazard? Environ Claims J 19:199–205. https://doi.org/10.1080/10406020701557958

Shoults-Wilson WA, Reinsch BC, Tsyusko OV, Bertsch PM, Lowry GV, Unrine JM (2011) Effect of silver nanoparticle surface coating on bioaccumulation and reproductive toxicity in earthworms (Eisenia fetida). Nanotoxicology 5:432–444. https://doi.org/10.3109/17435390.2010.537382

Singh A, Singh N, Hussain I, Singh H (2017) Effect of biologically synthesized copper oxide nanoparticles on metabolism and antioxidant activity to the crop plants Solanum lycopersicum and Brassica oleracea var. botrytis. J Biotechnol 262:11–27. https://doi.org/10.1016/j.jbiotec.2017.09.016

Souza IR et al (2020) Visible-light reduced silver nanoparticles' toxicity in Allium cepa test system. Environ Pollut 257:113551. https://doi.org/10.1016/j.envpol.2019.113551

Stuart EJ, Compton RG (2015) Nanoparticles-emerging contaminants. In: Environmental analysis by electrochemical sensors and biosensors. Springer, New York, pp 855–878. https://doi.org/10.1007/978-1-4939-1301-5_8

Tourinho PS, Van Gestel CA, Lofts S, Svendsen C, Soares AM, Loureiro S (2012) Metal-based nanoparticles in soil: fate, behavior, and effects on soil invertebrates. Environ Toxicol Chem 31:1679–1692. https://doi.org/10.1002/etc.1880

Tripathi DK et al (2017a) An overview on manufactured nanoparticles in plants: uptake, translocation, accumulation and phytotoxicity. Plant Physiol Biochem 110:2–12. https://doi.org/10.1016/j.plaphy.2016.07.030

Tripathi DK et al (2017b) Nitric oxide alleviates silver nanoparticles (AgNps)-induced phytotoxicity in Pisum sativum seedlings. Plant Physiol Biochem 110:167–177. https://doi.org/10.1016/j.plaphy.2016.06.015

Trujillo-Reyes J, Majumdar S, Botez CE, Peralta-Videa JR, Gardea-Torresdey JL (2014) Exposure studies of core–shell Fe/Fe3O4 and Cu/CuO NPs to lettuce (Lactuca sativa) plants: are they a potential physiological and nutritional hazard? J Hazard Mater 267:255–263. https://doi.org/10.1016/j.jhazmat.2013.11.067

Usman M et al (2020) Nanotechnology in agriculture: current status, challenges and future opportunities. Sci Total Environ 721:137778. https://doi.org/10.1016/j.scitotenv.2020.137778

Vance ME, Kuiken T, Vejerano EP, McGinnis SP, Hochella MF Jr, Rejeski D, Hull MS (2015) Nanotechnology in the real world: redeveloping the nanomaterial consumer products inventory. Beilstein J Nanotechnol 6:1769–1780. https://doi.org/10.3762/bjnano.6.181

Vittori Antisari L, Carbone S, Gatti A, Vianello G, Nannipieri P (2013) Toxicity of metal oxide (CeO2, Fe3O4, SnO2) engineered nanoparticles on soil microbial biomass and their distribution in soil. Soil Biol Biochem 60:87–94. https://doi.org/10.1016/j.soilbio.2013.01.016

Wagner S, Gondikas A, Neubauer E, Hofmann T, von der Kammer F (2014) Spot the difference: engineered and natural nanoparticles in the environment—release, behavior, and fate. Angew Chem Int Ed 53:12398–12419. https://doi.org/10.1002/anie.201405050

Waychunas GA, Kim CS, Banfield JF (2005) Nanoparticulate iron oxide minerals in soils and sediments: unique properties and contaminant scavenging mechanisms. J Nanopart Res 7:409–433. https://doi.org/10.1007/s11051-005-6931-x

WHO (2010) FAO/WHO expert meeting on the application of nanotechnologies in the food and agriculture sectors: potential food safety implications: meeting report. World Health Organization

Tan X-m, Lin C, Fugetsu B (2009) Studies on toxicity of multi-walled carbon nanotubes on suspension rice cells. Carbon 47:3479–3487. https://doi.org/10.1016/j.carbon.2009.08.018

Yao K-M, Habibian MT, O'Melia CR (1971) Water and waste water filtration. Concepts and applications. Environ Sci Technol 5:1105–1112

You T, Liu D, Chen J, Yang Z, Dou R, Gao X, Wang L (2018) Effects of metal oxide nanoparticles on soil enzyme activities and bacterial communities in two different soil types. J Soils Sediments 18:211–221. https://doi.org/10.1007/s11368-017-1716-2

Zheng X, Chen Y, Wu R (2011) Long-term effects of titanium dioxide nanoparticles on nitrogen and phosphorus removal from wastewater and bacterial community shift in activated sludge. Environ Sci Technol 45:7284–7290. https://doi.org/10.1021/es2008598

# Chapter 13
# Mechanism of Toxicity of Engineered Nanomaterials and Defense by the Crop Plants

**Ranjana Singh and Kajal Patel**

**Abstract** Rapid progress and broad applications of nanotechnology has resulted into a considerable increase in the number of engineered-nanomaterials that are inevitably entering in the environment and agriculture fields and have become a critical environmental threat. The production of engineered nanomaterials is increasing very rapidly due to their wide applications. The predicted global market value for nanomaterials would be about 11.3 billion USD by 2022 and almost 9–37% of engineered nanomaterials are directly or indirectly emitted into the atmosphere. It is apparent that due to technological upgradation, unsustainable use products and anthropogenic activities, contamination of agricultural soil with engineered nanomaterials has recently become a severe problem because it may lead to some unidentified issues related to food safety and quality. Finding a sustainable and ecological benign approach is a major challenge to contest engineered contamination of agriculture soil as well as crops grown there.

In spite of availability of huge toxicological data on engineered nanomaterials, systematic studies on impact of engineered nanomaterials and the underlying mechanism of their toxicity on crops are still not fully revealed. Therefore, this chapter comprehensively summarizes the current knowledge on exposure pathway of engineered nanomaterials, to agricultural soil, their interaction with crops and phytotoxicity at morphological, physiological and molecular level. Eventually, it highlights the mechanisms of engineered nanomaterials detoxification in crop plants in order to increase tolerance and sustainability. A clear understanding of engineered nanomaterials impacts on crops will help in optimizing their application in agriculture to improve crop production and to meet unforeseen demand of food in safe and sustainable way. The present chapter would provide a comprehensive current knowl-

R. Singh (✉)
Department of Botany, Government Model Degree College, Arniya-Bulandsahr, UP, India
e-mail: vermaranjana09@gmail.com

K. Patel
Natural Resources Management Laboratory, Department of Botany, University of Delhi, New Delhi, India



V. Kumar Singh et al. (eds.), *Sustainable Agriculture Reviews 50*, Sustainable Agriculture Reviews 50, https://doi.org/10.1007/978-3-030-63249-6_13

edge on the issue and a clear understanding of engineered nanomaterials impacts on crops plants.

**Keywords** Bioavailability · Detoxification mechanism · Phytotoxicity · Plant interaction · Tolerance · Transformation

## Abbreviations

| | |
|---|---|
| Ag-NPs | silver-nanoparticles |
| $Al_2O_3$-NPs | aluminum oxide-nanoparticles |
| Au-NPs | gold-nanoparticles |
| $CeO_2$-NPs | cerium oxide-nanoparticles |
| CuO-NPs | copper oxide-nanoparticle |
| GSSG | reduced glutathione |
| ICP-OES/MS | inductively coupled plasma optical emission spectrometry/mass spectrometry |
| $In_2O_3$-NPs | indium oxide-nanoparticles |
| MWCNTs | multi-walled carbon nanotubes |
| NADPH | nicotinamide adenine dinucleotide |
| NiO-NPs | nickel oxide-nanoparticles |
| $nTiO_2$ | nano-titanium dioxide |
| Pt-NPs | platinum-nanoparticles |
| SWCNTs | single-walled carbon nanotubes |
| $TiO_2$-NP | titanium dioxide-nanoparticle |
| WWTP | wastewater treatment plant |
| ZnO-NPs | zinc oxide- nanoparticles |
| $\gamma$-$Fe_2O_3$-NPs | gamma-iron oxide-nanoparticle |
| $\mu$-XANES | micro X-ray absorption near edge structure |
| $\mu$-XRF | micro X-ray fluorescence |

## 13.1 Introduction

Engineered-nanomaterials contamination is one of the prime environmental concerns that threaten the living organisms including human, as well as the ecological functioning of the environment. The rise in their uses in a wide range of products, their production at a mass-level and mishandling causes their exposure to atmosphere. Moreover, rise in application of agrochemicals, sewage sludge, organic waste manure, and industrial byproducts in modern agriculture adversely affect different food crops and hence the food chains (Antisari et al. 2013; Ma et al. 2016; Servin and White 2016).

Because of their unexpected applications, persistent nature, presence in every component of ecosystem (i.e. air, water and soil) and acute toxicity, engineered nanomaterials have emerged as a potent contaminant of soil (Yan et al. 2010). Engineered-nanomaterials found in contaminated soils are silica-based, carbon-based, quantum dots, and metal or metal-oxide based nanoparticles with different properties and sources of origin (Table 13.1). Among all of them, the metal or metal-oxide based nanoparticles are the most common and abundantly present in contaminated soil. Owing to their unique physiochemical properties like size ($\leq$100 nm), high surface to volume ratio, and improved reactivity, engineered nanoparticles are extensively used in agriculture to monitor soil properties and fertility and to manage diseases in crops by preventing and controlling pests and pathogens (Asli and Neumann 2009; Aslani et al. 2014). Toxicological studies of various types of engineered nanoparticles like carbon nano-tubes, metal and metal oxide nanoparticles (AgNPs, AuNPs, $Al_2O_3$, ZnO, CuO, $TiO_2$ and others), have been conducted both *in vivo* and *in vitro* in a broad range of living organisms including algae, plants, vertebrates and invertebrate (Schrand et al. 2010; Liu et al. 2013; Ma et al. 2016).

The soil is regarded to be the primary sink for engineered nanomaterials accumulation. It is evident from many studies that soil physiochemical properties like pH, ionic strength, natural organic matter and presence of other nanoparticles and biomacromolecules have significant impacts on nanoparticles aggregation and transformation (Cornelis et al. 2014; Tou et al. 2017; Ma et al. 2018). The bioavailability of engineered nanomaterials to plants depends on the mobility and stability of transformed/modified-engineered nanomaterials that ultimately determine the uptake and toxicity to crop plants (Lowry et al. 2012; Cornelis et al. 2014; Deng et al. 2014; Wang et al. 2016a, b; Tou et al. 2017). Thus, the bioavailable engineered nanomaterials should be consider as the exact engineered nanomaterials concentration to which a plant interact with and that may affect plant growth and development (International Organization for Standardization 2011).

Our understanding on engineered nanomaterials-crop interactions in soil is very narrow and imperfect. Very few toxicological assessment studies of engineered nanomaterials have been conducted in natural soils (Du et al. 2011; Dimkpa et al. 2012; Priester et al. 2012; Khodakovskaya et al. 2013). The majority of research has been conducted to understand the toxic nature of engineered nanomaterials to plants and both detrimental and beneficial effects have been documented (Yan et al. 2010; Du et al. 2011; Priester et al. 2012; Dimpka et al. 2013; Aslani et al. 2014; Gonzalez-Garcia et al. 2019). For example, carbon nanotubes treatment significantly increase the germination rate of rice seeds (Wang et al. 2012a), while $Al_2O_3$-NPs treatment negatively affects the seed germination and root elongation in some crops, including cucumber, *Zea mays*, *Glycine max*, carrot, and cabbage (Yamamoto et al. 2001; Aslani et al. 2014). ZnO-NPs are found to be the most toxic engineered nanomaterials that could negatively affect the root growth of the plants. Similarly, toxicological studies carried out on the *Arabidopsis thaliana*, with various engineered

**Table 13.1** Types of engineered nanomaterials, their sources, applications and properties

| Features | Metal and metal-oxide based | Quantum dots | Silica-based | Carbon-based | Dendrimers |
|---|---|---|---|---|---|
| Shells | None, metal-oxide, silica | None, other metal sulfide, polymers, silica | None | None | Generation, 0–7 (metal chalcogenide) |
| Cores | Metal and metal-oxides | Metal sulphides | Crystalline silica | Solid, hollow | 1000's of potential cores |
| Shapes | Sphere, rod, hexagonal, mono-,bi-,tri- and tetra-pod star etc. | Sphere, cube, rod, tetrahedral, octahedral etc. | Sphere, rod, fibre, sheet | Sphere, tubes | Sphere |
| Size | 1–100 nm | 1–10 nm | >10 nm | 1, 10–20, 30–50, >50 nm | 2–100 kDa |
| Specific properties | Photocatalysis, electrical conductivity, biocompatibility, metal-based with zero-valency | Semi conductance, optical and electrical properties, absorption of white and UV-light and reemission | Modification in mechanical, physical and biological properties of surface | High tensile strength, electrical and thermal conductivity | Multifunctional, flexible, molecular weight, surface modifications catalysis |
| Applications | Sensors, electronics, antimicrobial agents in food packaging and fabrics, soil, air, water remediation, fuel cells, paints, solar cells, sunscreen, cosmetics and coatings. | In biomedical field for cellular imaging/labelling, diagnostic tools and solar batteries | In dentistry to make silica matrix, skeletal application, repairment, improvement and construction of bones. | Plastics, automobiles, aircrafts, electronics, batteries, sensors, coatings, water purification | Microcapsules, nanolatex, coloured glass, sensors, electrodes, DNA chips, medicines |
| Examples | Metal based- Ag, Au, Ti, Zn, Cu, Fe etc. and metal oxide based- $TiO_2$, SiO2, $Al_2O_2$, $CeO_2$ etc. | Semiconductors like CdSe, CdTe, CdSeTe, ZnSe, ZnS etc. | Silica-cement and silica-nanoparticles | Fullerenes, MWCNT, SWCNT | Dendrimers |
| Emission via | Agriculture wastes, household wastes, waste water treatment and industrial wastes | Industrial wastes | Industrial wastes, waste water, wear and tear of polymers. | Burning, dry and wet drilling from polymers, abrasion and sanding from epoxy | Waste water and industrial water |

*MWCNT* Multi-walled carbon nanotube, *SWCNT* Single-walled carbon nanotube nanomaterials ($Al_2O_3$, $SiO_2$, ZnO, and $Fe_3O_4$) also reported that ZnO-NPs (at 400 mg $l^{-1}$) inhibited the germination (Huang et al. 2002; Ma et al. 2009). It was clear from the reported studies that nanotoxicity to plants is dependent on concentration and size of engineered nanomaterials, plant species, and experimental conditions (Aslani et al. 2014; Deng et al. 2014; Rizwan et al. 2017). Besides, surface reactivity is another factor that contributes to the toxicity of engineered nanomaterials. Engineered nanomaterials, when used at appropriately low concentrations it can support crop growth by improving nutrient sources. However, at excessive concentrations, engineered nanomaterials exposure to crops negatively affect plant growth, biomass reduction, and yield, photosynthetic system, and defense mechanism by affecting the expressions of a group of genes (Thuesombat et al. 2014; Jain et al. 2016; Ma et al. 2016; Anderson et al. 2017, Rizwan et al. 2017; Santos Filho et al. 2019).

A large amount of literature is available suggesting phytotoxicity of engineered nanomaterials in plants (Table 13.1), but very few studies describe the biotransformation, possible transfer to the next generation and bio-magnifications of engineered nanomaterials in food crops. Some previous studies, have reported in detail about the fate and behavior of engineered nanomaterials in crops through studying physiological, biochemical and molecular alterations, and their accumulation within the plant cell (Thuesombat et al. 2014; Ma et al. 2016) and provided contrasting evidences. Moreover, plants show a tolerance towards the engineered nanomaterials on continuous exposure. This could be possible by generation of reactive oxygen species which affects the physiological and biochemical cycles that ultimately activate defense pathway (Dimpka et al. 2012; Rico et al. 2013a; Da Costa and Sharma 2016; Abdel Latef et al. 2017; Gonzalez-Garcia et al. 2019). Some studies revealed that engineered nanomaterials provide benefits to plants by enhancing their resistance and alleviating toxicity caused by various biotic (pests and pathogens) and abiotic (cold, heat, saline, drought, heavy metals etc.) stresses (Mohammadi et al. 2014; Singh and Lee 2016; Abdel–Latef et al. 2017; Perez-Labrada et al. 2019). These engineered nanomaterials can cause oxidative stress in plants by inducing reactive oxygen species and reactive nitrogen species which may lead to oxidation of lipids, protein and deoxyribonucleic acid (DNA) in plant cells. It affects membrane integrity and increase the phytoavailability of engineered nanomaterials within cell which in turn may cause breaks in DNA strands, cross linkage mutations and affect the growth and development of crop plants (Fig. 13.1). In order to cope up with different engineered nanomaterials, plants possess defense systems to manage the engineered nanomaterials-induced oxidative stress. However, upregulation of these mechanisms and biomolecules may depend on plant species, level of their tolerance, engineered nanomaterials type and plant growth (Lopez- Moreno et al. 2010; Ma et al. 2016; Wang et al. 2016c, 2019).

Therefore in this chapter, we have summarized the results from the literatures related to engineered nanomaterials-crop interactions to attain an inclusive understanding of the:

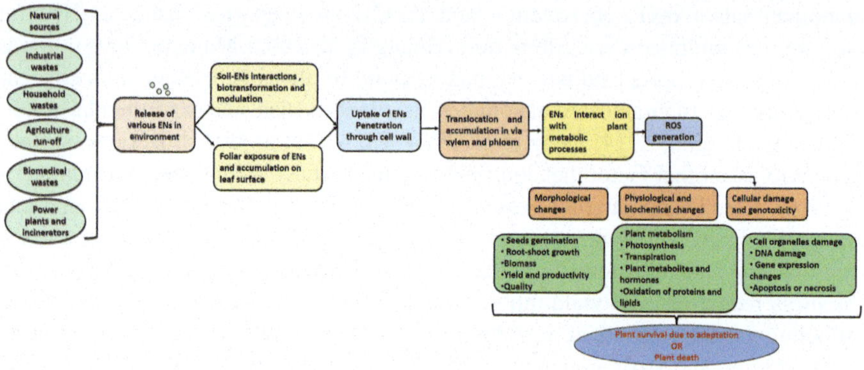


**Fig. 13.1** Schematic representation of complete cycle of engineered nanomaterials from their point of origin to their exposure in soil and then uptake by plants and phytotoxicity at various levels which ultimately results in either plant survival due to positive impacts of engineered nanomaterials leads to adaptations in plants or cause plant death due to their negative impacts. *ENs* engineered nanomaterials, *ROS* reactive oxygen species

1. diverse sources of engineered nanomaterials contamination in agro-ecosystems in relation to anthropogenic factors and processes,
2. morphological, physiological, and genetic/epigenetic responses of crop plants to engineered nanomaterials stress, and
3. strategies/defense mechanisms activated in plants in order to promote tolerance against engineered nanomaterials stress.

The present chapter would provide a comprehensive current knowledge on the issue and a clear understanding of engineered nanomaterials impacts on crops plants that will help in optimizing their application in crop production to meet continuously rising demand of food in safe and sustainable way.

## 13.2 Engineered Nanomaterials – Types, Sources and Exposure Pathway

Nanomaterials are the unique chemical substances or materials which could be natural, incidental or manufactured composed of free, aggregated or agglomerated particles of size ranges from 1 nm to 100 nm i.e. of nanoscale ($10^{-9}$ m). Due to specific size distribution, the nanomaterials exhibit unique properties compared to their bulk forms such as high surface area, refractive index, chemical reactivity, photocatalysis etc. (Cornelis et al. 2014; Ma et al. 2018). Their novel characteristics escalated their application in various sectors, and thus, utilized in the mass manufacturing of a wide range of products, for example, in cosmetics, medicines, materials, papers, food products, paints, solar cells, nanofertilizers, nanosensors etc. This growing demand of engineered nanomaterials in myriad of products and utilization, led to increased nanomaterials production with an added value which is estimated to be raised to about 30 billion USD by the end of 2020 in global markets (Wang et al. 2013). This increased growth in nanomaterials production may consequently increase their emission into the environment (Gottschalk and Nowack 2011). According to conservative estimations by Mordor Intelligence the predicted global value of nanomaterials will be expected to rise >22% growth rate during 2017–2022 from 4.1 billion USD (in 2015). The predicted global market value for nanomaterials would be about 11.3 billion USD by 2022 (Inshakova and Inshakov 2017). According to Keller et al. (2013), almost 9–37% of engineered nanomaterials are directly or indirectly emitted into the atmosphere. Nanomaterials are considered as an emerging environmental contaminant (Yan et al. 2010). Due to persistent nature, nanomaterials can stay in air, water and soil for a long-term period without degradation when discharged into the environment. On the basis of their point of sources or synthesis, nanomaterials can be categorized in two broad classes i.e. natural and anthropogenic (Phogat et al. 2016; Ha et al. 2019) (Fig. 13.2).

Among the various nanomaterials, engineered nanomaterials (are synthesized and manufactured at mass level for their utilization in order to achieve definite goals like improvement and remediation of water, air, agricultural soil, crops and ecosystems. According to Peralta-Videa et al. (2011) engineered nanomaterials can be further categorized into organic and inorganic, where organic engineered nanomaterials basically comprise of fullerenes and carbon nanotubes and metal-oxide based, metal-based and quantum dots constitute the class of inorganic engineered nanomaterials. The complete classification of nanomaterials is summarized in Fig. 13.2. According to Aslani et al. (2014), there are more than 1300 commercial engineered nanomaterials available with wide scale potential applications, they are summarized with their properties, sources and applications in Table 13.1.

The promising applications, progress in their utilization, their unabated mass scale production and release in environment raise concerns for the agriculture

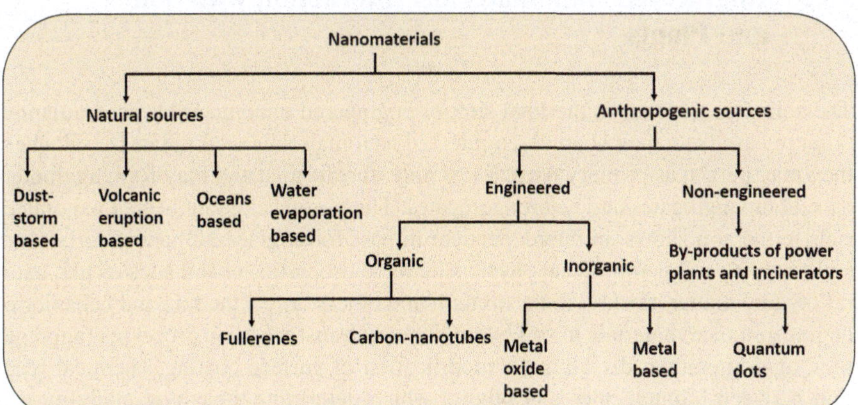


**Fig. 13.2** Flow chart of classification of engineered nanomaterials on the basis of their sources of origin and chemical nature

sector, which is facing innumerable challenges in terms of global food security and climate change. In recent years, use of nanotechnology in agriculture has gained a good momentum with wide scale usage by public contributions in terms of social and economic support. In modern agricultural practices, engineered nanomaterials are being applied as agrochemical agents, nanosensors, nanodevices, nanofertilizers etc. to improve crop production by improving nutrients, crop protection against various pathogens and abiotic stresses and by managing postharvest products (Ghidan and Al Antary 2019). Despite of beneficial applications, wide application of engineered nanomaterials in agriculture may further add engineered nanomaterials to soil that may cause deleterious impacts on crop yield and productivity. Engineered nanomaterials possibly contaminate the soil directly via handling (for example nanofertilizers, nanopesticides, nanoherbicides, nanobactericides, nanosensors etc.) to enhance crop productivity, growth, protection against both biotic and abiotic stresses, and to remediate soil to increase soil fertility and health improvement (De Rosa et al. 2010) or accidental release. The use of biosolids and sludge from wastewater treatment plant (WWTP) and wastewater in agriculture also increase the probability of release of engineered nanomaterials in the agricultural fields (Gottschalk et al. 2009).

Soil is considered to act as ultimate site of engineered nanomaterials accumulation in environment. After discharge these engineered nanomaterials may persist in all components (air, water and soil) of environment for a long time and interact with plants. In soils, roots of plants interact with engineered nanomaterials, while in air, aerial parts of plants like leaves and stem will be in direct contact with the atmospheric engineered nanomaterials. There is great risk of crop contamination with different types of engineered nanomaterials, which may affect food quality and human health.

## 13.3 Engineered Nanomaterials Interaction with Soil and Plants

The soil is considered as the final sink of engineered nanomaterials accumulation. After release from their sources, engineered nanomaterials accumulate in soil where they may persist as primary particles or may transformed and may form agglomerates, homoaggregates and heteroaggregates. The presence of engineered nanomaterials in the soil shows an interdependent impact for engineered nanomaterials and soil properties. It is found that physicochemical properties of soil such as pH, natural organic matter, texture, ionic strength and others, affect the fate and behavior of engineered nanomaterials in soil by influencing transformation processes (aggregation, agglomeration, dissolutions, modification of surface coating, chemical reaction between biomolecules and others) which determine their bioavailability and toxicity (Tou et al. 2017; Ma et al. 2018). Some studies also reported a vice-versa effect that engineered nanomaterials shows alterations in various soil properties including its physico-chemical and biological (enzymatic) properties (Peralta-Videa et al. 2011; Shah et al. 2014). Over the past several years, research has been focused on the engineered interaction with soil, plants and their impact on ecology, food chain and human health (Cox et al. 2016; Ali et al. 2017; Goswami et al. 2017; Ziental et al. 2020).

Surface modification or coating, affects the mobility and stability of engineered nanomaterials, and thus, determines bioavailability of engineered nanomaterials to plants, uptake and toxicity (Cornalis et al. 2014; Chai et al. 2015; Ma et al. 2018). In rhizosphere, interaction of engineered nanomaterials and biological molecules present in root exudates (carbohydrates, amino acids, organic acids) and microbial secretion, chelators (biosurfactants), organic acids (glucuronic acid and citric acid) hormones and enzymes (indol acetic acid/IAA and aminocyclopropane-carboxylate/ACC) take place that play a crucial role in engineered nanomaterials transformation as these substances/biomolecules modify surface characteristics of engineered nanomaterials. Root exudates may biotransform engineered nanomaterials or ions released from engineered nanomaterials and affect their bioavailability and uptake by plants. The microorganisms present in soil also play a crucial role in engineered nanomaterials biotransformation as they alter physico-chemical properties of rhizosphere near plant roots by introducing enzymes and chelating agents and contribute to engineered nanomaterials bioavailability to plants (El Badawy et al. 2012; Deng et al. 2014; Chai et al. 2015). Hence it could be concluded that engineered nanomaterials toxicity can be altered by the interaction of soil, root exudates and microbial communities present around the plants.

## 13.4 Uptake, Translocation and Accumulation of Engineered Nanomaterials in Plants

It is clear now that after being discharged into the soil, engineered nanomaterials are not found in their original form. Various abiotic and biotic factors of soil as well as their own properties determine their fate in soil and finally the uptake by plants (Cornelis et al. 2014; Tan et al. 2018). Biotransformation of engineered nanomaterials in soil as well as on plant roots may affect their uptake, translocation and accumulation within plants which ultimately determine plant growth and productivity (Shah et al. 2014; Chai et al. 2015) (Fig. 13.1). engineered nanomaterials when interact with plant roots then either in individual or in aggregated form, engineered nanomaterials show accumulation or adherence to tissues of rhizodermis (Lin and Xing 2008; Zhao et al. 2012). Some specific forces like electrostatic adsorption, mechanical adhesion or hydrophobic affinity of engineered nanomaterials determine the initial interaction (Zhang et al. 2012; Deng et al. 2014).

### 13.4.1 Uptake of Engineered Nanomaterials

The mechanism of engineered nanomaterials uptake by roots is not clear. But it is evident from the previous studies that uptake of engineered nanomaterials in plants depends on their size, chemical nature and surface properties. The entry of engineered nanomaterials into root cells depends on their size. Various carrier proteins, aquaporins, ion channels and endocytosis have been reported to involve in uptake process depending upon engineered nanomaterials size. Besides their transport through plasmodesmata, or their entry may facilitate via presence of biomolecules on surface coating through the formation of new pores on the root cell (Rico et al. 2011). The biotransformed nanoparticles got adhered to root surface where they are taken by penetration through the cell wall and membrane of root (Serag et al. 2013; Tripathi et al. 2018). Further, the engineered nanomaterials are uptaken passively with the non-metabolic processes of diffusion or mass-flow (Judy 2013). The matrix found to be responsible for the passage of engineered nanomaterials into plant cells via several pores and channels in intact cell membrane; sometimes it restricts their passage due to size incompatibility which shows the size specificity as a crucial factor for the uptake of engineered nanomaterials in plants (Ma et al. 2010; Rico et al. 2011). This movement leads the engineered nanomaterials to stelar region of root which is responsible of their translocation via two different routes including the apoplastic and symplastic (Kurepa et al. 2010; Larue et al. 2012). González-Melendi et al. (2008) reported the presence of engineered nanomaterials in extracellular space and within some cells in case of *Cucurbita*. Consequently, the restriction of some engineered nanomaterials also reported during apoplastic movement due to presence of suberised barrier known as casparian strip leads to accumulation of aggregates of engineered nanomaterials in endodermis (Larue et al. 2012; Patrick et al. 2015). These engineered nanomaterials eventually enter into symplastic route for efficient translocation to vascular system (Deng et al. 2014).

In case of uptake via roots, translocation and accumulation are served by xylem as the most important carrier involved in the unidirectional flow and distribution of engineered nanomaterials from bottom to up in aerial parts (Aslani et al. 2014). In addition to roots, the other exposure pathway that is via leaves in case of foliar application of engineered nanomaterials where they accumulate on the leaf surface and uptake of engineered nanomaterials are facilitated by cuticular and stomatal routes (Buick et al. 1993). Engineered nanomaterials which accumulate on stomata show translocation via phloem, and thus, bidirectional distribution (Tripathi et al. 2017). The engineered nanomaterials uptake in plants generally follows the active transport mechanism along with other processes including signaling, recycling and regulation of membrane (Etxeberria et al. 2009). Besides cell pores, rhizodermal, stomatal, cuticular uptake of engineered nanomaterials, the endocytic pathways are also found as an alternative for their uptake which can be both clathrin dependent and clathrin independent as studied in *Nicotiana tabaccum* for uptake of Au-NPs (Onelli et al. 2008; Iversen et al. 2011). Serag et al. (2011) reported a completely different endocytic pathway for internalization or uptake of multi-walled carbon nanotubes (MWCNTs) in *Catharanthus roseus.* Judy (2013) reported that the translocation and accumulation might be influenced by some physical and chemical properties of plants such as hydraulic conductivity, pore size of cell wall etc.

In addition to root and foliar applications, there is one more application to seeds which is the most commonly used protocol to study the engineered nanomaterials toxicity where the uptake of engineered nanomaterials from the suspension media is achieved via parenchymatous intercellular spaces by involving diffusion method (Lee et al. 2010; Ma et al. 2010). In seed coat, another channel for the uptake of engineered nanomaterials has found to be upregulation of aquaporins as observed in case of tomato (Khodakovskaya et al. 2009). These engineered nanomaterials reported to have both stimulatory as well as inhibitory impacts on seeds germination depends on characteristics of both plants and engineered nanomaterials (Stampoulis et al. 2009; Azimi et al. 2013). Hence based on exposure (root, foliar and seed), mechanism of uptake of engineered nanomaterials to plants also vary and depends on particle's size, shape and surface chemical nature.

### 13.4.2 Translocation and Accumulation of Engineered Nanomaterials

During uptake process via roots and leaves, engineered nanomaterials enter into the vascular system followed by their distribution and their accumulation to different parts of the plants (Tan et al. 2018). In the vascular system, both xylem and phloem play remarkable role in translocation. However, Wang et al. (2012b) studied that xylem shows unidirectional translocation i.e. movement of engineered nanomaterials from roots (bottom) to shoot (top) in case of soil amendments with engineered nanomaterials, while phloem helps in bidirectional translocation i.e. from leaves (top) to other parts of plants (bottom) in case of foliar application of CuO-NPs in *Zea mays* plants. The advanced technologies like micro X-ray absorption near edge structure (μ-XANES), micro X-ray fluorescence (μ-XRF), inductively coupled plasma optical emission spectrometry/mass spectrometry (ICP-OES/MS), optical microscopy and electron microscopy help to study the details for uptake and translocation of engineered nanomaterials in plants (Tan et al. 2018). Antisari et al. (2015) determined the fate, transport and possible toxicity of different nanoparticles ($CeO_2$, $Fe_3O_4$, $SnO_2$, $TiO_2$ and metallic Ag, Co, Ni) in tomato tissues from soils by using ICP-OES and found that the metals ions are accumulated in the roots of tomato plant; while in case of Ag, Co, Ni-NPs treatment, these nanoparticles (not ions) were present in higher concentrations in both above as well as below ground organs (including Ag in fruits) than that of untreated plants.

Direct visualization and whole plant mapping of engineered nanomaterials signaling gave direct evidences of engineered nanomaterials transport within stems, leaves, petioles and fruits in many studies. Some of the most basic conclusion withdrawn from these studies include the accumulation of engineered nanomaterials in shoots near or within vascular tissues, long range transport of small sized engineered nanomaterials from root to sub-apical tissues, leaves found to have higher accumulated concentration of engineered nanomaterials in comparison to stems and finally the engineered nanomaterials distribution to leaf periphery and trichomes also play role in detoxification strategies (Cifuentes et al. 2010; Ghafariyan et al. 2013; Deng et al. 2014). Engineered nanomaterials can translocate within the plants through various routes, but most of the engineered nanomaterials cannot move to sub-apical tissues because of their size. For example, $Fe_3O_4$NPs of 20 nm was found to penetrate and translocate, while the nanoparticles of 25 nm were unable to translocate within the *Cucurbita mixta* (pumpkin) plants (Zhu et al. 2008; Wang et al. 2011). A study conducted to evaluate the effect of different size of $nTiO_2$ (Anatase 14, 25, 140 nm and Rutile 22, 36, 655 nm) in *Triticum aestivum* demonstrated that engineered nanomaterials of size 36 nm could move to the steler region in roots, whereas bigger engineered nanomaterials of sizes (range 36–140 nm) could only reach to cortical region of root (Laure et al. 2012). Similarly, MWCNTs of micrometer lengths range adsorbed onto root surfaces and penetrated root epidermal tissues by inducing physical injuries and therefore caused stress without being internalized into the plant (Miralles et al. 2012).

Surface modification of engineered nanomaterials is another factor which determines their uptake and translocation in plants. Surface coating of engineered nanomaterials significantly enhance their uptake and translocation due to biocompatibility and higher affinity of biological compound or functional group with plasma membranes. However, there are very few studies for the large-scale patterns of engineered nanomaterials distribution in plants but it is still emerging. The accumulation and distribution of engineered nanomaterials to plants show diverse impacts which can be understood by studying the responses of plants to various engineered nanomaterials.

## 13.5 Impacts of Engineered Nanomaterials on Plants

The typical role of plants in maintaining the environmental balance as the most important part in the form of producers in the food chain makes them superior in order to study the impact and toxicity caused by engineered nanomaterials (Ma et al. 2010; Tan et al. 2018). Besides this, the food safety and security raises concern for engineered nanomaterials toxicity in crop plant species (Deng et al. 2014). Plants exposed to engineered nanomaterials may experience changes in its morphology physiology and genetic characters which may consequently influenced biomass, production and quality of crops. It is reported that engineered nanomaterials exposure can affect a plant system both directly (due to foliar uptake) or subtly (uptake from soil and water) (Ma et al. 2018). Based on knowledge from the existing literature, it is clear that plants can uptake engineered nanomaterials. Few studies have observed the beneficial impact on crop production, nonetheless, there are a huge number of studies addressing that low concentrations of engineered

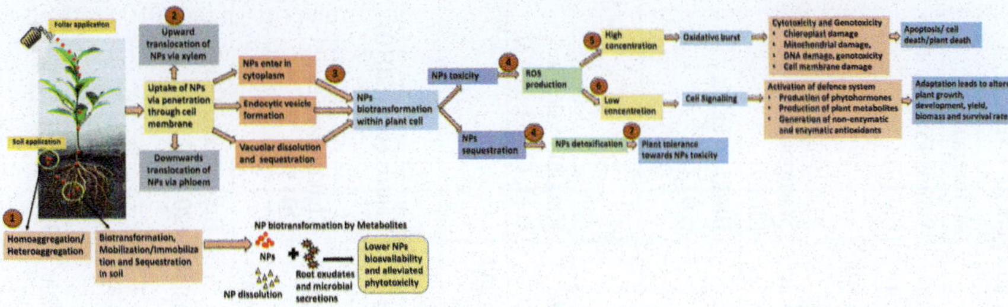


**Fig. 13.3** Engineered nanomaterials interaction with soil and plants and the mechanism of their toxicity and detoxification in plants (1). Biotransformation of nanomaterials in rhizosphere due to root exudates and microbial secretion leads to altered nanomaterials bioavailability to plants, and thus, their toxicity in plant. (2). Uptake of engineered nanomaterials followed by foliar and soil application through cell membrane and their translocation via xylem and phloem. (3). Nanonmaterials biotransformation within plant cell leads to either nanomaterials toxicity or their sequestration. (4). Nanomaterials toxicity induces ROS production and nanomaterials sequestration shows detoxification mechanism. (5). High concentration of nanomaterials causes oxidative burst which results in cytotoxicity and genotoxicity ultimately leading to cell/plant death (6). Low concentration of ROS helps in cell signalling and results in activation of defence system that ultimately helps in plant tolerance to engineered nanomaterials. (7). Nanomaterials detoxification leads to plant tolerance. *ROS* reactive oxygen species

nanomaterials may be detrimental to various physiological and biochemical processes of crops which in turn adversely affect their yield and productivity by altering growth and development (Deng et al. 2014; Cox et al. 2016; Mattiello and Marchiol 2017) (Figs. 13.1 and 13.3). Engineered nanomaterials caused phytotoxicity preliminary by clogging of pores and interruption in apoplastic stream which ultimately affect the uptake and translocation of various essential nutrients, reduce photosynthesis, reactive oxygen species generation and DNA damage (Asli and Neumann 2009; Aghdam et al. 2016; Santos Filho et al. 2019). Studies related to impact of engineered nanomaterials on crops at morphological, physiological, biochemical and cellular levels are summarized in Table 13.2.

### 13.5.1 Phytotoxicity at Physiological and Biochemical Level

To study the phytotoxicity of engineered nanomaterials at physiological levels, several standard parameters such as photosynthesis, transpiration, stomatal conductance etc. are kept under observation, while parameters like photosynthetic pigments, osmotic balance, reactive oxygen species generation and antioxidants production are evaluated as biochemical measures. Photosynthesis is known to be a good measure of plants performance under stress because it is the only source of energy in plants and hence affects all aspects of plant metabolism and physiology. Therefore, the evaluation of PS-II activity and photosynthetic rates are found as significant measures to evaluate the impacts of various stress factors.

**Table 13.2** Summary of phytotoxicity of various engineered nanomaterial at morphological, physiological, biochemical, cellular and genetic levels

| ENMs | Plants | Size of ENMs | Treatment | Concentration | Impacts on plants | References |
|---|---|---|---|---|---|---|
| $TiO_2$ | *Allium cepa* | 21 nm, anatase | Suspended in distilled water | 10,100, 1000 mg $l^{-1}$ | Reduced seed germination rate and root development | Santos Filho et al. (2019) |
| | | | | | Increased lytic vacuoles, oil bodies, nucleolar damage | |
| Si | *Cucurbita pepo* | 10 nm, powder | Suspended in media | 1000 mg $l^{-1}$ | Inhibited seed germination | Stampoulis et al. (2009) |
| $Al_2O_3$ | *Nicotiana tabacum* | Nanoparticles | Suspended in distilled water | 0, 0.1, 0.5 and 1% | Reduced root growth, biomass, leaf count and overall growth | Burklew et al. (2012) |
| | | | | | Alteration in miRNAs expression | |
| MWCNTs | *Solanum lycopersicon* | 125–250 nm, irregular | Suspended in media | 0, 50 and 200 μg $ml^{-1}$ | Accelerated plant growth | Khodakovskaya et al. (2009, 2013) |
| | | | | | More number of flowers and fruits | |
| | | | | 10, 20, 40 μg $ml^{-1}$ | Increased seed germination | |
| CdSe/ZnS QDs | *Medicago sativa* | 4–6 nm, nanocrystals | Suspended in media | 100 nM | Promoted cyto-oxidative stress | Santos et al. (2010) |
| Graphene quantum dots | Coriander and garlic | 5–10 mm | Suspended in distilled water | | Growth rates of the plants increased | Chakravarty et al. (2015) |
| | | | | | Enhanced yield and biomass | |
| MWCNT | *Amaranthus sp.* | 11 nm | Suspended in distilled water and media | 125, 250, 500 and 1000 mg $l^{-1}$ | Inhibition of stomatal opening or induces stomatal closing | Begum and Fugetsu (2012) |
| | | | | | Generation of ROS such as superoxides and $H_2O_2$ | |
| $TiO_2$ | *Phaseolus vulgaris* | <25 nm | Suspended in media | 10–30 ppm | Altered chlorophyll content | Jacob et al. (2013) |

| | | | | | | |
|---|---|---|---|---|---|---|
| | | | | | Modified antioxidant enzymatic activities | |
| CuO | *Oryza sativa* | <50 nm | Suspended in media | 0, 2.5, 10, 50, 100, and 1000 mg $l^{-1}$ | Germination rate, root and shoot length, and biomass decreased, | Da Costa and Sharma (2016) |
| | | | | | Decreased chloroplasts with lower number of thylakoids per granum | |
| | | | | | Photosynthetic and transpiration rate declined | |
| | | | | | Oxidative and osmotic stress | |
| | | | | | Elevated levels of ascorbate peroxidase and superoxide dismutase | |
| Ag | *Ricinus communis* | <100 nm | Suspended in distilled water | 0 (control), 100, 200, 500, 1000, 2000, and 4000 mg $l^{-1}$ | Seed germination rate not affected | Yasur and Rani (2013) |
| | | | | | Shoot and root growth most affected | |
| | | | | | Increased oxidative stress | |
| | | | | | Increase in antioxidants | |
| $Fe_2O_3$ | *Glycine max* | | Foliar spray | 0, 0.25, 0.5, 0.75 and 1 g $l^{-1}$ | Increased leaf and pod dry weight | Sheykhbaglou et al. (2016) |
| | | | | | Increase in grain yield | |
| $CeO_2$ | *Oryza sativa* | 231 ± 16 nm | Soil amendment by suspension | 0 and 500 mg $kg^{-1}$. | Grain quality compromised | Rico et al. (2013b) |
| | | | | | Reduced all antioxidant, except flavonoids | |
| Au | *Brassica juncea* | 10–20 nm | Foliar spray | 10, 25, 50 and 100 ppm | Increased growth and yield | Arora et al. (2012) |

(continued)

**Table 13.2** (continued)

| ENMs | Plants | Size of ENMs | Treatment | Concentration | Impacts on plants | References |
|---|---|---|---|---|---|---|
| | | | | | Imbalanced ethylene and auxin production | |
| | | | | | Oxidative damage increased | |
| | | | | | Increased sugars content | |
| Fullerol | *Allium cepa* | 18.17–24.36 nm | Suspended in distilled water | 10–110 mg $l^{-1}$ | Generation of ROS led to loss of membrane integrity | Chen et al. (2010) |
| | | | | | Necrosis observed | |
| ZnO | *Pisum sativum* | 10 nm | Suspended in water to amend soil | 125–500 ppm | Reduction in chlorophyll contents | Mukherjee et al. (2014) |
| | | | | | Inclined oxidative stress and lipid peroxidation | |
| | | | | | Increased root and stem elongation | |
| | | | | | Increased $H_2O_2$ accumulation and reduced stress enzymes activities | |
| SWCNT | *Oryza sativa* | 1–2 nm | Suspended in buffer | 25 mg $l^{-1}$ | Increased chromatin condensation and DNA damage | Shen et al. (2010) |
| | | | | | Affected viability of cells | |
| Ag | *Allium cepa* | <100 nm | Suspended in distilled water | 25, 20, 75, and 100 ppm | Reduced mitotic index, fragmented and chromosomes | Kumari et al. (2009) |
| MWCNT | *Lycopersicon esculentum* | 10–35 nm | Suspended in media | 50 μg $ml^{-1}$ | Up-regulated stress-related genes | Khodakovskaya et al. (2011) |
| | | | | | Altered expression of genes | |

| | | | | | | |
|---|---|---|---|---|---|---|
| $TiO_2$ | *Vician arbonensis* | <100 nm, mixture of rutile and anatase | Suspended in distilled water | 0.2, 1.0, 2.0, and 4.0% | Aberration of chromosomes and loss of gene expression | Castiglione et al. (2011) |
| | | | | | Possible alterations in mitotic activity, chromosomal aberrations, and micronuclei release | |
| $TiO_2$ | *Nicotiana tabacum* | 100 nm | Suspended in distilled water | 0, 2, 4, 6, 8 and 10 mM | Damaged DNA structure, micronuclei and chromosomal aberrations | Ghosh et al. (2010) |
| | | | | | Reduction in root growth and increased lipid peroxidation | |

*MWCNT* Multi-walled carbon nanotube, *SWCNT* Single-walled carbon nanotube, *CdSe/ZnS QDs Cadmium selenide/zinc sulphide quantum dots*

Various reports revealed that the engineered nanomaterials influence the chlorophyll concentration and its function in plants (Zheng et al. 2005; Smith et al. 2015; Pérez-Labrada et al. 2019). Some initial studies showed increase in photosynthesis, high rate of chlorophyll synthesis and higher Rubisco activity after $TiO_2$-NP exposure in spinach (Zheng et al. 2005). Additionally, it is found that very high concentration of NPs can affect photosynthesis severely which ultimately may cause suppression of plant growth or plant death. Engineered nanomaterials stress leads to alternation in stomatal conductance, transpiration, unavailability to $CO_2$ and the photosynthetic process which consequently induces oxidative stress by overproduction and accumulation of reactive oxygen species and/or reactive nitrogen species.

According to Wang et al. (2019), the engineered nanomaterials toxicity in plants depends on dose applied, exposure time and plant species. Begum et al. (2011) reported growth inhibition and increased reactive oxygen species generation in seedlings of cabbage, tomato and red spinach treated with doses of 500–2000 mg $l^{-1}$ of graphene for 20 days. Cucumber plants were shown to uptake $Fe_3O_4$-NPs which were eventually accumulated in various plant tissues (Zhu et al. 2008). Moreover, $Fe_3O_4$-NPs treatment reduced the chlorophyll pigment and photosynthetic efficiency in soybean plants and ultimately altered both biochemical and enzymatic processes (Ghafariyan et al. 2013). $\gamma$-$Fe_2O_3$-NPs (20–100 nm) treatment (50 and 100 mg $l^{-1}$ in hydroponic) found to decrease macronutrients (viz. Ca, K, Mg, and S) concentrations in the shoot, reduce photosynthetic pigment content as well as functioning of roots in *Helianthus annuus* L. (Martínez-Fernández et al. 2016). The CuO-NPs exposure through hydroponics to 6-day old rice seedlings caused major toxicity in terms of declined photosynthesis due to decreased chlorophyll pigments and transpiration due to decreased hydraulic conductivity.

The decline in these physiological processes led to oxidative as well as osmotic stress which in turn activates plant defense machinery by triggering various enzymatic antioxidants system (Da Costa and Sharma 2016). According to study by Santos et al. (2010), the quantum dots application through suspension in media led to promoted cyto-oxidative stress. The MWCNTs treatment under hydroponic condition induced stomatal closing in red spinach. An increased level of reactive oxygen species such as superoxides and peroxides was observed in concentration dependent manner which might eventually causes damage to cell and its death (Begum and Fugetsu 2012). An alteration in chlorophyll content and antioxidant enzymatic activities was observed in $TiO_2$-NPs treated *Phaseolus vulgaris* plant (Jacob et al. 2013). Yasur and Rani (2013) also reported a concentration dependent increase in oxidative stress and antioxidants in order to protect *Ricinus communis* from Ag-NPs stress.

The foliar application of Au-NPs in 30 days old *Brassica juncea* seedlings resulted in imbalanced ethylene and auxin production, increased oxidative damage and increased sugars content (both reducing and total) in order to maintain osmotic balance (Arora et al. 2012). Chen et al. (2010) demonstrated the phytotoxicity of fullerol in case of leaves sample of *Allium cepa* when immersed in solution of fullerol. Fullerol showed harmful impacts as they caused over generation of reactive oxygen species which finally led to loss of membrane integrity and necrosis and ultimately caused collapse of plant structure. The soil amendment with bulk ZnO and ZnO-NPs solutions posed some physiological and biochemical alterations in *Pisum sativum* such as reduction in chlorophyll contents, increased lipid peroxidation and hydrogen peroxide accumulation and reduced activities of stress enzymes catalase and ascorbate peroxidase (Mukherjee et al. 2014). In a study, Alkhatib et al. (2019) reported a size and concentration dependent impact of $Fe_3O_4$-NPs on physiological, biochemical, and ultrastructural properties of tobacco.

Although a plentiful of literature on phytotoxicity of engineered nanomaterials on different plants species are available, but all were conducted either in laboratory condition or in greenhouse facilities. The systematic studies in real field situation are still lacking and hence further research is needed to elucidate the mechanism of engineered nanomaterials toxicity in plant species. The mechanism of engineered nanomaterials toxicity has been discussed separately in next section.

### *13.5.2 Phytotoxicity at Cellular and Genetic Level*

The entry of nanoparticles in plants takes place by targeting some specific genes which helps in their smart delivery in plant system (Siddiqui et al. 2015). In addition to this, the engineered nanomaterials toxicity regulates the expression of diverse genes that are crucial for various physiological and biochemical processes. Thus, perhaps the molecular analysis might be the best means to study engineered nanomaterials phytotoxicity. Likewise, Ma and Yamaji (2015) observed activation of some specific genes (*Lsi1, Lsi2 and Lsi6*) to facilitate entry of silica- nanoparticles in *Oryza sativa*. Yan et al. (2013) observed single-walled carbon nano-tubes (SWCNTs) induced induction of various genes associated with root growth in *Zea mays*. Microarray-based analysis has been conducted on *Arabidopsis thaliana* roots to understand the mechanisms of ZnO-NPs, $TiO_2$-NPs, and fullerene soot toxicity (Landa et al. 2012). In an advanced study, Khodakovskaya et al. (2011) amalgamated genetic, photothermal, and photoacoustic strategies to detect engineered nanomaterials in different parts of tomato plants, especially in fruit. Carbon nanotubes treatment increases the expression of genes related to stress and water channel in tomato plant. A transcriptional analysis revealed that engineered nanomaterials interaction to roots regulates the gene expression which may determine the growth and development of root. Similar to transcriptome analysis, lesser number of studies focused on proteome analysis to investigate level of different proteins and their function in response to engineered nanomaterials stress in plants.

Besides, Santos Filho et al. (2019) observed increased lytic vacuoles, oil bodies and nucleolar damage in *Allium cepa* cells when treated with anatase $TiO_2$-NPs solution. The increased chromatin condensation caused damage to DNA and affected viability of cells was observed in 10 days old *Oryza sativa* seedlings exposed to SWNCT which finally caused apoptosis, and thus, cell death (Shen et al. 2010). Exposure of CuO-NPs to rice seedling caused cytotoxicity which is observed in the form of damaged chloroplasts and a lesser number of thylakoids per granum and found to be increased with engineered nanomaterials concentrations in solution (Da Costa and Sharma 2016). Burklew et al. (2012) reported that an alteration in miRNAs expressions takes place in cells of tobacco plants when treated with $Al_2O_3$-NPs. Ag-NPs treatment to onion roots showed cytotoxicity and genotoxicity in concentration dependent manner. In these results decrease in mitotic index consequently led to inhibition of cell division at metaphase causing formation of Chromatin Bridge and chromosomal breaks that ultimately caused DNA damage (Kumari et al. 2009).

Khodakovskaya et al. (2011) observed the impacts of MWCNT treatment on *Lycopersicon esculentum* plants which caused upregulation of stress related genes including water channel *LeAqp2* gene. Castiglione et al. (2011) observed the phytotoxicity of $TiO_2$-NPs for *Vician arbonensis* and *Zea mays* for which their seeds were soaked in NPs solution overnight. It was observed that seed germination got significantly affected and in addition to this, concentration dependent increase in the mitotic index resulted in chromosomal aberrations, altered mitotic activity and loss of gene expression. Thus, $TiO_2$-NPs showed negative impacts on both of the plant system in terms of reduced seed germination, mitosis in root cells and their growth and development. In another study conducted on *Allium cepa* and *Nicotiana tobacum* to elucidate the toxicity mechanism, $TiO_2$-NPs exposure was found to cause toxicity in these plants at cellular and genetic level. $TiO_2$-NPs exposure damaged DNA structure, and caused chromosomal aberrations which reduced the root growth (Ghosh et al. 2010). Further omics-based studies can give a more realistic and precise information to understand the genetic and molecular approach of engineered nanomaterials induced toxicity in various agricultural crops.

### 13.5.3 Phytotoxicity at Morphological Level

The engineered nanomaterials plants interactions cause several morphological alterations which can be observed by measuring various endpoints including seed germination, plant root and shoot growth, biomass and yield (Deng et al. 2014; Tripathi et al. 2017). Engineered nanomaterials found to have contrasting impacts on these morphological parameters depending on its function and concentration (Siddiqui et al. 2015). Engineered nanomaterials when applied to low concentration, are beneficial for plant growth and seed germination. Engineered nanomaterials exposure enhanced germination rate, broke seed dormancy, and enhanced growth and productivity in some plants (Asli and Neumann 2009; Santos Filho et al. 2019). Contrary to this, engineered nanomaterials application also posed some adverse impacts and inhibited seed germination and reduced growth and productivity of crop plants (Stampoulis et al. 2009; Azimi et al. 2013; Raliya et al. 2015).

The effect of five different types of engineered nanomaterials (MWCNT, aluminium (Al), alumina ($Al_2O_3$), Zn and ZnO) on the seed germination were examined in some plant species and found that engineered nanomaterials exhibit different toxicity pattern in all plants (Thuesombat et al. 2014). ZnO-NPs treatment adversely affected the seedling and decreased the biomass by causing shrinkage in root tips and collapsing the root epidermis of perennial ryegrass (*Lolium perenne*) (Lin and Xing 2007). Contrary to this, Zheng et al. (2005) reported that size dependent response of $TiO_2$-NPs treatment in seed germination and germination index. An exposure of small sized $TiO_2$-NP (0.25–4% w/v) enhanced the seed germination and its index while big sized nanoparticles at same concentration were unable to induce seed germination. Santos Filho et al. (2019) reported a dose dependent reduction in seed germination and root growth in $TiO_2$-NP treated *Allium cepa.* The reduction in root growth, biomass, leaf count and overall growth was observed in *Nicotiana tabacum* plants grown in hydroponic suspension of $Al_2O_3$-NPs (Burklew et al. 2012). The germination rate of seeds was not much affected but the root shoot growth got declined due to accumulation of CuO-NPs in tissues and negatively affected physiological processes (Da Costa and Sharma 2016). The silver NPs suspended in distilled water used for soaking of seeds of *Ricinus communis* was found to be ineffective for seed germination but the shoot and root growth were adversely affected; this might be because of toxicity at biochemical levels (Yasur and Rani 2013). Soil amendment with $CeO_2$-NPs suspension was also reported to degrade the quality of grains of rice (Rico et al. 2013a, b).

Besides, some studies also reported the delaying in flowering and fruit setting in *Oryza sativa* plants exposed to SWCNTs (Liu et al. 2010). Zhao et al. (2013) observed ZnO-NPs induced necrosis and collapsing of cell structures and epidermis in the roots of cucumber plant. Contrary to this, Khodakovskaya et al. (2009) reported increased seed germination of tomato on application of MWCNTs but still the impacts were found to be dependent on concentration. In a study, Chakravarty et al. (2015) demonstrated the positive role of graphene quantum dots in terms of increased plant growth rates, yields and biomass in both coriander and garlic plants whose seeds were soaked in solution of quantum dots. Similarly, the foliar application of $Fe_3O_4$-NPs improved the grain yield and dry weight of fruit in *Glycine max* (Sheykhbaglou et al. 2010). The positive impacts promote use of various engineered nanomaterials in agriculture in the form of nanofertilizers but still large scale scrutinization is required.

## 13.6 Mechanisms of Engineered Nanomaterials Toxicity and Plant Tolerance to Engineered Nanomaterials Stress

Engineered nanomaterials contamination in soil causes extensive losses to agricultural production globally by inducing stress in crop plants. Engineered nanomaterials stress causes alterations at morphological to molecular level that negatively affect crop yield and quality (Deng et al. 2014; Ma et al. 2016). Engineered nanomaterials may hamper plant metabolism in a number of ways, such as, by providing micronutrients, regulation of genes or inducing oxidative stress which results in oxidative burst (Nair and Chung 2014; Hossain et al. 2015). Therefore, this section is aimed to overview the mechanisms of engineered nanomaterials toxicity and tolerance strategies in plants with special reference to reactive oxygen species and its role in oxidative stress and in activation of antioxidative machinery (Fig. 13.3).

## *13.6.1 Phytotoxicity Mechanisms*

Generally, the over production of reactive oxygen species and damages induced by them are two main paradigms responsible for engineered nanomaterials phytotoxicity (Deng et al. 2014) mainly due to large surface area to volume ratio and oxidative potential of engineered nanomaterials.

### 13.6.1.1 Destructive Role of Reactive Oxygen Species and Oxidative Stress

In response to engineered nanomaterials mediated alteration in various metabolic processes including photosynthesis, oxidative stress is mediated in plants by the excessive generation and accumulation of reactive oxygen species and/or reactive nitrogen species. Reactive oxygen species are by-products of aerobic metabolism which are commonly produced in the chloroplast, peroxisome, and mitochondria, by the reduction of cellular oxygen (Hossain and Dietz 2016; Mittler 2017) in low concentrations. Various reactive oxygen species produced in these sites, include superoxide radical, hydrogen peroxide, singlet oxygen, hydroxyl, superoxide anion, peroxide and alkoxyl. The major reactive nitrogen species produce are radicals of nitric oxide and nitrogen dioxide, nitrous acid, dinitrogen tetroxide, peroxynitrite and nitroxyl anion (del Rio et al. 2015). These reactive oxygen species and reactive nitrogen species are considered as stress markers because they activate an antioxidative defense system that in turn allows the plant to tolerate engineered nanomaterials stress.

Under normal condition, reactive oxygen species and reactive nitrogen species are generally in cellular homeostasis, and quickly scavenged by antioxidants (Mittler 2002). It is well known that reactive oxygen species and reactive nitrogen species exhibit dual role i.e. destructive (induce oxidative damages) and signaling (helps in developing tolerance) that depends on the concentration of these highly reactive species and their scavenging. In order to control their action, plant cells have a complex antioxidant system that usually resides very close to the production site of reactive oxygen species so they can be immediately scavenged. This has been proven to be an additional mechanism of plant protection through modulating reactive oxygen species as signaling molecules (Corpas et al. 2015).

The entered engineered nanomaterials interfere with the cell organelles and disturb the equilibrium between their production and scavenging which lead to over accumulation of reactive oxygen species in plant cells (Wang et al. 2016c). It has been noted that engineered exposure in plants limits the rate of carbon fixation which triggers the overproduction of superoxide anions and hydrogen peroxide by enhancing photo-inhibition of photosystem (Foyer and Noctor 2005). Overproduction of singlet oxygen during abiotic stress in chloroplasts leads to yellowing of leaves due to oxidation of carotenoids followed by bleaching of chlorophyll molecules.

The excess production and accumulation of ROS in plants in response to engineered nanomaterials exposure can lead to an oxidative burst (Dimkpa et al. 2013; Deng et al. 2014; Pakrashi et al. 2014) because they interact with almost all components of cells viz. proteins, amino acids chains and lipids of membrane and impose deleterious impacts by inducing damage to DNA and RNA, oxidation of protein and lipids, inhibition of enzyme activity, electrolyte leakage, lipid peroxidation and membrane damage (Sharma et al. 2012; Deng et al. 2014; Wang et al. 2019) subsequently causing cell death either by inducing apoptosis or necrosis (Faisal et al. 2013; Yang et al. 2017) (Fig. 13.3). Engineered nanomaterials like $TiO_2$-NPs, carbon nanotubes, ZnO-NPs, CuO-NPs, Ag-NPs, $CeO_2$-NPs, Au-NPs etc. are found to induce oxidative stress in various crop plants (Dimpka et al. 2013; Faisal et al. 2013; Rico et al. 2013a, b; Da Costa and Sharma 2016; González-García et al. 2019; Perez-Labrada et al. 2019; Wang et al. 2016a, b, 2019). High concentration of $CeO_2$-NPs found to be increasing the membrane damage and photosynthetic inhibition in rice seedlings (Rico et al. 2013c). Phytotoxicity of ZnO-NPs and Ag-NPs was observed in ryegrass and *Lemna* sp. due to the formation of reactive oxygen species and lipid peroxidation (Lin and Xing 2008; Oukarroum et al. 2013). Similarly, the $CeO_2$-NPs treatment (800 mg $kg^{-1}$) increased about 10 times more accumulation of hydrogen peroxide i.e.upto 35 μM in comparison to that of control in *Zea mays* (2012). Hydrogen peroxide could be converted into hydroxyl radical that is the most reactive and toxic species among all of reactive oxygen species, therefore, it interacts quickly with the cell components and causes cellular damages and ultimately premature death of cell.

Lipid peroxidation is the most damaging process that takes place in the cell and membrane. An enhanced level of peroxidation of unsaturated fatty acids in phospholipids has been observed when reactive oxygen species exceed its threshold level in plants exposed to engineered nanomaterials that produces malondialdehyde as a by-product and leads to the cell membrane damage (Halliwell and Gutteridge 2015). Malondialdehyde is the best indicator of oxidative damage under stress. Lipid peroxidation affects the membrane integrity that subsequently increases electrolyte leakage followed by inactivation of receptors bound to membrane, protein channels, and enzymes structure and function.

Besides, reactive oxygen species and by-products are known to modify proteins and their charges by inducing oxidation of amino acids and fragmentation of peptide chain which lead to proteins proteolysis (Sharma et al. 2012). Protein oxidation of a number of amino acids (proline, lysine, arginine, threonine, histidine and tryptophan) leads to the production of free carbonyls groups that in turn inactivate protein cross linkages (Moller et al. 2011). Recently, Ag-NPs (10 mg $l^{-1}$) induced phytotoxicity and genotoxicity in germinating wheat seedlings was observed that was found to be associated with the alteration of proteins associated with cell metabolism (Vannini et al. 2014). A proteomic study conducted to evaluate the phytotoxicity of $Al_2O_3$, ZnO and Ag-NPs in soybean seedlings, revealed modification of 16 common proteins that were mainly associated with protein degradation and photosynthesis (PS-I and PS-II) (Hossain et al. 2016).

In addition to protein oxidation, reactive oxygen species like ˙hydroxyl radicals also cause oxidative damages in DNA of nucleus, chloroplasts and mitochondria by oxidizing purine, pyrimidine and deoxyribose backbone. Engineered nanomaterials induced DNA degradation has been observed in different plants species (López-Moreno et al. 2010; Shen et al. 2010; Khodakovskaya et al. 2011, 2013; Dimkpa et al. 2013; Santos Filho et al. 2019). López-Moreno et al. (2010) reported that $CeO_2$-NPs (at high concentrations) adversely affected the DNA in soybean plants.

Likewise, genotoxic effects of $TiO_2$-NPs have also been investigated (Pakrashi et al. 2014; Ghosh et al. 2015; Santos Filho et al. 2019). Moreover, generation of reactive oxygen species which directly or indirectly affects the various processes in plants, is determined by various factors such as morphology (size and shape), particles' solubility and dissolution, biotransformation, light and plant species (Dimkpa et al. 2012; Rui et al. 2015; Zhang et al. 2015). Zhang et al. (2015) studied the toxicity of different types of $CeO_2$-NPs to different species of *Lactuca*. Responses of various plant species to engineered nanomaterials exposure of size and shape have been studied (Hawthorne et al. 2012; Syu et al. 2014; Zhang et al. 2015; Ma et al. 2016). For example, small sized $CeO_2$-NP (7 nm) treatment significantly enhanced malondialdehyde, while no significant difference in malondialdehyde level was observed in 25 nm and its bulk counterpart treatment in root cells, indicating reactive oxygen species induced membrane damage in root cells.

Nevertheless, there are some other factors (viz. the composition, concentration, surface and morphology of engineered nanomaterials also attributing to phytotoxicity to engineered nanomaterials. Biotransformation (dissolution and surface modification) of engineered nanomaterials in soil as well as inside the plants play a critical role in phytotoxicity as observed in several studies (Xia et al. 2008; Zhang et al. 2012; Oukarroum et al. 2013; Cui et al. 2014; Rui et al. 2015; Ma et al. 2015). For example, in contrast to more soluble nanoparticles (ZnO-NPs and Ag-NPs), $CeO_2$-NPs are generally stable and that is why used as model nanoparticles in toxicity mechanism study. However, it has been observed that $CeO_2$-NPs transformed to $CePO_4$ and $Ce(CH_3COO)_3$ by the chemical reaction and dissolution of $CeO_2$-NPs to $Ce^{3+}$ ions was found to play an important role in imparting toxicity of $CeO_2$-nanoparticles (Cui et al. 2014; Zhang et al. 2015). Moreover, some phytotoxicity studies conducted on metal or metal-based engineered nanomaterials, observed that dissolution of engineered nanomaterials to metal ions was not attributed much in phytotoxicity in plants (Lin and Xing 2008; Zhang et al. 2015). A microarray analysis on *Arabidopsis thaliana* revealed that the $Cu^{2+}$ ions discharged from dissolution of CuO-NPs was responsible for half of the toxicity in response to CuO-NPs treatment (Tang et al. 2016). Similarly, Zn as Zn (II)-phosphate or Zn-citrate complexes forms was found to induce toxicity in size dependent manner in ZnO-NPs treated *Triticum aestivum* and *Glycine max* plants as there no elemental particles were found in plant tissue (Dimkpa et al. 2012, 2013; Hernandez-Viezcas et al. 2013). In another study, Ag-NPs were found to show more toxicity in the root elongation in *A. thaliana* as compared to $Ag^+$ ions. It was also noticed that seedlings adsorbed very few amounts of $Ag^+$. Ag-NPs may disturb the cellular homeostasis between reactive oxygen species and antioxidants and osmosis. The possible mechanism of toxicity in this case is that adsorbed Ag-NPs caused damage in thylakoid membrane of chloroplasts and decreased the chlorophyll content which in turn adversely affected the growth and development of plant.

Besides, the photocatalytic activity of some engineered nanomaterials (ZnO-NPs and $TiO_2$-NPs) may also attribute to phytotoxicity because they produce reactive oxygen species upon irradiation of light having energy above their band gap energy (Feizi et al. 2013; Ma et al. 2013a, b). Thus, it can be concluded that engineered nanomaterials do not show any accurate mechanisms of toxicity. It may vary from plant to plant and may change with physicochemical factors, engineered nanomaterials types and their transformation and so on, therefore suggesting further research in this direction.

### 13.6.2 Reactive Oxygen Species as Signaling Molecule

In addition to the deleterious activity, at low concentration, reactive oxygen species act as a second messenger and regulates various cellular processes including activation of intrinsic antioxidant machinery under different abiotic and biotic stresses (Foyer and Noctor 2003; Baxter et al. 2014). Besides, reactive oxygen species and reactive nitrogen species also control the activities of a number of cellular components and enzymes (Cheng and Song 2006; You and Chang 2014), therefore, it is important for the cell to regulate reactive oxygen species concentration which determines its mode of action (Mittler 2002).

To deal with the engineered nanomaterials stress, plants show different strategies and adaptive mechanisms. The following section overviews the different strategies and mechanisms of plant tolerance to engineered nanomaterials stress.

### 13.6.3 Defense Mechanisms of Plants Against Engineered Nanomaterials

Plants can uptake engineered nanomaterials from soil to root cells; however, the level of engineered nanomaterials concentrations may differ between plant species. Plants exhibit different defense strategies to avoid engineered nanomaterials stress. The first line of defense is prohibited adsorption of bioavailable engineered nanomaterials at the root surface by secreting organic acids which bind to the engineered nanomaterials (Fig. 13.3). Some transmission electron microscopy-based studies reported that plants can tolerate and store bioavailable engineered nanomaterials in plant tissues. They accumulate the uptaken engineered nanomaterials within the vacuoles of root cells to prevent them interfering in important physiological processes (Gonzalez-Garcia et al. 2019; Santos Filho et al. 2019). However, when these defense strategies fail, plants employ antioxidant defense mechanism to deal with engineered nanomaterials stress.

The presence of engineered nanomaterials in soil in excess quantity leads to abiotic stress in plants. In response to abiotic stress, plants have evolved a complex regulatory network on the basis of reactive oxygen species production, scavenging, and signaling regulation (Sharma et al. 2012; You and Chan 2015). The impact of oxidative stress on plant metabolism is complex, it causes a considerable imbalance in cell homeostasis with damage in plant tissues (Sharma et al. 2012). As discussed in above section, reactive oxygen species (hydrogen peroxide, and hydroxyl radicals) and/or reactive nitrogen species are over produced during engineered nanomaterials stress and are highly reactive causing oxidative damages to protein, lipids and DNA nucleic acid which result in cellular death (You and Chan 2015) subsequently causing losses in crop yield and productivity globally (Mittler et al. 2002). Moreover, over production of reactive oxygen species in stress conditions are known to act as signaling molecules and activate stress response pathways (Baxter et al. 2014; You and Chan 2015). Plants have a potential intrinsic antioxidant defense system to defend themselves from the reactive oxygen species and oxidative damages caused by them. Besides, they modulate low levels of reactive oxygen species for signal transduction and interrupt the cascades of uncontrolled oxidation by scavenging reactive oxygen species.

#### 13.6.3.1 Enzymatic Antioxidants of the Antioxidant Defense System

Plants have various antioxidant enzymes such as superoxide dismutase, catalase, monodehydroascorbate reductase, dehydroascorbate reductase, glutathione reductase, peroxidase, ascorbate peroxidase, glutathione S-transferase, peroxiredoxin and guaiacol peroxidases that scavenge reactive oxygen species and reactive nitrogen species (Sharma et al. 2012). These are found in various sites within the plant cells and worked collaboratively to scavenge reactive oxygen species (You and Chan 2015). Superoxide dismutase is the most important metalloenzyme that convert highly toxic superoxide radicals into less toxic hydrogen peroxide. Several reports have documented that an increment of superoxide dismutase improves the plants tolerance against engineered nanomaterials stress (Ma et al. 2016; Zhang et al. 2017; González-García et al. 2019; Pérez-Labrada et al. 2019). Faisal et al. (2013) recorded a significantly increased level of superoxide dismutase in tomato exposed to NiO-NPs (0–1000 mg $l^{-1}$). Ma et al. (2016) have investigated that $CeO_2$-NPs treatment enhanced superoxide dismutase activity only at low concentration (250 mg $l^{-1}$), whereas no noticeable change in superoxide dismutase activity was recorded at a high concentration in rice. After over production of hydrogen peroxide either through different cellular processes or in response to superoxide dismutase activity, other antioxidant enzymes like catalase, ascorbate peroxidase and oxidized glutathione would be activated to detoxify the hydrogen peroxide immediately in the cell that otherwise may lead to lipid peroxidation.

Besides, catalase converts hydrogen peroxide into oxygen ($O_2$) and water molecule ($H_2O$) directly. In addition to catalase, ascorbate peroxidase is considered as the most important reactive oxygen species scavenger and shows a greater affinity for hydrogen peroxide as compared to catalase and peroxidase. Its role is crucial in regulating ascorbic acid-glutathion cycle. In the process of ascorbic acid oxidation into dehydroascorbate and malondialdehyde, ascorbate peroxidase converts hydrogen peroxide to $H_2O$. In addition, glutathione reductase which is nicotinamide adenine dinucleotide (NADPH) dependent enzyme, take part in the oxidation of glutathione to glutathione disulfide (GSSG) and help in maintaining a high glutathion/GSSG ratio in plant cells. Glutathione peroxidase also shows positive role in plants against abiotic stress which catalyzes the reduction of hydroperoxides using glutathion. Ascorbic acid-glutathion cycle helps in regenerating reduced form of ascorbate and glutathion in cell and maintains the redox status of cell.

There are numbers of studies reporting the increased production of the antioxidants in the plant in response to engineered nanomaterials exposure (Faisal et al. 2013; Da Costa and Sharma 2016; González-García et al. 2019), however, the level of their activation depends on engineered nanomaterials types and concentration and exposure time. Various engineered nanomaterials (i.e. $Fe_3O_4$-NPs, $Co_3O$ -NPs, $CeO_2$-NPs, Au-NPs, $MnO_2$-NPs, CuO-NPs, fullerene and Pt-NPs) can activate catalase, superoxide dismutase and glutathione peroxidase in plants (Wei and Wang 2013). An increased activity of antioxidant enzymes and malondialdehyde level was reported in rice seedlings in response to CuO-NPs treatment (Shaw and Hossain 2013). Similarly, study on effect of CuO-NPs treatment on antioxidative defense and photosynthetic performance of *Hordeum vulgare* revealed decline in root and shoot growth due to decreased photosynthetic performance and inefficiency of antioxidant system (Shaw et al. 2014). Nair and Chung (2014) investigated the changes in growth, redox level, and genes expression in *Cicer arietinum* plants treated with CuO-NPs (50–500 $mgl^{-1}$). They reported a steady increase in malondialdehyde level with increase in peroxidase activity in concentration dependent manner. Similarly, mRNA expression levels of Cu/Zn superoxide dismutase, catalase, and ascorbate peroxidase also found to increase in CuO-NPs treated plants.

Karami Mehrian et al. (2015) noted an improvement in plant performance under saline stress on Ag-NPs exposure. Ag-NPs treatment induced the antioxidant enzymes which in turn reduced lipid peroxidation and reduced the level of the malondialdehyde and $Na^+$ content in stressed plants. Similarly, application of Cu-NPs (250 mg $l^{-1}$) on tomato plants also enhanced the antioxidant activity and accumulation of bioactive compounds in the fruits under salt stress (Hernández-Fuentes et al. 2017). $CeO_2$-NPs exposure caused phytotoxicity in *Phaseolus vulgaris* by disturbing its antioxidant defense system (Majumdar et al. 2014). In a study, catalase activity reduced significantly in roots of $CeO_2$-NPs treated plants suggesting that ascorbic acid-glutathion or glutathione peroxidase cycle may be involved in reactive oxygen species detoxification (Rico et al. 2013c).

Contrary to this, Servin et al. (2013) observed a significant rise (250 to 750 mg $l^{-1}$) in catalase activity in $TiO_2$-NPs treated cucumber plants with no noticeable difference in ascorbate peroxidase activity. Likewise, Ma et al. (2016) observed a significant decrease in catalase, superoxide dismutase, peroxidase and ascorbate peroxidase activities in *Arabidopsis thaliana* under $CeO_2$-NPs exposure. Recently, González-García et al. (2019) studied the antioxidant system of carbon nanomaterials (10, 50, 100, 250, 500, and 1000 mg $l^{-1}$) treated tomato seedlings and found that carbon nanomaterials application increased the enzymatic activity of catalase, glutathione peroxidase and ascorbate peroxidase and phenylalanine ammonia lyase. Abdel-Latef (2017) reported that $TiO_2$-NP (0.01%) application to *Vicia faba* plants grown under salinity stress, alleviated salinity stress by increasing the enzymatic activity of catalase, superoxide dismutase, and ascorbate peroxidase as well as the retention of $K^+$ and elimination of hydroxyl radicals (Pérez-Labrada et al. 2019).

Beside the fact that engineered nanomaterials activate the antioxidant system in plants, Wei and Wang (2013) evaluated the antioxidant potential of engineered nanomaterials and mechanisms that imitated the activity like the natural enzymes. Their review suggested the various enzyme like activities of engineered nanomaterials ($CeO_2$, $Fe_3O_4$, cobalt oxide NPs) imitating catalase like; $CeO_2$, iron, cobalt, manganese, CuO and Au-NPs showing peroxidase like, and $CeO_2$-NP and Pt-NPs showing superoxide dismutase like activities.

#### 13.6.3.2 Non-enzymatic Antioxidant Defense System

Plants have non-enzymatic antioxidant system which have many low molecular weight molecules such as thiols, ascorbic acid, glutathion, a-tocopherol, carotenoids and flavonoids (Sharma et al. 2012; You and Chan 2015). Ascorbic acid acts as the first barrier of antioxidant system in plants. Shaw and Hossain (2013) found a consistent increase in ascorbic acid activity in CuO-NPs treated rice seedlings. Similarly, Rico et al. (2013a) observed an altered ascorbic acid and free thiols concentration on 500 mg $l^{-1}$ $CeO_2$-NPs treatment which in turn enhanced oxidative damage in membrane and caused photosynthetic stress.

Glutathion is a tripeptide known for its key role in intercellular defense against reactive oxygen species. Besides, it is a part of the Ascorbic acid-glutathion cycle and regenerates other antioxidants such as Ascorbic acid. Recently, it was reported that $CeO_2$ and $In_2O_3$-NPs induced biosynthesis of glutathione in *Arabidopsis* (Majumdar et al. 2014; Ma et al. 2016). Glutathion under engineered nanomaterials stress is an alternative pathway to assess its role in detoxification processes. Engineered nanomaterials induced enhancement in glutathione disulfide concentration was observed in plants (Dimkpa et al. 2012, 2013), however, there is no direct relation in increased level of glutathione disulfide and the decline in glutathion, a process in turn converts hydrogen peroxide into $H_2O$ (Faisal et al. 2013).

Phenolic compounds (phenolic acids, flavonoids and many others are secondary metabolites that are found ubiquitously in plants and act as chemical and physical barriers in plant protection under biotic and abiotic stress (Zhou et al. 2019). They protect plants from damaging effects of engineered nanomaterials because of their ability to scavenge reactive oxygen species, chelate metal ions coming from dissolution of engineered nanomaterials and trap lipid alkoxyl (–$OCH_3$) polyphenols.

Carotenoids and tocopherols are lipophilic antioxidants which detoxify different free radicals. They protect the membrane structure and PS-II by preventing the lipids and other components of cell membrane to react with oxygen chemically and by quenching the chloroplasts physically (Ivanov and Khorobrykh 2003). However, our knowledge about the antioxidant potential of these compounds in plant protection in response to engineered nanomaterials stress is very scarce and imprecise.

### *13.6.4 Plant Hormones and Stress Tolerance*

Besides antioxidants (enzymatic and non-enzymatic), plants hormones (auxin, gibberellic acid, salicylic acid, abscisic acid etc.) play a significant role in signaling during abiotic stress including engineered nanomaterials. The plant development and adaptation to stress is a result of synergistic and antagonistic interactions of these phytohormones. The reactive oxygen species at low concentration act as secondary molecules which control the secretion and modulate each of their activity in response to engineered nanomaterials stress including others. An enhancement in cytokinin concentration and reduction in indol acetic acid and abscisic acid was observed in *Capsicum annuum* and cotton plant in two separate studies in response to Ag-NPs and CuO-NP, respectively. It indicates that engineered nanomaterials mediate hormonal balance through reactive oxygen species which in turn cause phytotoxicity by altering metabolic processes (Rastogi et al. 2017).

Thus, it can be concluded that plants can modulate their intrinsic antioxidant defense system and stress tolerance by inducing mild oxidative stress and phytohormone signaling. Engineered nanomaterials exposure to plants, over expresses the genes involved in stress signaling. As a result of which various antioxidant compounds including ascorbic acid, flavonoids, polyphenols are produced which in turn improve plant defense by decreasing oxidative stress (Gonzalez-Garcia et al. 2019). Moreover, a hormetic response can be observed. Low concentrations of engineered nanomaterials stimulate the plant growth and development but high concentrations negatively affect the metabolic processes and induce opposite effect. Besides, exposure to plant with higher concentrations may again show positive response. For example, graphene oxide treatment at 0.10 $mg^{-1}$ significantly increased the proline content in carrot plants (Gonzalez-Garcia et al. 2019) that was found to be beneficial to plant because it helped in mitigating oxidative stress by inducing the generation of free radicals and activation of antioxidant defense system.

Further, plants alleviate the toxic effects of engineered nanomaterials by limiting their uptake, prohibiting their movement by formation of complex and chelation and through their sequestration in vacuole. In addition to this, plants must regulate secondary defects produced in response to engineered nanomaterials stress either by avoiding or fix them. Nevertheless, various plant species have developed different strategies to combat engineered nanomaterials stress.

## 13.7 Conclusions and Future Perspectives

Excessive accumulation of engineered nanomaterials in agricultural soils has resulted in an unforeseen impact on quantity as well as quality of crops grown in contaminated soil. Thus, food safety and security remain a major challenge at global scale. Some of the major future recommendations based on conclusions of various studies are:

(i) A systematic knowledge about the engineered nanomaterials fate and behavior, their actual concentrations in soil and crops and successive trophic levels, as well as transformation will be decisive for accurate engineered nanomaterials risk assessment.
(ii) Depending on its size, chemical composition, surface coating, shapes and transformation, engineered nanomaterials could show different toxicity pattern in plants. Hence, a systematic research is needed to reveal the mechanisms of uptake kinetics and engineered nanomaterials interaction within the cells.
(iii) The effect of root exudates, soil properties such as pH, natural organic matter ionic strength and microbial community in transformation, stability, bioavailability and finally uptake of engineered nanomaterials is poorly understood and elusive. Hence, studies should be conducted to investigate engineered nanomaterials crop interactions under conditions that simulate environmental factors in rhizosphere.
(iv) Our understanding on the effects of reactive oxygen species and the oxidative damages caused by them in response to engineered nanomaterials stress on plants is still at preliminary stage. The lack of knowledge in context to proper mechanism of toxicity and plant-engineered nanomaterials interactions, defense strategies by plants in response to engineered nanomaterials stress still open the options for future research for which 'omics' like metabolomics and proteomics could be prove a best and appropriate approach to investigate plants responses under engineered nanomaterials stress.

Overall, a clear understanding of engineered nanomaterials impacts on crops plants at a wider-scale will help in optimizing their application in agriculture to improve crop production and to meet continuously rising demand of food in safe and sustainable way. Careful examination of engineered nanomaterials concentration in soil as well as in crops is necessary as these may possibly reach the hazardous level in animal or human being via the affected food chain.

**Acknowledgements** Authors are grateful to Professor K.S. Rao, Head, Department of Botany, University of Delhi, New Delhi, India for his kind support. Financial support by the UGC, New Delhi, is gratefully acknowledged.

## References

Abdel Latef AAH, Srivastava AK, El-sadek MSA, Kordrostami M, Tran LSP (2017) Titanium dioxide nanoparticles improve growth and enhance tolerance of broad bean plants under saline soil conditions. Land Degrad Dev 29:1065–1073. https://doi.org/10.1002/ldr.2780

Aghdam MTB, Mohammadi H, Ghorbanpour M (2016) Effects of nanoparticulate anatase titanium dioxide on physiological and biochemical performance of *Linum usitatissimum* (Linaceae) under well-watered and drought stress conditions. Braz J Bot 39:139–146. https://doi.org/10.1007/s40415-015-0227-x

Ali T, Tripathi P, Azam A, Raza W, Ahmed AS, Ahmed A, Muneer M (2017) Photocatalytic performance of Fe-doped $TiO_2$ nanoparticles under visible-light irradiation. Mater Res Express 4:015022. https://doi.org/10.1088/2053-1591/aa576d

Alkhatib R, Alkhatib B, Abdo N, Laith AE, Creamer R (2019) Physio-biochemical and ultrastructural impact of ($Fe_3O_4$) nanoparticles on tobacco. BMC Plant Biol 19:253. https://doi.org/10.1186/s12870-019-1864-1

Anderson A, McLean JE, Jacobson AR, Britt D (2017) CuO and ZnO nanoparticles modify inter kingdom cell signaling processes relevant to crop production: a review. J Agric Food Chem 66:6513–6524. https://doi.org/10.1021/acs.jafc.7b01302

Antisari LV, Carbone S, Gatti A, Vianello G, Nannipieri P (2013) Toxicity of metal oxide ($CeO_2$, $Fe_3O_4$, $SnO_2$) engineered nanoparticles on soil microbial biomass and their distribution in soil. Soil Biol Biochem 60:87–94. https://doi.org/10.1016/j.soilbio.2013.01.016

Antisari LV, Carbone S, Gatti A, Vianello G, Nannipieri P (2015) Uptake and translocation of metals and nutrients in tomato grown in soil polluted with metal oxide ($CeO_2$, $Fe_3O_4$, $SnO_2$, $TiO_2$) or metallic (ag, co, Ni) engineered nanoparticles. Environ Sci Pollut Res Int 22:1841–1853. https://doi.org/10.1007/s11356-014-3509-0

Arora S, Sharma P, Kumar S, Nayan R, Khanna PK, Zaidi MGH (2012) Gold-nanoparticle induced enhancement in growth and seed yield of *Brassica juncea*. Plant Growth Regul 66:303–310. https://doi.org/10.1007/s10725-011-9649-z

Aslani F, Bagheri S, MuhdJulkapli N, Juraimi AS, Hashemi FS, Baghdadi A (2014) Effects of engineered nanomaterials on plants growth: an overview. Sci World J 2014:641759. https://doi.org/10.1155/2014/641759

Asli S, Neumann M (2009) Colloidal suspensions of clay or titanium dioxide nanoparticles can inhibit leaf growth and transpiration via physical effects on root water transport. Plant Cell Environ 32:577–584. https://doi.org/10.1111/j.1365-3040.2009.01952.x

Azimi R, Feizi H, Hosseini MK (2013) Can bulk and nanosized titanium dioxide particles improve seed germination features of wheatgrass (*Agropyron desertorum*). Not Sci Biol 5:325–331. https://doi.org/10.15835/nsb539072

Baxter A, Mittler R, Suzuki N (2014) ROS as key players in plant stress signaling. J Exp Bot 65:1229–1240. https://doi.org/10.1093/jxb/ert375

Begum P, Ikhtiari R, Fugetsu B (2011) Graphene phytotoxicity in the seedling stage of cabbage, tomato, red spinach, and lettuce. Carbon N Y 49:3907–3919. https://doi.org/10.1016/j.carbon.2011.05.029

Buick RD, Buchan GD, Field RJ (1993) The role of surface tension of spreading droplets in absorption of a herbicide formulation via leaf stomata. Pestic Sci 38:227–235. https://doi.org/10.1002/ps.2780380218

Burklew CE, Ashlock J, Winfrey WB, Zhang B (2012) Effects of aluminum oxide nanoparticles on the growth, development, and micro RNA expression of tobacco (*Nicotiana tabacum*). PLoS One 7:1–9. https://doi.org/10.1371/journal.pone.0034783

Cao C, Chen XP, Ma ZB, Jia HH, Wang JJ (2016) Greenhouse cultivation mitigates metal-ingestion-associated health risks from vegetables in wastewater-irrigated agroecosystems. Sci Total Environ 560:204–211. https://doi.org/10.1016/j.scitotenv.2016.04.044

Castiglione MR, Giorgetti L, Geri C, Cremonini R (2011) The effects of nano-$TiO_2$ on seed germination, development and mitosis of root tip cells of *Vicia arbonensis* L. and *Zea mays* L. J Nanopart Res 13:2443–2449. https://doi.org/10.1007/s11051-010-0135-8

Chai H, Yao J, Sun J, Zhang C, Liu W, Zhu M, Ceccanti B (2015) The effect of metal oxide nanoparticles on functional bacteria and metabolic profiles in agricultural soil. B Environ Contam Toxicol 94:490–495. https://doi.org/10.1007/s00128-015-1485-9

Chakravarty D, Erande MB, Late DJ (2015) Graphene quantum dots as enhanced plant growth regulators: effects on coriander and garlic plants. J Sci Food Agric 95:2772–2778. https://doi.org/10.1002/jsfa.7106

Chen R, Ratnikova TA, Stone MB, Lin S, Lard M, Huang G, Hudson JS, Ke PC (2010) Differential uptake of carbon nanoparticles by plant and mammalian cells. Small 6:612–617. https://doi.org/10.1002/smll.200901911

Cheng Y, Song C (2006) Hydrogen peroxide homeostatis and signaling in plant cells. Science in China Series c Life Sciences-English Edition 49(1):1. https://doi.org/10.1007/s11427-005-0066-2

Cifuentes Z, Custardoy L, de la Fuente JM, Marquina C, Ibarra MR, Rubiales D, Pérez-de-Luque A (2010) Absorption and translocation to the aerial part of magnetic carbon-coated nanoparticles through the root of different crop plants. J Nanobiotechnology 8:26. https://doi.org/10.1186/1477-3155-8-26

Cornelis G, Hund-Rinke K, Kuhlbusch T, Van den Brink N, Nickel C (2014) Fate and bioavailability of engineered nanoparticles in soils: a review. Crit Rev Environ Sci Technol 44:2720–2764. https://doi.org/10.1080/10643389.2013.829767

Corpas FJ, Gupta DK, Palma JM (2015) Production sites of reactive oxygen species (ROS) in organelles from plant cells. In: Gupta DK, Palma JM, Corpas FJ (eds) Reactive oxygen species and oxidative damage in plants under stress. Springer, Cham, pp 1–22. https://doi.org/10.1007/978-3-319-20421-5_1

Cox A, Venkatachalam P, Sahi S, Sharma N (2016) Silver and titanium dioxide nanoparticle toxicity in plants: a review of current research. Plant Physiol Biochem 107:147–163. https://doi.org/10.1016/j.plaphy.2016.05.022

Cui D, Zhang P, Ma Y, He X, Li Y, Zhang J, Zhao Y, Zhang Z (2014) Effect of cerium oxide nanoparticles on asparagus lettuce cultured in an agar medium. Environ Sci Nano 1:459–465. 1010.1039/x0xx00000x

Da Costa MVJ, Sharma PK (2016) Effect of copper oxide nanoparticles on growth, morphology, photosynthesis, and antioxidant response in *Oryza sativa*. Photosynthetica 54:110–119. https://doi.org/10.1007/s11099-015-0167-5

De Rosa MC, Monreal C, Schnitzer M, Walsh R, Sultan Y (2010) Nanotechnology in fertilizers. Nat Nanotechnol 5:91. https://doi.org/10.1038/nnano.2010.2

Del-Rio LA, Sandalio LM, Corpas FJ, Palma JM, Barroso JB (2006) Reactive oxygen species and reactive nitrogen species in peroxisomes. Production, scavenging, and role in cell signaling. Plant Physiol 141:330–335. https://doi.org/10.1104/pp.106.078204

Deng Y, White JC, Xing B (2014) Interactions between engineered nanomaterials and agricultural crops: implications for food safety. J Zhejiang Univ Sci A15:552–572. https://doi.org/10.1631/jzus.A1400165

Dimkpa CO, McLean JE, Latta DE, Manangón E, Britt DW, Johnson WP, Boyanov MI, Anderson AJ (2012) CuO and ZnO nanoparticles: phytotoxicity, metal speciation, and induction of oxidative stress in sand-grown wheat. J Nanopart Res 14:1125. https://doi.org/10.1007/s11051-012-1125-9

Dimkpa CO, Latta DE, McLean JE, Britt DW, Boyanov MI, Anderson AJ (2013) Fate of CuO and ZnO nano- and microparticles in the plant environment. Environ Sci Techno 47:4734–4742. https://doi.org/10.1021/es304736y

Du W, Sun Y, Ji R, Zhu J, Wu J, Guo H (2011) $TiO_2$ and ZnO nanoparticles negatively affect wheat growth and soil enzyme activities in agricultural soil. J Environ Monitor 13:822–828. https://doi.org/10.1039/c0em00611d

El Badawy AM, Scheckel KG, Suidan M, Tolaymat T (2012) The impact of stabilization mechanism on the aggregation kinetics of silver nanoparticles. Sci Total Environ 429:325–331. https://doi.org/10.1016/j.scitotenv.2012.03.041

Etxeberria E, Gonzalez P, Pozueta J (2009) Evidence for two endocytic transport pathways in plant cells. Plant Sci 177:341–348. https://doi.org/10.1016/j.plantsci.2009.06.014

Faisal M, Saquib Q, Alatar AA, Al-Khedhairy AA, Hegazy AK, Musarrat J (2013) Phytotoxic hazards of NiO-nanoparticles in tomato: a study on mechanism of cell death. J Hazard Mater 250:318–332. https://doi.org/10.1016/j.jhazmat.2013.01.063

Feizi H, Kamali M, Jafari L, Moghaddam PR (2013) Phytotoxicity and stimulatory impacts of nanosized and bulk titanium dioxide on fennel (*Foeniculum vulgare* mill). Chemosphere 91:506–511. https://doi.org/10.1016/j.chemosphere.2012.12.012

Ghafariyan MH, Malakouti MJ, Dadpour MR, Stroeve P, Mahmoudi M (2013) Effects of magnetite nanoparticles on soybean chlorophyll. Environ Sci Technol 47:10645–10652. https://doi.org/10.1021/es402249b

Ghidan AY, Al Antary TM (2019) Applications of nanotechnology in agriculture. In: Niemeyer CM, Mirkin C (eds) Applications of nanobiotechnology. Intech Open. https://doi.org/10.5772/intechopen.88390

Ghosh M, Bandyopadhyay M, Mukherjee A (2010) Genotoxicity of titanium dioxide ($TiO_2$) nanoparticles at two trophic levels: plant and human lymphocytes. Chemosphere 81:1253–1262. https://doi.org/10.1016/j.chemosphere.2010.09.022

Ghosh M, Bhadra S, Adegoke A, Bandyopadhyay M, Mukherjee A (2015) MWCNT uptake in *Allium cepa* root cells induces cytotoxic and genotoxic responses and results in DNA hypermethylation. Mutat Res Fundam Mol Mech Mutagen 774:49–58. https://doi.org/10.1016/j.mrfmmm.2015.03.004

González-García Y, López-Vargas ER, Cadenas-Pliego G, Benavides-Mendoza A, González-Morales S, Robledo-Olivo A, Alpuche-Solís ÁG, Juárez-Maldonado A (2019) Impact of carbon nanomaterials on the antioxidant system of tomato seedlings. Int J Mol Sci 20:5858. https://doi.org/10.3390/ijms20235858

González-Melendi P, Fernández-Pacheco R, Coronado MJ, Corredor E, Testillano PS, Risueño MC, Marquina C, Ibarra MR, Rubiales D, Pérez-de-Luque A (2008) Nanoparticles as smart treatment-delivery systems in plants: assessment of different techniques of microscopy for their visualisation in plant tissues. Ann Bot 101:187–195. https://doi.org/10.1093/aob/mcm283

Goswami L, Kim KH, Deep A, Das P, Bhattacharya SS, Kumar S, Adelodun AA (2017) Engineered nano particles: nature, behavior, and effect on the environment. J Environ Manag 196:297–315. https://doi.org/10.1016/j.jenvman.2017.01.011

Gottschalk F, Nowack B (2011) The release of engineered nanomaterials to the environment. J Environ Monit 13:1145–1155. https://doi.org/10.1039/c0em00547a

Gottschalk F, Sonderer T, Scholz RW, Nowack B (2009) Modeled environmental concentrations of engineered nanomaterials ($TiO_2$, ZnO, ag, CNT, fullerenes) for different regions. Environ Sci Technol 43:9216–9222. https://doi.org/10.1021/es9015553

Ha SW, Weiss D, Weitzmann MN, Beck GR Jr (2019) Applications of silica-based nanomaterials in dental and skeletal biology. In: Ahmed W (ed) Nanobiomaterials in clinical dentistry. Elsevier, pp 77–112. https://doi.org/10.1016/b978-0-12-815886-9.00004-8

Halliwell B, Gutteridge JM (2015) Free radicals in biology and medicine. Oxford University Press, New York. https://doi.org/10.1093/acprof:oso/9780198717478.001.0001

Hawthorne J, Musante C, Sinha SK, White JC (2012) Accumulation and phytotoxicity of engineered nanoparticles to *Cucurbita pepo*. Int J Phytoremediation 14:429–442. https://doi.org/10.1080/15226514.2011.620903

Hernández-Fuentes AD, López-Vargas ER, Pinedo-Espinoza JM, Campos-Montiel RG, Valdés-Reyna J, Juárez-Maldonado A (2017) Postharvest behavior of bioactive compounds in tomato fruits treated with cu nanoparticles and NaCl stress. Appl Sci 7:980. https://doi.org/10.3390/app7100980

Hernandez-Viezcas JA, Castillo-Michel H, Andrews JC, Cotte M, Rico C, Peralta-Videa JR, Ge Y, Priester JH, Holden PA, Gardea-Torresdey JL (2013) In situ synchrotron X-ray fluorescence mapping and speciation of $CeO_2$ and ZnO nanoparticles in soil cultivated soybean (*Glycine max*). ACS Nano 7:1415–1423. https://doi.org/10.1021/nn305196q

Hossain MS, Dietz KJ (2016) Tuning of redox regulatory mechanisms, reactive oxygen species and redox homeostasis under salinity stress. Front Plant Sci 7:548. https://doi.org/10.3389/fpls.2016.00548

Hossain Z, Mustafa G, Komatsu S (2015) Plant responses to nanoparticle stress. Int J Mol Sci 16:26644–26653. https://doi.org/10.3390/ijms161125980

Huang FM, Tai KW, Chou MY, Chang YC (2002) Cytotoxicity of resin-, zinc oxide-eugenol-, and calcium hydroxide-based root canal sealers on human periodontal ligament cells and permanent V79 cells. Int Endod *J*35: 153–158., 2002. doi:https://doi.org/10.1046/j.1365-2591.2002.00459.x

Inshakova E, Inshakov O (2017) World market for nanomaterials: structure and trends. In: Bratan S, Gorbatyuk S, Leonov S, Roshchupk S (eds) MATEC web of conferences, vol 129, p. 02013. EDP Sciences. https://doi.org/10.1051/matecconf/201712902013

International Organization for Standardization (2011) Soil quality: Requirements and guidance for the selection and application of methods for the assessment of bioavailability of contaminants in soil and soil materials. ISO17402. https://doi.org/10.3403/30236055u

Ivanov B, Khorobrykh S (2003) Participation of photosynthetic electron transport in production and scavenging of reactive oxygen species. Antioxid Redox Signal 5:43–53. https://doi.org/10.1089/152308603321223531

Iversen TG, Skotland T, Sandvig K (2011) Endocytosis and intracellular transport of nanoparticles: present knowledge and need for future studies. Nano Today 6:176–185

Jacob DL, Borchardt JD, Navaratnam L, Otte ML, Bezbaruah AN (2013) Uptake and translocation of Ti from nanoparticles in crops and wetland plants. Int J Phytoremediation 15:142–153. https://doi.org/10.1080/15226514.2012.683209

Jain A, Ranjan S, Dasgupta N, Ramalingam C (2016) Nanomaterials in food and agriculture: an overview on their safety concerns and regulatory issues. Crit Rev Food Sci Nutr 6:1–21. https://doi.org/10.1080/10408398.2016.1160363

Judy JD (2013) Bioavailability of manufactured nanomaterials in terrestrial ecosystems. PhD Thesis. University of Kentucky. Lexington, Kentucky, USA

Karami Mehrian S, Heidari R, Rahmani FE (2015) Effect of silver nanoparticles on free amino acids content and antioxidant defense system of tomato plants. Indian J Plant Physiol 20:257–263. https://doi.org/10.1007/s40502-015-0171-6

Keller AA, McFerran S, Lazareva A, Suh S (2013) Global life cycle releases of engineered nanomaterials. J Nanopart Res 15:1692. https://doi.org/10.1007/s11051-013-1692-4

Khodakovskaya M, Dervishi E, Mahmood M, Xu Y, Li Z, Watanabe F, Biris AS (2009) Carbon nanotubes are able to penetrate plant seed coat and dramatically affect seed germination and plant growth. ACS Nano 3:3221–3227. https://doi.org/10.1021/nn900887m

Khodakovskaya MV, De Silva K, Nedosekin DA, Dervishi E, Biris AS, Shashkov EV, Galanzha EI, Zharov VP (2011) Complex genetic, photothermal, and photoacoustic analysis of nanoparticle-plant interactions. Proc Natl Acad Sci 108:1028–1033. https://doi.org/10.1073/pnas.1008856108

Khodakovskaya MV, de Silva K, Biris AS, Dervishi E, Villagarcia H (2012) Carbon nanotubes induce growth enhancement of tobacco cells. ACS Nano 6:2128–2135. https://doi.org/10.1021/nn204643g

Khodakovskaya MV, Kim BS, Kim JN, Alimohammadi M, Dervishi E, Mustafa T, Cernigla CE (2013) Carbon nanotubes as plant growth regulators: effects on tomato growth, reproductive system, and soil microbial community. Small 9:115–123. https://doi.org/10.1002/smll.201201225

Kurepa J, Paunesku T, Vogt S, Arora H, Rabatic BM, Lu J, Wanzer MB, Woloschak GE, Smalle JA (2010) Uptake and distribution of ultrasmall anatase $TiO_2$ alizarin red S nanoconjugates in *Arabidopsis thaliana*. Nano Lett 10:2296–2302. https://doi.org/10.1021/nl903518f

Landa P, Vankova R, Andrlova J, Hodek J, Marsik P, Storchova H, White JC, Vanek T (2012) Nanoparticle-specific changes in *Arabidopsisthaliana* gene expression after exposure to ZnO, $TiO_2$, and fullerene soot. J Hazard Mater 241–242:55–62. https://doi.org/10.1016/j.jhazmat.2012.08.059

Larue C, Laurette J, Herlin-Boime N, Khodja H, Fayard B, Flank AM, Brisset F, Carriere M (2012) Accumulation, translocation and impact of $TiO_2$ nanoparticles in wheat (*Triticum aestivum* spp.): influence of diameter and crystal phase. Sci Total Environ 431:197–208. https://doi.org/10.1016/j.scitotenv.2012.04.073

Lee CW, Mahendra S, Zodrow K, Li D, Tsai YC, Braam J, Alvarez PJ (2010) Developmental phytotoxicity of metal oxide nanoparticles to *Arabidopsisthaliana*. Environ Toxicol Chem Int J 29:669–675. https://doi.org/10.1002/etc.58

Lin D, Xing B (2008) Root uptake and phytotoxicity of ZnO nanoparticles. Environ Sci Technol 42:5580–5585. https://doi.org/10.1021/es800422x

Liu Q, Zhao Y, Wan Y, Zheng J, Zhang X, Wang C, Fang X, Lin J (2010) Study of the inhibitory effect of water-soluble fullerenes on plant growth at the cellular level. ACS Nano 4:5743–5748. https://doi.org/10.1021/nn101430g

Liu XY, Chen GX, Keller AA, Su CM (2013) Effects of dominant material properties on the stability and transport of $TiO_2$ nanoparticles and carbon nanotubes in aquatic environments: from synthesis to fate. Environ Sci Process Impacts 15:169–189. https://doi.org/10.1039/c2em30625e

López-Moreno ML, de la Rosa G, Hernández-Viezcas JÁ, Castillo-Michel H, Botez CE, Peralta-Videa JR, Gardea-Torresdey JL (2010) Evidence of the differential biotransformation and genotoxicity of ZnO and $CeO_2$ nanoparticles on soybean (*Glycine max*) plants. Environ Sci Technol 44:7315–7320. https://doi.org/10.1021/es903891g

Lowry GV, Espinasse BP, Badireddy AR, Richardson CJ, Reinsch BC, Bryant LD, Bone AJ, Deonarine A, Chae S, Therezien M, Colman BP, Hsu-Kim H, Bernhardt ES, Matson CW, Wiesner MR (2012) Long-term transformation and fate of manufactured ag nanoparticles in a simulated large scale freshwater emergent wetland. Environ Sci Technol 46:7027–7036. https://doi.org/10.1021/es204608d

Ma JF, Yamaji N (2015) A cooperative system of silicon transport in plants. Trends Plant Sci 20:435–442. https://doi.org/10.1016/j.tplants.2015.04.007

Ma H, Bertsch PM, Glenn TC, Kabengi NJ, Williams PL (2009) Toxicity of manufactured zinc oxide nanoparticles in the nematode *Caenorhabditis elegans*. Environ Toxicol Chem 28:1324–1330. https://doi.org/10.1897/08-262.1

Ma X, Geiser-Lee J, Deng Y, Kolmakov A (2010) Interactions between engineered nanoparticles (ENPs) and plants: phytotoxicity, uptake and accumulation. Sci Total Environ 408:3053–3061. https://doi.org/10.1016/j.scitotenv.2010.03.031

Ma R, Levard C, Michel FM, Brown GE, Lowry GV (2013a) Sulfidation mechanism for zinc oxide nanoparticles and the effect of sulfidation on their solubility. Environ Sci Technol 47:2527–2534. https://doi.org/10.1021/es3035347

Ma H, Williams PL, Diamond SA (2013b) Ecotoxicity of manufactured ZnO nanoparticles – a review. Environ Pollut 172:76–85. https://doi.org/10.1016/j.envpol.2012.08.011

Ma C, Liu H, Guo H, Musante C, Coskun SH, Nelson BC, White JC, Xing B, Dhankher OP (2016) Defense mechanisms and nutrient displacement in *Arabidopsis thaliana* upon exposure to $CeO_2$ and $In_2O_3$ nanoparticles. Environ Sci Nano 3:1369–1379. https://doi.org/10.1039/c6en00189k

Ma C, White JC, Zhao J, Zhao Q, Xing B (2018) Uptake of engineered nanoparticles by food crops: characterization, mechanisms, and implications. Annu Rev Food Sci Technol 9:129–153. https://doi.org/10.1146/annurev-food-030117-012657

Majumdar S, Peralta-Videa JR, Bandyopadhyay S, Castillo-Michel H, Hernandez-Viezcas JA, Sahi S, Gardea-Torresdey JL (2014) Exposure of cerium oxide nanoparticles to kidney bean shows disturbance in the plant defense mechanisms. J Hazard Mater 278:279–287. https://doi.org/10.1016/j.jhazmat.2014.06.009

Martínez-Fernández D, Barroso D, Komárek M (2016) Root water transport of *Helianthus annuus* L. under iron oxide nanoparticle exposure. Environ Sci Pollut Res 23:1732–1741. https://doi.org/10.1007/s11356-015-5423-5

Mattiello A, Marchiol L (2017) Application of nanotechnology in agriculture: assessment of $tio_2$ nanoparticle effects on barley. In: Janus M (ed) Application of titanium dioxide. InTech, London, pp 23–39. https://doi.org/10.5772/intechopen.68710

Miralles P, Johnson E, Church TL, Harris AT (2012) Multiwalled carbon nanotubes in alfalfa and wheat: toxicology and uptake. J R Soc Interface 9:3514–3527. https://doi.org/10.1098/rsif.2012.0535

Mittler R (2002) Oxidative stress, antioxidants and stress tolerance. Trends Plant Sci 7:405–410. https://doi.org/10.1016/s1360-1385(02)02312-9

Mittler R (2017) ROS are good. Trends Plant Sci 22:11–19. https://doi.org/10.1016/j.tplants.2016.08.002

Mohammadi R, Maali-Amiri R, Mantri NL (2014) Effect of $TiO_2$ nanoparticles on oxidative damage and antioxidant defense systems in chickpea seedlings during cold stress. Russ J Plant Physio 61:768–775. https://doi.org/10.1134/s1021443714050124

Moller IM, Rogowska-Wrzesinska A, Rao RSP (2011) Protein carbonylation and metal-catalyzed protein oxidation in a cellular perspective. J Proteome 74:2228–2242. https://doi.org/10.1016/j.jprot.2011.05.004

Mukherjee A, Peralta-Videa JR, Bandyopadhyay S, Rico CM, Zhao L, Gardea-Torresdey JL (2014) Physiological effects of nanoparticulate ZnO in green peas (*Pisum sativum* L.) cultivated in soil. Metallomics 6:132–138. https://doi.org/10.1039/c3mt00064h

Nair PM, Chung IM (2014) Impact of copper oxide nanoparticles exposure on *Arabidopsis thaliana* growth, root system development, root lignificaion, and molecular level changes. Environ Sci Pollut Res Int 21:12709–12022. https://doi.org/10.1007/s11356-014-3210-3

Onelli E, Prescianotto-Baschong C, Caccianiga M, Moscatelli A (2008) Clathrin-dependent and independent endocytic pathways in tobacco protoplasts revealed by labelling with charged nanogold. J Exp Bot 59:3051–3068. https://doi.org/10.1093/jxb/ern154

Oukarroum A, Barhoumi L, Pirastru L, Dewez D (2013) Silver nanoparticle toxicity effect on growth and cellular viability of the aquatic plant *Lemnagibba*. Environ Toxicol Chem 32:902–907. https://doi.org/10.1002/etc.2131

Pakrashi S, Jain N, Dalai S, Jayakumar J, Chandrasekaran PT, Raichur AM, Chandrasekaran N, Mukherjee A (2014) In vivo genotoxicity assessment of titanium dioxide nanoparticles by *Allium cepa* root tip assay at high exposure concentrations. PLoS One 9:e87789. https://doi.org/10.1371/journal.pone.0087789

Patrick JW, Tyerman SD, Bel AJE (2015) Long-distance transport. In: Buchanan BB, Gruissem W, Jones RL (eds) Biochemistry and molecular biology of plants, 2nd edn. Wiley, West Sussex, pp 658–710. https://doi.org/10.1002/cbf.1131

Peralta-Videa JR, Zhao L, Lopez-Moreno ML, De la Rosa G, Hong J, Gardea-Torresdey JL (2011) Nanomaterials and the environment: a review for the biennium 2008–2010. J Hazard Mater 186:1–15. https://doi.org/10.1016/j.jhazmat.2010.11.020

Perez-Labrada F, López-Vargas ER, Ortega-Ortiz H, Cadenas-Pliego G, Benavides-Mendoza A, Juárez-Maldonado A (2019) Responses of tomato plants under saline stress to foliar application of copper nanoparticles. Plan Theory 8:151. https://doi.org/10.3390/plants8060151

Phogat N, Khan SA, Shankar S, Ansary AA, Uddin I (2016) Fate of inorganic nanoparticles in agriculture. Adv Mater Lett 7:3–12. https://doi.org/10.5185/amlett.2016.6048

Priester JH, Ge Y, Mielke RE, Horst AM, Moritz SC, Espinosa K, Gelb J, Walker SL, Nisbet RM, An YJ, Schimel JP (2012) Soybean susceptibility to manufactured nanomaterials with evidence for food quality and soil fertility interruption. Proc Natl Acad Sci U S A 109:E2451–E 2456. https://doi.org/10.1073/pnas.1205431109

Raliya R, Nair R, Chavalmane S, Wang WN, Biswas P (2015) Mechanistic evaluation of translocation and physiological impact of titanium dioxide and zinc oxide nanoparticles on the tomato (*Solanum lycopersicum* L.) plant. Metallomics 7:1584–1594. https://doi.org/10.1039/c5mt00168d

Rastogi A, Zivcak M, Sytar O, Kalaji HM, He X, Mbarki S, Brestic M (2017) Impact of metal and metal oxide nanoparticles on plant: a critical review. Front Chem 5:78. https://doi.org/10.3389/fchem.2017.00078

Rico CM, Majumdar S, Duarte-Gardea M, Peralta-Videa JR, Gardea-Torresdey JL (2011) Interaction of nanoparticles with edible plants and their possible implications in the food chain. J Agric Food Chem 59:3485–3498. https://doi.org/10.1021/jf104517j

Rico CM, Morales MI, McCreary R, Castillo-Michel H, Barrios AC, Tafoya A, Lee WY, Ramirez AV, Peralta-Videa JR, Gardea-Torresdey JL (2013a) Cerium oxide nanoparticles modify the antioxidative stress enzyme activities and macromolecule composition in rice seedlings. Environ Sci Technol 47:14110–14118. https://doi.org/10.1021/es4033887

Rico CM, Morales MI, Barrios AC, McCreary R, Hong J, Lee WY, Nunez J, Peralta-Videa JR, Gardea-Torresdey JL (2013b) Effect of cerium oxide nanoparticles on the quality of rice (*Oryza sativa* L.) grains. J Agric Food Chem 61:11278–11285. https://doi.org/10.1021/jf404046v

Rico CM, Hong J, Morales MI, Zhao L, Barrios AC, Zhang JY, Peralta-Videa JR, Gardea-Torresdey JL (2013c) Effect of cerium oxide nanoparticles on rice: a study involving the antioxidant defense system and *in vivo* fluorescence imaging. Environ Sci Technol 47:5635–5642. https://doi.org/10.1021/es401032m

Rizwan M, Ali S, Qayyum MF, Ok YS, Adrees M, Ibrahim M, Zia-ur-Rehman M, Farid M, Abbas F (2017) Effect of metal and metal oxide nanoparticles on growth and physiology of globally important food crops: a critical review. J Hazard Mate 322:2–16. https://doi.org/10.1016/j.jhazmat.2016.05.061

Rui Y, Zhang P, Zhang Y, Ma Y, He X, Gui X, Li Y, Zhang J, Zheng L, Chu S, Guo Z, Chai Z, Zhao Y, Zhang Z (2015) Transformation of ceria nanoparticles in cucumber plants is influenced by phosphate. Environ Pollut (Barking Essex) 1987(198):8–14. https://doi.org/10.1016/j.envpol.2014.12.017

Santos Filho RD, Vicari T, Santos SA, Felisbino K, Mattoso N, Sant'Anna-Santos BF, Cestari MM, Leme DM (2019) Genotoxicity of titanium dioxide nanoparticles and triggering of defense mechanisms in *Allium cepa*. Genet Mol Biol 42:425–435. https://doi.org/10.1590/1678-4685-gmb-2018-0205

Santos AR, Miguel AS, Tomaz L, Malhó R, Maycock C, Patto MCV, Fevereiro P, Oliva A (2010) The impact of CdSe/ZnS quantum dots in cells of *Medicago sativa* in suspension culture. J Nanobiotechnology 8:24. https://doi.org/10.1186/1477-3155-8-24

Schrand AM, Rahman MF, Hussain SM, Schlager JJ, Smith DA, Syed AF (2010) Metal-based nanoparticles and their toxicity assessment. WIREs Nanomed Nanobiotechnol 2:544–568. https://doi.org/10.1002/wnan.103

Serag MF, Kaji N, Gaillard C, Okamoto Y, Terasaka K, Jabasini M, Tokeshi M, Mizukami H, Bianco A, Baba Y (2011) Trafficking and subcellular localization of multiwalled carbon nanotubes in plant cells. ACS Nano 5:493–499. https://doi.org/10.1021/nn102344t

Serag MF, Kaji N, Habuchi S, Bianco A, Baba Y (2013) Nanobiotechnology meets plant cell biology: carbon nanotubes as organelle targeting nanocarriers. RSC Adv 3:4856–4862. https://doi.org/10.1039/c2ra22766e

Servin AD, White JC (2016) Nanotechnology in agriculture: next steps for understanding engineered nanoparticle exposure and risk. Nano Impact 1:9–12. https://doi.org/10.1016/j.impact.2015.12.002

Servin AD, Morales MI, Castillo-Michel H, Hernandez-Viezcas JA, Munoz B, Zhao L, Nunez JE, Peralta-Videa JR, Gardea-Torresdey JL (2013) Synchrotron verification of $TiO_2$ accumulation in cucumber fruit: a possible pathway of $TiO_2$ nanoparticle transfer from soil into the food chain. Environ Sci Technol 47:11592–11598. https://doi.org/10.1021/es300955b

Shah V, Collins D, Walker VK, Shah S (2014) The impact of engineered cobalt, iron, nickel and silver nanoparticles on soil bacterial diversity under field conditions. Environ Res Lett 9:024001. https://doi.org/10.1088/1748-9326/9/2/024001

Sharma P, Jha AB, Dubey RS, Pessarakli M (2012) Reactive oxygen species, oxidative damage, and antioxidative defense mechanism in plants under stressful conditions. J Bot 2012:1–26. https://doi.org/10.1155/2012/217037

Shaw AK, Hossain Z (2013) Impact of nano-CuO stress on rice (*Oryza sativa* L.) seedlings. Chemosphere 93:906–915. https://doi.org/10.1016/j.chemosphere.2013.05.044

Shaw AK, Ghosh S, Kalaji HM, Bosa K, Brestic M, Zivcak M, Hossain Z (2014) Nano-CuO stress induced modulation of antioxidative defense and photosynthetic performance of Syrian barley (*Hordeum vulgare* L.). Environ Exp Bot 102:37–47. https://doi.org/10.1016/j.envexpbot.2014.02.016

Shen CX, Zhang QF, Li J, Bi FC, Yao N (2010) Induction of programmed cell death in *Arabidopsis* and rice by single-wall carbon nanotubes. Am J Bot 97:1602–1609. https://doi.org/10.3732/ajb.1000073

Sheykhbaglou R, Sedghi M, Shishevan MT, Sharifi RS (2010) Effects of nano-iron oxide particles on agronomic traits of soybean. Not Sci Biol 2:112–113. https://doi.org/10.15835/nsb224667

Siddiqui MH, Al-Whaibi MH, Firoz M, Al-Khaishany MY (2015) Role of nanoparticles in plants. In: Siddiqui MH, Al-Whaibi MH (eds) Nanotechnology and plant sciences. Springer, Cham, pp 19–35. https://doi.org/10.1007/978-3-319-14502-0_2

Singh J, Lee BK (2016) Influence of nano-$TiO_2$ particles on the bioaccumulation of cd in soybean plants (*Glycine max*): a possible mechanism for the removal of cd from the contaminated soil. J Environ Manag 170:88–96. https://doi.org/10.1016/j.jenvman.2016.01.015

Smith K, Ghoshroy K, Ghoshroy S (2015) Analysis of plant responses to titanium dioxide ($TiO_2$) nanoparticles. Microsc Microanal 21:871–872. https://doi.org/10.1017/s1431927615005152

Stampoulis D, Sinha SK, White JC (2009) Assay-dependent phytotoxicity of nanoparticles to plants. Environ Sci Technol 43:9473–9479. https://doi.org/10.1021/es901695c

Syu Y, Hung JH, Chen JC, Chuang H (2014) Impacts of size and shape of silver nanoparticles on *Arabidopsis* plant growth and gene expression. Plant Physiol Biochem 83:57–64. https://doi.org/10.1016/j.plaphy.2014.07.010

Tan W, Peralta-Videa JR, Gardea-Torresdey JL (2018) Interaction of titanium dioxide nanoparticles with soil components and plants: current knowledge and future research needs–a critical review. Environ Sci Nano 5:257–278. https://doi.org/10.1039/c7en00985b

Tang Y, He R, Zhao J, Nie G, Xu L, Xing B (2016) Oxidative stress induced toxicity of CuO nanoparticles and related toxicogenomic responses in *Arabidopsis thaliana*. Environ Pollut 212:605–614. https://doi.org/10.1016/j.envpol.2016.03.019

Thuesombat P, Hannongbua S, Akasit S, Chadchawan S (2014) Effect of silver nanoparticles on rice (*Oryza sativa* L. cv. KDML 105) seed germination and seedling growth. Ecotoxicol Environ Saf 104:302–309. https://doi.org/10.1016/j.ecoenv.2014.03.022

Tou F, Yang Y, Feng J, Niu Z, Pan H, Qin Y, Guo X, Meng X, Liu M, Hochella MF (2017) Environmental risk implications of metals in sludges from waste water treatment plants: the discovery of vast stores of metal-containing nanoparticles. Environ Sci Technol 51:4831–4840. https://doi.org/10.1021/acs.est.6b05931

Vannini C, Domingo G, Onelli E, De Mattia F, Bruni I, Marsoni M, Bracale M (2014) Phytotoxic and genotoxic effects of silver nanoparticles exposure on germinating wheat seedlings. J Plant Physiol 171:1142–1148. https://doi.org/10.1016/j.jplph.2014.05.002

Wang H, Kou X, Pei Z, Xiao JQ, Shan X, Xing B (2011) Physiological effects of magnetite ($Fe_3O_4$) nanoparticles on perennial ryegrass (*Lolium perenne* L.) and pumpkin (*Cucurbita mixta*) plants. Nanotoxicology 5:30–42. https://doi.org/10.3109/17435390.2010.489206

Wang X, Han H, Liu X, Gu X, Chen K, Lu D (2012a) Multiwalled carbon nanotubes can enhance root elongation of heat (*Triticum aestivum*) plants. J Nanopart Res 14:article 841. https://doi.org/10.1007/s11051-012-0841-5

Wang Z, Xie X, Zhao J, Liu X, Feng W, White JC, Xing B (2012b) Xylem-and phloem-based transport of CuO nanoparticles in maize (*Zea mays* L.). Environ Sci Technol 46:4434–4441. https://doi.org/10.1021/es204212z

Wang P, Menzies NW, Lombi E, McKenna BA, Johannessen B, Glover CJ, Kappen P, Kopittke PM (2013) Fate of ZnO nanoparticles in soils and cowpea (*Vigna unguiculata*). Environ Sci Technol 47:13822–13830

Wang F, Liu X, Shi Z, Tong R, Adams CA, Shi X (2016a) Arbuscular mycorrhizae alleviate negative effects of zinc oxide nanoparticle and zinc accumulation in maize plants—a soil microcosm experiment. Chemosphere 147:88–97. https://doi.org/10.1016/j.chemosphere.2015.12.076

Wang P, Menzies NW, Dennis PG, Guo J, Forstner C, Sekine R, Lombi E, Kappen P, Bertsch PM, Kopittke PM (2016b) Silver nanoparticles entering soils via the wastewater-sludge-soil pathway pose low risk to plants but elevated cl concentrations increase ag bioavailability. Environ Sci Technol 50:8274–8281. https://doi.org/10.1021/acs.est.6b01180

Wang Z, Xu L, Zhao J, Wang X, White JC, Xing B (2016c) CuO nanoparticle interaction with Arabidopsis thaliana: toxicity, parent-progeny transfer, and gene expression. Environ Sci Technol 50:6008–6016. https://doi.org/10.1021/acs.est.6b01017

Wang Q, Li C, Wang Y, Que X (2019) Phytotoxicity of graphene family nanomaterials and its mechanisms: a review. Front Chem 7:292. https://doi.org/10.3389/fchem.2019.00292

Wei H, Wang E (2013) Nanomaterials with enzyme-like characteristics (nanozymes): next-generation artificial enzymes. Chem Soc Rev 42:6060–6093. https://doi.org/10.1039/C3CS35486E

Yamamoto Y, Kobayashi Y, Matsumoto H (2001) Lipid peroxidation is an early symptom triggered by aluminum, but not the primary cause of elongation inhibition in pea roots. Plant Physiol 125:199–208. https://doi.org/10.1104/pp.125.1.199

Yan S, Subramanian SB, Tyagi RD, Surampalli RY, Zhang TC (2010) Emerging contaminants of environmental concern: source, transport, fate, and treatment. Practice Periodical of Hazardous, Toxic, and Radioactive Waste Management 14:2–20. https://doi.org/10.1061/(asce)hz.1944-8376.0000015

Yan S, Zhao L, Li H, Zhang Q, Tan J, Huang M, He S, Li L (2013) Single-walled carbon nanotubes selectively influence maize root tissue development accompanied by the change in the related gene expression. J Hazard Mater 246:110–118. https://doi.org/10.1016/j.jhazmat.2012.12.013

Yasur J, Rani PU (2013) Environmental effects of nanosilver: impact on castor seed germination, seedling growth, and plant physiology. Environ Sci Pollut Res 20:8636–8648. https://doi.org/10.1007/s11356-013-1798-3

You J, Chan Z (2015) ROS regulation during abiotic stress responses in crop plants. Front Plant Sci 6:1092. https://doi.org/10.3389/fpls.2015.01092

Zhang P, Ma Y, Zhang Z, He X, Zhang J, Guo Z, Tai R, Zhao Y, Chai Z (2012) Biotransformation of ceria nanoparticles in cucumber plants. ACS Nano 6:9943–9950. https://doi.org/10.1021/nn303543n

Zhang P, Ma Y, Zhang Z, He X, Li Y, Zhang J, Zheng L, Zhao Y (2015) Species-specific toxicity of ceria nanoparticles to *Lactuca* plants. Nanotoxicology 9:1–8. https://doi.org/10.3109/17435390.2013.855829

Zhang P, Ma Y, Liu S, Wang G, Zhang J, He X, Zhang J, Rui Y, Zhang Z (2017) Phytotoxicity, uptake and transformation of nano-$CeO_2$ in sand cultured romaine lettuce. Environ Pollut 220:1400–1408. https://doi.org/10.1016/j.envpol.2016.10.094

Zhao L, Peralta-Videa JR, Ren M, Varela-Ramirez A, Li C, Hernandez-Viezcas JA, Aguilera RJ, Gardea-Torresdey JL (2012) Transport of Zn in a sandy loam soil treated with ZnO NPs and uptake by corn plants: Electron microprobe and confocal microscopy studies. Chem Eng J 184:1–8. https://doi.org/10.1016/j.cej.2012.01.041

Zheng L, Hong F, Lu S, Liu C (2005) Effect of nano-$TiO_2$ on strength of naturally aged seeds and growth of spinach. Biol Trace Elem Res 104:83–91. https://doi.org/10.1385/bter:104:1:083

Zhou J, Ma Y, Jia Y, Pang M, Cheng G, Cai S (2019) Phenolic profiles, antioxidant activities and cytoprotective effects of different phenolic fractions from oil palm (*Elaeis guineensis* Jacq.) fruits treated by ultra-high pressure. Food Chem 288:68–77. https://doi.org/10.1016/j.foodchem.2019.03.002

Zhu H, Han J, Xiao JQ, Jin Y (2008) Uptake, translocation, and accumulation of manufactured iron oxide nanoparticles by pumpkin plants. J Environ Monit 10:713–717. https://doi.org/10.1039/b805998e

Ziental D, Czarczynska-Goslinska B, Mlynarczyk DT, Glowacka-Sobotta A, Stanisz B, Goslinski T, Sobotta L (2020) Titanium dioxide nanoparticles: prospects and applications in medicine. Nano 10:387. https://doi.org/10.20944/preprints202002.0271.v1

# Part V
# Traditional Knowledge and Social Concerns About Emerging Contaminants

# Chapter 14
# Non-Judicious Use of Pesticides Indicating Potential Threat to Sustainable Agriculture

**Ajay Kumar Mishra, Rahul Arya, Rakhi Tyagi, Dipti Grover, Jyotiprakash Mishra, Shobhit Raj Vimal, Sarita Mishra, and Sheetal Sharma**

**Abstract** Worldwide, pesticides have proved to be a boon for the farmers by increasing agricultural yield and providing innumerable benefits to society directly and indirectly. For instance, India witnessed the green revolution during the 1960s that redefined the low input to high input more intense agriculture system that contributed to transforming Indian agriculture from food-deficit to food-surplus nation. Nevertheless, the issue of hazards posed by pesticides to human health and the environment cannot be ignored. Non-judicious overuse of pesticides is acting as the source of emerging contaminants in agro-ecosystems. It has been linked to a wide range of human health hazards, ranging from short-term impacts such as headaches and nausea to chronic impacts like cancer, reproductive disorders, and endocrine disruption. Moreover, contamination of air, water and soil that end up adversely affecting the survival of other organisms such as non-targeted plants, birds, and aquatic flora and fauna has also been reported.

A. K. Mishra (✉)
Terrestrial Ecosystem Management Laboratory, Kyoto University, Kyoto, Japan

International Rice Research Institute, New Delhi, India
e-mail: a.k.mishra@irri.org

R. Arya · R. Tyagi · D. Grover
Institute of Environmental Studies, Kurukshetra University, Kurukshetra, Haryana, India

J. Mishra
Department of Agronomy, OUAT, Bhubaneswar, Odisha, India

S. R. Vimal
Department of Environmental Microbiology, Babasaheb Bhimao Ambedkar University, Lucknow, Uttar Pradesh, India

S. Mishra
Department of Chemistry, UdaiPratap College, Varanasi, Uttar Pradesh, India

S. Sharma (✉)
International Rice Research Institute, New Delhi, India
e-mail: sheetal.sharma@irri.org



V. Kumar Singh et al. (eds.), *Sustainable Agriculture Reviews 50*, Sustainable Agriculture Reviews 50, https://doi.org/10.1007/978-3-030-63249-6_14

The present study aimed to give a brief history of pesticide use globally, case studies of pesticide use consequences particularly in the Indian context and strategies to mitigate the ill effects. In general, this study also focuses on direct and indirect impacts of pesticide exposure on air quality, water contamination, soil pollution and human health. Besides, we highlighted detailed case studies of over usage of pesticides states of Punjab, Haryana and Kerala in India. In all the case studies, hyper-accumulation of pesticides led to detrimental effects in all age classes of human beings and their surroundings. Integrated pest management (IPM) should be developed and extended, for effective, affordable, and environmentally-sound control strategies for detrimental biotic agents. The integration of pesticides with natural enemies has more significant potential than is utilized at present and it could help to mitigate the negative consequences of pesticide overuse. Judicious use with IPM, ecological engineering and biological control is highly considerable and beneficial in the long term to sustain agriculture productivity without compromising the environment and human health.

**Keywords** Biomagnification · Biopesticides · Cancer · Ecological engineering · Environmental impact · Health impacts · Integrated pest management

## Abbreviations

| | |
|---|---|
| DDT | Dichloro Diphenyl Trichloroethane |
| DNA | Deoxyribonucleic Acid |
| EPA | Environment Protection Agency |
| HCH | Hexachlorocyclohexane |
| HYVP | High Yielding Variety Programme |
| IPM | Integrated Pest Management |
| OCP | Organo-Chlorine Pesticides |
| PPE | Personal Protection Equipment |
| USA | United States of America |
| USGS | United States Geological Survey |
| WHO | World Health Organization |

## 14.1 Introduction

According to the World Health Organization (WHO) each year more than 3,000,000 cases of pesticide poisoning and 22,000 deaths are reported in developing countries (Lah 2011). Moreover, the inefficient and extensive use of pesticides during the last

5 to 6 decades inadvertently has led to the serious deterioration of environmental quality with increased health risks to living organisms, including humans (Khalid et al. 2020). One major challenge encountered globally is emerging contaminants in agriculture mainly due to hasty growth of the pesticide industry that increased the chemical burden on the natural environment in various ways. For instance, about 75% of pesticide use in the USA was in agriculture (Calvert et al. 2008) and in Europe, despite international efforts to promote the sustainable use of pesticides in agriculture. Overall pesticide use did not decline substantially in the WHO European Region during the period of 1990s (Robertson et al. 2004). Worldwide, contamination of soil and water with various types of pesticides, resulting from accidental spills at agrochemical manufacturing, formulation, and distribution facilities, farm loading/wash-down sites, or abandoned waste sites, is a serious environmental problem (Abbas and Al-Madhhachi 2020). Md Meftaul et al. (2020) provides updated knowledge on the sources, nature and extent of pesticide pollution in the urban environment, and the ecological and human health effects of pesticides and their residues that hinders sustainable livelihood. However, the urgent need is to understand and act holistically on emerging contaminants for a more ecological sustainable approach with promising innovative ideas that potentially transforms agriculture and food production systems.

Indian economy chiefly depends on the agriculture sector as it provides employment and livelihood to more than 70% of the population. As per the current trend of population explosion around the globe, it is predictable that India will surpass China and become the most populous nation in the world by 2022 (BBC News Asia 2015). At present, India supports 17.84% (1.32 billion) of the world population, with 2.4% land resource and 4% of water resources (Yadav and Dutta 2019). Vegetables and fruits are considered a healthy choice to fulfill the essential nutritional requirement of the increasing population. There are many reasons for cultivating vegetables as it has good market value, short cultivation time (45–60 days) and a new earning source for farmers. Though fulfilling the nutritional requirements, cultivation of fruits and vegetables heavily rely on use of pesticides for controlling diseases and pests. Pesticides have played an essential role in reducing losses from the weeds, diseases and insect pests that can markedly reduce the amount of loss and increase the profit (Kaur et al. 2019). However, almost 62% of vegetables and 82% of fruits have been found to be contaminated with pesticides (FDA Pesticide monitoring report 2019). According to the Food and Agriculture Organization (2018) "*Pesticides are any substance or blend of substances that diminish the harmful population of pests of bugs.*" According to the United States Environmental Protection Agency (2018) "*Pesticides intend to control pests, as well as weeds and protect plant health from harmful biotic agents*". The pattern of pesticide application in the world and India, are shown in Fig. 14.1. In India, insecticides alone constitute 76% of the total pesticides, whereas this figure is only 44% at the world level (Mathur 1999).

When pesticides are applied, plant tissues absorb them. Not only the absorption but adsorption also takes place (Kaur et al. 2019). Adsorption is the physical binding of pesticide molecules to soil particles or any other surface. The physical and

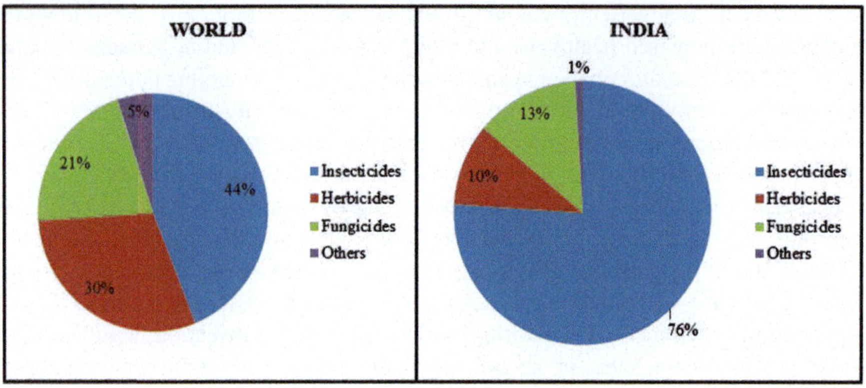


**Fig. 14.1** The categories wise application of pesticides in the world and India during cultivation

chemical properties of pesticides depend on the pH of the soil, its concentration, and the composition of the soil. Some soluble pesticides bind strongly with soil as a soil containing clay has more surface area and high water holding capacity (Das et al. 2019). The more clay particles and organic matter in the soil, the more the pesticide is held by the soil and becomes immobile and unlikely to leach down unless the soil particles to which they were adsorbed move with water (Takeshita et al. 2019). The longer the molecules of a pesticide are held, the more microbiological degradation will occur, which reduces the risk of leaching and erosion. Erosion is the movement of soil particles from the application site by heavy rains or excess irrigation. If the pesticide is not adsorbed to the soil particle, the pesticide is also being moved off-site and may contaminate the groundwater. This chapter highlights pesticide application in global and Indian agriculture and its impact on air, water, soil and human health. The pesticides hazard cases especially from Punjab, Haryana and Kerala states of India have been explored in detail in the present chapter.

## 14.2 Classification of Pesticides

The classification of pesticides, based on their chemical composition and nature of ingredients, is depicted in Fig. 14.2. The classification is based on the efficacy, physical and chemical properties of the respective pesticides (Kaur et al. 2019).

## 14.3 Pathways of Pesticides Contamination

In natural soil conditions, there may be cocktails of pesticides together or other contaminants in addition to pesticides at a time in a particular location. The co-occurrence of different pesticides and other contaminants might have different

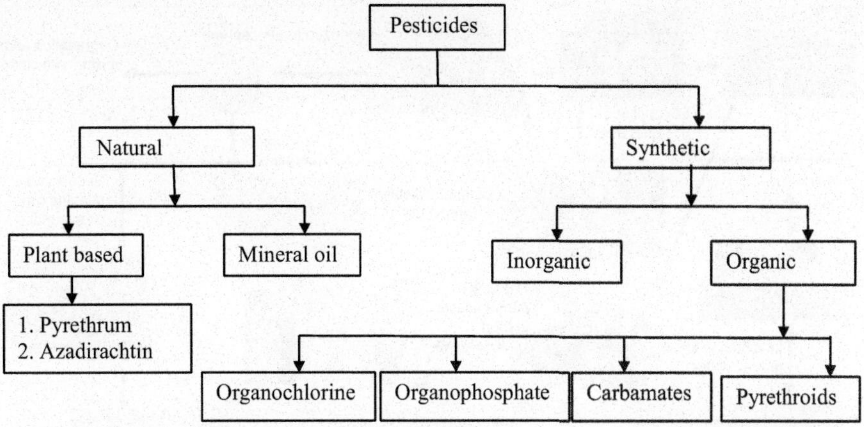


**Fig. 14.2** Demonstration of the classification of the pesticides in natural and synthetic categories, with their principal chemical constitutes

properties and effects on microbial activity, biomass, and diversity. The pesticides application may be harmful to the microbial community, disturb soil ecosystem, and thus, may affect human health by entering in the food chain (Hussain et al. 2009). There are many ways through which pesticides can enter the human body and can cause hazardous effects on human health. For example:

(a) Breathing- Pesticides are present in vapors form that can be easily inhaled during their application time.
(b) Ingestion- Ingestion of these chemicals occurs through crops or through direct accidental ingestion.
(c) Dermal- Absorption through the skin.
(d) Eye exposure- Pesticides can enter through eyes in the human body when these are used in spray form.

Over dose of pesticides act as pollutants and can also enter the food chain through contaminated food and may induce many health hazards (Fig. 14.3). Morphological and biochemical changes in *Vigna radiata* and *Spinacia oleracea* induced by fluoride contamination in soils from various sources were reported by Tyagi et al. (2019). Several studies have reported that in many developing countries including India, the pesticide application is indiscriminate and beyond safe limits (Dey 2016).

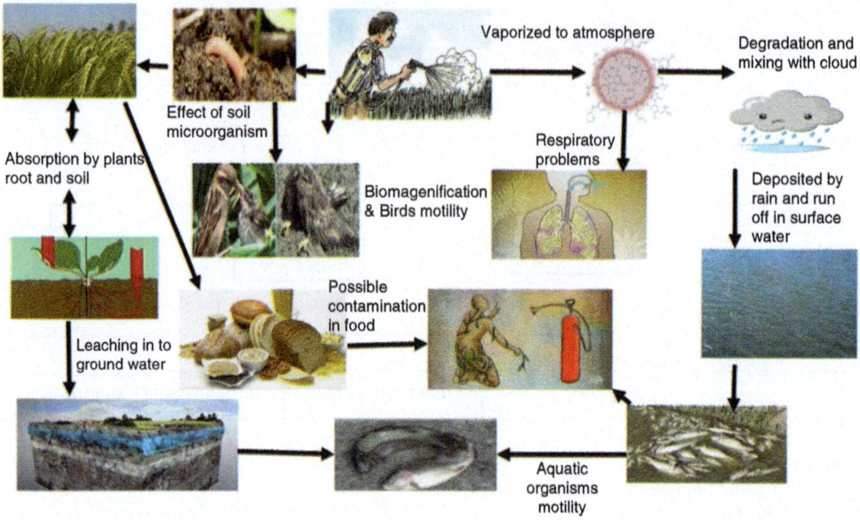


**Fig. 14.3** Pictorial representation of pesticides application on crops and their transmission into atmosphere, terrestrial and aquatic ecosystems to animals and human beings which leads lethal effect both on human and environment through biomagnification, consequently inter- and intra-transmission in different components of ecosystem, human and animal and its harmful effects

## 14.4 Outcomes of Pesticide Application

### *14.4.1 The Global Perspectives*

Globally pesticide consumption has increased drastically for crop/food security to feed the increasing population (Zhang et al. 2011). Total pesticides (tonnes) applied worldwide in agriculture is presented in Fig. 14.4. In history, pesticides may be divided into three phases; the first phase of the pesticide history was before the 1870s in ancient Greece when sulfur was used as natural pesticides to control pests (Zhang et al. 2011). The era of inorganic synthetic pesticides was the second phase (1870s-1945), wherein natural material and inorganic compounds were used for the controlling pests. The third phase which is well known as the era of organic synthetic pesticides started in 1945, in which man made organic pesticides like DDT, 2,4-D, HCH, Dieldrin were used. In the earlier era, only three organically synthesized pesticides viz. Carbamate, Organophosphorus, and Organochlorine were used but after the introduction of herbicides, fungicides, and bactericides significant changes in uses of these pesticides were noticed (Zhang et al. 2011). Pesticides consumption endures substantial changes since the 1960s, it increased from 20% in 1960 to 48% in 2005 (Zhang et al. 2011).

The largest consumer of pesticides is Europe followed by Asia, while China, the US, France, Brazil and Japan are the largest pesticide producers, consumers or traders in the world (Zhang et al. 2011). A portion of the credit of high yields and

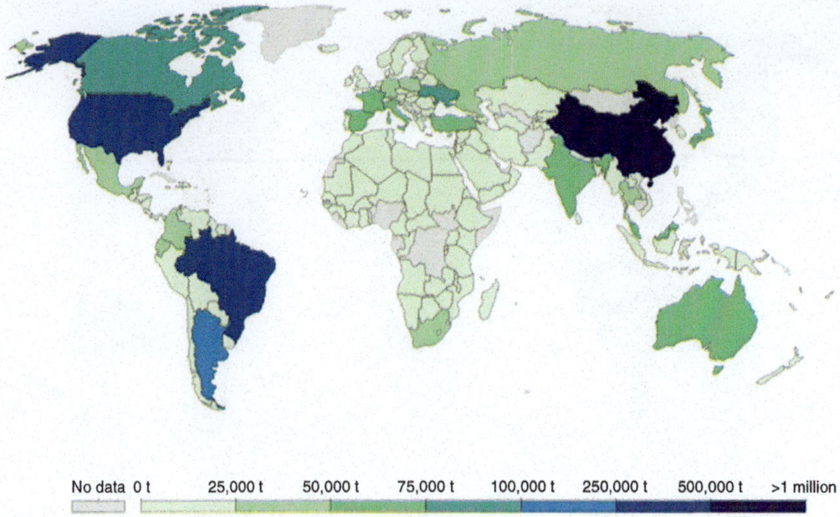


**Fig. 14.4** Details of the total pesticides measured in tones applied in global agriculture. Dark geographical area and light geographic area demonstrates higher and lower measurement of pesticides, respectively (FAO 2017)

production, ensuring food security goes to pesticide application while they have several negative consequences on air, water, and soil by acting as environmental contaminants and causing hazards to human and other life forms exerting toxic effects.

Worldwide approximately 10,000 insects and mite's species, 50,000 plant pathogens species, and 8000 weeds species are documented to damage plant health and hence crop productivity (Dhaliwal et al. 2010). Insect pests cause approximately 14% loss, plant pathogens cause nearly 13% losses, and weeds around 13% loss (Pimentel 2009). In a farmer's perspective, pesticide is crucial for enhancing agricultural production. Studies have even reported that around one-third of the agricultural products are produced by using pesticides (Liu et al. 2002), while without using pesticide, the loss of fruits, vegetables, and cereals due to pest injury would reach 78%, 54%, and 32%, respectively (Cai 2008). The pesticides application significantly reduces 35–42% crop loss due to various pests (Pimentel 1997; Liu and Liu 1999). Because of the world's limited croplands and growing population (Zhang et al. 2006; Zhang 2008), it is necessary to take all measures to increase crop production. Increased pesticides application (average) worldwide was to ensure reduction in a range of pests infesting agricultural crops (Fig. 14.5).

Developing strong consciousness among the policymakers about Integrated pest management (IPM) as an ecosystem-based approach and strategy that focuses on long-term prevention of pests or their damage through a combination of techniques such as biological control, habitat manipulation, modification of cultural practices, and use of resistant varieties. Ecological engineering holistically adopted to increase

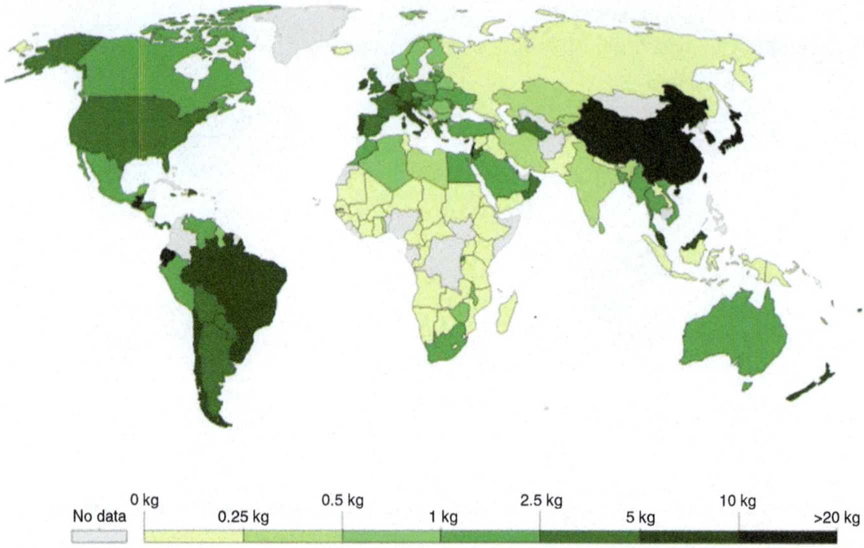


**Fig. 14.5** Details of the average pesticide application per unit of cropland, measured in kilograms per hectare in global agriculture. Dark geographical area and light geographic area demonstrates the higher and lower use of pesticides, respectively (FAO 2017)

natural predators of pests that not only controls pest population but also enhances aesthetic value. In the next section, we focused on pesticide use and ill effects in Indian context with some case studies from different regions of India.

## *14.4.2 The Indian Perspectives*

Pesticide application as a trend started after the green revolution, and since the green revolution, there has been an enormous increase in food grain (Niti.gov.in 2017). It has made a significant contribution in transforming India from food-deficit to food-sufficient nation. To increase agricultural productivity, the Government of India initiated a new 'agriculture strategy', the High-Yielding Varieties Programme (HYVP). This depended crucially on high yielding varieties of seeds, along with regular and adequate irrigation, fertilizers, pesticides, insecticides, and agricultural equipment. However, an enormous amount of contaminant loaded fertilizers and pesticides usage leads to various deformities in bodies, mental problems, hormonal problems, cancer and lots of new types of diseases (Zhang et al. 2011). Despite several achievements, the green revolution has resulted in the above listed effects in India.

In India, pesticide production started in 1950 and now the country has become the 2nd largest producer of pesticide. According to the Directorate of Plant Protection and Quarantine (2011), India grew steadily from 5000 MT in 1959 to 85,000 MT in

2009–2010, which makes it 2% of the total market of the world in revenue generation (Statistical Yearbook of India 2011). Organochlorine dichlorodiphenyltrichloroethane (DDT) was the first agrochemical which was used in India in 1948 as an insecticide (Bhardwaj and Sharma 2013). After the green revolution period, the consumption of pesticides increased from $0.29 \times 10^6$ MGT (1960–61) to $12.15 \times 10^6$ MGT (1992–1993) (Singh 2000). Though, the consumption of pesticides in India is less from other countries like Japan and Germany, non-judicious usage of pesticides in India increased due to poor agricultural management strategies and practices, lack of knowledge about hazardous behaviour of pesticides among farmers. As a tropical country, India differed in the consumption of pesticides from the rest of the world, which skewed toward the usage of insecticides (Indian Pesticides Industry 2011), which account for 76% of total usage. Crop specific consumption of pesticides reveals that cotton accounts for highest consumption followed by paddy at 37% and 20%, respectively. Other crops such as wheat, vegetables, and other plant crops consumed 9%, 2% and 7% of pesticides, respectively (Ministry of Agriculture, Govt. of India 2009). Andhra Pradesh ranked first followed by Punjab and Maharashtra in the consumption of pesticides (Bhardwaj and Sharma 2013).

## 14.5 The Potential Threat of Pesticides in Air, Water and Soil Environment

Excessive dose of agrochemicals (pesticides) with several emerging contaminants affect the air quality, water purity, and soil health by their hazardous transformations (Fig. 14.6). Limited research is available on contamination of air by pesticides

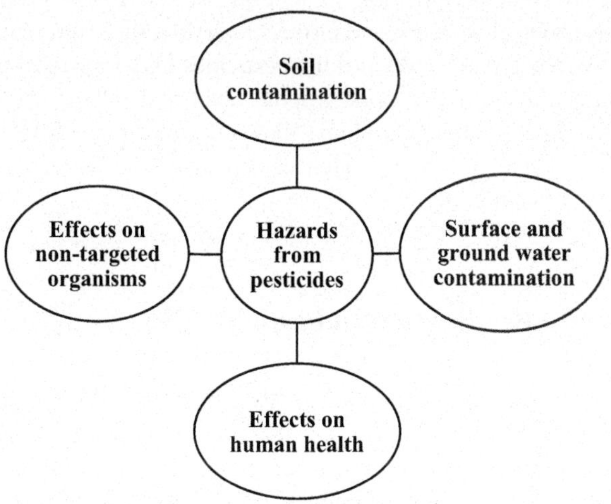


**Fig. 14.6** The hazardous impact of massive use of pesticides on soil, water, air, other living creatures and human health

application than other pollutants. Pesticide sprays may drift and volatilize and can form aerosol in the target area, which can contaminate the ambient air of that area (Socorro et al. 2016). The drift of sprays losses 2–25% of applied agrochemical which can sink into the air and spread >100 miles and 80–90% of applied treatment can volatilize within couple of the day after application and secondary pollutants may be resultant from the same as the precursor (Socorro et al. 2016).

Surface, as well as groundwater contamination, is a worldwide problem due to pesticide application. Such contamination of both surface and groundwater affects aquatic fauna and flora, as well as human health (Cerejeira et al. 2003; Hossain et al. 2015). Aquatic organisms are directly exposed to chemicals resulting from agricultural production via surface run-off or indirectly through trophic chains. Pesticides can reach surface water like river streams etc. by run-off and into groundwater by leaching (Aktar et al. 2009). Pesticides leaching and run-off contribute >50% to the surface water pollution (EPA 2009). According to the United States Geological Survey (USGS), at least 143 different pesticides and 21 transformation products have been found in groundwater (Aktar et al. 2009). Organochlorine Pesticides (OCPs), DDT, Endosulfan, Endrin, Aldrin, Dieldrin, etc. were reported in drinking water supply in Delhi and river Yamuna (Agrawal et al. 2015). Endosulfan and Di-chloro-diphenyl-trichloro-ethane (DDT) detected in Bihar and Uttar Pradesh (Agarwal et al. 2015). Total 58% of samples of drinking water were observed to be contaminated with pesticides in Bhopal, Madhya Pradesh, India (Kole and Bagchi 1995). Eleven states found nitrate contamination in India due to pesticides application (Kumar and Shah 2006).

Excessive application of agrochemicals leads to changes in soil reaction, poor soil fertility, improper nitrogen fixation, changes in biogeochemical cycles, loss in carbon content, etc. problems in agriculture (Aktar et al. 2009; Arya et al. 2018). Exposure to pesticides may affect humans and other life forms as an unwanted side effect to the environment (Igbedioh 1991; Forget et al. 1993). Producer workers, distributors, pest control workers, and farmers may directly or indirectly get affected due to exposure to pesticides. Most of the pesticides may transfer through the food chain by biomagnification from one trophic level to another next trophic level. Pesticides, hazardous in chemical nature like neurotoxins, can affect liver, kidney, and reproductive cycle in humans. Hence, pesticides are toxic, risky and health hazardous to the human being.

## 14.6 Case Studies Related to Pesticide Use

Indiscriminate use of numerous pesticides like insecticides, herbicides/weedicides, fungicides, rodenticides, etc. lead to accumulation of diverse emerging contaminants in the soil-plant-water continuum and adversely affect the environment and human health. All the farmers and communities who are directly or indirectly linked to exposure as well as biomagnification through the food chain have to face the effect and repercussions of using herbicides and pesticides. At that time Punjab state being a wealthy state, and literacy rate being high, they accepted and adopted the modern technologies of green revolution in 1965 (Luwanda 2018) but in the present scenario it is subjected to the severe repercussions of excessive pesticide use and is worst affected.

In Rangareddy district, Andhra Pradesh, India, a study was conducted by Rao et al. (2009), in 2007 and they found pesticide residues were revealed in the vegetable like okra, tomato, cucumber, brinjal, etc. and also in the collected water samples from Kotapally Adarsha watershed. Approximately, the quantity of chlorpyrifos, cypermethrin and endosulfan was 0.001 to 0.352 mg $kg^{-1}$, 0.001 to 5.154 mg $kg^{-1}$ and 0.001 to 0.784 mg $kg^{-1}$, respectively. Endosulfan residues were found to be below residual limit while chlorpyrifos and cypermethrin residues were above the limit in 59, 4 and 2 samples, respectively. In water samples residues were also found but below the residual limit. This shows that pesticides, whether in high or low concentrations, are always served in plates in the form of vegetables.

Kumari and Sharma (2012) investigated the consequences of pesticide usage on the health of farmers of Kullu and Shimla districts of Himachal Pradesh, India. Interviews of 100 farmers were conducted in each district by using a pre-tested questionnaire. Most of the farmers were directly exposed to pesticides when spray was done by themselves in high-value cash crops. While doing spray, only 22% and 8% of the farmers were using polythene to cover their nose and mouth in Kullu and Shimla, respectively. Personal Protective Equipment (PPE) was not used by farmers while spraying pesticides. The majority of the farmers were not adopting IPM in Kullu (80%) and Shimla (86%). Due to the indiscriminate use of pesticides, farmers suffered from several toxicity related diseases. In Kullu, farmers responded that 86% were suffering from eye irritation, 81% from fatigue, 66% from skin irritation, 59% from back pain and headache, 56% from vomit and 22% from dizziness. While in Shimla, 78% farmers were suffering from the symptoms of eye irritation and back pain followed by fatigue (77%), headache (77%), skin irritation and nausea (41%) and eye flu (31%). 90% of the farmers in Kullu and 62% in Shimla reported that pesticides also killed beneficial insects and bees. This implies that the usage of excessive pesticides without protective measures adversely affect human health and the environment (Kumari and Sharma 2012). Therefore, there is a need to make farmers more aware through authorities responsible regarding the usage of IPM and protective gear, while handling pesticides. The present study compiles the effect of pesticides in four different regions of India.

### *14.6.1 A Case Studies of Punjab, India*

The present case study was an attempt to compile scattered pieces of literature that evidenced the adverse effects of pesticides in Punjab, India. Punjab that witnessed green revolution is now majority ruined because of adoption of massive and non-judicious use of pesticide, herbicide and fertilizer overuse which had weakened the immunity of local population and made them susceptible to diseases (Mittal et al. 2014).

In Malwa region of Punjab nearly 75% consumption of the pesticide was reported in comparison to the whole state. With an enormous amount of pesticide usage, it is suspected to affect the whole environment such as soil, water and air which ultimately enter the food chain and affect the local population and nearby population. Within two decades, many cases of mental disabilities and reproductive disorders were reported. According to the study of Kumar (2005), 7441 deaths were recorded in the period 1993–2003 because of cancer in Talwandi region and maximum cases were of female reproductive cancer. Out of 11 districts of Malwa region, 4 districts namely Muktsar, Mansa, Bathinda, and Faridkot were reported to be badly affected by cancer cases of various types. In this region, 2 villages Giana and Jajjal of Bathinda district have been declared as "*cancer-stricken villages*" (Thakur et al. 2008) and Malwa region is known as 'Cancer Capital' of India. A passenger train from Bohar to Jodhpur passengers known as 'cancer train' because of approximately 60 cancer patients per day were carried by it to the cancer treatment hospital and research institute in Bikaner, Rajasthan, India. Kaur et al. (2011) studied the DNA of exposed farmers and control (210 samples) which reveals that 1/3rd of the samples was genetically damaged due to pesticide exposure. Thereby, a high risk of DNA damage due to pesticide exposure has been observed. Another study of the past year reported cancer and other life-threatening ailments due to contaminated drinking water with heavy metals and pesticides (Thakur et al. 2008).

### *14.6.2 A Case Study from Faridabad, Haryana, India*

Study revealed that continuous and long-term exposure of chemical pesticides in various forms and conditions leads to chronic health effects which were admitted by the respondents. The most common symptoms observed in respondents were headache and dizziness (27%), followed by eye problems like itching and redness (23%). Other most widespread problems were allergy to hands, face, neck, feet, and skin of other exposed body parts (18%), and vomiting and nausea (16%). More than 11% of farmers also reported diarrhea and stomach-ache. Serious health problems such as asthma, migraine, development of permanent skin patches, etc. were found in almost 16% of farm workers (Tyagi et al. 2015). Knowledge dissemination, awareness campaign and provision of alternatives to fares are the key to reduce ill effects of pesticide to human health.

### 14.6.3 A Case Study from Kasaragod, Kerala, India

Kasaragod tragedy due to Endosulfan aerial spraying is well known after the episode of 'Satyamev Jayate' where they publicise the hazardous after effects of pesticide usage. There, aerial spraying of Endosulfan was started on cashew cropland continuously for 20 years (Venugopal 2008). Effects of such exposure were revealed when cases of mental disorder, chemical neurotoxicity, physical deformities, reproductive failure, and delayed sexual maturity came into front very often. One out of five houses was affected by any one of the above problems due to pesticide exposure. In 2011, the Supreme Court banned its production and distribution. Over the years, other studies confirmed these findings, and the health hazards associated with endosulfan are now widely known and accepted (Patocka et al. 2016). The chemical, biochemical, environmental, and toxicological data of endosulfan reviewed by Patocka et al. (2016) showed its toxicological potential risk to human health that had been banned globally by the Persistent Organic Pollutants Review Committee in 2010.

## 14.7 Technological Intervention and Advancement in Development of Less Toxic Pesticides

Ecological engineering, biopesticides, nano-based smart pesticide formulations, stimuli-responsive nano-based pesticides are the emerging options to control pests with least effect to environment and human health. Ecological engineering for pest management is a new paradigm to enhance the natural enemies of pests in an agro-ecosystem and it relies on use of cultural techniques to bring about habitat manipulation and enhance biological control (Gurr et al. 2004). The key objective in ecological engineering is to make a conducive environment for the better survival of natural enemies of pests in the agro-ecosystem. In ecological engineering, habitat manipulation targets to provide natural enemies of pests with nectar, pollen, physical refuge, alternate prey, alternate hosts and living sites (Manage 2019). This can be achieved through plantation of appropriate companion plants like floral trap crops and repellent crops, through which the population of pollinators, predators and parasitoids that can be enhanced to manage the herbivorous insect pests (Manage 2019). As ecological engineering is knowledge intensive package of practice, lack of awareness and information sources to farmers about ecological engineering possess obstacles in adoption and scaling of this eco-friendly technology.

The microbiome or consortium of microorganisms not only act as a biological pesticide but also offers diverse plant growth-promoting attributes such as biological nitrogen fixation. Such microbiomes showed promising results as an alternative to chemical pesticides and are known to play a pivotal role against insects and pathogens. Still the efficacy and time to reduce pest attack is not convincing. *Moringa* leaf and stem extract may be used as a low cost and easily available biopesticide (Kumar et al. 2019). Further strengthening research on bio-pesticides especially for emerging contaminants is the need of the hour to move forward sustainably in an eco-friendly manner.

In recent years, the application of nanotechnology to formulate nano-based pesticide has shown boundless potential for diminishing the indiscriminate use of pesticides and proposing environmentally safer alternatives. Smart nano-based pesticides are designed to efficiently deliver sufficient amounts of active ingredients in response to biotic and/or abiotic stressors that act as triggers, employing targeted and controlled release mechanisms (Camara et al. 2019). Presently, although stimuli-responsive systems are well developed in the area of medicine, their applications in agriculture principally in the pesticide field remain limited. Therefore, continued systematic and strategic research for the development of improved eco-friendly responsive, targeted, controlled-release pesticide formulations are required to excel a healthy and safer food production system (Camara et al. 2019).

Photocatalysis is one of the promising advanced oxidation technologies to alleviate pesticide induced contamination. Titanium dioxide ($TiO_2$) is the most popular photocatalyst due to its low cost, nontoxicity, high oxidizing abilities, and easy immobilization on various surfaces. Kanan et al. (2020) highlighted recent advancements in photocatalytic degradation of pesticides and major organic pollutants using $TiO_2$-based photocatalysts. Such research need to be more emphasised for the developing countries.

## 14.8 Conclusions and Future Thrust

Pesticides have proved to be a boon for the farmers by increasing agricultural yield manifold and providing innumerable benefits to society directly and indirectly in terms of food security. Nevertheless, the issue of hazards posed by pesticides to human health and the environment cannot be ignored. The outcomes of the case studies on ill-effects of pesticides in India is alarming and indicate a paradigm shift towards the knowledge-intensive farming system with a decreased footprint to the environment and human health. The studies reviewed in this paper have raised concerns about the safety of pesticides to use without affecting the non-target species and the environment. Moreover, the detrimental consequences of non-judicious over use of pesticides had been witnessed worldwide, and worsened the health and environment condition in developing countries particularly in India. Several case studies suggesting adverse effect of pesticide overuse on soil-plant-human-environment nexus in different regions of India was discussed throughout this review. Although the threats associated with pesticides cannot be eliminated, they could be circumvented in one way or the other to at least make it user and environment friendly. Integration of pesticides with its natural enemies has greater potential than is utilized at present and it could help in mitigating negative consequences of pesticide overuse. Low cost, efficient, target specific, eco-friendly integrated pest management package of practice needs to be developed based on site specific pest problems to sustain the food production system sustainably. Indeed, strict policy framework, systematic and strategic responsible research, and ground-breaking awareness pathways are the demand of time in order to develop and implement for the betterment of society.

Recommendations for research in pesticides management to combat ill effects on soil, air, water and human health nexus:

- For now, pesticide free agriculture is nearer to impossible keeping in view of the food security and population growth, therefore, judicious use of pesticide is the need of the hour.
- Integrated pest management coupled with ecological engineering based interventions should be site specifically standardized, tested and disseminated to avail the benefits.
- Biopesticides, nano based eco-friendly encapsulation/formulation, and photocatalysis are the future research areas to strengthen pesticide free soil, air, water and human nexus.
- Development and use of a low-cost pesticide detection kit available to end-users to find out the permissible limit of residual pesticide in day to day consumable cereals, pulses, fruits and vegetables.

## References

Abbas MN, Al-Madhhachi A-ST (2020) Pesticide remediation from soil and water accessing to zero residue level. Quantum J Environ Stud 1(1):5–21. http://myjms.moe.gov.my/index.php/qjoes/article/view/8263

Agarwal A, Prajapati R, Singh OP, Raza SK, Thakur LK (2015) Pesticide residue in water–a challenging task in India. Environ Monit Assess 187(2):54

Aktar MW, Sengupta D, Choudhary A (2009) Impact of pesticides use in agriculture: their benefits and hazards. Interdisc Toxicol 2(1):1–12

Arya R, Mishra AK, Chaudhry S (2018) Variation in soil properties and carbon stocks under roadside plantation and rice-wheat cropping system in North-Western Haryana, India. Int J Curr Microbiol App Sci7(04): 1939–1949

BBC News Asia (2015) India 'to overtake China's population by 2022' UN. https://www.bbc.com/news/world-asia-33720723. Access on 12 June 2020

Bhardwaj T, Sharma JP (2013) Impact of pesticides application in agricultural industry: an Indian scenario. Int J Agric Food Sci Technol 4(8):817–822

Cai DW (2008) Understand the role of chemical pesticides and prevent misuse of pesticides. Bull Agric Sci Technol 1:36–38

Calvert GM, Karnik J, Mehler L, Beckman J, Morrissey B, Sievert J, Barrett R, Lackovic M, Mabee L, Schwartz A, Mitchell Y, Moraga-McHaley S (2008) Acute pesticide poisoning among agricultural workers in the United States, 1998–2005. Am J Ind Med 51:883–898

Camara MC, Campos EVR, Monteiro RA et al (2019) Development of stimuli-responsive nanobased pesticides: emerging opportunities for agriculture. J Nanobiotechnol 17:100. https://doi.org/10.1186/s12951-019-0533-8

Cerejeira MJ, Viana P, Batista S, Pereira T, Silva E, Valerio MJ (2003) Pesticides in Portuguese surface and ground waters. Water Res 37(5):1055–1063

Das R, Sahoo S, Singh H, Suman SN, Singh P (2019) Humus pesticide interaction: the fate of pesticide in soil environment: an overview. Int J Chem Stud 7(3):3117–3123

Dey D (2016) Use of essential oils for insect Pest management – a review. Innovative Farming 1(2):21–29. Retrieved from http://www.innovativefarming.in/index.php/innofarm/article/view/126

Dhaliwal GS, Vikas J, Dhawan AK (2010) Insect pest problems and crop losses: changing trends. Indian J Ecol 37(1):1–7

Doubling the farmer's income (niti.gov.in) (2017) https://niti.gov.in/writereaddata/files/document_publication/DOUBLING%20FARMERS%20INCOME.pdf

EPA-Environmental protection Agency (2009) What is a pesticide? http://www.epa.gov/opp00001/about/

FDA 2017, Pesticide monitoring report (2019). https://www.fda.gov/food/pesticides/pesticide-residue-monitoring-2017-report-and-data

Forget G, Goodman T, de Villiers A (1993) Impact of pesticide use on health in developing countries. IDRC, Ottawa: 2

Gurr GM, Wratten SD, Altieri MA (2004) Ecological engineering: a new direction for agricultural pest management. Aust Farm Bus Manag J 1(1):23–28

Hossain MS, Chowdhury MAZ, Pramanik MK, Rahman MA, Fakhruddin ANM, Alam MK (2015) Determination of selected pesticides in water samples adjacent to agricultural fields and removal of organophosphorus insecticide chlorpyrifos using soil bacterial isolates. Appl Water Sci 5:171–179. 10.1007/s13201-014-0178-6, 10.1016/j.scitotenv.2019.134645

Hussain S, Siddique T, Saleem M, Arshad M, Khalid A (2009) Chapter 5: Impact of pesticides on soil microbial diversity, enzymes, and biochemical reactions. Adv Agron 2(102):159–200

Igbedioh SO (1991) Effects of agricultural pesticides on humans, animals, and higher plants in developing countries. Arch Environ Health 46(4):218–240. https://doi.org/10.1080/00039896.1991.9937452

Kanan S, Moyet MA, Arthur RB, Patterson HH (2020) Recent advances on $TiO_2$-based photocatalysts toward the degradation of pesticides and major organic pollutants from water bodies. Catal Rev 62(1):1–65. https://doi.org/10.1080/01614940.2019.1613323

Kaur R, Kaur S, Lata M (2011) Evaluation of DNA damage in agricultural workers exposed to pesticides using single cell gel electrophoresis (comet) assay. Indian J Hum Genet 17:179–187

Kaur R, Mavi GK, Raghav S (2019) Pesticides classification and its impact on environment. Int J Curr Microbiol App Sci 8(03):1889–1897. https://doi.org/10.20546/ijcmas.2019.803.224

Khalid S, Shahid M, Murtaza B, Bibi I, Natasha MAN, Niazi NK (2020) A critical review of different factors governing the fate of pesticides in soil under biochar application. Sci Total Environ 711:134645

Kole RK, Bagchi MM (1995) Pesticide residues in the aquatic environment and their possible ecological hazards. J Inland Fish Soc India 27(2):79–89

Kumar R (2005) An epidemiological study of cancer cases reported from villages of Talwandi Sabo Block. Department of Community Medicine, Postgraduate Institute of Medical Education and Research, District Bhatinda, Punjab. http://umendradutt.blogspot.in/2007/09/final-report-of-pgimer-epidemiological.html

Kumar V, Kumar P (2019) Pesticides in agriculture and environment: impacts on human health. Contaminants in agriculture and environment: health risks and remediation. https://doi.org/10.26832/AESA-2019-CAE-0160-07

Kumar MD, Shah T (2006) Groundwater pollution and contamination in India: the emerging challenge. https://www.researchgate.net/publication/241760985

Kumar G, A Giri A, Arya R, Tyagi R, Mishra S, Mishra AK (2019) Multifaceted applications of different parts of Moringa species: review of present status and future potentials. Int J Chem Stud 7(2):835–842

Kumari S, Sharma H (2012) The impact of pesticides on Farmer's health: a case study of fruit bowl of Himachal Pradesh. Int J Sci Res 2319-7064(3):358. https://doi.org/10.21275/23101402

Lah K (2011) Effects of pesticides on human health. In: Toxipedia. http://www.toxipedia.org/display/toxipedia/Effects+of+Pesticides+on+Human+Health
Liu ZJ, Liu ZB (1999) The status of the world agrochemicals. Chem Technol Market 22(12):14–17
Liu CJ, Men WJ, Liu YJ (2002) The pollution of pesticides in soils and its bioremediation. System Sci Comprehen Stud Agric 18(4):295–297
Luwanda L (2018) The unintended health consequences of the green revolution in Punjab, India. GRIN Verlag, Munich. https://www.grin.com/document/432637
Manage (2019) Ecologically sustainable strategies for pest management. Extension Digest 3(1) https://www.manage.gov.in/publications/edigest/jun2019.pdf
Mathur SC (1999) Future of Indian pesticides industry in next millennium. Pesticide Inform 24(4):9–23
Md Meftaul I, Venkateswarlu K, Dharmarajan R, Annamalai P, Megharaj M (2020) Pesticides in the urban environment: a potential threat that knocks at the door. Sci Total Environ 711:134612. https://doi.org/10.1016/j.scitotenv.2019.134612
Mittal S, Kaur G, Vishwakarma GS (2014) Effects of environmental pesticides on the health of rural communities in the Malwa region of Punjab, India: a review. Human Ecol Risk Assess Int J 20(2):366–387. https://doi.org/10.1080/10807039.2013.788972
Patocka J, Wu O, França TCC, Ramalho TC, Kuca PR (2016) Clinical aspects of the poisoning by the pesticide endosulfan. Quim Nova 39(8):987–994. https://doi.org/10.5935/0100-4042.20160102
Pimentel D (1997) Techniques for reducing pesticide use: environmental and economic benefits. Wiley, Chichester
Pimentel D (2009) Pesticides and pest control. In: Rajinder P, Dhawaneds A (eds) Integrated pest management: innovation-development process, vol 1. Springer, pp 83–87
Rao GV, Sahrawat K, Ch S, Das B, Reddy K, Bharath BS, Rao V, Murthy KVS, Wani S (2009) Insecticide residues in vegetable crops grown in Kothapalli watershed, Andhra Pradesh, India: a case study. Indian J Dryland Agric Res Dev 24(2):21–27
Robertson A, Tirado C, Lobstein T, Jermini M, Knai C, Jensen JH, Ferro-Luzzi A, James WPT (2004) Food and health in Europe: a new basis for action, European series, No. 96. WHO Regional Publications, Geneva, p 385
Singh RB (2000) Environmental consequences of agricultural development: a case study from the Green Revolution state of Haryana, India. Agric Ecosyst Environ 82:97–103
Socorro J, Durand A, Temime-Roussel B, Gligorovski S, Wortham H, Quivet E (2016) The persistence of pesticides in atmospheric particulate phase: an emerging air quality issue. Sci Rep 6:33456. https://doi.org/10.1038/srep33456
Statistical yearbook of India (2011). http://www.mospi.gov.in/statistical-year-book-india/2011
Takeshita V, Mendes KF, Alonso FG, Tornisielo VL (2019) Effect of organic matter on the behavior and control effectiveness of herbicides in soil. Planta Daninha 37:e019214401. Epub October 17, 2019. https://doi.org/10.1590/s0100-83582019370100110
Thakur JS, Rao BT, Rajwanshi A, Parwana HK, Kumar R (2008) Epidemiological study of high cancer among rural agricultural community of Punjab in Northern India. Int J Environ Res Public Health 5(5):399–407. https://doi.org/10.3390/ijerph5050399
Tyagi H, Gautam T, Prashar P (2015) Survey of pesticide use patterns and farmers'perceptions: a case study from cauliflower and tomato cultivating areas of district Faridabad, Haryana, India. Int J MediPharm Res 3:139–146
Tyagi R, Luhach J, Mishra AK, Chaudhry S (2019) Morphological and biochemical changes in Vigna radiata and Spinacia oleracea induced by fluoride contamination in soils. Int J Curr Microbiol App Sci 6(5):395–402
U.S. Geological Survey (1998) National Water-Quality Assessment. Pesticide national synthesis project. Pesticides in surface and ground water of the United States; Summary of results of the National Water Quality Assessment Program. http://water.wr.usgs.gov/pnsp/allsum/fi g02.gif
Venugopal PN (2008) Endosulfan victims still suffer in India, Article February 2008. http://southasia.oneworld.net/Article/endosulfan-victims-still-suffer-in-india/?

Yadav S, Dutta S (2019) A study of pesticide consumption pattern and farmer's perceptions towards pesticides: A case of Tijara Tehsil, Alwar (Rajasthan). Int J Curr Microbiol App Sci 8(04):96104. https://doi.org/10.20546/ijcmas.2019.804.012

Zhang WJ (2008) A forecast analysis on world population and urbanization process. Environ Dev Sustain 10:717–730

Zhang WJ, Qi YH, Zhang ZG (2006) A long-term forecast analysis on worldwide land uses. Environ Monit Assess 119:609–620

Zhang W, Jiang F, Ou J (2011) Global pesticide consumption and pollution: with China as a focus. Proc Int Acad Ecol Environ Sci 1(2):125–144

# Chapter 15
# Indigenous Agricultural Knowledge Towards Achieving Sustainable Agriculture

**Anwesha Borthakur and Pardeep Singh**

**Abstract** With modernization of agriculture comes the downside of intense application of pesticides, weedicides, fertilizers etc. It is widely accepted that such practices in the long run do pose major challenges to sustainable agriculture. Indigenous agricultural knowledge has the potential to provide a viable alternative here. This knowledge is the product of hundreds of years of experiences of the farmers and their experiments with nature. Such knowledge evolves gradually over the years through constant engagement with the natural processes, passes across generations, and thus, integrates the agro-climatic factors of a particular geographical area.

In this chapter, we primarily attempt to rationalize the significance of indigenous agricultural knowledge in the present-day context. The concept of indigenous people, indigenous knowledge and indigenous agricultural knowledge are reviewed. The significance of indigenous agricultural knowledge in achieving the much talked about sustainable development goals are discussed. Its role in managing the contaminants from agro-ecosystems and ensuring sustainable yield are assessed in detail with a few empirical evidences of successful cases. We propose that an integration of indigenous agricultural knowledge with modern scientific knowledge has immense potential towards ensuring sustainable agriculture.

**Keywords** Indigenous agricultural knowledge · Sustainable agriculture · Sustainable development goals · SDG-2 · Agricultural contaminants

A. Borthakur (✉)
Leuven International and European Studies (LINES), KU Leuven, Leuven, Belgium
e-mail: anwesha.borthakur@kuleuven.be

P. Singh
Department of Environmental Science, PGDAV College, University of Delhi, New Delhi, India
e-mail: psingh.rs.apc@itbhu.ac.in



V. Kumar Singh et al. (eds.), *Sustainable Agriculture Reviews 50*, Sustainable Agriculture Reviews 50, https://doi.org/10.1007/978-3-030-63249-6_15

## 15.1 Introduction

Indigenous agricultural knowledge is gaining increasing attention in the contemporary global environmental and agricultural discourse. It has been argued that it has the potential to substantially contribute towards achieving the Sustainable Development Goals (SDG) proposed by the United Nations (UN). Before exclusively focusing on the indigenous agricultural knowledge, it is essential to define 'indigenous knowledge', the superset of indigenous agricultural knowledge. Further, it is imperative to understand who are essentially the 'indigenous people' owning the indigenous knowledge. According to the United Nations Educational, Scientific, and Cultural Organization (UNESCO), indigenous knowledge refers to "the understandings, skills and philosophies developed by societies with long histories of interaction with their natural surroundings. For rural and indigenous people, local knowledge informs decision-making about fundamental aspects of day-to-day life".[1] The UN defines indigenous people as the "inheritors and practitioners of unique cultures and ways of relating to people and the environment. They have retained social, cultural, economic and political characteristics that are distinct from those of the dominant societies in which they live".[2] They are essentially the people who had settled in a country preceding colonization or who acknowledge themselves as successors of these very peoples (OECD 2017). Indigenous knowledge is conveyed both informally and formally amongst the kinfolks through ritual practices, oral traditions, social encounters and other such processes which comprise of methods of sowing and harvesting; specific understandings of the native ecosystems; decorative and symbolic approaches of communiqué; cosmological observations, oral accounts towards narrating human histories; creation of specific tools and technologies etc. (Bruchac 2014).

Indigenous people today represent approximately 5% of the global population (OECD 2017). Although represents only a fraction of the global populace, their knowledge in the form of indigenous science are imperative in addressing present-day global crisis, whether it is towards ensuring environmental and economic sustainability or towards confirming equal social opportunities. In general, indigenous communities share unique knowledge on their direct environs, verbally passed on in what could be called as diverse forms of indigenous science (Zidny and Eilks 2020). Within the broader concept of indigenous knowledge, indigenous agricultural knowledge is attracting considerable attention in the recent time. There is a growing cognizance that agricultural research and development (R&D) interventions ought to be based on a number of traditional practices that are rooted on comprehensive, rational and scientific principles (Anchirinah et al. 2001). Thus, indigenous agricultural knowledge has the potential to contribute in improved understanding in the usage and sustenance of agro-ecosystems and agricultural production (Anchirinah

[1] http://www.unesco.org/new/en/natural-sciences/priority-areas/links/related-information/what-is-local-and-indigenous-knowledge/

[2] https://www.un.org/development/desa/indigenouspeoples/about-us.html

et al. 2001). According to the Center for International Earth Science Information Network (CIESIN) at Columbia University, reports such as the Brundtland Commission's Our Common Future (World Commission on Environment and Development (WCED) 1987) and World Conservation Strategy (International Union for the Conservation of Nature and Natural Resources (IUCN)1980) along with the United Nations Conference on Environment and Development (UNCED) in 1992 had a pivotal role in the global recognition of indigenous environmental and agricultural knowledge.[3] Today, indigenous agricultural knowledge is a major focus area for the research and policy community towards achieving the objectives of sustainable agriculture.

In this chapter, we primarily attempt to rationalize the significance of indigenous agricultural knowledge in the present-day context. After an introduction of the concept of indigenous people, indigenous knowledge and indigenous agricultural knowledge, in the subsequent sections the significance of indigenous agricultural knowledge in achieving the much talked about SDGs will be discussed. Its role in managing the contaminants from agro-ecosystems and ensuring adequate and sustainable yield will be assessed in detail with a few empirical evidences of successful cases. The sustainability aspect of indigenous agricultural knowledge will be appraised taking into consideration its contributions in maintaining and restoring agro-ecosystems. Overall, the chapter endevours to provide a comprehensive understanding of the indigenous knowledge associated with agriculture and its relevance in assuring sustainable agricultural practices in today's world.

## 15.2 Indigenous Agricultural Knowledge, Sustainable Agriculture and Sustainable Development Goals

Indigenous knowledge, as a fundamental constituent of a country's knowledge system, refers to the local knowledge responsible for the local decision makings regarding natural resource management, agriculture, food preparation, health care, education, and a range of other such activities especially in the rural settings (Rao 2006). For instance, it has been anticipated that traditional medicine alone uses between 25,000 and 75,000 plant species among which a mere 1% is recognized by the scientific community, and thus, acknowledged towards commercial intentions (Grethel 2001). Traditional knowledge, indigenous knowledge, and traditional environmental knowledge are the three major foundations of 'intellectual capital' (Che and Korota'aini 2012). All these knowledge have been primarily shared or communicated verbally throughout generations. A few such forms are articulated via folklore, stories, songs, legends, proverbs, myths, dances, agricultural practices, beliefs, cultural values, community laws, rituals, taxonomy, local language, materials, equipments, animal breeds, plant species etc. (Acharya and Shrivastava 2008; Rao

[3] See: http://www.ciesin.org/TG/AG/iksys.html

2006). In the recent years, an increased international appreciation has been observed of the activities of indigenous people as innovative, creative, and for their significant contributions to research and development, science and towards overall human progress (Kamau 2012). Indigenous people, comprising of approximately 5000 distinct and varied groups, represent about one third of world's exceptionally deprived rural people and reside in over 70 countries with 70% of them residing in the different countries across Asia (Rao 2006).

- In 2015, all member states of the United Nations adopted the 2030 Agenda for Sustainable Development. The agenda primarily offers a common outline for prosperity and peace of both the planet and its people today and into the future. Accordingly, in a global corporation, 17 Sustainable Development Goals (SDGs) are set calling for urgent actions to be taken by all the countries irrespective of developing or developed.[4] Experts have observed that indigenous agricultural knowledge has immense potential to contribute towards achieving the SDGs. For instance, the most closely associated SDG goal with indigenous agricultural knowledge is SDG – 2. It can aid in realizing the SDG – 2 which is aimed at 'Zero Hunger'. The comprehensive goal is to accomplish improved nutrition, food security and promote sustainable agricultural practices. As defined by the UN, "It is time to rethink how we grow, share and consume our food. If done right, agriculture, forestry and fisheries can provide nutritious food for all and generate decent incomes, while supporting people-centered rural development and protecting the environment".[5] As in the case of most of the indigenous knowledge-based practices, the agricultural practices preserved and performed by the indigenous communities are often sustainable, both environmentally and economically. Today's environmental advocates argue that the world must focus on sustainable agriculture in the coming years in order to avoid detrimental consequences of erroneous capitalist agricultural modes. Sustainable agriculture primarily aims to integrate three key objectives into its practices – economic and social justice, economic viability and environmental protection.[6] Unlike several modern agricultural practices which may result in increasing productivity, but at the same time, also are responsible for causing irreparable damage to the environment, indigenous knowledge-based practices maintain a harmony with the nature. As argued by Pretty (2008) in the context of agriculture:

  *New approaches are needed that will integrate biological and ecological processes into food production, minimize the use of those non-renewable inputs that cause harm to the environment or to the health of farmers and consumers, make productive use of the knowledge and skills of farmers, so substituting human capital for costly external inputs, and make productive use of people's collective capacities to work together to solve common agricultural and natural resource problems, such as for pest, watershed, irrigation, forest and credit management.*

---

[4] See: https://sustainabledevelopment.un.org/

[5] https://www.un.org/sustainabledevelopment/hunger/

[6] https://asi.ucdavis.edu/programs/ucsarep/about/what-is-sustainable-agriculture

As suggested by the World Wildlife Foundation (WWF), agriculture today is the largest industry in the world responsible for providing employment to over 1 billion people, annually generates food worth trillions of dollars with cropland and pasture occupying approximately 50% of the Earth's habitable land.[7] Accordingly, it is essential to shift towards the sustainable mode of agriculture which ensures social, economic and environmental wellbeing of the people and the planet. Indigenous agricultural knowledge has the potential to be the starting point in this context. As indigenous systems of knowledge develop over centuries and inherently incorporate the knowhow of generations, these systems are typically ecologically sustainable and sensitive to the surrounding ecosystems. These systems are local-specific and developed considering the native geographical circumstances. Therefore, it is essential to provide sufficient importance to these systems of knowledge towards achieving the goals of sustainable agriculture.

## 15.3 Significance of Indigenous Agricultural Knowledge in Sustainable Agriculture

### *15.3.1 Indigenous Agricultural Knowledge in Ensuring Sustainable Agricultural Production*

Diverse and local-specific agricultural systems have been developed by indigenous farmers for centuries with an aim towards conserving agrobiodiversity and community food security through original practices of management (Altieri 2004). The scientific knowledge base of the indigenous communities, which could be called as indigenous science, has been playing a fundamental role in agricultural sustainability. With the objective of realizing the requirements of agriculture, soil and water conservation along with medicine, forest and land management, indigenous science practices typically concentrate on responsible natural resources utilization (Zidny and Eilks 2020). Nevertheless, the modern agricultural practices are moving away from the indigenous methods and techniques, putting a question mark on their sustainability both in terms of economic and environmental perspectives. A good example is the seed production and preservation paradigm which is largely controlled by the multinational seed companies in the world today. As argued by Fernandez (1994), a shift towards sustainable agriculture necessitates essential transformations in the paradigm of seed production, for which, it is imperative that the farmers hold control over their seeds, especially those farmers in the Third World countries. The author further argues that, in this context, an acknowledgement and knowledge of alternative technologies is crucial for crop improvement, genetic conservation, seed and crop production, and other associated issues.

[7] See: https://www.worldwildlife.org/industries/sustainable-agriculture

Indigenous agricultural systems encompassing seeds and related practices are an affluent source of such alternatives.

In particular, indigenous communities, still very much upholding their diverse and unique cultural traditions, are common in Asia and a few other regions in the world (Zidny and Eilks 2020). For instance, India is country with presence of large indigenous communities throughout its the length and breadth. Diverse sets of indigenous agricultural knowledge are omnipresent among these communities which largely guide them in their agricultural practices. As a country whose economy is predominantly based on agriculture, these practices play a vital role towards ensuring sustainable agriculture. Indigenous seed protection and conservation techniques are common in the country. One such example is the practices in Uttarakhand Himalayas. In this region, farmers often use ash, plant-based parts, oil etc. as a measure to protect grains and seed materials and these are precisely done by learning and identifying the resources through their ancestors (Mehta et al. 2012). As a seed dresser, wooden and cow dung ash and red baked soil are used because of the presence of silica in these materials which prevent the larvae feeding and the egg formation (Mehta et al. 2012). Both ash and soil act as indigenous pesticides too. The authors Mehta et al. (2012) further argue that unlike many other farming communities where indigenous techniques are almost dying, the farmers in the Himalayan Uttarakhand are still very much reliant on their indigenous seed conservation techniques which form the backbone of their cropping in the next season. A major characteristic of indigenous agriculture is that it upholds a subsistence orientation wherein a significant section of the populace is occupied in agricultural activities with roughly all the production are meant for local utilization (Schroeder 1985).

Not only the concerns associated with the direct agricultural practices and processes, indigenous communities own a rich source of knowledge about the assessment of soil quality parameters and measures. In the recent years, there has been an increasing consent that, by means of participatory methods and conventional scientific methods, indigenous people have the potential to generate sets of data which are as accurate as those derived professionally (Fry 2011). For instance, a study conducted by Gosai et al. (2011) correlates the scientific and traditional knowledge in assessing the nutrient availability status of the agricultural soil as practiced by the 'Nyishi' tribe of Arunachal Pradesh in North Eastern India. Nyishi tribe, an indigenous community, assesses the soil fertility since historical past with the help of their own set of techniques and methods. As described by the authors, the soil is characterized by means of their yield, texture, and visual appearance, for example, colour and topographic positioning of the terrain or land. Further, the 'Nyishis' possess excellent knowledge on plot variants and the tribe has the expertise in distinguishing soil in terms of cropping potential too. Finally, the authors argue that the indigenous knowledge of the 'Nyishis' associated with soil is at par with the scientifically appropriate land use patterns and scientific physico-chemical analysis of the soil samples for soil characterization (Gosai et al. 2011). Mostly, it has been observed that the locale specific agricultural and environmental monitoring systems are economical than modern specialized monitoring systems involving complicated resources and manpower. These cost effective monitoring systems are financially sustainable than pricey professional schemes. Moreover, enhancing its longevity, the locally rooted approaches involve the indigenous communities in every single step such as data collection, analysis, planning, and decision-making, which in turn creates local support and ownership for a particular monitoring programme (Fry 2011).

### 15.3.2 Indigenous Agricultural Knowledge in Managing Agricultural Contaminants

- Management of agricultural contaminants is a major concern for the agricultural communities across the world today. The improved attention of the soil scientists on the significant potential of 'indigenous' soil knowledge towards facilitating, informing, or modifying the relevance of traditional scientific research is comparatively recent (Payton et al. 2003). In this context, 'ethnopedology' has emerged as a discipline gaining momentum in the recent years. The term 'ethnopedology' was originally coined by Williams and Ortiz-Solorio in 1981 and it is associated with the study of folk taxonomies (Krasilnikov and Joe 2003). The aim of ethnopedology is to comprehend and document the indigenous approaches to soil classification, perception, use, appraisal, and overall management (Barrera-Bassols and Zinck 2003). It is essentially a discipline that deals with the reflection with which pedodiversity is robustly associated; not simply to soil taxonomic properties and physical-chemical behaviour, but also to cultural practices, historical uses, anthropic management, and the entire indigenous knowledge that exemplifies an explicit area (Capra et al. 2015). The last decade of the twentieth century especially observed the emergence of ethnopedology (Krasilnikov and Joe 2003). As further described by Barrera-Bassols and Zinck (2003: 171–172):

  > *Ethnopedology, a hybrid discipline nurtured by natural as well as social sciences, encompasses the soil and land knowledge systems of rural populations, from the most traditional to the modern. Despite thousands of years of applied ethnopedology practiced by indigenous peoples, local soil knowledge has not been historically reflected in soil science research. Over the last few decades, remarkable examples of soil knowledge from non-western civilizations living mainly in tropical and subtropical areas have been described, and their current strength has been assessed.*

Traditionally, several indigenous plant species have been employed by the local farmers of a particular region to remove contaminants from the soil, in a process today known and recognized as 'phytoremediation'. Modern research shows that indigenous plant species across continents have widely been successful in removing contaminants from the soil including heavy metal. As these species are locally available, they provide great ecological services along with their phytoremediation characteristics. An excellent example in this context is the removal of arsenic from contaminated soil by native or indigenous plants. Arsenic is considered as one of the highly toxic metalloids and of immense environmental concerns owing to its widespread contamination of hundreds and thousands of sites across the globe consequential of a range of human activities (Antosiewicz et al. 2008). A study was conducted by Visoottiviseth et al. (2002) in two areas in Thailand, namely Ron Phibun District (Nakorn Si Thammarat province) and Bannang Sata District (Yala province), having arsenic pollution histories from mine tailings towards assessing the prospective of native plant species for possible phytoremediation. Upon selecting the plants through criteria such as high arsenic tolerance, short life cycle, high bioaccumulation factor and propagation rate, large shoot biomass and wide distribution, a herb (*Mimosa pudica*), two species of ferns (*Pityrogramma calomelanos* and *Pteris vittata*), and a shrub (*Melastoma malabrathricum*) were observed to be apposite for phytoremediation. These are the native plants widely available in that particular geographic area and thus, have the potential to immensely contribute towards environmental remediation through the removal of contaminants from soil.

Similarly, in a study carried out in a region of copper mining tailings situated in Southern Brazil towards evaluating spontaneously occurring plants for their phytoremediation potential for heavy metal concentrations illustrates that 11 plant species have copper concentrations exceeding 100 mg $kg^{-1}$ among which seven species highlighted concentrations between 321 and 586 mg $kg^{-1}$ (Afonso et al. 2020). The same species (*S. viarum* Dunal and *B. trimera* being the most effective) demonstrated chromium concentrations up to 440 mg $kg^{-1}$ (Afonso et al. 2020). In Umuahia in eastern-Nigeria, another study was conducted to discover indigenous plants growing in petroleum hydrocarbon contaminated sites for their ability to phytoremediate soils under a tropical monsoon climate setting (Oriebe and Atagana 2018). Among the total number of 28 native plant species belonging to diverse families budding in and around petroleum hydrocarbon-impacted soil, species such as *Chromolaena odorata, Aspilia africana, Chloris barbata, Pasparlum vaginatum, Bryophyllum pinnatum, Paspalum scrobiculatum, Cosmos bipinnatus, Eragrostis atrovirens, Cyperus rotundus,* and *Uvaria chamae* showed considerable tendencies for phytoremediation of contaminated soil (Oriebe and Atagana 2018). Yet another study was conducted Indian Sundarban wetlands to assess the phytoremediation potential for trace metal contamination by selected mangrove plants native to that particular area (Chowdhury et al. 2016). The study found that the specie *S. apetala* is effective in the accumulation of trace metals, and thus, has the potential to protect the vulnerable Sundarban ecosystem by acting both as biogeochemical and physical barriers to trace metal mobility (Chowdhury et al. 2016). Thus, throughout climatic conditions across continents, indigenous plants show great potential in the removal of agricultural contaminants. These plants are known to the indigenous communities since ages and they have probably been using these plants in some way or the other. Overall, an integration of indigenous and modern knowledge could successfully aid in removing contaminations, including highly toxic heavy metals, from the soil.

### *15.3.3 Indigenous Agricultural Knowledge in Maintaining/ Restoring the Agroecosystems*

Indigenous agricultural knowledge has been playing a pivotal role in maintaining and restoring agro-ecosystems since ages. The practices of the indigenous communities are typically in sync with nature and natural processes. A major characteristic of the indigenous knowledge systems is that these systems emerged from years of knowledge accumulation and translations by the local communities in view of the ecological processes typical in a particular geography. Over the time, indigenous knowledge systems were enriched and advanced by the experiences of the native people. Indigenous agricultural systems are no exceptions. Indigenous agricultural systems are known for providing the communities with self-sustenance and sufficiency with respect to food and nutrition, while at the same time conserving the agro-ecosystems and associated biodiversity. An excellent example of such an indigenous system of agriculture is the '*Bari System of Farming*', popular in Assam, a North-East Indian state. A *bari* is precisely a sustainable system of farming practiced by the native people of North Eastern India and recognized from the perspectives of consumption, conservation, and biodiversity management.

- *Bari* indicates an operational unit towards meeting the essential necessities of a rural household wherein several crops (including trees) are grown along with livestocks such as poultry and fish production (Barooah and Pathak 2009). A study carried out by the authors Barooah and Pathak (2009) on the indigenous knowledge and practices of Thengal-Kachari women, a small ethnic community with rich cultural diversity, illustrates interesting observations. As documented by Barooah and Pathak (2009), an all-purpose *bari* structure with substantial flexibility and diversity towards facilitating the production of key livelihood requirements has been developed by the women of the community through several years of experimentations and observation. Their expertise lies in selecting crops that are co-adapted and results in giving cumulative advantages. One of the major attributes of a bari has been that they are designed towards permitting most advantageous harvest of solar energy "through the strategy of fitting phenological classes and life forms together in space and time, and through niche diversification techniques". As further states by the authors:

> *Multiple crops are present in a multi-tier canopy configuration. The leaf canopies of the components are arranged in such a way that they occupy different vertical layers with the tallest components having foliage tolerant to strong light and high evaporation demand and the shorter components having foliage requiring or tolerating shade and high humidity. Although the baris exhibit a general pattern, each garden is unique in its spatial and temporal structure, crop mix and arrangement, and overall design. Some crops are always planted in regular patterns, while others are planted wherever space is available.*

Dugout fish pond is an important component in the *bari* system of farming. So are the livestock. Kitchen waste from the rural households is used for composting purposes and also to feed the fishes in the pond. Further, the aquatic plants present in the ponds too act as a source of food to the fishes. Excreta of the livestock serve

**Fig. 15.1** Preparation of a dugout pond in a *bari system of farming* in Assam. (Source: Collected by the authors through Ms. Nizara Phukon (2019))

as manures for the vegetable production and for the fruit trees. Overall, *bari* is an ecologically sustainable way of farming technique with almost zero waste production. It is a self-sustaining unit which is both productive and environment- and resource- friendly (Figs. 15.1 and 15.2).

## 15.4 Conclusion

Indigenous knowledge and techniques play a significant role in addressing major socio-economic, cultural, health and environmental concerns ranging from medicine and agriculture to the issues related to climate change and mitigation of natural disasters. These knowledge and technologies have, since long, been providing the indigenous communities with self-sufficiency and comfort. It has been observed that involving local or indigenous communities and incorporating their indigenous knowledge as agricultural and environmental management measures result in

**Fig. 15.2** A well-functioning dugout pond in a *bari* in Assam with plants and crops at the bank. (Source: Collected by the authors through Ms. Nizara Phukon (2019))

effective planning and forecasting strategies intended for addressing key issues. Indigenous knowledge bearers, for instance, are the best people having knowledge about any locale specific agricultural issue ranging from crop selection, yield up to harvesting and further. Any risks to the agricultural and natural resources on which the local communities depend and interact on a daily basis are often rapidly detected by these communities. As they have been interacting with these resources for generations, the indigenous communities have paramount knowledge about the history of the resources and their gradual evolution. Therefore, the issues related to these resources could be addressed through local or indigenous management actions and strategies.

In the contemporary world, it is becoming increasingly essential to shift towards the sustainable mode of agriculture which ensures social, economic and environmental wellbeing of the people and the planet. Modern agricultural practices are proven to have intensified the use of chemicals in the forms of pesticides, herbicides, weedicides etc. These practices may result in an increasing yield, nonetheless, also by making compromises with the planet's ecology and environment. Thus, it is imperative to shift our focus towards sustainable agriculture and indigenous agricultural knowledge has the potential to be a focal point in this context. For instance, indigenous agriculture has immense potential to contribute to the sustainable development goals, aimed to be achieved by the year 2030. In particular, SDG – 2 which aims for 'zero hunger' could only be achieved with a major focus on sustainable agriculture. Indigenous agricultural knowledge certainly helps in realizing the goal where no one remains hungry and food deprived.

In this chapter, we attempted to assess the role of indigenous agricultural knowledge not only in maintaining and restoring a dynamic agro-ecosystem, but also in managing the contaminants from these agro-ecosystems and ensuring adequate and sustainable yield. We assessed the scenario in detail with a few empirical evidences of successful cases. We observed that the scientific knowledge base of the indigenous communities, which could be called as indigenous science, has been playing a fundamental role in agricultural sustainability. The knowledge of the indigenous communities on different aspects of agriculture such as soil characteristics, crop and seed selections, cropping patterns and so on are not only ecologically sustainable, but also cost-effective and socially inclusive.

We conclude that indigenous knowledge, if integrated appositely with modern scientific knowledge, has immense potential to address key environmental issues encountered in the current world. In the contemporary world, there is an emerging need for integrating indigenous and modern knowledge systems in order to derive maximum benefits towards addressing a wide range of agricultural and environmental concerns such as poverty, nutrition, soil degradation, pollution, climate change among others. Many success stories of such integration have been recorded in the recent past from different parts of the globe. Nevertheless, such stories are still a rarity in many regions in spite of having an astounding indigenous knowledge base. Scientific communities should engage themselves in documenting the indigenous agricultural knowledge presently in practice throughout the world for the larger benefit of the ecology and overall environment. Much of such knowledge is at the verge of extinction. Unless conserved, some remarkable, unique and extensive knowledge sources in the form of indigenous agricultural knowledge across continents could be lost.

**Acknowledgement** The authors are grateful to Ms. Nizara Phukon for the photos of the '*Bari*' system of farming.

## References

Acharya D, Shrivastava A (2008) Indigenous herbal medicines: tribal formulations and traditional herbal practices. Aavishkar Publishers Distributor, Jaipur

Afonso TF, Demarco CF, Pieniz S, Quadro MS, Camargo FAO, Andreazza R (2020) Bioprospection of indigenous flora grown in copper mining tailing area for phytoremediation of metals. J Environ Manag 256:109953

Altieri MA (2004) Linking ecologists and traditional farmers in the search for sustainable agriculture. Front Ecol Environ 2(1):35–42

Anchirinah VM, Yiridoe EK, Bennett-Lartey SO (2001) Enhancing sustainable production and genetic resource conservation of Bambara groundnut: a survey of indigenous agricultural knowledge systems. Outlook on Agriculture 30(4):281–288

Antosiewicz DM, Escudě-Duran C, Wierzbowska E, Skłodowska A (2008) Indigenous plant species with the potential for the phytoremediation of arsenic and metals contaminated soil. Water Air Soil Pollut 193(1):197–210. https://doi.org/10.1007/s11270-008-9683-2

Barrera-Bassols N, Zinck JA (2003) Ethnopedology: a worldwide view on the soil knowledge of local people. Geoderma 111(3):171–195. https://doi.org/10.1016/S0016-7061(02)00263-X

Barooah M, Pathak A (2009) Indigenous knowledge and practices of Thengal Kachari women in sustainable management of bari system of farming. Indian J Tradit Knowl 8:25–40

Bruchac M (2014) Indigenous knowledge and traditional knowledge. In: Smith C (ed) Encyclopedia of global archaeology. Springer, New York, pp 3814–3824

Capra GF, Ganga A, Buondonno A, Grilli E, Gaviano C, Vacca S (2015) Ethnopedology in the study of toponyms connected to the indigenous knowledge on soil resource. PLoS One 10:e0120240

Che SMB, Korota'aini OS (2012) Small is big: the charms of indigenous knowledge for sustainable livelihood. Procedia Soc Behav Sci 36:602–610

Chowdhury R, Lyubun Y, Favas PJC, Sarkar SK (2016) Phytoremediation potential of selected mangrove plants for trace metal contamination in Indian Sundarban wetland. In: Ansari AA, Gill SS, Gill R, Lanza GR, pp NL (eds) Phytoremediation: Management of Environmental Contaminants, vol 4. Springer, Cham, pp 283–310

Fernandez PG (1994) Indigenous seed practices for sustainable agriculture. Indigenous Knowledge and Development Monitor 2(2):9–12

Fry BP (2011) Community forest monitoring in REDD+: the 'M' in MRV? Environ Sci Pol 14:181–187

Gosai K et al (2011) Indigenous knowledge of soil fertility management in the humid tropics of Arunachal Pradesh. Indian J Tradit Knowl 2:508–511

Grethel A (2001) Access to genetic resources and protection of traditional knowledge in the territories of indigenous peoples. Environ Sci Pol 4:241–256

Kamau MC (2012) Traditional knowledge management and preservation: intersections with library and information science. Int Inf Libr Rev 44:13–27

Krasilnikov PV, Joe AT (2003) Perspectives on utilitarian ethnopedology. Geoderma 111(3):197–215. https://doi.org/10.1016/S0016-7061(02)00264-1

Mehta PS, Negi KS, Rathi RS, Ojha SN (2012) Indigenous methods of seed conservation and protection in Uttarakhand Himalaya. Indian J Tradit Knowl 11(2):279–282

OECD (2017). Available at http://www.oecd.org/regional/regional-policy/indigenous-communities.htm

Oriebe AR, Atagana HI (2018) Profiling of plants at petroleum contaminated site for phytoremediation. Int J Phytoremediation 20(4):352–361. https://doi.org/10.1080/15226514.2017.1393386

Payton RW, Barr JJF, Martin A, Sillitoe P, Deckers JF, Gowing JW, Hatibu N, Naseem SB, Tenywa M, Zuberi MI (2003) Contrasting approaches to integrating indigenous knowledge about soils and scientific soil survey in East Africa and Bangladesh. Geoderma 111(3):355–386. https://doi.org/10.1016/S0016-7061(02)00272-0

Pretty J (2008) Agricultural sustainability: concepts, principles and evidence. Philos Trans R Soc Lond Ser B Biol Sci 363(1491):447–465. https://doi.org/10.1098/rstb.2007.2163

Rao SS (2006) Indigenous knowledge organization: an Indian scenario. Int J Inf Manag 26:224–233

Schroeder RF (1985) Himalayan subsistence systems: indigenous agriculture in rural Nepal. Mt Res Dev 5(1):31–44. https://doi.org/10.2307/3673221

Visoottiviseth P, Francesconi K, Sridokchan W (2002) The potential of Thai indigenous plant species for the phytoremediation of arsenic contaminated land. Environ Pollut 118(3):453–461. https://doi.org/10.1016/S0269-7491(01)00293-7

Zidny R, Eilks I (2020) Integrating perspectives from indigenous knowledge and Western science in secondary and higher chemistry learning to contribute to sustainability education. Sustain Chem Pharm 16:100229

The manufacturer's authorised representative in the EU is Springer Nature Customer Service Centre GmbH, Europaplatz 3, 69115 Heidelberg, Germany. If you have any concerns regarding our products, please contact ProductSafety@springernature.com

Printed and bound by CPI Group (UK) Ltd, Croydon, CR0 4YY
30/09/2026
02225491-0014